Smart Hydrogel Modelling

Hua Li

Smart Hydrogel Modelling

Springer

Prof. Hua Li
Nanyang Technological University
College of Engineering
School of Mechanical & Aerospace Engineering
50 Nanyang Ave.
Singapore 639798
Singapore

ISBN 978-3-642-42492-2 ISBN 978-3-642-02368-2 (eBook)
DOI 10.1007/978-3-642-02368-2
Springer Heidelberg Dordrecht London New York

Softcover re-print of the Hardcover 1st edition 2009

Cover design: WMXDesign GmbH, Heidelberg

Printed on acid-free paper

Springer is part of Springer Science+Business Media (www.springer.com)

Dedicated first and foremost to my motherland,
and to Duer, Anne and my parents

Preface

The science of mathematical modelling and numerical simulation is generally accepted as the third mode of scientific discovery (with the other two modes being experiment and analysis), making this field an integral component of cutting edge scientific and industrial research in most domains. This is especially so in advanced biomaterials such as polymeric hydrogels responsive to biostimuli for a wide range of potential BioMEMS applications, where multiphysics and multiphase are common requirements. *These environmental stimuli-responsive hydrogels are often known as smart hydrogels.* In the published studies on the smart or stimuli-responsive hydrogels, the literature search clearly indicates that the vast majority are experimental based. In particular, although there are a few published books on the smart hydrogels, none is involved in the *modelling* of *smart* hydrogels.

For the few published journal papers that conducted mathematical modelling and numerical simulation, results were far from satisfactory, and showed significant discrepancies when compared with existing experimental data. This has resulted in ad hoc studies of these hydrogel materials mainly conducted by trial and error. This is a very time-consuming and inefficient process, and certain aspects of fundamental knowledge are often missed or overlooked, resulting in off-tangent research directions. Thus it is absolutely necessary to publish a book on the modelling and simulation of the smart hydrogels with real multidisciplinary requirement for establishment of a theoretical platform by developing the correct mathematical models and also the powerful numerical techniques required to solve these challenging highly nonlinear and coupled models.

Polymeric hydrogels are form of matters that possess both the properties of solid and liquid. Their structural framework chains are formed from networks of randomly crosslinked polymers that embody three different phases in general, namely the three-dimensional solid polymeric matrix network, interstitial fluid and ion species. Depending on the component characteristics and synthesis methods, the hydrogels can be designed or tailored to demonstrate the unique property of undergoing discrete or continuous volume transformation in response to infinitesimal changes of external environment stimuli, such as solution pH, electric field, temperature, solvent composition, glucose/carbohydrates, salt concentration/ionic strength, light/photon, pressure, coupled magnetic and electric fields. These magnificent features make the hydrogel better known as smart or stimuli-responsive hydrogels.

Due to their unique properties that include swelling/deswelling behaviour, sorption capacity, mechanical property, permeability and surface property, the hydrogels provide the instrumentation for creating functional materials for broad spectrum of applications as they can sense the environmental changes and eventually induce structural changes without a need for an external power source. Artificial muscle, microfluidic control, sensor/actuator, separation process and chromatographic packing are just few examples of the successful applications of hydrogels. Another exceptional promise of the hydrogels is their biocompatibility and biostability potentials, suggesting that the hydrogels are also an excellent substitution for the human body tissues or biomimetic applications. There are also extensive explorations of the hydrogels in the medical and pharmaceutical applications, such as drug delivery system, articular cartilage, biomaterial scaffold, corneal replacement and tissue engineering. As such, the multi-state characteristics of the smart polymer hydrogels and their wide-range multiphysics applications make the multidisciplinary and multiphase the basic requirements for the mathematical models. For example, these models are required to be highly multi-disciplinary and at least to take into consideration the coupled chemo-electro-mechanical multi-fields and multiphase deformation of polymeric network solid matrices with flow of ions and interstitial fluid. This results in several mathematical challenges, in which usually the models consist of coupled nonlinear partial differential equations with requirements of moving boundary and localized high gradient.

A comprehensive study through modelling and simulation is thus warranted for theoretical understanding of the response behaviour of the smart hydrogels in BioMEMS devices subject to different environmental stimuli, due to the advantages of the material characteristics of the smart polymer hydrogels and their wide range of multiphysics applications. However, as mentioned above, there is a lack of open publications on modelling and simulation of the smart hydrogels, and this monograph is thus written to systematically document the response behaviour of the smart hydrogels to various environmental stimuli. A complete theoretical platform detailing of the fundamental theory for the smart polymer hydrogels is established. It is composed of several novel mathematical models which are already successfully developed. Response of the smart hydrogels to surrounding environment is examined in detail for the basic stimuli within common BioMEMS devices such as solution pH, externally applied electric voltage, temperature, solvent composition, glucose/carbohydrates and salt concentration/ionic strength. The effects of various material properties and environmental conditions on the responsive performance of the smart hydrogels, including Young's modulus, initially fixed charge density, effective crosslink density, ionic strength and valence of bath solution, initial volume fraction of polymeric network, initial geometry, are also investigated in various parametric studies. In addition, an analysis of drug delivery system for the controlled drug release from non-swellable micro-hydrogel particles is presented with consideration of drug dissolution and diffusion through the continuous matrices of spherical micro-hydrogel particles.

This is the first monograph of its kind, which primarily meets the needs of scientists and engineers in the broad areas of polymer materials science, biomaterials engineering, biomedical engineering, sensor/actuator, micro-electro-mechanical system (MEMS) and BioMEMS, physics, chemistry, biophysics, biochemistry and bioengineering. It is especially useful for them as a reference source, and also if they wish to conduct further studies so as to extend their work to practical application. Other important primary readers are postgraduate students in the area of polymer materials science and biomedical engineering, especially those with involvement in the computational aspects such as the modelling and simulation of stimuli-responsive soft materials. Possible secondary readers include undergraduate students taking the advanced mechanical and electric engineering courses which involve sensor or actuator, MEMS and BioMEMS. The chapters on the fundamental theoretical development are especially useful to these students. Correspondingly, the course lecturers will also find this book a good reference source. This book provides both the casual and interested reader with insights into the special features and intricacies of smart polymer hydrogels when environmental stimuli are involved. It is also invaluable to design engineers in the polymer sensor/actuator, MEMS and BioMEMS industry and biomedical engineering, serving as a useful reference source with benchmark results to compare and verify their experimental data against.

The author would like to thank Professor Justin Hanes for his constant encouragement over the years, and especially for writing the Foreword to this book. Special thanks also go to Professors N.R. Aluru, Erik Birgersson, Khin-Yong Lam, Teng-Yong Ng and Zhigang Suo for their strong support and useful advice. Finally the author is very grateful to Drs J. Chen, J.Q. Cheng, J. Fu, R.M. Luo, Q.X. Wang, X.G. Wang, Z.J. Wang, S.N. Wu, G.P. Yan, Y.K. Yew and Z. Yuan for their invaluable contributions to this research.

Hua Li, Ph.D.
School of Mechanical & Aerospace Engineering
Nanyang Technological University
Singapore
Singapore, May 2009

Author's Brief Biography

Dr. Hua Li received his BSc and MEng degrees in engineering mechanics from Wuhan University of Technology, P.R.C., in 1982 and 1987, respectively. He obtained his PhD degree in mechanical engineering from the National University of Singapore in 1999. From 2000 to 2001, Dr. Li was a postdoctoral associate at the Beckman Institute for Advanced Science and Technology, University of Illinois at Urbana-Champaign. At the end of 2005, he was a visiting scientist (on invitation) at the Department of Chemical and Biomolecular Engineering of Johns Hopkins University. From 2001 to 2006, he was a research scientist in the A∗STAR Institute of High Performance Computing. Dr. Li is currently an assistant professor in School of Mechanical & Aerospace Engineering at Nanyang Technological University. His research interests include the modelling and simulation of MEMS focusing on the use of smart hydrogels in BioMEMS applications, the development of advanced numerical methodologies and the dynamics of high-speed rotating shell structures. He has co-authored a book on "*Rotating Shell Dynamics*" published by Elsevier and two book chapters, one on MEMS simulation and the other on hydrogel drug delivery system modelling, and authored/co-authored over 90 articles published in top international peer-reviewed journals. He is also extensively funded by agencies and industries, including the principal investigator of a Computational BioMEMS project awarded under A*STAR's strategic research programme in MEMS.

Foreword

Since the author told me that he planned to write a book about the mathematical modelling of smart hydrogels, I have looked forward to its completion with great anticipation. I have known the author and have followed his research for years; this work is a consummation of his extraordinary contributions in the area of soft active materials (SAMs). To write this book, the author has exhausted his knowledge, experience and free time. I believe this book will be of great value to both experts and also those with a casual interest in hydrogels, soft materials, drug delivery, BioMEMs and related fields of active scientific investigation. Its author, professor Hua Li, is among the most highly respected scientists in the world in the area of BioMEMs hydrogel theory. It is an honour to write the Foreword for his book.

Smart hydrogels have wide-ranging applications in bioengineering, such as in the development of soft sensors/actuators, controlled drug delivery systems and stimuli-responsive BioMEMS devices. The majority of relevant studies in these areas are experiment based; considerably less attention has been paid to theoretical aspects. This monograph provides a comprehensive and systematic study for modelling smart polymer hydrogels in the BioMEMS environment. It covers development of the models characterized in chemo-electro-mechanical multi-energy coupled domains and expressed in form of nonlinear partial differential governing equations for smart hydrogels. It also documents benchmark results, namely simulating the performance and predicting the characteristics of smart hydrogels responding to solution pH, externally applied electric voltage, environmental temperature, glucose/carbohydrates and salt concentration/ionic strength, which are basic stimuli in common BioMEMS devices.

This book is written in a straightforward manner without losing depth, such that it makes informative reading for a graduate student working or intending to work in this area. It will also undoubtedly serve as a rich reference source for experts in the field. I congratulate professor Hua Li for a tremendous achievement, and I hope and expect that the reader will benefit greatly from it.

Justin Hanes, Ph.D.
Professor of Chemical & Biomolecular Engineering
Director of Therapeutics, The Institute for NanoBioTechnology
The Johns Hopkins University
3400 N. Charles Street,
221 MD Hall, Baltimore, MD 21218

Contents

Chapter 1
Introduction

1.1 Definition and Application of Hydrogel

Hydrogel is a form of materials generally constructed by hydrophilic multiphase polymer mixture that may exhibit both solid-like and liquid-like properties. Its structural framework is formed from three-dimensional networks of randomly crosslinked polymeric chains that embody three different phases, namely solid polymer network matrix, interstitial water or biological fluid and ion species. A schematic drawing of the microscopic structure of charged hydrogel is shown in Fig. 1.1.

The solid portion of the hydrogel is a network of crosslinked polymer chains where their three-dimensional structure is usually described as a mesh, with the interstitial space filled up with fluid. The meshes of networks hold the fluid in place and also impart rubber-like elastic force that can compete with the expansion or contraction of thc hydrogcl, thus providing thc solidity of thc hydrogcl. Thc

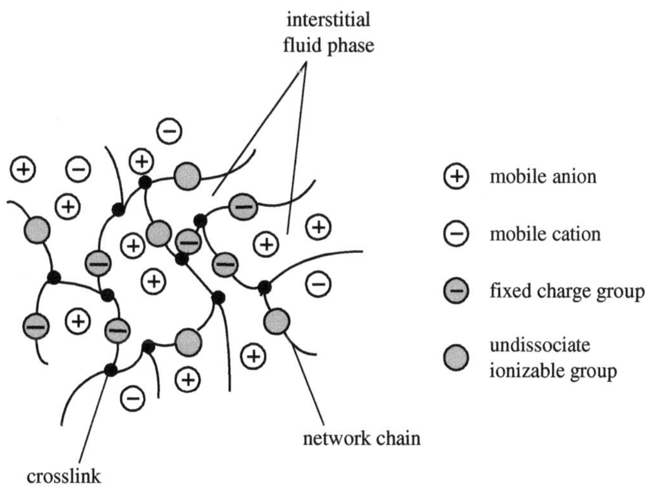

Fig. 1.1 Schematic microscopic structure of charged hydrogel

H. Li, *Smart Hydrogel Modelling*, DOI 10.1007/978-3-642-02368-2_1,

crosslinked polymer network may be formed physicochemically, for instance, by hydrogen bonding, van der Waals interaction between chains, covalent bond, crystalline, electrostatic interaction or physical entanglements. For the fluid phase of the hydrogel, it fills up the interstitial pores of the network and makes the hydrogel wet and soft, which is similar to biological tissues in some respects. Regarding the ionic phase of the hydrogel, generally it is composed of ionizable groups bound onto the polymer chains and a number of mobile ions which include counterions and co-ions due to the presence of electrolytic solvent that surrounds the hydrogel. The ionizable groups may dissociate in solution completely for strong electrolyte or partially for weak polyelectrolyte groups, and the network is left with the charged groups fixed to its chains.

There is a considerable variety of materials in either naturally existing or synthesizing, as examples of swellable hydrated polymeric gels. Crosslinked guar gum and collagens are the examples of the natural polymer that are modified to produce hydrogels. Examples of the synthetic hydrogels include *N*-isopropylacrylamide (NIPA), poly(acrylic acid) (PAA), poly(acrylonitrile) (PAN), poly(acrylamide) (PAM), poly(acrylonitrile)/poly(pyrrole) (PAN/PPY), poly(vinyl alcohol) poly(acrylic acid) (PVA–PAA), poly(hydroxyethyl methacrylate) (PHEMA). Depending on the preparation history of the hydrogel and the physical and chemical characteristics of the polymer, hydrogels may be categorized further into subclasses. For example, hydrogels can be synthesized to be either neutral or ionic, determined by the chemical characteristics of the pendant groups fixed onto the polymer matrix. From the point of physical mechanism, if the overall structure of hydrogels is homogeneous, the polymer chains have a high degree of mobility. If it is heterogeneous, there is a great deal of internal polymeric interactions and the polymer chains are virtually immobile at the molecular level. However, the eventual stability of the hydrogel depends on the interaction between the polymer matrix network and the aqueous medium where the hydrogel is immersed.

A smart or called environmental stimuli-responsive hydrogel is often synthesized when an ionic monomer is incorporated into the hydrogel network. The resulting charged group is generally termed the fixed charge since its mobility is much less than that of freely mobile ions within the interstitial fluid. The fixed charge groups produce electrostatic repulsion force among themselves, which influences greatly the expansion or contraction of hydrogel network. It is therefore known that the fixed charge density has an important effect on the electrostatic force and is able to play a substantial role in the change of degree of swelling/deswelling of the hydrogel.

The smart hydrogel is increasingly attracting more and more attention due to its great promise for a wide range of bioengineering applications, especially in bio-micro-electro-mechanical systems (BioMEMS). This kind of hydrogels is able to demonstrate the unique property of undergoing discrete or continuous volume transformation or volume phase transition in response to infinitesimal changes in external environment conditions, such as solution pH, electric field, temperature, solvent composition, glucose/carbohydrates, salt concentration or ionic strength, light/photon, pressure, coupled magnetic and electric fields. Usually the

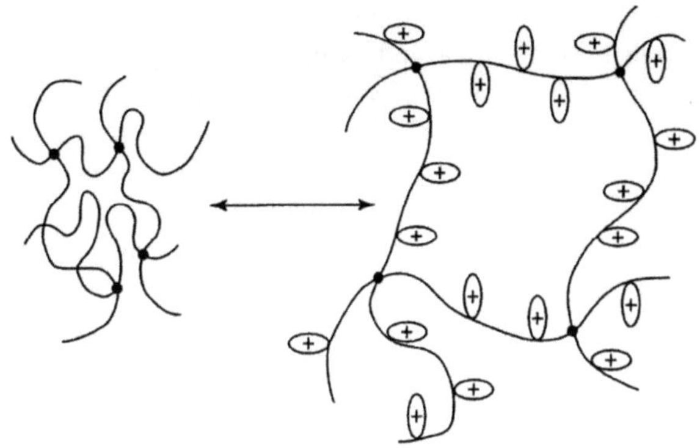

Fig. 1.2 Reversible expansion or contraction of the smart hydrogel due to environmental changes

volume transformation or volume phase transition due to environmental changes is reversible when the external environmental stimuli disappear (Fig. 1.2). These magnificent features make the hydrogel better known as smart (intelligent) or environmental stimuli-responsive hydrogels.

Due to the wonderful properties of the smart hydrogels, such as reversible swelling/deswelling behaviour, sorption capacity, novel mechanical property, high ionic conductivity, high environmental sensitivity, permeability and surface property, the hydrogels provide platform for creating novel smart materials for a wide range of BioMEMS applications as the highly manoeuvrable smart and adaptive hydrogels are able to respond to environmental changes and eventually induce structural changes without requirement of external power source. Examples of the successful BioMEMS applications of the smart hydrogels include microfluidic control, biomimetic biosensor/bioactuator, separation process and artificial muscle. Another exceptional promise of the hydrogels is their biocompatibility and biostability potentials, by which the hydrogels become excellent substitution for the human body tissues or biomimetic applications. There are also extensive explorations of the hydrogels in the medical and pharmaceutical applications, such as drug delivery system, articular cartilage, biomaterial scaffold, corneal replacement and tissue engineering.

1.2 Historical Development of Modelling Hydrogel

As mentioned above, the smart hydrogels as three-dimensional crosslinked hydrophilic polymer networks are capable of swelling/deswelling reversibly in water and retaining large volume of liquid in swollen state. They may perform dramatic volume transition in response to a variety of physical and chemical stimuli, where the physical stimuli include temperature, electric or magnetic field, light, pressure and sound, while the chemical stimuli include pH, solvent composition, ionic strength and molecular species. Owing to their smart or stimulus-responsive property, the hydrogels have been used for applications in numerous areas such as mechano-chemical actuators and sensors, environmental remediation, microfluidic control and separation, drug delivery and BioMEMS, artificial muscles and biomaterials for biomedical and tissue engineering.

The characteristics of volume transition of the smart hydrogels have drawn considerable attentions. The literature search reveals that numerous studies have been done. However, the majority of them are experimental based. For the few published studies that conducted mathematical modelling and numerical simulation, results were far from satisfactory, where comparisons with experimental data show significant discrepancies. This has resulted in ad hoc studies of these hydrogel materials mainly conducted by trial and error. This is a very time-consuming and inefficient process, and certain aspects of fundamental knowledge can often be missed or overlooked, resulting in off-tangent research directions. Thus it is absolutely necessary to establish a theoretical platform through development of better mathematical models with excellent capability of characterizing the multiphase hydrogels in multiphysics environmental conditions, and also the powerful numerical simulations for solution of these challenging highly nonlinear coupled models, where multiphysics and multiphase are common requirements.

In addition, with increment of biomedical applications, the smart hydrogels with more complex shapes are increasingly required and the accurate dimensional measurement of their volume transitions is becoming a challenge in experimental analysis. Therefore, computational modelling and simulation to predict the performance of the smart hydrogels become a critical tool for understanding of the characteristics of the smart hydrogels. On the other hand, when the optimization of hydrogel characteristics is required for specified applications, ready modelling and simulation will prove indispensable. In the past several years, some encouraging progresses were achieved in computational prediction of volume transition behaviours of the hydrogels in various case studies.

In this section, we summarize and review several typical existing models that have been widely used, and then categorize them roughly into two groups, the steady-state models for equilibrium simulation and the transient models for kinetics simulation, respectively. Comments on the models and several key parameters are also given for full understanding. It should be pointed out here that the present category is not done exactly since some models can be employed for both equilibrium and kinetics simulations. However, all we do here is to provide readers a clear outline of historical development and technological application of mathematical models and numerical simulations of the smart hydrogels. We believe that the review on the achievements and challenges is greatly instrumental to further study in this aspect.

1.2.1 Steady-State Modelling for Equilibrium of Smart Hydrogels

The present steady-state models focus on the responsive behaviours of the smart hydrogels at different equilibrium states. We provide an overview of the developed models in both macroscopic and molecular levels for simulation of volume transition behaviours of the smart hydrogels. The laws of energy, mass and momentum conservations provide good starting points for developing mathematical models for the hydrogel volume transition. Thermodynamic models, transport models, multiphase mixture theory and molecular simulation are discussed in details. Thermodynamic models provide the qualitative description of equilibrium volume transition of the hydrogels. Both the transport models and multiphase mixture theory are able to predict the equilibrium and kinetics volume transitions of the hydrogels. Molecular simulation provides the mechanism for understanding the hydrogel volume transition. Several key parameters are summarized for future model development.

1.2.1.1 Mathematical Models and Simulations

The volume transition of the smart hydrogel involves intricate chemical process, hydrophilic, hydrophobic and electrostatic interactions and mechanical deformation. In order to accurately characterize the volume transition behaviour of the hydrogel, it is absolutely necessary to develop computational models incorporating the multidiscipline complexity. Conservation laws, including the energy conservation law, the mass conservation law and the momentum conservation law, are good starting points for developing mathematical models simulating the volume transitions of the smart hydrogel. Four types of the typical existing steady-state models are thus reviewed here.

1. *Thermodynamic Models*

   Energy balance arises from the energy conservation law, which may be characterized by thermodynamics theory. Thermodynamic models offer simple approaches in order to describe the volume transition behaviours of neutral or ionic hydrogels in equilibrium. The equilibrium state of volume transition of the smart hydrogel in solvent is reached when the solvent inside the hydrogels is in thermodynamic equilibrium with that outside. This can be characterized in terms of the free energy, the chemical potential or the related osmotic pressure.
2. *Transport Models*

   Mass conservation law can be transferred into a mathematical form of diffusion equations to describe the concentration distributions of mobile ions inside and outside the hydrogels, where the transport models account for osmotic pressure, electrical potential, hydrogel stress, and thus the hydrogel volume transition associated with diffusive flux of mobile ions. The models can provide both transient and equilibrium simulations for description of the volume transition behaviour of the ionizable hydrogels.
3. *Multiphase Mixture Theory*

   Momentum conservation law can describe the stress due to friction in a shear flow and in a compression or expansion flow. The multiphase mixture theory supposes that the volume transition of the elastic hydrogel is controlled by the friction force in the hydrogel network. It encompasses the flow-dependent mechano-electro-chemical behaviours to balance the frictional force in hydrogels for description of volume transition behaviours of the hydrogel.

4. *Molecular Simulation*

   In order to deeply understand the volume transition behaviours of the hydrogel, an ideal approach is the modelling of various hydrophilic, hydrophobic and electrostatic interactions in molecular or even atomistic level, which is involved in the volume transition processes of the hydrogel. This can be achieved by molecular simulation, which incorporates various interactions with different energy expressions.

Thermodynamic Models

The most extensively used thermodynamic model was derived by Flory (1953) for description of the equilibrium volume transition of the hydrogels. The hydrogel at equilibrium state is treated as a special solution system, where the mesh chains cannot move freely to each other relatively, but the chains can become elongated in a non-interacting way during the volume transition process. The free energy of the neutral hydrogels is assumed to be a sum of the mixing free energy due to the mixing process and the elastic free energy due to the rubber elasticity. The osmotic pressure acting on the hydrogel network can be calculated according to the change of free energy.

The osmotic pressure associated with the change of mixing free energy, π_{mix}, is given by the Flory–Huggins polymer solution theory as follows (Flory 1953):

$$\pi_{\text{mix}} = -\frac{RT}{V_1}\left(\ln(1-\phi) + \phi + \chi\phi^2\right) \tag{1.1}$$

where R is the universal gas constant, T is the absolute temperature, V_1 is the molar volume of solvent and χ is the polymer–solvent interaction parameter. ϕ is the polymer volume fraction of the hydrogel, thus $(1-\phi)$ is the solvent volume fraction.

The osmotic pressure due to the change of elastic free energy, π_{el}, is obtained by the rubber elasticity theory, which in turn is derived from the Gaussian chain model as follows:

$$\pi_{\text{el}} = -\frac{RT}{V_1}N_c^{-1}\left(\phi^{1/3}\phi_0^{2/3} - \phi/2\right) \tag{1.2}$$

where ϕ_0 is the polymer volume fraction at the relaxed state which may be taken as preparation state and N_c is average number of segments in the network chain.

For highly crosslinked hydrogels, when the polymer chains contain fewer than 20 monomers between crosslinks, it is appropriate to use an expression for π_{el} that assumes the non-Gaussian chain statistics (Peppas, 1986). Several expressions describing the volume transition of the non-Gaussian polymer networks were proposed by Kovac (1977) and Galli (1982).

When the hydrogel volume transition reaches equilibrium, the total osmotic pressure of the hydrogel system, which is the sum of Eqs. (1.1) and (1.2), is equal to zero.

According to the equations, the polymer volume fraction at equilibrium state ϕ_e can be calculated by the following equation:

$$\ln(1-\phi_e)+\phi_e+\chi\phi_e^2+N_c^{-1}\left(\phi_e^{1/3}\phi_0^{2/3}-\phi_e/2\right)=0 \tag{1.3}$$

To account for the steep or discrete volume transition of the hydrogel, the χ parameter is assumed to be concentration dependent in form of $\chi = \chi_1 + \phi\chi_2$ (Erman and Flory, 1986). Both the parameters χ_1 and χ_2 are functions of temperature. Critical conditions for collapse of the neutral hydrogels require $\chi_1 = 1/2$ and $\chi_2 = 1/3$, which are easily met for a polymer–solvent system through adjustment of the crosslink density and/or attachment of charged ions onto the network chains.

The attachment of charged ions brings about the translational entropy of the counterions, which gives rise to the change of free energy of ions. The osmotic pressure due to ions, π_{ion}, is given as follows:

$$\pi_{\text{ion}} = \left(fRT/\overline{V}_r\right)\phi \tag{1.4}$$

where f is the charge density of the hydrogel, i.e. the fraction of segments bearing ionic groups. $\overline{V}_r$ is the molar volume of the polymer monomer.

Based on Eqs. (1.1), (1.2) and (1.4), the well-known Flory–Rehner relationship is obtained (Kovac, 1977):

$$\ln(1-\phi)+\phi+\chi\phi^2+N_c^{-1}\left[\phi^{1/3}\phi_0^{2/3}-\phi/2\right]-V_1\frac{f}{\overline{V}_r}\phi=0 \tag{1.5}$$

Adding salt with concentration C^s_{salt} in external solution results in variational distribution of mobile ions inside and outside the hydrogel. Influence of the electrolyte can be evaluated by the ideal Donnan equation, which gives rise to

$$\ln(1-\phi)+\phi+\chi\phi^2+N_c^{-1}\left[\phi^{1/3}\phi_0^{2/3}-\phi/2\right]-2(K-1)V_1C^s_{\text{salt}}-V_1\frac{f}{\overline{V}_r}\phi=0 \tag{1.6}$$

where K is the Donnan coefficient of the mobile ions inside and outside the hydrogel, and it is calculated by

$$K\left(K+\frac{f\phi}{\overline{V}_rC^s_{\text{salt}}}\right)-1=0$$

When the concentration of external electrolyte is larger than that of counterions belonging to the polymer, and the concentration difference of the mobile ions between the external electrolyte and the internal hydrogel is comparable in magnitude to the concentration, π_{ion} of the hydrogel with ionizable groups in electrolyte solution can be expressed in term of ionic strength as follows:

$$\pi_{\text{ion}} = RTV_1 \left(\frac{i^2 c_2^2}{4I} \right) \tag{1.7}$$

where i is the degree of ionization, c_2 is the concentration of ionizable polymer and I is the ionic strength of the medium for volume transition of the hydrogel.

Based on Flory's pioneering work, various modified models for predicting the volume transition behaviours of the hydrogel are proposed with the following assumptions:

- The elastic, mixing and ionic contributions to the free energy are independent of each other
- The chains possess the Gaussian statistics in approximation of the elastic contribution
- The end-to-end distance of a chain scales linearly with the linear extension of the hydrogel, namely so-called affine assumption
- The ionic contribution represents the coupled effect of the network being charged with the presence of mobile ions. The osmotic pressure originates from the electrostatically confined ions inside the hydrogel and the repulsive electrostatic interactions between the charged groups

Based on Eqs. (1.1) and (1.7) as well as the phantom network model, Caykara et al. (2003) derived the following equation for predicting the equilibrium swelling behaviour of copolymeric hydrogels containing monoprotic acid moieties:

$$\begin{aligned} &\left[\frac{K_a}{10^{-\text{pH}} + K_a} \right]^2 \frac{V_1 f_i^2}{4I\overline{V}_r^2} - \phi^{-2} \ln(1-\phi) - \phi^{-1} \\ &= \chi + \frac{\left(1 - 2/\Omega\right) V_1 \rho \phi_0^{2/3} \phi^{-5/3}}{\overline{M}_c} \end{aligned} \tag{1.8}$$

where $\overline{M}_c$ is the average molecular weight between the crosslinks, K_a is the dissociation constant of the anionic polymer, f_i is the molar fraction of the ionic unit in the hydrogel system, ρ is the density of the polymer network, pH is the pH value of medium surrounding the hydrogel and Ω is the functionality at the crosslinking site. The equation can be used to predict the equilibrium volume transition behaviours of the copolymeric hydrogels containing diprotic and triprotic acid moieties and the amphiphilic hydrogels after proper modification of the expression for i in Eq. (1.7) (Sen et al., 1998, 2000; Caykara et al., 2000). The above equation is used to determine the $\overline{M}_c$ and χ parameters of the copolymeric hydrogels.

Additional contribution of electrostatic interaction to the free energy for ionizable hydrogels was taken into account by Fomenko et al. (2002). Kramarento et al.

(2000) expressed the free energy of weakly charged hydrogel as a sum of six terms: (1) the elastic free energy, (2) the free energy of interaction of the monomer units, (3) the free energy gained from ion pairing, (4) the free energy associated with the translational entropy of mobile counterions inside the hydrogel and the outer solution, (5) the free energy of Coulombic interaction and (6) the free energy related to the combinatorial entropy of distribution of counterions between three possible states. The free energy was also described in terms of probability distribution function of different configurations (Huang et al. 2002). Hong and Bae (2002) combined the modified double-lattice model with Flory–Erman's elastic model and the ideal Donnan theory to capture the swelling behaviours of the electrolyte-bounded hydrogels. A network parameter κ ranging from 0 to ∞ is employed to describe the constraints on fluctuations of junctions from the surrounding chains in which they are embedded (Erman and Flory, 1986; Hong and Bae, 2002). $\kappa = 0$ when the network approaches a "phantom network" in the limit of high volume transition, while $\kappa \rightarrow \infty$ if the network is an "affine network". Contribution of hydrogen bonding to the free energy of hydrogels was also taken into account by several researchers. Annaka et al. (2000) incorporated a numerical constant Z to quantify the alteration of hydrogel osmotic pressure. The Z value depends on the chemical structure of hydrogen-bonded unit. The hydrogen bonding fraction was integrated into the mixing free energy term by Varghese et al. (2000).

However, the separation of the free energy of hydrogels into independent terms, i.e. the additivity of the different free energy contributions, is questionable. These terms all depend on the configuration of mesh chains. Therefore, they should be connected by the nature of hydrogel structure. This assumes the random mixing of polymer and solvent in hydrogel system, but this cannot hold owing to the hydrogen bonding between water and hydrophilic polymer and possible hydrophobic hydration at small water content. It also regards polymer monomers as unconnected, and the conformational entropy of polymer chains and their interactions with the solvent molecules are separated. This underestimates the monomer correlation of polymer chains and hence overestimates the intra-chain monomer repulsion in good solvent. On the other hand, the Gaussian chain assumption, which is the basis of the elastic free energy for most thermodynamic models, overestimates the elastic free energy for stretched chains. This is attributed to the two effects. First, the chains in most hydrogels do not behave in a Gaussian way. Second, the chains are not long enough to be taken as Gaussian way since the number of monomers between two crosslinks, N_c, is often less than 100.

In general, the parameters in the thermodynamic models are difficultly ascertained. Usually the estimated or best-fitted parameters are employed. This restricts the prediction accuracy of the thermodynamic models. Since they cannot incorporate temporal term into the system, the thermodynamic models cannot provide transient description for the kinetics volume transition behaviours of the hydrogel. This means that the thermodynamic models can be employed only for steady-state modelling of equilibrium of the smart hydrogels. The thermodynamic models provide a qualitative prediction of overall volume transition of the hydrogel. Acceptable agreement between the computational results and the experimental data may be obtained if some parameters such as f and χ are taken as adjustable parameters. In applications of the thermodynamic models, the adjustment of these parameters is critical for simulation of volume transition behaviours of some hydrogels. Two of these parameters, the effective charge density and the polymer–solvent interaction parameter, are of most importance, and they are further discussed in detail as follows:

1. *Effective Charge Density*

   Polyelectrolyte hydrogels with counterions in solution are known to perform different degrees of counterion association with the polyelectrolyte. Ion pairing with subsequent aggregation of ion pairs into multiplet structure can lead to the appearance of a so-called super-collapsed state of hydrogels, where practically all counterions form ion pairs (Khokhlov and Kramarenko, 1994). This results in the reduction of the hydrogel charge density and the alteration of solution properties. Therefore, the volume transition of ionizable hydrogels is not always a monotonous increasing function with increasing ionic group contents (Durmaz and Okay, 2000; Melekaslan and Okay, 2000; Okay and Durmaz, 2002). The Flory–Rehner theory including the ideal Donnan equilibrium reveals that the existence of counterions inside the hydrogel is responsible for the observed volume transition behaviours. The fixed charge density calculated from the ionic group content cannot accurately describe the contribution of ionic groups to the free energy of the hydrogel. Kramarenko et al. (2000) proposed that three possible states of counterions could be distinguished, namely the free counterions in the solution and within the hydrogels and bound counterions forming ion pairs with the charged units of polymer chains. The fraction of bound counterions depends essentially on the dielectric constant of the medium inside the hydrogel, which changes during the hydrogel volume transition (Khokhlov and Yu, 1994).

   The effective charge density is used to take into account the reduction of the charge density of hydrogels. It is obvious that the effective charge density is dependent on the hydrogel composition. For example, the effective charge density f_{eff} of poly(acrylamide(AAm)-*co*-sodium acrylate) hydrogels swollen in aqueous NaCl solution is a function of sodium acrylate mole fraction f_{SA} in form of $f_{\text{eff}} = 1 - 19f_{\text{SA}} + 155{f_{\text{SA}}}^2$ (Okay and Sariisik, 2000). Simulation indicates that not all the fixed charges inside the hydrogel have influence on the hydrogel volume transition. The charge distribution of hydrogels also strongly affects the effective charge density and thus the volume transition behaviours. In the *N*-isopropylacrylamide (NIPAAm)/acrylic acid (AAc) hydrogels, the localized carboxyl ions are ineffective in preventing the thermally induced collapse at temperature above the volume phase transition temperature, compared with the randomly distributed carboxyl ions (Ogawa et al., 2002).

   The effective charge density of hydrogels changes with the alteration of volume transition medium. The effective charge density of the NIPAAm/2-acrylamido-2-methylpropane-sulfonic acid (AMPS) hydrogel in aqueous poly(ethylene glycol) (PEG)-300 solution was reported to vary with the volume fraction of PEG-300 ϕ_{PEG} in addition to the mole fraction of AMPS

f_{AMPS} according to $f_{eff} = (31/N_s) f_{AMPS}^{0.54}(1 - A\phi_{PEG})$, where A is a constant describing the fraction of trapped counterions that form ion pairs in pure PEG and N_s is the number of segments on the network chains (Melekaslan and Okay, 2000). Electrokinetic studies also manifest that the effective charge density of poly(NIPAAm) varies with temperature (Makino et al., 2000a, b). The variation may be caused by the structural changes of the polymer chains and the water molecules in the hydrogels.

However, there is still certain degree of ambiguity as to the state of the associated counterions, which result in the decrease of effective charge density.

2. *Polymer–Solvent Interaction Parameter*

The polymer–solvent interaction parameter χ is a normalized interaction enthalpy. It is supposed that the nature of a pair interaction is not altered when the concentration of a component in the system changes. However, one must concede that the interaction of a given water molecule with another water molecule or with one of the polymer units in the hydrogel might be different. The interaction depends on the intensity of the molecule's link to a polymer chain or on the degree to which the polymer unit is screened by surrounding water molecules. According to the regular solution theory, the Flory–Huggins interaction parameter χ is calculated from Hildebrand solubility parameters (Madkour, 2001) as follows:

$$\chi = \frac{v}{RT}(\delta_1 - \delta_2)^2 \tag{1.9}$$

where v is the volume of a polymer segment, δ_1 and δ_2 are the Hildebrand solubility parameters of the polymer and the solvent, respectively. Three forces, namely the dispersion force, the polar force and the hydrogen bonding effect, make contribution to Hildebrand solubility parameter. However, their contributions are unequal (Hansen, 2000; Lindvig et al., 2002). Equation (1.9) clearly shows that the χ value decreases with the increase of temperature. This is observed in poly(methyl methacrylate (MMA))-methanol system, where $\chi = -1.985 + 919.23/T \pm 0.2$ is obeyed with 95% confidence limit (Hassan and Durning, 1999). However, it is found that χ of protein-based hydrogels increases when temperature is below their phase transition temperature (Lee et al., 2001). It may be due to specific interaction, e.g. hydrophobic hydration, and free volume change with temperature.

The ionization of dissociable groups in the hydrogels significantly changes the dispersion force, the polar force and the hydrogen bonding effect in the system, which has remarkable influence on the polymer–solvent interaction parameters. The polymer–solvent interaction parameter of AAc/methacrylated dextran hydrogels changes with varying ionization of AAc subject to different external pH conditions (Chiu et al., 2002), formulated by $\chi = 0.517 - 0.060 f_{AAc} - 0.061 f_{AAc}^2 - 0.310 f_{AAc}^3$ at pH 7.4, in which f_{AAc} is the volume fraction of ionizable AAc in hydrogels (Chiu et al., 2002a, b).

Hydrogel conformation and the hydrophobic hydration of the polymer chains may change during the volume transition process. Consequently, the polymer–solvent interaction parameters of the hydrogels vary with the polymer concentration at different degrees of volume transitions. It was reported that $\chi \approx 1/2 + \phi/3$ for poly(NIPAAm-*co*-AAc) hydrogel (Xue et al., 2001a, b). By considering the dependence of polymer–water interaction parameter on the polymer concentration as $\chi = 0.3875 + 0.518\phi$, the reentrant volume transitions of NIPAAm/AMPS hydrogels in aqueous PEG solution are successfully described (Melekaslan and Okay, 2001).

Addition of organic solvent may change the structure of water surrounding the hydrophobic polymers and the hydrogen bonding network. This complicates the dependence of polymer–water interaction parameter with the polymer volume fraction. Chuang et al. (2000) used a ternary interaction parameter $\chi_T = -1.0 + 0.02\phi_{\text{ethanol}} + 1.3\phi$, depending on both polymer and ethanol concentrations, in which ϕ_{ethanol} is the volume fraction of ethanol. The volume transition behaviour of poly(ethylene-*co*-vinyl alcohol) is reasonably predicted in the presence of this ternary interaction parameter.

Dependence of polymer–water interaction parameter on both temperature and polymer concentration is also proposed for poly(*N*-*t*-butylacrylamide (TBA)-*co*-AAm) hydrogels (Ozturk and Okay, 2002). Incorporating this sensitive dependence, the thermodynamic model may predict the volume transition of the hydrogels in satisfactory agreement to experimental data.

Transport Models

It is understood for the transport models that diffusion is a driving source to make the volume transition of the ionic smart hydrogels. Ion diffusive flux controls the osmotic pressure, the electrical potential of the network, the hydrogel stress, and thus the volume transition of hydrogel. Osmotic pressure is the directly driving force for volume transition of the smart hydrogels. Two expressions of the osmotic pressure lead to two different transport models.

A transport model for prediction of the volume transition of the ionic smart hydrogels was employed by Wallmersperger et al. (2001), which couples the Nernst–Planck equations with the Poisson equation and the mechanical equilibrium equation to simulate the diffusive ionic concentration, the electric potential and the hydrogel volume transition. The Nernst–Planck equation is given as follows:

$$\overline{D}_k \frac{\partial^2 c_k}{\partial x^2} + \mu_k z_k \frac{\partial c_k}{\partial x}\frac{\partial \psi}{\partial x} + \mu_k z_k c_k \frac{\partial^2 \psi}{\partial x^2} = 0 \tag{1.10}$$

where $\overline{D}_k$, μ_k, c_k and z_k are the effective diffusivity, the mobility, the concentration and the valence of the *k*th ionic species inside the hydrogel, respectively. The index *k* (1, 2, 3, ..., *N*) describes the *N* ionic species in the solution. ψ is the electric potential and it is modelled by the Poisson equation as

$$\frac{\partial^2 \psi}{\partial x^2} = -\frac{F}{\varepsilon\varepsilon_0}\left(\sum_{k=1}^{N} z_k c_k + z_f c_f\right) \tag{1.11}$$

where ε_0 is the dielectric constant of vacuum, ε is the relative dielectric constant of the solvent, F is the Faraday constant, c_f and z_f are the concentration and the valence of the fixed charge in the hydrogel, respectively. And the fixed charge concentration c_f is related to the hydrogel volume transition by (Grimshaw et al., 1990)

$$c_f = \frac{1}{H} \frac{c_f^0 K}{(K + c_{\mathrm{H}})} \tag{1.12}$$

where H is the local hydration state of the smart hydrogel, K is the dissociation constant of the fixed charge group, c_f^0 is the concentration of ionizable groups in the hydrogel before volume transition and c_{H} is the concentration of hydrogen ions within the hydrogel.

The mechanical equation governs the equilibrium of the hydrogel with the osmotic pressure, which is given as

$$\nabla \cdot \sigma = 0 \tag{1.13}$$

where σ is the stress tensor.

The osmotic pressure π is calculated by the concentration difference of mobile ions between inside and outside the hydrogels, which is controlled by the diffusion process

$$\pi = RT \sum_{k=1}^{N} \left(c_k^h - c_k^s \right) \tag{1.14}$$

where superscripts h and s represent the hydrogel and bulk solution, respectively.

The first transport model thus consists of Eqs. (1.10), (1.11), (1.12), (1.13) and (1.14) which are nonlinearly coupled. A self-consistent solution is thus required by iteratively solving the Nernst–Planck equation, the Poisson equation and the mechanical equilibrium equation. This model simplifies the interactions in the hydrogel system. The required parameters in the formulation are easily determined. It is a convenient way for simulation of both equilibrium and kinetics of volume transition behaviours of ionic smart hydrogels in aqueous solutions. However, the model neglects the interactions between the polymer and water in the hydrogels, which include the hydrophilic and hydrophobic interactions and the hydrogen bonding effects. These forces may not be negligible in certain ionic hydrogels, compared with the osmotic pressure due to the concentration difference of ionic species. Their contributions should be incorporated for accurate prediction of the volume transition behaviours of the smart hydrogels.

The second transport model considers the contributions of mixing, elastic and electrostatic free energies to the osmotic pressure, which was developed by Achilleos et al. (2000, 2001) for predicting the volume transition of polyelectrolyte hydrogels. The negligible inertia, electroneutrality and constant equal partial species densities are assumed in this model. The conservation laws for species mass and polymer mass, the mechanical balance and the overall charge balance for a mixture of four components ($k\equiv$ 1–4, $1\equiv H_2O$, $2\equiv Na^+$, $3\equiv Cl^-$, $4\equiv p$ the polymer) are given as follows:

$$\rho\frac{d_p w_k}{dt} - \rho\frac{w_k}{w_p}\frac{d_p w_p}{dt} + \nabla\cdot\left(j_k - \frac{w_k}{w_p}j_p\right) = 0 \tag{1.15}$$

$$\nabla\cdot v_p + \frac{1}{\rho}\frac{d_p\rho}{dt} + \frac{1}{w_p}\frac{d_p w_p}{dt} = 0 \tag{1.16}$$

$$\nabla\cdot\sigma - \nabla P = 0 \tag{1.17}$$

$$\sum_{k=1}^{p} z_k w_k \frac{M_1}{M_k} = 0 \tag{1.18}$$

where ρ, v_P and P are the mixture density, the polymer velocity and the hydrostatic pressure, respectively. w_k, j_k and M_k are the weight fraction, diffusive flux and molecular weight of the kth species, respectively. d_P/dt is the substantial derivative following the polymer velocity. The diffusive flux of the kth component is given as

$$j_k = -\overline{D}_k w_k\left(\nabla\pi_k + \frac{M_k}{M_1}\nabla\cdot\sigma + z_k\nabla\psi\right) \tag{1.19}$$

where π_K is the osmotic pressure of the kth component and it is derived from the free energy, which sums up the mixing, elastic and electrostatic contributions by

$$\begin{aligned}\pi_k = {} & (\ln w_k + \chi_{kp}w_p + 1) - w_P\sum_{\beta=1}^{p-1}\frac{\chi_{\beta p}w_\beta M_k}{M_\beta} - \sum_{\beta=1}^{p-1}\frac{w_\beta M_k}{M_\beta} \\ & + C_{DH}\left(I^{1.5}\frac{M_k}{M_1} - 3z_k I^{1.5}\right) + C_{mf}i^2 w_p\left(-\frac{z_k^2 M_1}{IM_k} + 1\right)\end{aligned} \tag{1.20}$$

where C_{DH} and C_{mf} are two determined constants.

The network stress is related to the polymer velocity v_P and d_P/dt and given as

$$\sigma = \frac{d_P\sigma}{dt} - \sigma\cdot\nabla v_P - \nabla v_P^T\cdot\sigma + \sigma\nabla\cdot v_P = G_0\frac{\phi_p}{\phi_p^0}\left(\nabla v_P + \nabla v_P^T\right) \tag{1.21}$$

where G_0 is the hydrogel modulus, ϕ_p and ϕ_p^0 are the polymer weight fraction in swelling and at gelation, respectively.

The second transport model thus consists of Eqs. (1.15), (1.16), (1.17), (1.18), (1.19), (1.20) and (1.21) for simulation of both equilibrium and kinetics of volume transition behaviours of the polyelectrolyte hydrogels. This model may be understood as a transport model combined with the modified thermodynamic model. The volume transition of the hydrogels is still considered to be driven by the ionic diffusion, but the osmotic pressure is contributed by the mixing, elastic and electrostatic free energies. Though it takes into account the interactions between the polymer and solution in the hydrogels, it needs additional parameters to be determined. In general, it is accurate enough for the polyelectrolyte hydrogels with low degree of ionization. The transport model creates the relationship between the hydrogel volume transition behaviours and the ion diffusion process. It can thus provide both the equilibrium and kinetics descriptions of the volume transition behaviours of hydrogels, which is an advantage over the thermodynamic models. The model is able to achieve quantitatively the simulation of volume transition behaviours of the hydrogels when a suitable expression of osmotic pressure is employed.

As well known, the parameters required by the transport models include mainly the diffusive coefficients of the mobile species and the elastic material properties of the hydrogels. Constant or concentration-dependent diffusive coefficients are employed. The elastic material properties of the polymer network may become difficult to be determined when the geometrical sizes of hydrogels are very small. Accurate values of the elastic material properties are important when the transport models are used. However, the elastic material properties are observed to change during the volume transition process. Therefore, it is necessary to make a further discussion of the relationship between the elastic material properties and the volume transition behaviours of hydrogels.

The elastic property of hydrogels falls between two idealized limits, namely the affine and phantom. They relate the network structure to the elastic deformation subject to applied stress. In the affine model, crosslinks are assumed to be affine with respect to macroscopic strain and the fluctuations of the junctions are completely suppressed. In the phantom model, chains are considered immaterial or phantom and can freely cross each other. The chains exert forces on the crosslinks, which can move independently from the applied stress.

The crosslink density has a primary effect on the elastic property of hydrogels. The crosslinking process of the polymer network matrix partially replaces the weak van der Waals intermolecular bonds by the strong covalent bonds, which evidently increases the mechanical strength of the polymeric network. The increase of crosslinking temperature is observed to monotonically decrease the elastic modulus due to the spatial heterogeneity of the crosslinked hydroxypropylcellulose (HPC) hydrogel (Valente et al., 2002).

Theoretically, the increase of charge density should monotonically decrease the elastic modulus of hydrogels. However, the elastic modulus of the AAm/AMPS hydrogels is found to increase first with increasing the charge density and decrease subsequently (Okay and Durmaz, 2002). This demonstrates two opposite effects of charged groups on the elastic modulus of hydrogels: (1) the formation of multiplets to increase the elastic modulus of the ionic hydrogels and (2) the effect of electrostatic interaction of charged groups to decrease the elastic modulus. The initial increase of the modulus with increasing the charge density is connected with the condensation of counterions to ion pairs. The multiplets act as additionally physical crosslinks in the hydrogel and contribute to the elastic modulus by increasing the effective crosslink density of the hydrogel. The decrease of the elastic modulus at higher ionic group content is associated with the contribution of electrostatic interaction to the conformational change of the hydrogel chains.

The medium surrounding the hydrogels has remarkably influences on the elastic property of the hydrogels. Young's modulus of pH-sensitive hydrogel is dependent on the medium pH (Johnson et al., 2002), while the negative dependence of the elastic modulus to temperature is observed in the crosslinked PAAm hydrogel (Muniz and Geuskens, 2001). In the collapsed state, the elasticity of poly(NIPAAm) hydrogel depends on the concentration and composition of salt in the aqueous solutions (Ikehat and Ushiki, 2002), probably due to the viscoelasticity from the shrinkage of the polymeric hydrogel.

The elastic behaviour of hydrogels varies with the state of volume transition of the hydrogel. Young's moduli of the hydrogels at swelling and deswelling states are remarkably different (Johnson et al., 2002). The modulus is inversely proportional to the swelling degree (Whiting et al., 2001). The relation between the shear modulus and the swelling degree for covalently crosslinked poly(aldehyde guluronate) hydrogels shows a linear logarithm relation (Lee et al., 2000). Three different regions of variation of Young's modulus were reported with the degree of volume transition of a chitosan–poly(ethylene oxide) (PEO) hydrogel (Khalid et al., 2002). The first region is a slight decrease of Young's modulus, which may be explained by the peripheral swelling of the hydrogel due to the uniform diffusion of the aqueous medium towards the centre of the hydrogel. The second region is a distinct decrease of Young's modulus, which is ascribed to the arrival of the swelling border in the centre part of the hydrogel. Finally, a steady Young modulus of the polymeric network is described with a stationary hydration of the hydrogel. The relations between the solution fraction and the equilibrium shear modulus are derived with a statistical theory (Franse and Nijenhuis, 2000).

Multiphase Mixture Theory

In the multiphase mixture theory, it is assumed that the volume transition of the hydrogels is driven by the gradient of chemical or electrochemical potentials, which is balanced by the frictional force between the phases when one phase flows through the other (Zhou et al., 2002). The hydrogel system is divided into three phases, which are solid phase (denoted by s), water phase (denoted by w) and ion phase containing both anion (denoted by $-$) and cation (denoted by $+$). It is also assumed that the three phases are intrinsically incompressible, whereas the hydrogel as a whole is compressible through exudation of water. If both the body and the inertial forces are neglected, the momentum balance equations are given as

$$\nabla \cdot \sigma = 0 \tag{1.22}$$

$$-\rho^w \nabla \mu^w + f_{ws}(v^s - v^w) + f_{w+}(v^+ - v^w) + f_{w-}(v^- - v^w) = 0 \tag{1.23}$$

$$-\rho^+ \nabla \tilde{\mu}^+ + f_{+s}(v^s - v^+) + f_{+w}(v^w - v^+) + f_{+-}(v^- - v^+) = 0 \tag{1.24}$$

$$-\rho^- \nabla \tilde{\mu}^- + f_{-s}(v^s - v^-) + f_{-w}(v^w - v^-) + f_{-+}(v^+ - v^-) = 0 \tag{1.25}$$

where $\tilde{\mu}^k$ is the electrochemical potentials for the kth species, μ^w is the chemical potential of fluid phase, $f_{\alpha\beta}$ is the frictional coefficient per unit tissue volume between the inter-diffusing α and β components and it is assumed to be symmetric, i.e. $f_{\alpha\beta} = f_{\beta\alpha}$.

The saturation condition is given as

$$\sum_{j=s,w,+,-} \phi^j = 1 \tag{1.26}$$

where $j = s$, w, $+$ and $-$. $\phi^j = dV^j/dV$ is the volume fraction of the jth component.

The continuity equation, electroneutrality condition and continuity equation of fixed charge group are given as follows:

$$\frac{\partial \rho^j}{\partial t} + \nabla \cdot (\rho^j v^j) = 0 \quad (j = w, s, +, -) \tag{1.27}$$

where v^j is the velocity of component j and ρ^j its mass density

$$\sum_{j=+,-} z_j c_j + z_f c_f = 0 \tag{1.28}$$

$$\frac{\partial(\phi^w c_f)}{\partial t} + \nabla \cdot (\phi^w c_f v^s) = 0 \tag{1.29}$$

For an isotropic hydrated charged hydrogel with infinitesimal deformation, the constitutive equations are given as follows:

$$\sigma = PI + \lambda_s tr(E) I + 2\mu_s E \tag{1.30}$$

$$\mu^w = \mu_0^w + \frac{1}{\rho_T^w}[P - RT\xi(c_+ + c_-)] \tag{1.31}$$

$$\tilde{\mu}^+ = \mu_0^+ + \frac{RT}{M_+}\ln(\gamma_+ c_+) + \frac{z_+ F\psi}{M_+} \tag{1.32}$$

$$\tilde{\mu}^- = \mu_0^- + \frac{RT}{M_-}\ln(\gamma_- c_-) + \frac{z_- F\psi}{M_-} \tag{1.33}$$

where I and E are the identity tensor and the elastic strain tensor of the solid phase, respectively; λ_s and μ_s are the Lamé coefficient and the shear modulus of solid phase, respectively; μ_0^k is the chemical potential at reference; ρ_T^w is the true mass density of water; γ_+ and γ_- are the activity coefficients of cation and anion, respectively; M_+ and M_- are the molecular weights of the cation and anion, respectively and ξ is the osmotic coefficient.

The multiphase mixture model is applicable to the hydrogels with linear volume transition. It simplifies the interactions between the phases of the water, mobile ion and the polymer network, without consideration of variation of the hydrogel elastic properties in volume transition process. At the same time, the model assumes the constant frictional coefficients between the phases. This directly restricts the applicability of the model. However, this model can describe both the equilibrium and kinetics of volume transition behaviours of the hydrogels.

Molecular Simulation

Unfortunately, the three kinds of the models discussed above, the thermodynamic models, the transport models and the multiphase mixture models, cannot evaluate the hydrophilic, hydrophobic and hydrogen bonding interactions involved in the hydrogels. However, the dynamic and structural characteristics of the hydrogels are accessible via molecular simulations. Over the past years, it is demonstrated that the molecular simulation is a uniquely useful tool for investigating the properties of polymer networks since it allows the construction and investigation of near-perfect model networks with well-characterized structures. It is able to examine the interaction between the electrolyte and the polymer matrix, which is much less understood. Fermi resonance perturbed Raman spectroscopy reveals that there exists weak polymer–electrolyte interaction between the poly(MMA) and the solvent molecules, ethylene and propylene carbonate (Ostrovskii et al., 2003). Ab initio molecular orbital calculation confirms that the hydrogen bonding hydration of the ion–water clusters to the polymers is destabilized by anions but stabilized by cations via ionic hydration in most cases (Muta et al., 2001a, b). It is understood that the volume transition of the hydrogels is based on the electron pair acceptance and electron pair donation change of water through the ionic hydration.

Molecular simulation can provide examination of the mechanism of the volume transition. Using the Monte Carlo and molecular dynamics techniques, the swelling of weakly charged polyelectrolyte hydrogel in poor solvent is simulated (Lyulin et al., 1999). Splitting of the spherical polyelectrolyte globule into a dumbbell-type structure accompanied by a sharp increase in interaction occurs in the chain's radius of gyration upon increasing in the Coulombic repulsion. By simultaneously treating the electrostatics and chain connectivity, the swelling of the crosslinked polyelectrolyte hydrogels is simulated by Monte Carlo technique (Schroder and Oppermann, 2002). This reveals that the assumption of Gaussian statistics in the hydrogel theories holds for the polymeric hydrogels but not for the polyelectrolyte hydrogels, whereas the affine assumption seems to be reasonable only for the polyelectrolyte hydrogels in equilibrium with pure solvent. Molecular dynamics simulation at an atomistically detailed level is carried out to predict the polymer volume transition by a two-step method in combination of the thermodynamic integration approach with Widom's particle insertion method (Nick and Suter, 2001). The quality of simulation depends only on the energy model used, i.e. on the force field for the atom–atom interactions. The simulation indicates the concentration dependence of the polymer–water interaction parameter, expressed by $\chi_{\text{BPA-PC}} = 4.7 - 6.6\phi_w$ for the bisphenol-A-polycarbonate (BPA-PC) and $\chi_{\text{PVA}} = 2.6 - 5.9\phi_w$ for the poly(vinyl alcohol) (PVA), respectively, where ϕ_w is the volume fraction of water in the network. A statistical mechanics combined with the molecular dynamics simulation evaluates the influence of the volume transition on the entropy of mixing

(Hansen, 2000). The loss in entropy due to the deformation of the chains takes on a negative value, and the final value of the mixing entropy decreases in both coiling and swelling.

Molecular dynamics is also combined with other approaches to simulate the volume transition behaviours of the hydrogels. The swelling of athermal networks in an athermal hard sphere solvent is simulated by a discontinuous molecular dynamics combined with Monte Carlo simulation technique (Kenkare et al., 2000). An elastohydrodynamic approach is employed to model the hydrogel undergoing a large volume change during solvent diffusion (Barriere and Leibler, 2003). It couples the diffusion process with the large elastic deformation undergone by the polymer network. The solvent–polymer friction is described as a liquid flowing through a porous medium. The network elasticity over the range of volume fraction is divided into the isotropic and non-isotropic parts. Employing a Gibbs ensemble molecular dynamics method, the dynamic and structural results for equilibrium swelling of a model network in contact with different chain-like solvents are simulated by Lennard–Jones non-bonded interaction (Aydt and Hentschke, 2000). The swelling behaviour of model polymer networks in different solvents subject to the controlled thermodynamic conditions is simulated by a "two-box particle-transfer" molecular dynamic simulation method (Lu and Hentschke, 2002a, b). The solvent chemical potentials are calculated via the Widom's test particle method for one-site solvents or the Rosenbluth sampling method for chain-like solvents, in which the simulation results are in qualitative agreement.

Although the molecular simulation techniques provide the better understanding of the mechanism of both polymer–solvent interaction and polymer network conformation, the simulation is computationally intensive and time consuming. The selection of energy description models is also crucial to the simulation accuracy. It is limited to the hydrogels with relatively small structures. Therefore, it is difficult to practically predict the volume transition behaviour of the smart hydrogels with complex structures.

Remarks

We have discussed four types of models that are developed and based on different assumptions and approximations. With basic assumptions of the different contributions into the osmotic pressure, the thermodynamic models provide a tool for qualitative estimation of overall volume transition of the hydrogels in different equilibrium states. These thermodynamic models establish the relationship between the volume transition and the important material properties for the hydrogels. Transport models and multiphase mixture theory can quantitatively achieve both the equilibrium and kinetics simulations of volume transition behaviours of ionic smart hydrogels with complex shapes. The transport models suppose that the volume transition of the ionic smart hydrogels is driven by ionic diffusion and the osmotic pressure is the directly driven force for the volume transition of the hydrogel. These transport models establish the relationship between the volume transition behaviours of the hydrogels and the ion diffusion process and elastic stress. By formulation of the osmotic pressure in the thermodynamic models, the transport models can also take into account important material parameters of the hydrogels. The multiphase mixture theory assumes that the hydrogels are composed of intrinsically incompressible solid phase, incompressible water phase and ion phase, and the volume transition of the hydrogels results from exudation of water driven by the gradients of chemical potential and electric potential. The theory generates the relationship between water flow and frictional forces. By integrating various hydrophobic, hydrophilic and electrostatic interactions involved, the molecular simulation makes it instrumental in investigate the mechanisms of volume transition of the hydrogels and to evaluate the influences of various environmental and material parameters. However, the requirement of intensive computation restricts its application to the small scale of the hydrogels.

1.2.1.2 Key Parameters in Steady-State Modelling for Equilibrium of Hydrogels

It is known that there is a variety of parameters having influences on the volume transition behaviours of the smart hydrogels. For example, the ionic additives affect electrostatic interactions, the surfactants implicate the hydrophilic and hydrophobic forces and the organic solvents uphold the involvement of hydrogen bonding. Even the same additives may have remarkably different influences in different hydrogels. Successful mathematical models should be able to capture the influences of various parameters that significantly affect the volume transition of the smart hydrogel, which is definitely a challenge for modelling development and simulation. Several key parameters are summarized below.

Hydrogel Composition

The swelling/deswelling behaviour of the smart hydrogel is distinctly influenced by the monomer composition. Incorporating more hydrophilic or hydrophobic monomers in hydrogel composition is a useful approach to regulate the volume transition behaviour of the hydrogel. For example, the copolymerization of charged monomers is a well-tested strategy to increase hydrophilicity in polymeric chains and thus obtain the polyelectrolyte hydrogels. Among the charged monomers, AMPS and [(methacrylamido)-propyl]trimethyl ammonium chloride (MAPTAC) have received more attention due to their strongly ionizable sulfonate groups and complete dissociation in the overall pH range. It is demonstrated that the hydrogels incorporated with AMPS and MAPTAC can exhibit pH-independent swelling behaviour (Varghese et al., 2000; Ozmen and Okay, 2003).

In general, the increase of charged monomer contents of the hydrogels increases the degree of volume transition (Katime and Rodriguez, 2001). One of the reasons is the simultaneous increase of counterions inside the hydrogels, which generates an additional osmotic pressure that expands the hydrogels. However, the degree of volume transition of the hydrogel is not always a monotonically increasing function of the charged monomer contents of the hydrogel. It may increase first with increasing ionic group concentration but then level off after passing a critical concentration due to the finite extensibility of the network chains and due to the ion pair formation (Durmaz and Okay, 2000; Melekaslan and Okay, 2000; Okay and Durmaz, 2002). Some hydrogels show a maximum degree of volume transition at certain ionic group concentrations. For example, the swelling ratio of poly(NIPAAm-*co*-AAc) hydrogel reaches to the largest value at 20 mol% of AAc content (Yoo et al., 2000), the poly(NIPAAm-*co*-methacrylic acid (MAAc)) hydrogel shows the largest swelling at 30 mol% of MAAc content (Diez-Pena et al., 2002) and the swelling ratio of poly(AAm-*co*-AAc) hydrogel reaches to the maximum value at 50 mol% AAc (Bouillot and Vincent, 2000). In exception, a slightly decreased swelling ratio with increasing content of charged monomer is observed in poly(NIPAAm)/chitosan hydrogels (Lee and Chen, 2001). It is proposed that addition of chitosan increases the hydrogel crosslinking and makes the network structure denser and more hydrophobic. When the zwitterionic monomer is appropriately introduced, the volume transition of the hydrogels may also be enhanced. The swelling ratio of thermo-sensitive poly(NIPAAm-*co*-1-vinyl-3-(3-sulfopropyl) imidazolium betaine (VSIB)) hydrogels in salt solution increases with increasing zwitterionic VSIB content (Lee and Yeh, 2000).

Incorporating more hydrophilic neutral monomers may also increase the degree of the volume transition of corresponding hydrogels. The hydrophilic PEG and PEO polymers are often incorporated in the hydrogels to remarkably increase the volume transition (Cho et al., 2000; Bajpai and Shrivastava, 2002). The increase of the hydrophilic AAm content in the NIPAAm/AAm hydrogels gives rise to a higher equilibrium swelling ratio (Yildiz et al., 2002). The swelling ratio of the PVA/poly(NIPAAm) interpenetrating network (IPN) hydrogels increases with the increase of the hydrophilic poly(NIPAAm) content in the IPNs (Kim et al., 2003). However, the swelling ratio of the poly(2-hydroxyethyl methacrylate)(HEMA)/PEG IPN hydrogels increases first with increasing PEG concentration, but then decreases monotonously after PEG concentration increases beyond 0.33 g (Bajpai and Shrivastava, 2002).

In general, the volume transition of the hydrogels decreases with integrating more hydrophobic comonomers (Bajpai and Giri, 2002). The swelling ratio of the poly(NIPPAm-*co*-AAm-*co*-HEMA) hydrogels decreases with increasing the content of hydrophobic AAm or HEMA (Yildiz et al., 2002). The formation of intermolecular hydrogen bonding between hydroxyl and amino groups decreases the hydrophilic group content of the hydrogels and thus the affinity towards water decreases. A more interesting observation is that, with increase of the hydrophobic monomer content of the di-*n*-propylacrylamide (DPAM), di-*n*-octylacrylamide (DOAM) and didodecylacrylamide (DDAM), the swelling ratio increases in the poly(NIPAAm) hydrogels (Xue and Hamley, 2002).

The hydrogels even with the same composition may display different volume transition behaviours with different polymer architectures. The hydrogels made of random copolymer chains swell much larger than those composed of the triblock copolymer chains (Simmons et al., 2000; Triftaridou et al., 2002), which is attributed to the absence of microphase separation in the former type of network and the presence in the latter.

Preparation History

Variation of the synthesis methods or formation conditions of the smart hydrogels provide another approach to control the volume transition of the hydrogel. The hydrogel microstructure and degree of topological constraint depend on the preparation conditions. The hydrophobic interactions in the temperature stimulus-responsive hydrogels also change with formation conditions (Suetoh and Shibayama, 2000).

The monomer concentration has the most important effect, relatively compared with other preparation conditions. The swelling ratio can be controlled by preparing certain monomer concentration (Furukawa, 2000). The mechanism of volume transition of the poly(AAc) hydrogel is dependent on the preparing concentration of AAc in solution during γ-irradiation in addition to pH of the medium of the volume transition (Jabbari and Nozari, 2000).

Solvent in the hydrogel preparation can change the porosity and crosslinking of the formed product. Ethanol as solvent is more effective than water for creating the porous structure of the hydrogels (Pradas et al., 2001). The primary cyclization in the hydrogels is facilitated by increasing the solvent concentration. The primary cyclization decreases the crosslinking density and increases the molecular weight between crosslinks. Consequently, this increases the degree of the volume transition of the hydrogels. On the other hand, the hydrogels prepared in bulk conditions exhibit much poorer volume transition behaviour than those in solution (Loh et al., 2001).

Variation of physical preparing conditions, such as the heating time, radiation dose and applied electric field, also changes the volume transition of the hydrogel. The degree of the volume transition of poly(AAm-sodium acrylate-PVA) hydrogel increases to 1200 g water per 1 g swollen polymeric network after heat treatment at 120°C (Vashuk et al., 2001). The increasing porosity of the polymeric network results in the decrease of elastic modulus and thus the increase of the volume transition. Schmidt et al. (2003) reported that the poly(vinyl methyl ether) (PVME) structures formed by radiation crosslinking of PVME in aqueous solution can be microgel particles, porous hydrogels, branched molecules and homogeneous hydrogels, which are influenced by preparation parameters including radiation dose and irradiation temperature. The poly(NIPAAm-*co*-AAc) hydrogels prepared under constant electric current show gradual phase transition in response to temperature, which is in contrary to the general overall phase transition in aqueous solution (Shin et al., 2003). The copolymeric hydrogels start the phase transition from the section polymerized on the anode side and gradually change to the section polymerized on the cathode side when the temperature is increased. The gradual phase transition arises from the protonation of AAc near the anode and the ionization of AAc near the cathode by the electric current.

The conditions of hydrogel pretreatment after preparation also have influences on the volume transition behaviours. When the nonionic NIPAAm hydrogel is dried before placed in the aqueous ethanol solution, a single phase transition occurs when the ethanol concentration increases, whereas three phase transitions are observed when the hydrogel is immersed directly in the aqueous solution without drying process (Huther et al., 2002). Washing the prepared hydrogels results in a considerable reduction of volume transition of the NIPAAm/sodium methacrylate hydrogels in aqueous ethanol solutions, which is ascribed to the exchange of hydrogen ions with sodium ions and the conversion of completely dissociated sodium methacrylate groups into weakly dissociated methacrylic acid groups (Huther et al., 2002).

Electrolytes

Addition of electrolyte in the medium of volume transition of the hydrogels may alter the interactions of polymer–polymer, polymer–water, polymer–ion and water–ion. The addition of salt ions may enhance the polymer–water mixing conditions (salting-in) to enhance the volume transition or may impair the conditions (salting-out) to suppress the volume transition. It is observed that the salts of bromide and chloride exert the salt-out effect, whereas the iodide produces the salting-in effect on the volume transition of the copolymeric hydrogels of *N*-acryloyl-*N*′-alkyl piperazine and MMA (Gan et al., 2001). Usually the cations exhibit the salting-out effect and the volume transition decreases in the well-known Hofmeister order (Muta et al., 2001c). The decrease of degree of the volume transition of the hydrogels in aqueous salt solution is dependent on the counterion species, irrespective of the co-ions (Atta, 2002). However, the relative order of cations may reverse with different anions due to ion pair effects. The Hofmeister order for anions is also observed in the salting-out property. For example, the order of effectiveness in the depression of the degree of the volume transition of the HEMA/PEG hydrogel in aqueous salt solutions follows the sequence $Cl^- > Br^- > I^-$ (Bajpai and Shrivastava, 2002a, b, c). The Hofmeister order demonstrates the promotion of hydrophobic associations that arise from the alteration of water structure and thus the free energy of interaction between polymer and water. The low-frequency Raman scattering experiments reveal that the presence of alkali metal chloride salts tends to reduce the chemical potential of water, which causes the exclusion of water molecules from the NIPAAm hydrogel network chains, probably the proximity of isopropyl groups (Annaka et al., 2002). The Hofmeister order shows that Cit^{3-} is the water structure breakers, whereas SCN^- is the water structure maker. Therefore, Cit^{3-} efficiently suppresses the volume transition of poly(NIPAAm-*co*-dimethylaminoethyl methacrylate) hydrogel, whereas SCN slightly increases the swelling in the concentration range from 10^{-5} to 10^{-3} M (Zha et al., 2002).

The salt ions may change the electrostatic interactions within the hydrogels. They may shield the ionized groups attached onto the polymeric network chains and screen the ion pair interactions. Shielding of ionized 16-acryloylhexadecanoic acid (AHA) groups results in the disorder–order transition of poly(AHA-*co*-AAc) hydrogel (Matsuda et al., 2000). The shielding effect caused by the counterions drastically decreases the swelling of poly(MAAc) in NaCl solution when the ionic strength increases (Zhang and Peppas, 2000). Screening of the intra-chain and intra-group interactions within the poly(1-(3-sulfopropyl)-2-vinyl-pyridinum-betaine) hydrogel remarkably increases the swelling of the hydrogel in dilute aqueous salt solution (Xue et al., 2001). In particular, the anti-polyelectrolyte behaviour of the amphoteric hydrogels is observed in poly(*N*-vinylimidazole-*co*-sodium styrene sulfonate) hydrogels (Valencia and Pierola, 2001). The swelling ratio increases slightly upon increasing the ionic strength of the aqueous salt solution. This is due to the reduction of intra-group and intra-chain aggregations of zwitterionic monomer units by the external electrolytes. The electrostatic interaction in electrolyte solution significantly changes the swelling behaviour of the smart hydrogel subject to externally applied electric field due to the movement of the ions towards their counter-electrodes. The increase of NaCl concentration enlarges the bending angle of the PVA/chitosan IPN hydrogel under electric field (Kim et al., 2002). It is noted that the PVA/poly(diallyldimethylammonium chloride) IPN hydrogel performs the largest bending angle in 0.2 wt% NaCl aqueous solution (Kim et al., 2003a, b), which may be attributed to the coupled effect of increasing free ion movement from the surrounding solution towards their counter-electrodes or into the hydrogel and the shielding effect of the poly-ions by the ions in the electrolytic solution.

Ion-specific interactions with hydrophobic sites of the hydrogels complicate the salt influences on the swelling of hydrogels. The vibrational Raman spectroscope study reveals that the addition of lithium perchlorate into poly(AAm) hydrogel strongly disturbs the hydrogen bonding of both the polymer network and the interstitial water due to the specific interaction of lithium with amide groups and causes the structural breakdown by the perchlorate anions (da Costa and Amado, 2000). The amide groups become more accessible to chemical transformations and the hydrogel flexibility increases. It is known that a specific divalent–polymer interaction reduces the effective charge per poly(AAc) polymeric chain at various degrees and the interaction seems to be related to the energy of hydration of the cations (Boisvert et al., 2002). The ion–polymer binding increases with increasing the electrolyte concentration, which leads to the increase of hydrophilicity and thus the degree of swelling of the hydrogels (Lobo et al., 2001). However, the further increase of salt concentration beyond an optimum binding of the ions with polymer chains might reduce the volume transition of the hydrogel (Bajpai and Shrivastava, 2000). In the presence of the thiocyanate ions (CNS^-) in the concentration range from 0.01 to 0.40 M, the swelling ratio of the poly(HEMA)/PEG IPN hydrogels increases with increasing the salt concentration due to the formation of the thiocyanate–polymer complexes. The polymer-bound thiocyanate ions render the hydrogels negatively charged (Bajpai and Shrivastava, 2002). Formation of electrolyte–polymer complexes decreases the swelling degree of the poly(*N*-vinyl-2-pyrrolidone/itaconic acid) hydrogels due to the exclusion of water molecules by adsorbed electrolytes in the hydrogel system (Sen and Güven, 2002).

Ionic osmotic pressure can be used to quantify the salt effect without ion-specific interactions. In general, the increase of ionic strength decreases the osmotic pressure since the higher salt concentration reduces the concentration gradient between the inside and the outside of hydrogels. The greater size the diffusible ion has, the lesser osmotic pressure difference occurs between the interior hydrogel and the external solution, consequently, the lesser volume transition the hydrogel behaves. For halide anions, the sequence in the increasing order of ion sizes is $Cl^-<Br^-<I^-$. The same order of the degree of the volume transition is observed in the poly(AAc-styrene) hydrogels in potassium halide solution with concentrations ranging from 0.01 to 0.1 M (Bajpai and Shrivastava, 2001).

Crosslinkers and Crosslink Density

The structure and elasticity of the hydrogels depend on the nature of crosslinking agent as well as on the crosslinking degree. The crosslinking density is the critical indicative parameter that controls the volume transition behaviours and mechanical properties of the smart hydrogels. The osmotic pressure of the crosslinked polymeric network can differ from that of the non-crosslinked one with the same chemical nature. Different crosslinking methods can distinctly influence the prepared polymer network structure and subsequent volume transition behaviours (Martens and Anseth, 2000). Crosslinking by γ-ray irradiation randomly introduces the crosslinks in the hydrogels, whereas the chemically crosslinked hydrogels exhibit more inhomogeneous distribution of crosslinking points due to the difference in the reactivity ratios of monomers and crosslinkers (Pradas et al., 2001). The inhomogeneity of the chemically crosslinked hydrogels increases with increasing crosslinker concentration due to the diffusive nature of the conventional polymerization reaction. The small angle neutron scattering studies demonstrate that the poly(NIPPAm) hydrogels crosslinked by γ-ray irradiation are more homogeneous than the hydrogels crosslinked by conventional polymerization with BIS (Norisuye et al., 2002). The differences of the hydrogel structures can be interpreted well by the statistical-mechanical Panyukov–Rabin theory (Panyukov and Rabin, 1996).

The structure of the hydrogels is also determined by the functionality and hydrophilicity of crosslinkers. Hexafunctional crosslinkers generate more rigidly crosslinked polymer networks, compared with tetrafunctional crosslinkers. Therefore, it greatly reduces the volume transition of hydrogels when the tetrafunctional crosslinker is replaced by the hexafunctional crosslinker (Karadag and Saraydin, 2002). When a hydrophilic crosslinking agent is employed, the hydrogel behaves a high affinity towards the aqueous solution owing to the presence of polar groups (Castelli et al., 2000). The degree of the volume transition of hydrogels increases with increasing the polarity of the polar groups in the hydrophilic crosslinker (Martens and Anseth, 2000). By replacing bisacrylamide with more hydrophilic glyoxal bis(diallyacetal) (GLY) as crosslinker, the swelling of the crosslinked poly(NIPAAm) hydrogel is enhanced (Xue et al., 2001; Xue and Hamley, 2002). The pronounced increase of the polymer–water interaction parameter is also observed with increasing GLY content.

The swelling degree of the hydrogel is reduced obviously as the amount of crosslinker increases. The degree of swelling is observed to decrease with the crosslink density at exponent of 0.5 (Alvarez-Lorenzo and Concheiro, 2002). However, at low crosslink density, the increase of amount of hydrophilic crosslinker may increase the degree of swelling. The poly(HEMA) hydrogels show an exceptionally large increase of swelling with increasing the amount of crosslinker tripropyleneglycol diacrylate at pH 12.0 (Ferreira et al., 2000). Hydrolysis of HEMA or tripropyleneglycol diacrylate in the highly crosslinked hydrogels is supposed to contribute to the enhancement of swelling.

Surfactants

The interaction between the hydrogel and surfactant is of great importance for understanding the fundamentals of the volume transition of hydrogels because of the amphoteric property of surfactants. The hydrogels can be ionized upon binding of ionic surfactant molecules to the polymer network through hydrophobic interactions. The polymer–surfactant complex is formed when the surfactant is above the critical aggregation concentration (cac) (Gan et al., 2001). In general, cac is lower than the critical micelle concentration (cmc) of the respective surfactant. The charged head groups of the surfactant molecules bound to the polymer chains give rise to the electrostatic repulsion among the polymer chains, which results in the increase of the volume transition and the lower critical solution temperature (LCST) of the hydrogels. The addition of sodium dodecyl sulfate (SDS) at concentration up to 0.5 wt% causes the significant increase of the swelling ratio of NIPAAm/DPAM copolymeric hydrogel and the LCST value (Xue and Hamley, 2002). When the SDS concentration exceeds 0.5 wt%, there are two volume phase transitions at 36–40 and 70°C, respectively. However, the cationic surfactant in very dilute solution behaves like simple electrolyte and exerts salting-out effect on the hydrogels, hence the LCST values decrease initially (Gan et al., 2001).

Solvents

Addition of other solvents into water may affect the conformation of the hydrogels and the hydrogen bonding between the hydrogels and water (Tokuhiro, 2001), which changes the interactions of polymer–polymer, polymer–solvent, water–solvent and solvent–solvent. Depending on acetone concentration, two transition regions with different swelling ratios are observed due to formation of a two-phase structure (Ilavsky et al., 2002). When the dimethylsulfoxide (DMSO) content in the DMSO–water mixtures is continuously increased, the poly(TBA-*co*-AAm) hydrogel exhibits the reentrant volume transition behaviour (Ozturk and Okay, 2002). The poly(TBA-*co*-AAm) hydrogel also exhibits the reentrant conformational transition in the ethanol–water mixtures (Ozmen and Okay, 2003). When the alcohol concentration is increased from 0 to 100 vol%, the poly(acryloyl-l-proline methyl ester) hydrogels show the two swelling phases, the first swelling at 0–10 vol% and the second swelling at 50–80 vol%, and the two shrinking phases, the first shrinking at 10–50 vol% and the second shrinking at 80–100 vol%, in all the aqueous alcohol solutions (Hiroki et al., 2001). The reentrant phase transition of the poly(NIPAAm) hydrogels is also observed in the polymeric solvent PEG and water mixture (Melekaslan and Okay, 2001). The competing attractive interactions between water–solvent and polymer–solvent are responsible to the reentrant phase transition.

Addition of polymeric solvent into the medium may selectively increase the external osmotic pressure and consequently decrease the volume transition of the hydrogel, which is observed in the swelling of the NIPAAm hydrogel in aqueous PEG solution (Huther et al., 2002). The degree of swelling of the hydrogel decreases with increasing the PEG concentration in solution, and a phase transition to a shrunken state occurs. The swelling of the Na-gel in HPC solution decreases (Evmenenko and Budtova, 2000). Unlike the organic solvents, the polymeric solvents may form complexes with the polymer network, and thus change the volume transition behaviours of the hydrogels. The formation of complex between HPC and poly(AAc) hydrogel decreases the swelling of the poly(AAc) hydrogel with the increase of HPC concentration in aqueous solution (Evmenenko and Budtova, 2000).

Co-nonsolvency effect is observed upon addition of any solvent to the poly(NIPAAm)–water system (Costa and Freitas, 2002). Co-nonsolvency is referred to poor solubility in mixture of solvents that are good solvents individually for the polymer. In a general manner, the poly(NIPAAm) solubility is reduced within the range of intermediate solvent concentrations in binary aqueous solutions due to the preferential hydrophobic hydration of solvent molecules in water-rich region, which results in the weakening of the poly(NIPAAm)–water interactions.

Metal Ions

Unlike simple cations, heavy metal ions have strong affinity to polar groups. Addition of metal ions may decrease the swelling of the hydrogel (Katime and Rodriguez, 2001; Vashuk et al., 2001). The adsorption of heavy metal ions by the hydrogels leads to more physical crosslinking formed by the inter-chain metal complex formation (Loh et al., 2001). However, the capacity of metal adsorption decreases with increasing the crosslinker density due to the more rigid polymer networks. The metal binding in ionic poly(MAAc-*co*-AAc) hydrogels is proposed to be entropy driven (Eichenbaum et al., 2000). Therefore, there is a distinct difference of the metal ion binding affinity between ionic hydrogels.

Hydrogen Bonding Effect

Hydrogen bonding effect has significant influence on the volume transition behaviours of the hydrogels in water, where water forms strong hydrogen bonding with hydrophilic polymer chains (Qu et al., 2000). The mechanism of the volume transition behaviours of the hydrogel is partially controlled by rearrangement of hydrogen bonding. The AMPS/*N*-*t*-butylacrylamide hydrogel performs a discontinuous phenomenon of volume transition as the hydrophilic–hydrophobic balance is changed (Varghese et al., 2000), because of the alteration of the hydrogen bonding in the hydrogel–water system and temperature-dependent hydrophobicity of the polymer. Formation of hydrogen bonding makes the hydrogel more hydrophobic, which leads the hydrogel to a collapsed state, which is observed in the poly(MAAc-*co*-methacryloxyethyl glucoside) hydrogels in acidic media (Kim and Peppas, 2002). Deswelling of the poly(MAAc) hydrogel is the result of the hydrogen bonding formation arising from the protonation of the carboxylic acid groups. The hydrogen bonding increases with the MAAc content in the copolymer networks. Intermolecular hydrogen bonding intermediated by water molecules is also found to play an important role in the mechanism of deswelling of the poly(*N*-vinyl-2-pyrrolidone) (PVP) hydrogels (Muta et al., 2002).

On the other hand, the destruction of hydrogen bonding may increase the hydrophilicity of the system and thus increase the degree of swelling. The dissociation of the hydrogen bonding between the chitosan and PEO as well as the ionization of the amine functions in the chitosan–PEO semi-IPN hydrogels leads to the relaxation of the macromolecular chains when pH is in acidic condition (Khalid et al., 2002).

The ionic hydration can stabilize or destabilize the hydrogen bonding network (Muta et al., 2001a, b). The hydrogen bonding of water in the hydrogels is strengthened in the presence of strongly hydrated cations or weakened in the presence of strongly hydrated anions (Muta et al., 2001c). Integrating hydrophobic groups to the hydrogels decreases the hydrogen bonding of water in the hydrogels (Yasunaga et al., 2002).

Water State

In general, water in the hydrogels exists in three physical states: (1) the free water which can freeze at the usual freezing point, (2) the intermediate water which is less strongly associated due to the existing hydrophobic interactions and can freeze at a temperature lower than the usual freezing point and (3) the bound water which is strongly associated with the hydrophilic segments of hydrogels and cannot freeze at the usual freezing point (Dibbern-brunelli and Atvars, 2000; Li et al., 2000; Ito et al., 2000; McConville and Pope 2001; Manetti et al., 2002; Meakin et al., 2003). The swelling characteristics of the hydrogels are dominated by the nature of the polymer and the state of water. The poly(*N*-vinyl 2-pyrrolidone-*graft*-citric acid) hydrogels exhibit continuous changes in water content as a function of temperature, which is attributed to the fact that water remaining within the hydrogels consists mostly of the binding water (Caykara et al., 2000). The oscillation in volume transition behaviour is related at least to the three types of water molecule movement in the hydrogels (Makino et al., 2000).

The content of non-freezing water is affected by the chain mobility or crosslink density of the hydrogels. Various hydrogels are reported to have quite different contents of non-freezing water, for example, the 23% non-freezing water in the chitosan–PEO hydrogel (Khalid et al., 2002), the 24–28% non-freezing water in a chitosan hydrogel (Qu et al., 2000), the 35% non-freezing water in the HEMA hydrogels and as high as the 43% non-freezing water in the 2,3-dihydroxypropyl methacrylate (DHPMA) hydrogels (Gates et al., 2003). The molecules of non-freezing water are hydrogen bonded to the hydrophilic groups of the hydrogels. Therefore, the content of non-freezing water is proportional to the amount of hydrophilic groups in the hydrogels. The content of non-freezing water increases with the increase of the ratio of hydrophilic groups (Katime and de Apodaca, 2000) and decreases with the increase of the ratio of hydrophobic groups (Kim et al., 2003). Obviously, the proportion of free water becomes preponderant as the swelling ratio increases.

1.2.2 Transient Modelling for Kinetics of Smart Hydrogels

The present transient models focus on the responsive kinetics of the smart hydrogels subject to different environmental stimuli, especially in the rates of swelling or shrinking process of fast response hydrogels.

Currently the fast response hydrogels are increasingly attracting attention. Usually their fast response can be achieved in five ways: (1) reducing the dimension of the hydrogels due to the swelling rate being inversely proportional to the square of the geometrical size of the hydrogels; (2) forming the heterogeneous network structure by phase separation technique; (3) introducing the porosity into the hydrogels; (4) graft copolymerization where the free ends of the grafts act to accelerate the dehydration rate and (5) using the silane crosslinking agent. The extensively studied phase separation technique is an approach for preparation of the fast response hydrogel. The poly(*N*-isopropylacrylamide-*co*-methyl methacrylate) hydrogels with capability of fast response can be prepared through the phase separation technique in water/glacial acetic acid as the mixed solvent (Zhang and Zhuo, 2002). Using water/acetone as the mixed solvent during the redox polymerization/crosslinking reaction, the fast responsive poly(*N*-isopropylacrylamide) hydrogel may also be prepared (Zhang and Zhuo, 2000a). This is attributed to the fact that the polymer chains produced in the mixed solvent are very soluble and get widely expanded during polymerization. Owing to the widely expanded chains, the polymer system is more unstable, and thus it exhibits the fast deswelling rate. The electron beam irradiation provides another technique for phase separation. The thermo-sensitive poly(*N*-isopropylacrylamide) hydrogels crosslinked by electron beam irradiation are prepared (Panda et al., 2000), which shrink about 100 times faster than the conventionally crosslinked gels. The mechanism of the fast response of the hydrogels may be attributed to the two effects: (1) the polymerization and crosslinking at extremely high dose rates may result in intra-molecularly crosslinked gels and (2) the temperature during electron beam irradiation may rise above the lower critical solution temperature (LCST), resulting in the phase separation in polymerization during crosslinking reaction. In addition, it is noted that the porous structure enlarges the surface area to volume ratio and allows the solute to diffuse more rapidly into the hydrogel. The porous hydrogels may be made by several methods, such as the porogen leaching, phase separation, particulate crosslinking, microemulsions, gas blowing and freeze drying. Addition of polyethyleneglycol (PEG)-400 as the pore-forming agent to the monomer solution of *N*-isopropylacrylamide (NIPAAm) is found to dramatically increase the swelling kinetics of the prepared hydrogels (Zhang and Zhuo, 2000b). The higher the PEG-400 contents are in the monomer solution, the larger fraction the big pore sizes are in the network of the hydrogel.

By adding a gas-forming agent Pluronic® F127 during polymerization, the superporous sucrose hydrogels are synthesized (Chen and Park, 2000). Compared with the sucrogels, the superporous hydrogels swell faster with higher ratio of swelling and degrade faster in both acidic and basic conditions. The gas blowing technique is also employed to prepare the superporous poly(acrylamide-*co*-acrylic acid) hydrogels whose response is the orders of magnitude faster than that expected for nonporous hydrogels with the same dimension (Gemeinhart et al., 2000). The superporous hydrogels have low bulk density of 0.30±0.03 g/ml and water molecules are taken up into the superporous hydrogels by capillary forces. By reducing the dimension of the hydrogels as well as introducing the porous structure into the hydrogels, the fast pH- and ionic strength-sensitive hydrogels are prepared in microchannels by photopolymerization of monomer mixtures containing water and surfactant (Zhao and Moore, 2001). The mechanism of the fast response of the hydrogels is attributed to the porous structure generated by the surfactant phase template and phase separation, irrespective of the network crosslinking density or acrylic acid (AAc) composition in the hydrogel. The swelling of the hydrogels follows the second order of kinetics (Quintana et al., 1999). A two-step polymerization method is proposed to prepare the fast response thermo-sensitive poly(NIPAAm-*co*-di-*n*-propylacrylamide) hydrogels (Xue and Hamley, 2002), where the polymerization is conducted at 20°C for a short time and the polymerization is then completed at −28°C. The responsive kinetics can be adjusted according to the polymerization time at 20°C, which seems to influence the porosity of the network structure of the hydrogels.

The literature search reveals that the transient models for simulation of kinetics of the smart hydrogels have so far been limited probably due to the limited experimental data. The fast response hydrogels have found wide-range promising applications in various areas, such as the mechano-chemical actuator and sensor, drug delivery device and especially in the bio-micro-electro-mechanical system (BioMEMS), as well as biomaterials for biomedical and tissue engineering applications. The ready mathematical models with capability of predicting and simulating the responsive kinetics of the fast response smart hydrogels are undoubtedly crucial for the applications. The fast response hydrogels provide an attractive routine to investigate the kinetics of volume variation. Currently the modelling development of the volume variation of the fast response hydrogels is becoming a popular research area. A review on the developed models could greatly facilitate this effort. In this section, three typical models developed with different complexities are discussed to investigate the kinetics of the volume variation of the hydrogels. A phenomenal model based on the second order of reaction kinetics provides a simple method to correlate the experimental data of kinetics. A power law model can be employed to evaluate different mechanisms of the volume variation according to the diffusion exponential coefficients. The multi-field model couples the mechanical deformation with the diffusion of ion species. The rate of volume variation is dependent on both hydrogel molecular design and operation conditions. Several key parameters are summarized for simulation of kinetics of the fast response smart hydrogels.

1.2.2.1 Mathematical Models and Simulations

The kinetics process of swelling of the smart hydrogels is a complicated phenomenon, which generally involves three successive steps: (1) the diffusion of water molecules into the polymer system; (2) the relaxation of the hydrated polymer chains and (3) the expansion of the polymer network in surrounding aqueous solution. The swelling kinetics of the hydrogel may be further divided into more detailed steps. For example, the kinetics process of the swelling or shrinking of poly(*N*-vinylimidazole-*co*-sodium styrene sulfonate) hydrogels includes the following simultaneous steps (Valencia and Pierola, 2002): (1) the diffusion of water towards the hydrogel; (2) the chain disentanglement; (3) the sodium–proton interchange through the external bath boundary; (4) approaching of chains to allow the interaction of sulfonate groups with neighbouring protonated imidazole moieties and (5) the diffusion of water outside the hydrogel. On the other hand, the wetting process of polymer dry gel with cylindrical shape is divided into three steps (Ji and Ding, 2002): (1) swelling of the gel in the radial direction with cusp-like patterns evolving on the surface; (2) shrinking of the gel in the radial direction and swelling in the axial direction and (3) the re-swelling of the gel to the final stage and eventually disappearing of patterns.

Generally, two processes are employed to describe the diffusion of water into the polymer matrix, namely the diffusion of the solvent into the swollen matrix and the advancement of the swollen–unswollen boundary as a result of polymer relaxation. When the rate-determining step is the diffusion of the solvent into the swollen matrix, there is a linear relation between the water uptake and time$^{1/2}$, the system is regarded to exhibit the Fickian behaviour. In contrast, if the advancement of the swollen–unswollen boundary is slower than the diffusion of the solvent in the swollen polymer, the zero order of kinetics is achieved with the water uptake time, and the system is regarded to exhibit the non-Fickian behaviour. In pure water condition, the diffusion of water generally approaches to the Fickian behaviour. However, the diffusion of water into the hydrogels under solution conditions follows the non-Fickian behaviour due to the dominance of polymer relaxation. The swelling of the poly(NIPAAm) hydrogel in the gelated corn starch aqueous solution is determined mainly by the relation of hydrated polymer chains, instead of the water diffusion rate (Zhang and Zhuo, 2000c). The starch may act as the long graft-like chains of the poly(NIPAAm) hydrogels, which provides the channels for water to be released during the shrinking process. It is claimed that the molecular relaxation controls the velocity of water uptake of hydroxypropylmethyl cellulose (HPMC), carboxymethylcellulose-sodium (NaCMC) and mixed HPMC/NaCMC hydrogels irrespective of pH of the aqueous phase (Michailova et al., 2000), in which there is an inversely proportional dependence between the viscosity and the water uptake velocity.

Three typical transient models developed with different complexities are reviewed here for simulation of the kinetics of the volume variation of the smart hydrogels. They are the phenomenal model, the power law model and the multi-component diffusion model.

Phenomenal Model

The phenomenal model is based on experimental observation for describing the extensive polymer swelling (Schott, 1992) and is written as follows:

$$\frac{t}{W} = A + Bt \qquad (1.34)$$

where W is the water uptake at time t, $B = 1/W_\infty$ is the inverse of the maximum swelling W_∞ and $A = 1/(dW/dt)_0$ is the reciprocal of the initial swelling rate.

Equation (1.34) implies the second order of kinetics of swelling for volume variation of the hydrogels, which is expressed as $dW/dt = k_r\,(W_\infty - W)^2$. The specific rate k_r is a constant related to the parameter A as $k_r = 1/(AW_\infty^2)$. By substituting the swelling data into Eq. (1.34), it is found that the plots of t/W against t can give straight lines with good correlation coefficients for experimental observations.

Obviously, the phenomenal model cannot provide accurate information about the mechanism of the volume variation of hydrogel. In most cases, the initial phase of the volume variation of hydrogel follows the zero order of kinetics of swelling, instead of the second order of kinetics. This means that the model is suitable for simulation of kinetics of the hydrogel volume variation after the initially fast swelling process. However, at least it provides a simple approach to correlate experimental data.

Power Law Model

As well known, the diffusion of water into the hydrogel system results in the volume variation of the hydrogels. The water transport in the hydrogels can be characterized either by Fick's second law or by a more advanced equation of anomalous diffusion, which incorporates a diffusive term coupled with a pseudo-convective velocity term. A simple method of describing water transport in polymers is given via the power law (Astarita, 1989)

$$M_t/M_\infty = kt^n \qquad (1.35)$$

where M_t is the mass of water gained or lost at time t, M_∞ is the initial mass of water contained in the polymer, k is a constant and n is the diffusion exponent.

Usually the mechanism of the water transport in the hydrogels may be characterized by the diffusion exponent n and also by two limited cases of this power law model, $n = 0.5$ and $n = 1$. When $n = 0.5$, the water transport is controlled exclusively by the chemical potential gradient, little or no volume variation occurs during the transport. This is referred to as Case I diffusion or Fickian diffusion, and then the volume variation can be described by a diffusion coefficient. If $n = 1$, the rate of the volume variation of the hydrogels is proportional to time t, in which the stress relaxation controls the kinetics of the volume variation of the hydrogels and the volume variation counteracts the mechanical stresses produced by water transport. This is termed Case II diffusion, where the velocity of the waterfront describes the kinetics process of the volume variation. The intermediate case called the anomalous diffusion occurs when the water transport is proportional to t^n, where $0.5<n<1$.

Actually the water transport in the hydrogels may result from a mixture of Case I and Case II processes, in which both the waterfront velocity and diffusion coefficient are required to characterize the volume variation process (Chou et al., 2000). A model proposed can account for Case I, Case II and anomalous water transport processes (Harmon et al., 1987), in which the total flux J consists of two components, One is due to the diffusion with a concentration gradient and the other is due to the stress relaxation of polymer chains with a propagation speed v. For example, for one-dimensional volume variation of the hydrogels

$$J = -D\frac{\partial C}{\partial X} + v(C - C_0) \quad \text{for} \quad 0 \le |X| \le 1 \tag{1.36}$$

where $C = C(X)$ and $C_0 = C(0)$ are the concentrations at points X and $X = 0$, respectively. The diffusion coefficient D and the velocity v come from Case I and Case II water transport processes. According to the law of mass conservation, one can have

$$\frac{\partial C}{\partial t} = D\frac{\partial^2 C}{\partial X^2} - v\frac{\partial C}{\partial X} \quad \text{for} \quad 0 \le |X| \le 1 \tag{1.37}$$

After integrating the concentration over the domain with the boundary condition of constant surface concentration C_0 at $X = \pm 1$, the variation of water mass M_t at time t associated with the equilibrium water mass M_∞ is written as follows (Harmon et al. 1987):

$$\frac{M_t}{M_\infty} = 1 - 2\sum_{n=1}^{\infty} \frac{\lambda_n^2\left(1 - 2\cos\lambda_n \exp\left(-\frac{vl}{2D}\right)\right)}{\beta_n^4\left(1 - \frac{2D}{vl}\cos^2\lambda_n\right)} \exp\left(-\frac{\beta_n^2 Dt}{l^2}\right) \tag{1.38}$$

where

$$\lambda_n = \frac{vl}{2D}\tan\lambda_n \tag{1.39}$$

$$\beta_n^2 = \frac{v^2\, l^2}{4D^2} + \lambda_n^2 \tag{1.40}$$

where the roots λ_n (n= 1, 2, 3, ..., ∞) of Eq. (1.39) are used in Eqs. (1.38) and (1.40), respectively.

For a special case where v is equal to zero, Eqs. (1.39) and (1.40) are simplified to $\lambda_n = (n + 1/2)\pi$ and $\beta_n = \lambda_n$, respectively. Equation (1.38) is then reduced to

$$M_t/M_\infty = 1 - \sum_{n=1}^{\infty}\left[8\Big/(2n+1)^2\pi^2\right]\exp\left[-(2n+1)^2\pi^2\left(Dt\big/l^2\right)\right] \tag{1.41}$$

The short-time limited expression for Eq. (1.41) is thus obtained as

$$\frac{M_t}{M_\infty} = \left(\frac{4}{\pi^{0.5}}\right)\left(\frac{Dt}{l^2}\right)^{0.5} \tag{1.42}$$

It is observed from the above equation that the relation between M_t/M_∞ and $t^{0.5}$ yields a straight line. Based on the slope of the graph, the value of diffusion coefficient D is calculated. As a good example of applications, the experiment of the water transport in crosslinked 2-hydroxyethyl methacrylate (HEMA) hydrogels is in excellent agreement with the prediction by the model (Chou et al., 2000).

In order to achieve better approximation, the exact solution of Eq. (1.37) is derived, termed the Berens–Hopfenberg model (Enscore et al., 1977; Berens and Hopfenberg, 1978), and it is expressed as follows:

$$M_t/M_\infty = \phi_F\left[1 - \sum_{n=1}^{n=\infty}\frac{8}{(2n+1)\pi^2}\exp\left(\frac{-D(2n+1)^2 t}{4l^2}\right)\right] + \phi_R\left(1 - \exp(-kt)\right) \tag{1.43}$$

where k is the first order of relaxation constant, ϕ_F and ϕ_R are the fractions of water uptake contributed by the Fickian diffusion and the chain relaxation, respectively. This heuristic model can be used for analysis of the overall water uptake in terms of the Fickian and non-Fickian contributions, which leads to the determination of both the diffusion coefficient D and the characteristic relaxation time τ that is defined as the reciprocal of the constant k.

It is noted that the model mentioned above does not take into account the interactions between the polymer and the solvent, which significantly influence the water diffusion and the stress relaxation of polymer chains. In addition, the diffusion coefficient D is assumed to be constant in the model. In fact, the hydrogel network structure always changes with the volume variation of the hydrogel, which results in a varying diffusion coefficient. However, at least when $0 < M_t/M_\infty < 0.60$, the model can predict well the kinetics process of volume variation of the smart hydrogels.

Multi-field Model

The kinetics of the volume variation of the hydrogels involves the water diffusion and the mechanical deformation simultaneously. For the ionic hydrogels, the volume variation of the hydrogels is strongly dependent upon the diffusions of all species and the variation of electrical potential. A good model for the volume variation of the hydrogels should be able to incorporate the multi-field effects.

Modelling of the ion transport could be conducted by many models within various theoretical frameworks. It is sufficient to model the ion transport as follows:

$$\frac{\partial c_k}{\partial t} = \frac{\partial}{\partial x}\left(D_k\frac{\partial c_k}{\partial x}\right) + \frac{\partial}{\partial x}\left(\frac{cZ_kFD_k}{RT}\frac{\partial \phi}{\partial x}\right) \tag{1.44}$$

where c_k is the concentration of the ion species k, D_k is the binary diffusion coefficient of the ion species k, c is the total concentration, z_k is the valence of the ion species k, ϕ is the electrostatic potential, F is the Faraday constant, R is the universal gas constant and T is the absolute temperature. The concentration of ionized pendant groups in the polymer network is always in equilibrium with the concentration of hydrogen ions in the polymer network.

The kinetics of swelling of the ionic hydrogels induced by pH change is modelled by considering the diffusion of hydrogen ions, which is governed by the chemical diffusion equation together with the mechanical equation (Chatterjee et al., 2003). This model assumes that the mechanical deformation of the polymer network occurs instantaneously with the diffusion of hydrogen ions. The mechanical equation takes into account the deformation of the polymer network that occurs during the diffusion of hydrogen ions into the hydrogel, namely

$$\Gamma_k = \phi \left[-\overline{D}_k \frac{\partial c_k}{\partial x} - \mu_k z_k c_k \frac{\partial \psi}{\partial x} \right] + c_k U \tag{1.45}$$

where Γ_k is the flux of the kth ion, ϕ is the gel porosity and U is the area-averaged fluid velocity relative to the polymer network. The diffusion rate inside the hydrogel is related to the diffusion in aqueous solution through the obstruction model as follows:

$$\frac{\overline{D}_k}{D_k} = \left(\frac{K}{2+H} \right)^2 \tag{1.46}$$

where H is the hydration of the hydrogel and it is defined as the ratio of fluid volume to polymer volume.

Continuity condition for the divergence of each ionic flux is given by

$$\frac{\partial}{\partial t} \left(HC_k + Hc_k^b \right) = -\frac{\partial \left(\alpha \Gamma_k \right)}{\partial X} \tag{1.47}$$

where c_k^b is the concentration of the ion k that can be reversibly bound to the polymeric fixed charge, X is the Lagrangian coordinate system associated with the hydrogel and α is the total hydrogel area normalized to its initial area.

The concentration of the ion k reversibly bound to the polymeric chains (c_k^b) in the presence of chemical reactions is calculated by (Grimshaw et al., 1990)

$$c_k^b = \frac{c_f^0}{H} \frac{c_k}{(K + c_k)} \tag{1.48}$$

where c_f^0 is the total concentration of ionizable groups within the hydrogel before the volume variation.

By rearrangement of the above equations, a nonlinear diffusion–reaction equation for the concentration of H^+ ions in the hydrogel is obtained as

$$\frac{\partial}{\partial t} \left[c_H \left(H + \frac{c_f^0}{K + c_H} \right) \right] = -\frac{\partial}{\partial X} \left[\alpha \phi \left(\overline{D}_H \frac{\partial c_H}{\partial x} + \mu_H z_H c_H \frac{\partial \psi}{\partial x} \right) - \alpha c_H U \right] \tag{1.49}$$

where $\bar{D}_{\mathrm{H}}$ is the diffusion rate of hydrogen ions within the hydrogel, c_{H} is the internal concentration of hydrogen ions and c_H^b is the concentration of hydrogen ions reversibly bound to the fixed charges of the hydrogel.

The presence of buffer in the solution increases the apparent diffusion rate of hydrogen by providing an alternate path for diffusing hydrogen ions between the hydrogel and bath solution. Buffer augmented transport of hydrogen ions under certain conditions may result in the apparent diffusion rates of hydrogen, which is several orders of magnitude higher than the diffusion coefficient of hydrogel alone (Ohs et al., 2001). It is found that the kinetics of the water uptake in the buffered solutions is markedly faster than that in the unbuffered solutions for the hydrophobic ionizable copolymer gels composed of PEG, AAc and styrene, though the water uptake is contributed by both the mechanisms of diffusion and polymer chain relaxation (Bajpai and Shrivastava, 2001). The influence of the buffer on the transport of hydrogen ions can be modelled by including additional terms in the continuity equation of hydrogen ions within the hydrogel

$$\frac{\partial}{\partial t}\left[Hc_{\mathrm{H}} + Hc_H^b + Hc_{\mathrm{HB}}\right] = -\frac{\partial(\alpha\Gamma_{\mathrm{H}} + \alpha\Gamma_{\mathrm{HB}})}{\partial X} \tag{1.50}$$

where c_{HB} is the concentration of hydrogen ions bound to the buffer, Γ_{H} is the flux of hydrogen ions and Γ_{HB} is the flux of hydrogen ions bound to the buffer.

$$c_{\mathrm{HB}} = \frac{c_{\mathrm{T}}c_{\mathrm{H}}}{(K_{\mathrm{B}} + c_{\mathrm{H}})} \tag{1.51}$$

where K_{B} is the dissociation constant of the buffer and c_{T} is the total buffer concentration.

The flux of the buffer is proportional to the flux of the hydrogen ions:

$$\Gamma_{\mathrm{HB}} = \left(\frac{\overline{D}_{\mathrm{HB}}}{\overline{D}_{\mathrm{H}}}\right)\left(\frac{c_{\mathrm{T}}}{K_{\mathrm{B}} + c_{\mathrm{H}}}\right)\Gamma_{\mathrm{H}} \tag{1.52}$$

where $\overline{D}_{\mathrm{HB}}$ is the diffusion rate of buffer in the hydrogel.

The Poisson equation is used to calculate the electrostatic potential

$$\frac{\partial^2\psi}{\partial x^2} = -\frac{F}{\varepsilon\varepsilon_0}\left(\sum_{k=1}^{N} z_k c_k + z_f c_f\right) \tag{1.53}$$

The chemical diffusion equation then becomes

$$\begin{aligned}\frac{\partial}{\partial t}&\left[Hc_{\mathrm{H}} + \frac{c_f^0 c_{\mathrm{H}}}{K + c_{\mathrm{H}}} + \frac{Hc_{\mathrm{T}}c_{\mathrm{H}}}{K_{\mathrm{B}} + c_{\mathrm{H}}}\right]\\ &= -\frac{\partial}{\partial X}\left[\alpha\left(\frac{H}{1+H}\right)\left(1 + \frac{\overline{D}_{\mathrm{HB}}}{\overline{D}_{\mathrm{H}}}\frac{c_{\mathrm{T}}}{KB + c_{\mathrm{H}}}\right)\left(\overline{D}_{\mathrm{H}}\frac{\partial c_{\mathrm{H}}}{\partial x}\right)\right]\end{aligned} \tag{1.54}$$

The change in the hydration of the hydrogel is calculated by the following mechanical equation that describes the forces and fluid flow:

$$\frac{\partial H}{\partial t} = \frac{\partial}{\partial X}\left[ak'\left(-\frac{\partial (M\varepsilon)}{\partial x} + z_f c_f F\frac{\partial \psi}{\partial x}\right)\right] \tag{1.55}$$

where k' is the hydraulic permeability of the hydrogel, M is the bulk modulus of the hydrogel and ε is the compressive strain.

The advantage of this model is the capability of providing a good evaluation of the volume variation of the hydrogel by integration of the chemical, electrical and mechanical interactions. It captures the variations of the concentration distributions of all ionic species with time. However, this model is valid only for the ionic hydrogels, and the deformation of the polymeric network matrix is characterized by the volume-based hydration H, instead of the displacement vectors. The mathematical complexity also limits its extensive applications.

1.2.2.2 Key Parameters in Transient Modelling for Kinetics of Hydrogels

Although there are many parameters having influences on the kinetics of volume variation of the smart hydrogels, basically they may be classified into two categories: (1) the designs of the hydrogels, which include the ionic group content, thickness, shape and strength and (2) the operating conditions, which include the intensity and function of environmental stimuli and the type and concentration of ionic species in outer electrolyte solution. A good transient model should be able to characterize the effects of various parameters that have significant influences on the kinetics of volume transition of the smart hydrogel. This is a challenge for modelling development and simulation. Several key parameters are summarized below.

Environmental Medium

The environmental medium of the hydrogels plays an important role in kinetics of the hydrogels with volume variation. The temperature and pH of solutions are the most widely studied parameters of operation conditions. It is found that, with increasing temperature, the full interpenetrating polymer networks (IPNs) of crosslinked gelation with polyacrylamide experiences an anomalous type of diffusion and the rate of swelling increases with the rise in ambient temperature (Rathna and Chatterji, 2001). For the gelation with poly(NIPAAm) above its LCST, the diffusion deviates from the anomalous to Fickian types, due to the predominance of the hydrophobic characteristics of NIPAAm, resisting the advancing of solvent molecules. The swelling behaviour of thermo-reversible NIPAAm hydrogels is related to their looser or denser structures and pore sizes, which is dependent on the surrounding temperature, polymerization temperature and polymerization media (Lee and Yeh, 2000).

However, the effect of solution pH is much more complicated. The kinetics of swelling of poly(N, N'-dimethylaminoethyl methacrylate) hydrogels is observed to be the non-Fickian type in lower pH and tend to the Fickian type at higher pH (Chen and Yi, 2001). There is a jump corresponding to the pH values and hysteresis loop on elongation or contraction behaviour for PVA and PAA hydrogels (Fei and Gu, 2002). The elongating process takes place much slower than that of contracting process. Time of swelling to equilibrium for the IPN hydrogels is much longer than that of shrinking to equilibrium (Lu et al., 2000), which is ascribed to the fact that the hydrogels shrink in the skin layer first and form a thick, dense and collapsed skin layer on the surface region. This dense layer is believed to be impermeable to water.

Composition of Hydrogels

The properties of hydrogels are determined by the composition of comonomers in the hydrogels. It is found that the water sorption by the hydrogels follows both the Fickian and non-Fickian swelling mechanisms, depending upon the composition of the hydrogel (Bajpai and Giri, 2002). The incorporation of hydrophobic comonomers is generally required for fast response of phase transition to pH change. For example, the incorporation of small amount of maleic acid into the nonionic polyacrylamide hydrogels results in the transition of swelling mechanism from the Fickian to non-Fickian swelling behaviours (Bajpai, 2000). Then the hydrogels require more time to swell. The increase of ionic strength causes the decrease of swelling, while the swelling increases with pH of the swelling medium. With lower acid contents, the swelling increases while higher concentrations of maleic acid cause the decrease of swelling. The hydrogels require more time to shrink, compared with that required for swelling. The water uptake of the hydrogel varies sensitively with various contents of the hydrophilic polymer, copolymer, crosslinking agent and the environmental medium temperature (Bajpai and Shrivastava, 2002a–c). The increasing proportion of the hydrophilic polymer PEG results in the shift of the water transport mechanism from the anomalous to Fickian types due to the network with a greater number of macromolecular chains, while the hydrophilic monomer acrylamide tends to move the swelling mechanism from the Fickian to non-Fickian ranges due to the increase of network density. On the other hand, the increasing content of the hydrophobic monomer (styrene) shifts the water-sorption mechanism from the Fickian to anomalous types, which is due to the steric hindrance and dispersion forces operating between the polystyrene chains in the hydrogels. The bending speed increases in proportion to the poly(acrylic acid) content in PVA/PAA semi-interpenetrating network hydrogel (Fei et al., 2002). The increase of the hydrophilicity by acrylic acid (AAc) decreases the response rate to pH variation (Chiu et al., 2001). It is claimed that the swelling ratio and rate increase with increasing the content of the hydrophilic monomer in the hydrogels (Martellini et al., 2002), and there is a significant variation in the diffusion coefficients for different temperatures. The corresponding swelling is classified as Fickian type. It is also observed that, with increasing the PEG content, the value of n increases in the non-Fickian region, while increasing HEMA tends to increase the value of n towards Fickian value in the Fickian range (Bajpai and Shrivastava, 2002a). It is found that the new crosslinked hydrogels based on *N*-acryloyl-*N′* -methyl piperazine (AcrNMP) and methyl methacrylate (MMA) are able to be responsive to pH and temperature (Gan et al., 2001). With increasing the hydrophilic content in the hydrogels or with rising temperature, the diffusion constants enlarge. The apparent activation energy for the diffusion process is independent of the composition of the hydrogel networks, and the enthalpy of mixing has a strong influence on the hydrophilic content of the hydrogel. It is emphasized that the balance between the two effects, (1) the maintenance of effective hydrophobic aggregation force between collapsing polymer chains and (2) facilitation of water channel formation in the gels to readily release water through the collapsing gel-skin layer, is very important to generate the fast shrinking of NIPAAm and 2-carboxyisopropylacrylamide (CIPAAm) copolymer gels (Ebara et al., 2001). The shrinking rate of NIPAAm–CIPAAm hydrogels increases with increasing CIPAAm content. In contrast, structurally analogue NIPAAm–AAc copolymer gels lose temperature sensitivity with introduction of a few mole percent of AAc. Additionally, the deswelling rates of NIPAAm–AAc gels decrease with increasing AAc content. It is found that the kinetics of swelling of the hydrogels decreases with increasing AAm and MBAAm concentration in the gel matrix (Yildiz et al., 2002).

The HEMA-*co*-sodium acrylate hydrogels may exhibit an overshooting phenomenon in the kinetics of swelling behaviour at higher HEMA content and lower temperature (Lee and Lin, 2001), where the phenomenon that the water uptake of the gel increases over the equilibrium value is called “overshooting”. On the other hand, the presence of impenetrable polymer molecules may lead to the increase of the path length for diffusion, a phenomenon called the obstruction effect.

Usually the hydrogel structure changes with the composition. In the study of water permeability in agarose–dextran gels, it is found that the dextran covalently linked to agarose by electron beam irradiation causes very significant reduction in the Darcy permeability, by as much as the order of magnitude (White and Deen, 2002). Modelling and simulation indicate that, at high agarose concentration, the dextran chains behave as fine fibres interspersed among coarse agarose fibrils, whereas the dextran molecules at low concentrations begin to resemble spherical obstacles embedded in agarose gels.

Salt

Basically salts have significant influences on the kinetics of volume variation of the hydrogels due to the change in the interactions of polymer–polymer, polymer–water, polymer–ion, water–ion and so on. It is observed that the bending speed of the PVA/chitosan IPN hydrogels increases with the increase of the applied voltage and the concentration of aqueous NaCl solution (Kim et al., 2002). However, the time required to reach the equilibrium bending angle depends on the mobility of the cations, the electric field and the gel thickness but not on the salt concentration (Yao and Krause, 2003).

In the study of the volume phase transition of poly(NIPAAm) hydrogels induced by aqueous hydroxides and chlorides (Dhara and Chatterji, 2000), the hydrogels exhibit the volume phase transition at additive concentration characteristics of the anion. It is presumed that the bubble formation observed during the anion induced by the deswelling of initially equilibrium swollen gels is the spontaneous separation of the partially collapsed gel into polymer-rich and solvent-rich domains. The permeability of the bubble wall and its stability may be osmotically driven.

Crosslink Density

The crosslink density directly affects the mechanical deformation of the hydrogels. It is easily understood that the hydrogels with higher crosslink density can have the larger resistance to the deformation during volume variation. The poly(aldehyde guluronate) hydrogels with higher crosslink density show very retarded degradation behaviour (Lee et al., 2000). The rate of hydrogen ion formation is significantly reduced when the crosslinking ratio increases in the gels (Podual et al., 2005). The response rate to the change in external temperature may be improved during the shrinking process and the oscillating shrinking–swelling cycles (Zhang et al., 2003). The interior network structure of the hydrogels becomes more porous with the decrease of the crosslinking level, which provides numerous water channels for the diffusion of water, while the water content decreases with increasing the level of crosslinking at room temperature. It is observed that the increase of *N,N′*-methylene bisacrylamide (MBA) concentration enhances the non-Fickian nature of the swelling process, which might be attributed to the slower relaxation of network chains (Bajpai and Shrivastava, 2002b). However, at higher concentration of ethylene glycol dimethacrylate (EGDMA), the equilibrium swelling is attained faster (Bajpai and Shrivastava, 2002a).

Diffusion Coefficient

Probably the diffusion coefficient has the most important effect on the characteristics of the kinetics of volume variation of the hydrogels. Previous studies utilize the concentration-dependent diffusion coefficient or time-dependent diffusion coefficient or time- and memory-dependent diffusion coefficient. The time- or position-dependent diffusion coefficients are also used to account for the shape change in solvent diffusivity across the swelling front between glassy and rubbery regions in the polymer.

The diffusion of water into the polymer matrix may be influenced by the copolymer composition and microstructure that are the features determined by the kinetics of the polymerization of the co-monomers, the polarity of the polymer segment, the glass transition temperature of the polymer, the flexibility of the polymer network, the crosslink density and inter-chain interaction, the molecular weight of the polymer, the degree of chain branching and the presence of bulky co-monomer pendent groups. For example, the diffusion coefficient may be controlled by addition of the non-polar monomers to the polar HEMA monomer. With increasing the proportion of *n*-butyl methacrylate or cyclohexyl methacrylate in the copolymers with

HEMA, the diffusion coefficients decrease, reflecting the influence of the compositional microstructure of the copolymers on the diffusion of water into the cylinders (Hill et al., 2000).

The thermal effect dominates at high temperature and the influence of water content is significant at low temperature. The temperature may affect the transport mechanism of swelling of the gels. For example, the diffusion coefficient of the NIPAAm and 1-vinyl-3-(3-sulfopropyl) imidazolium betaine copolymeric gels increases with increasing temperature (Lee and Yeh, 2000). The diffusion coefficients of IPN hydrogels increase with increasing temperature and decrease with increasing the content of poly(propylene glycol) (PPG) (Kim et al., 2003).

The diffusion coefficients of water may be concentration dependent. Based on the free volume theory, the relation for the effective water diffusion coefficient D_{eff} with the water concentration c_{w} is written as follows:

$$D_{\text{eff}} = D_0 \exp\left[-b_i\left(1 - \frac{c_{\text{w}}}{c_{\text{w,eq}}}\right)\right] \tag{1.56}$$

where D_0 is the limited water diffusion coefficient in the equilibrium swollen hydrogel and $c_{\text{w,eq}}$ is the water concentration in the equilibrium swollen hydrogel.

Based on the free volume theory suitable for rubbery polymers and considering the plasticization effect induced by the penetrant, a modified free volume model is proposed to predict the diffusion coefficients of small molecule solvent in glassy polymers (Wang et al., 2000). The solvent self-diffusion coefficient at temperature T is expressed as follows:

$$D_{\text{self}} = D_{\text{ex}} \exp\left(-\frac{E}{RT}\right) \exp\left(-\frac{\gamma(w_1 V_1^* + w_2 \xi V_2^*)}{V_{\text{FH}}}\right) \tag{1.57}$$

The solvent is denoted as component 1 and the polymer as component 2. w_i is the mass fraction of component i and V_i^* represents the specific critical hole-free volume of the component i required for a diffusion jump. D_{ex} is a pre-exponential factor, E is the critical energy that a molecule must possess to overcome the attractive force holding it to its neighbours and γ is an overlap factor that is introduced because the same free volume may be obtained for more than one molecule. ξ is the ratio of the molar volume of the jumping unit of the solvent to that of the polymer. V_{FH} is the average hole-free volume, which is estimated by addition of those for solvent and polymer with the assumption that all volumes associated with the solvent and polymer are additive at any concentration and temperature (Vrentas and Vrentas, 1994).

The mutual diffusion coefficient D_{m} for a solvent–polymer system may be calculated by

$$D_{\text{m}} = D_0(1 - v_1)^2(1 - 2\chi v_1) \tag{1.58}$$

where v_1 denotes the volume fraction of solvent in the solvent–polymer system and χ is the polymer–solvent interaction parameter, which is closely associated with the thermodynamics of polymer solution. The mutual diffusion coefficients with the polymer–solvent interaction parameter χ can be estimated by the group-contribution lattice fluid model (Lee and Danner, 1996).

The influence of moving phase boundary associated with the polymer swelling and diffusion-induced convection is considered to simulate the unsteady-state processes of water sorption by polymer (Alsoy and Duda, 2002). The effective diffusivity D_{eff} is defined as

$$D_{\text{eff}} = D_0 \left[(1 - \omega_1) \frac{\widehat{V}_2^0}{\widehat{V}} \right]^2 \tag{1.59}$$

where ω_1 is the mass fraction of the solvent, $\widehat{V}$ is the partial specific volume of the mixture, $\widehat{V}_2^0$ is the initial partial specific volume of the polymer. In the above formulation, the effective diffusivity results from the moving boundary, the convective flows induced by diffusion, as well as the conventional, mutual binary diffusion coefficient. The studies show that, for most polymer–solvent systems, the influence of the diffusion-induced convection associated with the volume change in mixing may be neglected in analysis of sorption process.

Free volume theories analyse the solute transport in terms of the free-volume-based topological analysis that incorporates the mesh size or correlation length of the three-dimensional structure.

There is a significant dependence of the diffusion coefficients on the water concentration, which increases with increasing the water activity (Pradas et al., 2001). Diffusion of water in a polymer gel generally decreases with decreasing the degree of swelling. A small amount of water in the dry gel may be regarded as plasticizer and thus enhances the chain relaxation or relatively depresses the water penetration. By the pulsed field gradient NMR method, the water self-diffusion coefficient is measured for a set of nine commercially available contact lens hydrogels, where water content itself is the predominant factor in determining the water diffusion coefficient by an exponential decay using the specific binding model (McConville and Pope, 2000). No significant dependence of the mobility is found on polymer composition, which could not be related to equilibrium water content (McConville and Pope, 2001). The average mobility of the water in the hydrogels is largely dependent on the water content in the gel and increases with increasing the water content towards that of free water in phosphate buffered saline.

From the long-time shrinking data of the DHPMA hydrogel, it is known that, as freezing water desorbs linearly with the square root of time, departure from linearity coincides with non-freezing water desorption (Gates et al., 2003). Additionally, $D_{\text{eff}}/D_0 = (1 - x_{\text{bw}})\,\xi$ is proposed (Manetti et al., 2002), in which the term ξ is the obstructive coefficient of the polymer material, x_{bw} is the fraction of water molecules bound to the polymer network.

Based on the Monte Carlo calculation, the following relation is obtained for the reduced diffusivities as a function of PVA volume fraction ϕ_{PVA} (Fergg and Keil, 2001):

$$\frac{D_{\text{eff}}}{D_0} = \frac{0.154 - 0.18\phi_{\text{PVA}}}{0.1562}(1 - \phi_{\text{PVA}}) \tag{1.60}$$

By taking into account the variation of thickness due to swelling or shrinking and assuming a concentration-dependent diffusion coefficient, a model is developed as follows (Vicente and Gottifredi, 2000):

$$M(\tau) = \left[\frac{C_{av}(\tau)}{1 + K(1 - C_{AV}(\tau))}\right] \tag{1.61}$$

This model predicts the anomalous behaviour by taking into account the simultaneous polymer shrinking with Fickian diffusion. It is unnecessary to assume the presence of relaxation phenomena to explain the observed sigmoidal effect of volume change due to the uptake.

1.2.3 A Theoretical Formalism for Diffusion Coupled with Large Deformation of Hydrogel

Apart from the above models discussed, a significant theoretical contribution to the modelling of polymer hydrogel has to be mentioned here. Recently Suo and his group developed a theoretical formalism for modelling of mass transport of small molecular species in three-dimensional polymeric network coupled with large deformation of the hydrogel (Hong et al., 2008), where there are two modes of large deformations. The first results from the fast process of local rearrangement of small molecules, allowing the change in shape but not volume of the hydrogel. The second results from slow process of long-range migration of the molecules, allowing the change in both shape and volume of the hydrogel. The present formalism is based on a nonequilibrium thermodynamic theory, where the network hydrogel swells as the small molecules mix with long polymeric chains so that the configurational entropy of the mixture increases while the configurational entropy of the network decreases. When the changes of the two entropies balance each other, the system of the hydrogel and solvent reaches equilibrium.

In the theoretical formalism, a three-dimensional inhomogeneous field is characterized by the concentration of small molecular species $C(X,t)$, deformation gradient F_{iK} and free energy function $W(F,C)$ as follows:

$$F_{iK} = \frac{\partial x_i(\mathbf{X},t)}{\partial X_K}; \quad W(\mathbf{F},C) = W_s(\mathbf{F}) + W_m(C) \tag{1.62}$$

where X is the coordinate in reference state where the dry gel is not subjected to any mechanical load. In current state at time t, the marker X moves to a position with coordinate $x(X, t)$. $W_s(F)$ is the free energy due to network stretching and $W_m(C)$ the free energy due to mixing of solvent and polymers.

If a nominal stress $s_{ik}(X,t)$ is defined as

$$\int s_{iK}\frac{\partial \xi_i}{\partial X_K}dV = \int B_i\xi_i dV + \int T_i\xi_i dA \tag{1.63}$$

one can have the equilibrium condition of the system as

$$\int \delta W dV = \int B_i\delta x_i dV + \int T_i\delta x_i dA + \int \mu\delta C dV \tag{1.64}$$

It leads to the equivalent equilibrium conditions as follows:

$$s_{iK} = \frac{\partial W(\mathbf{F},C)}{\partial F_{iK}};\ \mu = \frac{\partial W(\mathbf{F},C)}{\partial C};\ \frac{\partial s_{iK}}{\partial X_K} + B_i = 0;\ s_{iK}N_K = T_i \tag{1.65}$$

The theoretical formalism is based on the three assumptions as follows: (1) the volumetric change due to physical association is neglected; (2) the gel has no voids and (3) the individual solvent molecule and polymer are incompressible as

$$1 + \upsilon C = \det(\mathbf{F}) \tag{1.66}$$

where υ is the volume per solvent molecule and υC is the volume of the small molecules in the hydrogel divided by the volume of the dry polymers.

The above molecular incompressibility condition (1.66) can be enforced as a constraint by introducing a Lagrange multiplier $\prod(X,t)$, namely

$$W = W(\mathbf{F},C) + \Pi[1 + \upsilon C - \det(\mathbf{F})] \tag{1.67}$$

Then one can have the equations of state as

$$s_{iK} = \frac{\partial W(\mathbf{F},C)}{\partial F_{iK}} - \Pi H_{iK}\det(\mathbf{F});\ \mu = \frac{\partial W(\mathbf{F},C)}{\partial C} + \Pi\upsilon \tag{1.68}$$

and the true stress as

$$\sigma_{ij} = \frac{\partial W(\mathbf{F},C)}{\partial F_{iK}}\frac{F_{jK}}{\det(\mathbf{F})} - \Pi\delta_{ij} \tag{1.69}$$

where Π can be interpreted as the osmotic pressure that increases the chemical potential of the solvent entering the hydrogel.

This theoretical formalism represents a significant progression. It opens a general form of formulas for the free energy function $W(F,C)$ and kinetic law, both being material specific, such that it is applicable for modelling of solvent diffusion coupled with large deformation of various polymeric hydrogels.

1.2.4 Remarks

In terms of the steady-sate modelling for simulation of the equilibrium of smart hydrogels, three macroscopic models and a microscopic molecular simulation are overviewed for description of equilibrium behaviours of the volume transition of the hydrogel. The thermodynamic models establish the relationship between the hydrogel structure and the overall behaviours of the volume transition in terms of the free energy change, the chemical potential or the related osmotic pressure. They provide a simple and qualitative approach for prediction of equilibrium of the volume transition of various hydrogels. The transport models account for the osmotic pressure, the electrical potential, the hydrogel stress, and thus the hydrogel volume transition by the mobile ion diffusive flux. They characterize both kinetics and equilibrium of volume transition behaviours of the hydrogels by diffusion equations with incorporation of different osmotic pressure expressions. The multiphase mixture model predicts the kinetics and equilibrium of the volume transition of the hydrogels by considering the frictional forces among three phases. The molecular simulation gives further understanding of the mechanisms of the volume transition of the hydrogels by simulating various interactions involved in the volume transition process of the hydrogels. However, its intensive computational requirement restricts its application to the hydrogels with small structure. Furthermore, several key parameters for model development are summarized and reviewed. It is suggested that additional chemical and structural complexities must be considered to integrate these key parameters. The volume transition of the hydrogels can be predicted accurately with more detailed approaches. In terms of the transient modelling for simulation of the kinetics of smart hydrogels, three models are examined to investigate the kinetics of the volume variation of fast response hydrogels. The phenomenal model based on the second order of reaction kinetics provides a simple approach for experimental data correlation. The power law model suggests a method to investigate the kinetic mechanisms. The multi-component diffusion model successfully incorporates the mechanical deformation and ion transport in the hydrogels to explain the kinetics of the volume variation of the hydrogels.

In brief, the present survey of the developed models for investigation of volume variation of the smart hydrogels gives an outline of historical development of modelling smart hydrogels and provides a useful guide for further development of better mathematical models for simulation of the novel smart hydrogels subject to various environmental biostimuli.

1.3 About This Monograph

This monograph provides a comprehensive and systematic study of mathematical model development and numerical simulation of the smart polymer hydrogels in BioMEMS environment, and it is based on the author's works conducted over the past years. Several basic assumptions throughout this monograph are made as follows:

- the hydrogel has no voids;
- all the three phases (polymeric network matrix, interstitial fluid and ionic species) are incompressible, and thus it is understood that the swelling/deswelling of the hydrogels results mainly from the variation of volume fraction of the fluid phase by absorbing/exuding the fluid;
- the hydrogel is immersed in an unstirred solution in vibration-free experimental device, the bulk flow of fluid or hydrodynamic velocity can thus be eliminated and subsequently the convective flux is neglected; and
- the pore of the hydrogel is narrow enough so that the diffusion dominates the transmission of flux.

The monograph is divided into seven chapters and each chapter is further divided into sections to better organize the materials. Chapter 1 briefly introduces the definition and applications of the smart polymer hydrogels and provides a historical background of model development of the polymer hydrogels.

Chapter 2 presents the model development and systematic parametric studies for the smart hydrogels in response to the change in pH of surrounding solution. After stating the basic considerations and hypotheses, a novel mathematical model, termed the multi-effect-coupling pH stimulus (MECpH) model, is developed for simulation of the pH stimulus-responsive hydrogels. Numerical simulations are carried out for analysis of influences of various hydrogel material properties and environmental conditions on the responsive performance of the pH-sensitive hydrogels in equilibrium state.

In Chap. 3, a novel multiphysics model is developed for modelling and simulation of the smart hydrogels responsive to externally applied electric field, which is called the multi-effect-coupling electric stimulus (MECe) model. After imposing the boundary and initial conditions and numerically discretizing the nonlinear coupled partial differential governing equations, the model is validated well by comparison with experimental data extracted from published literature. Both the steady-state and transient simulations are conducted for analysis of the equilibrium and kinetics of the electric stimulus-responsive hydrogels. Correspondingly, parametric studies are done for further discussion of the electric-sensitive hydrogels in equilibrium and kinetics, respectively.

It is well known that many pH-sensitive hydrogels are often able to be responsive to externally applied electric voltage simultaneously. In Chap. 4, the multi-effect-coupling pH-electric stimuli (MECpHe) model is presented for simulation of the smart hydrogels responding to the pH-electric coupled stimuli, when the hydrogels are immersed in pH buffer solution subject to externally applied electric voltage. Several case studies are made for understanding the influences of important material properties and environmental conditions on the responsive behaviour of the smart hydrogels to the pH-electric coupled stimuli.

Chapter 5 focuses on the model development and numerical simulation of the smart hydrogels responding to environmental temperature. A novel multiphysics model is developed mathematically, consisting of nonlinear partial differential equations coupled with a transcendental equation, termed the multi-effect-coupling thermal stimulus (MECtherm) model. This is a steady-state model and the formulation is based on the thermodynamics equilibrium theory. After the numerical discretization and examination of the model for the temperature-sensitive hydrogels, the detailed parametric studies are conducted for several important material properties and environmental conditions. In addition, a transient model is presented for kinetics of the thermal stimulus-responsive neutral hydrogels and corresponding simulations are done for prediction of kinetics of the smart hydrogels.

Chapter 6 briefs the author's latest works on the development of two novel models for simulation of the smart hydrogels responsive to the glucose concentration and ionic strength of surrounding solutions, respectively.

In Chap. 7, a transient modelling is presented for simulation of the controlled drug release from a non-swellable micro-hydrogel particle as delivery carrier. The mathematical model directly couples the drug dissolution with drug diffusion as two driving source forces for controlled drug release. The drugs diffuse and dissolute through the continuous matrix of the micro-hydrogel particle. Computational result is compared numerically with published experiment data and very good agreement is achieved. It is followed by numerical simulations for wide-range parametric studies of various environmental conditions and the properties of the micro-hydrogel particle delivery systems.

References

E.C. Achilleos, K.N. Christodoulou, I.G. Kevrekidis. (2001). A transport model for swelling of polyelectrolyte gels in simple and complex geometries. *Computational and Theoretical Polymer Science*, 11, 63–80.

E.C. Achilleos, R.K. Prud'homme, K.N. Christodoulou, K.R. Gee, I.G. Kevrekidis. (2000). Dynamic deformation visualization in swelling of polymer gels. *Chemical Engineering Science*, 55, 3335–3340.

S. Alsoy, J.L. Duda. (2002). Influence of swelling and diffusion-induced convection on polymer sorption processes. *AIChE Journal*, 48, 1849–1855.

C. Alvarez-Lorenzo, A. Concheiro. (2002). Reversible adsorption by a pH- and temperature-sensitive acrylic hydrogel. *Journal of Controlled Release*, 80, 247–257.

M. Annaka, Y. Amo, S. Sasaki, Y. Tominaga, K. Motokawa, T. Nakahira. (2002). Salt effect on volume phase transition of a gel. *Physical Review E*, 65, 031805-(8).

M. Annaka, M. Tokita, T. Tanaka, S. Tanaka, T. Nakahira. (2000). The gel that memorizes phases. *The Journal of Physical Chemistry*, 112, 471–477.

G. Astarita. (1989). Heat and mass transfer in solid polymer system. In: Transport Phenomena in Polymeric Systems, R.A. Mashelkar, A.S. Mujumdar, R. Kamal (Eds.) New York: Wiley, pp. 339–351.

A.M. Atta. (2002). Swelling behaviors of polyelectrolyte hydrogels containing sulfonate groups. *Polymers for Advanced Technologies*, 13, 567–576.

E.M. Aydt, R. Hentschke. (2000). Swelling of a model network: A Gibbs-ensemble molecular dynamics study. *Journal of Chemical Physics*, 112, 5480–5487.

S.K. Bajpai. (2000). Swelling-deswelling behavior of poly(acrylamide-co-maleic acid) hydrogels. *Journal of Applied Polymer Science*, 80, 2782–2789.

A.K. Bajpai, A. Giri. (2002). Swelling dynamics of a macromolecular hydrophilic network and evaluation of its potential for controlled release of agrochemicals. *Reactive and Functional Polymers*, 53, 125–141.

A.K. Bajpai, M. Shrivastava. (2000). Dynamic swelling behavior of polyacrylamide based three component hydrogels. *Journal of Macromolecular Science: Pure and Applied Chemistry*, A37, 1069–1088.

A.K. Bajpai, M. Shrivastava. (2001). Water sorption dynamics of hydrophobic, ionizable copolymer gels. *Journal of Scientific and Industrial Research India*, 60, 131–140.

A.K. Bajpai, M. Shrivastava. (2002a). Enhanced water sorption of a semi-interpenetrating polymer network (IPN) of poly(2-hydroxyethyl methacrylate) (PHEMA) and poly(ethylene glycol) (PEG). *Journal of Macromolecular Science: Pure and Applied Chemistry*, A39, 1069–1088.

A.K. Bajpai, M. Shrivastava. (2002b). Water sorption dynamics of a binary copolymeric hydrogel of a-hydroxyethyl methacrylate (HEMA). *Journal of Biomaterials Science – Polymer Edition*, 13, 237–256.

A.K. Bajpai, M. Shrivastava. (2002c). Swelling kinetics of a hydrogel of poly(ethylene glycol) and poly(acrylamide-co-styrene). *Journal of Applied Polymer Science*, 85, 1419–1428.

B. Barriere, L. Leibler. (2003). Kinetics of solvent absorption and permeation through a highly swellable elastomeric network. *Journal of Polymer Science Part B: Polymer Physics*, 41, 166–182.

A.R. Berens, H.B. Hopfenberg. (1978). Diffusion and relaxation in glassy polymer powders: 2. separation of diffusion and relaxation parameters. *Polymer*, 19, 489–496.

J.P. Boisvert, A. Malgat, I. Pochard, C. Daneault. (2002). Influence of the counter-ion on the effective charge of polyacrylic acid in dilute condition. *Polymer*, 43, 141–148.

P. Bouillot, B.A. Vincent. (2000). Comparison of the swelling behaviour of copolymer and interpenetrating network microgel particles. *Colloid and Polymer Science*, 278, 74–79.

F. Castelli, G. Pitarresi, G. Giammona. (2000). Influence of different parameters on drug release from hydrogel systems to a biomembrane model. Evaluation by differential scanning calorimetry technique. *Biomaterials*, 21, 821–833.

T. Caykara, U. Bozkaya, O. Kantoglu. (2003). Network structure and swelling behavior of poly(acrylamide/crotonic acid) hydrogels in aqueous salt solution. *Journal of Polymer Science Part B: Polymer Physics*, 41, 1656–1664.

T. Caykara, C. Ozyurek, O. Kantoglu, O. Guven. (2000). Equilibrium swelling behavior of pH- and temperature-sensitive poly(*N*-vinyl 2-pyrrolidone-g-citric acid) polyelectrolyte hydrogels. *Journal of Polymer Science Part B: Polymer Physics*, 38, 2063–2071.

A.N. Chatterjee, Q. Yu, J.S. Moore, N.R. Aluru. (2003). Mathematical modeling and simulation of dissolvable hydrogels. *Journal of Aerospace Engineering*, 16, 55–64.

J. Chen, K. Park. (2000). Synthesis of fast-swelling, superporous sucrose hydrogels. *Carbohydrate Polymers*, 41, 259–268.

Y. Chen, M. Yi. (2001). Swelling kinetics and stimuli-responsiveness of poly(DMAEMA) hydrogels prepared by UV-irradiation. *Radiation Physics and Chemistry*, 61, 65–68.

H.C. Chiu, A.T. Wu, Y.F. Lin. (2001). Synthesis and characterization of acrylic acidcontaining dextran hydrogels. *Polymer*, 42, 1471–1479.

H.C. Chiu, Y.F. Lin, Y.H. Hsu. (2002a). Effects of acrylic acid on preparation and swelling properties of pH-sensitive dextran hydrogels. *Biomaterials*, 23, 1103–1112.

H.C. Chiu, Y.F. Lin, S.H. Hung. (2002b). Equilibrium swelling of copolymerized acrylic acid-methacrylated dextran networks: Effects of ph and neutral salt. *Macromolecules*, 35, 5235–5242.

C.S. Cho, Y.I. Jeong, S.K. Kim, J.W. Nah, M. Kubota, T. Komoto. (2000). Thermoplastic hydrogel based on hexablock copolymer composed of poly(-benzyl -glutamate) and poly(ethylene oxide). *Polymer*, 41, 5185–5193.

K.F. Chou, C.C. Han, S. Lee. (2000). Water transport in 2-hydroxyethyl methacrylate copolymer irradiated by γ rays in air and related phenomena. *Journal of Polymer Science Part B: Polymer Physics*, 38, 659–671.

W.Y. Chuang, T.H. Young, D.M. Wang, R.L. Luo, Y.M. Sun. (2000). Swelling behavior of hydrophobic polymers in water ethanol mixtures. *Polymer*, 41, 8339–8347.

R.O.R. Costa, R.F.S. Freitas. (2002). Phase behavior of poly(*N*-isopropylacrylamide) in binary aqueous solutions. *Polymer*, 43, 5879–5885.

A.M.A. Da Costa, A.M. Amado. (2000). Molecular interactions in polyacrylamide/lithium perchlorate hydrogel composites. *Polymer*, 41, 5361–5365.

D. Dibbern-brunelli, T.D.Z. Atvars. (2000). Thermal transitions of poly(vinyl alcohol) hydrogel sensed by a fluorescent probe. *Journal of Applied Polymer Science*, 76, 815–824.

S. Ji, J. Ding. (2002). The wetting process of a dry polymeric hydrogel. *Polymer Journal*, 34, 267–270.

D. Dhara, P.R. Chatterji. (2000). Phase transition in linear and cross-linked poly(*N*-isopropylacrylamide) in water: Effect of various types of additives. *Journal of Macromolecular Science, Part C: Polymer Reviews*, C40, 51–68.

E. Diez-Pena, I. Quijada-Garrido, J.M. Barrales-Rienda. (2002). On the water swelling behaviour of poly(*N*-isopropylacrylamide) [P(*N*-iPAAm)] poly(methacrylic acid) [P(MAA)] their random copolymers and sequential interpenetrating polymer networks (IPNs). *Polymer*, 43, 4341–4348.

S. Durmaz, O. Okay. (2000). Acrylamide/2-acrylamido-2-methylpropane sulfonic acid sodium salt-based hydrogels synthesis and characterization. *Polymer*, 41, 3693–3704.

M. Ebara, T. Aoyagi, K. Sakai, T. Okano. (2001). The incorporation of carboxylate groups into temperature-responsive poly(*N*-isopropylacrylamide)-based hydrogels promotes rapid gel shrinking. *Journal of Polymer Science Part A: Polymer Chemistry*, 39, 335–343.

G.M. Eichenbaum, P.K. Kiser, D. Shah, W.P. Meuer, D. Needham, S.A. Simon. (2000). Alkali earth metal binding properties of ionic microgels. *Macromolecules*, 33, 4087–4093.

D.J. Enscore, H.B. Hopfenberg, V.T. Stannett. (1977). Effect of particle size on the mechanism controlling n-hexane sorption in glassy polystyrene microspheres. *Polymer*, 18, 793–800.

B. Erman, P.J. Flory. (1986). Critical phenomena and transitions in swollen polymer networks and in linear macromolecules. *Macromolecules*, 19, 2342–2353.

G. Evmenenko, T. Budtova. (2000). Structure changes in hydrogels immersed in a linear polymer solution studied by SANS. *Polymer*, 41, 4943–4947.

J. Fei, Z. Zhang, L. Gu. (2002). Bending behavior of electroresponsive poly(vinyl alcohol) and poly(acrylic acid) semi-interpenetrating network hydrogel fibres under an electric stimulus. *Polymer International*, 51, 502–509.

F. Fergg, F.J. Keil. (2001). Diffusion and reactions of multicomponent electrolytes in poly(vinyl alcohol) hydrogels — modeling and experiment. *Chemical Engineering Science*, 56, 1305–1315.

L. Ferreira, M.M. Vidal, M.H. Gil. (2000). Evaluation of poly(2-hydroxyethyl methacrylate) gels as drug delivery systems at different pH values. *International Journal of Pharmaceutics*, 194, 169–180.

P.J. Flory. (1953). *Principles of Polymer Chemistry*, Ithaca, New York: Cornell University Press.

A. Fomenko, Z. Sedlakova, J. Plestil, M. Ilavsky. (2002). Phase transition in swollen gels: Part 32. Temperature transition in charged poly(*N*-isopropylmethacrylamide) hydrogels in water and aqueous NaCl solutions. *Physical Chemistry Chemical Physics*, 4, 4360–4367.

M.W.C.P. Franse, K. te Nijenhuis. (2000). Crosslinking index, molecular weight distribution and rubber equilibrium shear modulus during polyfunctional crosslinking of existing polymer. Part 5. Primary polymer with a discrete distribution of the molecular weight. *Journal of Molecular Structure*, 554, 1–10.

H. Furukawa. (2000). Effect of varying preparing-concentration on the equilibrium swelling of polyacrylamide gels. *Journal of Molecular Structure*, 554, 11–19.

L.H. Gan, G.R. Deen, X.J. Loh, X.Y. Gan. (2001). New stimuli-responsive copolymers of *N*-acryloyl-*N*'-alkyl piperazine and methyl methacrylate and their hydrogels. *Polymer*, 42, 65–69.

G. Gates, J.P. Harmon, J. Ors, P. Benz. (2003). 2, 3-dihydroxypropyl methacrylate and 2-hydroxyethyl methacrylate hydrogels: Gel structure and transport properties. *Polymer*, 44, 215–222.

A.J. Galli, W.H. Brumage. (1983). The freely jointed chain in expanded form. *Journal of Chemical Physics*, 79, 2411.

R.A. Gemeinhart, J. Chen, H. Park, K. Park. (2000). pH-Sensitivity of fast responsive superporous hydrogels. *Journal of Biomaterials Science*, Polymer Edition, 11, 1371–1380.

P.E. Grimshaw, J.H. Nussbaum, A.J. Grodzinsky. (1990). Kinetics of electricity and chemically induced swelling in polyelectrolyte gels. *Journal of Chemical Physics*, 93, 4462–4472.

C.M. Hansen. (2000). *Hansen Solubility Parameters: A User's Handbook*, Boca Raton: CRC Press.

J.P. Harmon, S. Lee, J.C.M. Li. (1987). Methanol treatment in PMMA: the effect of mechanical deformation. *Journal of Polymer Science Part A: Polymer Chemistry*, 25, 3215–3229.

M.M. Hassan, C.J. Durning. (1999). Effects of polymer molecular weight and temperature on case II transport. *Journal of Polymer Science Part B: Polymer Physics*, 37, 3159–3171.

D.J.T. Hill, N.G. Moss, P.J. Pomery, A.K. Whittaker. (2000). Copolymer hydrogels of 2-hydroxyethyl methacrylate with n-butyl methacrylate and cyclohexyl methacrylate synthesis characterization and uptake of water. *Polymer*, 41, 1287–1296.

A. Hiroki, Y. Maekawa, M. Yoshida, K. Kubota, R. Katakai. (2001). Volume phase transitions of poly(acryloyl-L-proline methyl ester) gels in response to water-alcohol composition. *Polymer*, 42, 1863–1867.

Y.P. Hong, Y.C. Bae. (2002). Phase behaviors of partially ionized hydrogels in aqueous salt solutions: Applicability of the modified double-lattice model. *Journal of Polymer Science Part B: Polymer Physics*, 40, 2333–2338.

W. Hong, X. Zhao, J. Zhou, Z. Suo. (2008). A theory of coupled diffusion and large deformation in polymeric gels, *Journal of the Mechanics and Physics of Solids*, 56, 1779–1793.

Y. Huang, I. Szleifer, N.A. Peppas. (2002). A molecular theory of polymer gels. *Macromolecules*, 35, 1373–1380.

A. Huther, B. Schafer, X. Xu, G. Maurer. (2002). Phase equilibria of hydrogel systems. *Physical Chemistry Chemical Physics*, 4, 835–844.

A. Ikehat, H. Ushiki. (2002). Effect of salt on the elastic modulus of poly(*N*-isopropylacrylamide) gels. *Polymer*, 43, 2089–2094.

M. Ilavsky, G. Mamytbekov, L. Hanykova, K. Dusek. (2002). Phase transition in swollen gels: 31. Swelling and mechanical behaviour of interpenetrating networks composed of poly(1-vinyl-2-pyrrolidone) and polyacrylamide in water/acetone mixtures. *European Polymer Journal*, 38, 875–883.

K. Ito, Y. Ujihira, T. Yamashita, K. Horie. (2000). Temperature dependence of free volume of polyacrylamide gels studied by positron lifetime measurements. *Radiation Physics and Chemistry*, 58, 521–524.

E. Jabbari, S. Nozari. (2000). Swelling behavior of acrylic acid hydrogels prepared by γ-radiation crosslinking of polyacrylic acid in aqueous solution. *European Polymer Journal*, 36, 2685–2692.

B.D. Johnson, D.J. Niedermaier, W.C. Crone, J. Moorthy, D.J. Beebe. (2002). Mechanical properties of a pH sensitive hydrogel, *Proceedings of the 2002 Annual Conference of Society for Experimental Mechanics,* Milwaukee, Wisconsin.

E. Karadag, D. Saraydin. (2002). Swelling of superabsorbent acrylamide/sodium acrylate hydrogels prepared using multifunctional crosslinkers. *Turkish Journal of Chemistry*, 26, 863–875.

I. Katime, E.D. de Apodaca. (2000). Acrylic acid/methyl methacrylate hydrogels. I. Effect of composition on mechanical and thermodynamic properties. *Journal of Macromolecular Science: Pure and Applied Chemistry*, A37, 307–321.

I. Katime, E. Rodriguez. (2001). Absorption of metal ions and swelling properties of poly(acrylic acid-*co*-itaconic acid) hydrogels. *Journal of Macromolecular Science: Pure and Applied Chemistry*, A38, 543–558.

N.R. Kenkare, C.K. Hall, S.A. Khan. (2000). Theory and simulation of the swelling of polymer gels. *Journal of Physical Chemistry*, 113, 404–418.

M.N. Khalid, F. Agnely, N. Yagoubi, J.L. Grossiord, G. Couarraze. (2002). Water state characterization, swelling behavior, thermal and mechanical properties of chitosan based networks. *European Journal of Pharmaceutical Sciences*, 15, 425–432.

A.R. Khokhlov, E. Yu. Kramarenko. (1994). Polyelectrolyte/Ionomer behavior in polymer gel collapse. *Macromolecular Theory and Simulations*, 3, 45–59.

S.J. Kim, S.J. Park, I.Y. Kim, M.S. Shin, S.I. Kim. (2002). Electric stimuli responses to poly(vinyl alcohol)/chitosan interpenetrating polymer network hydrogel in NaCl solutions. *Journal of Applied Polymer Science*, 86, 2285–2289.

S.J. Kim, S.J. Park, S.I. Kim. (2003a). Synthesis and characteristics of interpenetrating polymer network hydrogels composed of poly(vinyl alcohol) and poly(*N*-isopropylacrylamide). *Reactive and Functional Polymers*, 55, 61–67.

B. Kim, N.A. Peppas. (2002). Complexation phenomena in pH-responsive copolymer networks with pendent saccharides. *Macromolecules*, 35, 9545–9550.

S.J. Kim, G.Y. Yoon, Y.M. Lee, S.I. Kim. (2003b). Electrical sensitive behavior of poly(vinyl alcohol)/poly(diallyldimethylammonium chloride) IPN hydrogel. *Sensors and Actuators B: Chemical*, 88, 286–291.

J. Kovac. (1978). Modified Gaussian model for rubber elasticity. *Macromolecules*, 11, 362–365.

E. Yu. Kramarenko, A.R. Khokhlov, K. Yoshikawa. (2000). A three-state model for counterions in a dilute solution of weakly charged polyelectrolytes. *Macromolecular Theory and Simulations*, 9, 249–256.

K.Y. Lee, K.H. Bouhadir, D.J. Mooney. (2000). Degradation behavior of covalently cross-linked poly(aldehyde guluronate) hydrogels. *Macromolecules*, 33, 97–101.

W.F. Lee, Y.J. Chen. (2001). Studies on preparation and swelling properties of the *N*-isopropylacrylamide/chitosan semi-IPN and IPN hydrogels. *Journal of Applied Polymer Science*, 82, 2487–2496.

B.C. Lee, R.P. Danner. (1996). Application of the group-contribution lattice-fluid equation of state to random copolymer solvent systems, *Fluid Phase Equilibria*, 117, 33–39.

W.F. Lee, Y.H. Lin. (2001). pH-reversible hydrogels. IV. Swelling behavior of the 2-hydroxyethyl methacrylate-co acrylic acid-co-sodium acrylate copolymeric hydrogels. *Journal of Applied Polymer Science*, 81, 1360–1371.

J. Lee, C.W. Marcosko, D.W. Urry. (2001). Phase transition and elasticity of protein-based hydrogels. *Journal of Biomaterials Science – Polymer Edition*, 12, 229–242.

W.F. Lee, P.L. Yeh. (2000). Thermoreversible hydrogels. IX. Swelling behaviors of thermosensitive hydrogels copolymerized by *N*-isopropylacrylamide with 1-vinyl-3-(3-sulfopropyl) imidazolium betaine. *Journal of Applied Polymer Science*, 77, 14–23.

B. Li, D. Ding, P. Sun, Y. Wang, J. Ma, B. He. (2000). PGSE NMR studies of water states of hydrogel P(Am-NaA). *Journal of Applied Polymer Science*, 77, 424–427.

T. Lindvig, M.L. Michelsen, G.M. Kontogeorgis. (2002). A Flory–Huggins model based on the Hansen solubility parameters. *Fluid Phase Equilibria*, 5093, 1–14.

V.M.M. Lobo, A.J.M. Valente, A.Y. Polishchuk, G. Geuskens. (2001). Transport of non-associated electrolytes in acrylamide hydrogels. *Journal of Molecular Liquids*, 94, 179–192.

X.J. Loh, G.R. Deen, X.Y. Gan, L.H. Gan. (2001). Water-sorption and metal-uptake behavior of pH-responsive poly(*N*-acryloyl-*N*'-methylpiperazine) gels. *Journal of Applied Polymer Science*, 80, 268–273.

Z.Y. Lu, R. Hentschke. (2002a). Computer simulation study on the swelling of a model polymer network by a chainlike solvent. *Physical Review E*, 65, 041807.

Z.Y. Lu, R. Hentschke. (2002b). Swelling of model polymer networks with different cross-link densities: A computer simulation study. *Physical Review E*, 66, 041803.

X. Lu, M. Zhai, J. Li, H. Ha. (2000). Radiation preparation and thermo-response swelling of interpenetrating polymer network hydrogel composed of PNIPAAm and PMMA. *Radiation Physics and Chemistry*, 57, 477–480.

A.V. Lyulin, B. Dunweg, O.V. Borisov, A.A. Darinskii. (1999). Computer simulation studies of a single polyelectrolyte chain in poor solvent. *Macromolecules*, 32, 3264–3278.

T.M. Madkour. (2001). A combined statistical mechanics and molecular dynamics approach for the evaluation of the miscibility of polymers in good, poor and non-solvents. *Chemical Physics*, 274, 187–198.

K. Makino, H. Agata, H. Ohshima. (2000a). Dependence of temperature-sensitivity of poly(*N*-isopropylacrylamide-*co*-acrylic acid) hydrogel microspheres upon their sizes. *Journal of Colloid and Interface Science*, 230, 128–134.

K. Makino, J. Hiyoshi, H. Ohshima. (2000b). Kinetics of swelling and shrinking of poly(*N*-isopropylacrylamide) hydrogels at different temperatures. *Colloids and Surfaces B: Biointerfaces*, 19, 197–204.

C. Manetti, L. Casciani, N. Pescosolido. (2002). Diffusive contribution to permeation of hydrogel contact lenses theoretical model and experimental evaluation by nuclear magnetic resonance techniques. *Polymer*, 43, 87–92.

F. Martellini, L.H.I. Mei, J.L. Balino, M. Carenza. (2002). Swelling and water transport in temperature-sensitive hydrogels based on 2-methoxyethylacrylate. *Radiation Physics and Chemistry*, 63, 29–33.

P. Martens, K.S. Anseth. (2000). Characterization of hydrogels formed from acrylate modified poly(vinyl alcohol) macromers. *Polymer*, 41, 7715–7722.

A. Matsuda, Y. Katayama, T. Kaneko, J.P. Gong, Y. Osada. (2000). Ionization and order–disorder transition of hydrogels with ionizable hydrophobic side chain. *Journal of Molecular Structure*, 554, 91–97.

P. McConville, J.M. Pope. (2000). A comparison of water binding and mobility in contact lens hydrogels from NMR measurements of the water self-diffusion coefficient. *Polymer*, 41, 9081–9088.

P. McConville, J.M. Pope. (2001). ^{1}H NMR T_2 relaxation in contact lens hydrogels as a probe of water mobility. *Polymer*, 42, 3559–3568.

J.R. Meakin, D.W.L. Kukins, C.T. Imrie, R.M. Aspden. (2003). Thermal analysis of poly(2-hydroxyethyl methacrylate) (pHEMA) hydrogels. *Journal of Materials Science: Materials in Medicine*, 14, 9–15.

D. Melekaslan, O. Okay. (2000). Swelling of strong polyelectrolyte hydrogels in polymer solutions: Effect of ion pair formation on the polymer collapse. *Polymer*, 41, 5737–5747.

D. Melekaslan, O. Okay. (2001). Reentrant phase transition of strong polyelectrolyte poly(*N*-isopropylacrylamide) gels in PEG solutions. *Macromolecular Chemistry and Physics*, 202, 304–312.

V. Michailova, S. Titeva, R. Kotsilkova, E. Krusteva, E. Minkov. (2000). Water uptake and relaxation processes in mixed unlimited swelling hydrogels. *International Journal of Pharmaceutics*, 209, 45–56.

E.C. Muniz, G. Geuskens. (2001). Influence of temperature on the permeability of polyacrylamide hydrogels and semi-IPNs with poly(*N*-isopropylacrylamide). *Journal of Membrane Science*, 172, 287–293.

H. Muta, K. Ishida, E. Tamaki, M. Satoh. (2002). An IR study on ion-specific and solvent-specific swelling of poly(*N*-vinyl-2-pyrrolidone) gel. *Polymer*, 43, 103–110.

H. Muta, R. Kojima, S. Kawauchi, A. Tachibana, M. Satoh. (2001a). Ion-specificity for hydrogen-bonding hydration of polymer: An approach by *ab initio* molecular orbital calculations. *Journal of Molecular Structure: THEOCHEM*, 536, 219–226.

H. Muta, M. Miwa, M. Satoh. (2001b). Ion-specific swelling of hydrophilic polymer gels. *Polymer*, 42, 6313–6138.

H. Muta, T. Sin, A. Yamanaka, S. Kawauchi, M. Satoh. (2001c). Ion-specificity for hydrogen-bonding hydration of polymer: An approach by *ab initio* molecular orbital calculations II. *Journal of Molecular Structure: THEOCHEM*, 574, 195–211.

B. Nick, U.W. Suter. (2001). Solubility of water in polymers-atomistic simulations. *Computational and Theoretical Polymer Science*, 11, 49–54.

T. Norisuye, N. Masui, Y. Kida, D. Ikuta, E. Kokufuta, S. Ito, S. Panyukov, M. Shibayama. (2002). Small angle neutron scattering studies on structural inhomogeneities in polymer gels: irradiation cross-linked gels vs chemically crosslinked gels. *Polymer*, 43, 5289–5297.

K. Ogawa, Y. Ogawa, E. Kokufuta. (2002). Effect of charge inhomogeneity of polyelectrolyte gels on their swelling behavior. *Colloids and Surfaces A: Physicochemical and Engineering Aspects*, 209, 267–279.

R.R. Ohs, S.K. De, N.R. Aluru. (2001). Modelling of ionic hydrogel kinetics in buffered solutions, Technical Proceedings of the 2001 International Conference on Modeling and Simulation of Microsystems, 1, 7–10.

O. Okay, S. Durmaz. (2002). Charge density dependence of elastic modulus of strong polyelectrolyte hydrogels. *Polymer*, 43, 1215–1221.

O. Okay, S.B. Sariisik. (2000). Swelling behavior of poly(acrylamide-*co*-sodium acrylate) hydrogels in aqueous salt solutions: Theory versus experiments. *European Polymer Journal*, 36, 393–399.

D. Ostrovskii, M. Edvardsson, P. Jacobsson. (2003). Weak polymer-electrolyte interaction revealed by Fermi resonance perturbed Raman bands. *Journal of Raman Spectroscopy*, 34, 40–49.

M.M. Ozmen, O. Okay. (2003). Swelling behavior of strong polyelectrolyte poly(*N*-t-butylacrylamide-*co*-acrylamide) hydrogels. *European Polymer Journal*, 39, 877–886.

V. Ozturk, O. Okay. (2002). Temperature sensitive poly(*N*-t-butylacrylamide-*co*-acrylamide) hydrogels synthesis and swelling behavior. *Polymer*, 43, 5017–5026.

A. Panda, S.B. Manohar, S. Sabharwal, Y.K. Bhardwaj, A.B. Majali. (2000). Synthesis and swelling characteristics of poly (*N*-isopropylacrylamide) temperature sensitive hydrogels crosslinked by electron beam irradiation. *Radiation Physics and Chemistry*, 58, 101–110.

S. Panyukov, Y. Rabin. (1996). Statistical physics of polymer gels. *Physics Reports*, 269, 1–131.

N.A. Peppas. (1986). Hydrogels in Medicine and Pharmacy, Boca Raton, FL: CRC Press.

K. Podual, N.A. Peppas. (2005). Relaxational behavior and swelling-pH master curves of poly[(diethylaminoethyl methacrylate)-graft-(ethylene glycol)] hydrogels. *Polymer International*, 54, 581–593.

M.M. Pradas, J.L.G. Ribelles, A.S. Aroca, G.G. Ferrer, J.S. Anton, P. Pissis. (2001). Porous poly(2-hydroxyethyl acrylate) hydrogels. *Polymer*, 42, 4667–4674.

X. Qu, A. Wirsen, A.C. Albertsson. (2000). Novel pH-sensitive chitosan hydrogels swelling behavior and states of water. *Polymer*, 41, 4589–4598.

J.R. Quintana, N.E. Valderruten, N.E. Katime. (1999). Synthesis and swelling kinetics of poly(dimethylaminoethyl acrylate methyl chloride quaternary-co-itaconic acid) hydrogels. *Langmuir*, 15, 4728–4730.

G.V.N. Rathna, P.R. Chatterji. (2001). Swelling kinetics and mechanistic aspects of thermosensitive interpenetrating polymer networks. *Journal of Macromolecular Science: Pure and Applied Chemistry*, A38, 43–56.

T. Schmidt, C. Querner, K.F. Arndt. (2003). Characterization methods for radiation crosslinked poly(vinyl methyl ether) hydrogels. *Nuclear Instruments and Methods in Physics Research Section B: Beam Interactions with Materials and Atoms*, 208, 331–335.

H. Schott. (1992). Swelling kinetics of polymers. *Journal of Macromolecular Science: Physics*, B31, 1–9.

U.P. Schroder, W. Oppermann. (2002). Computer simulation of network formation via crosslinking copolymerization. *Macromolecular Theory and Simulations*, 6, 151–160.

M. Sen, O. Güven. (1998). Prediction of swelling behaviour of hydrogels containing diprotic acid moieties. *Polymer*, 39, 1165–1172.

M. Sen, O. Güven. (2000). Prediction of the swelling behavior of amphiphilic hydrogels and the determination of average molecular weight between cross-links. *Computational and Theoretical Polymer Science*, 11, 475–482.

M. Sen, O. Guven. (2002). Dynamic deswelling studies of poly(*N*-vinyl-2-pyrrolidone/itaconic acid) hydrogels swollen in water and terbinafine hydrochloride solutions. *European Polymer Journal*, 38, 751–757.

B.C. Shin, S.S. Kim, J.K. Ko, J. Jegal, B.M. Lee. (2003). Gradual phase transition of poly(*N*-isopropylacrylamide-*co*-acrylic acid) gel induced by electric current. *European Polymer Journal*, 39, 579–584.

M.R. Simmons, E.N. Yamasaki, C.S. Patrickios. (2000). Cationic amphiphilic model networks: Synthesis by group transfer polymerization and characterization of the degree of swelling. *Macromolecules*, 33, 3176–3179.

Y. Suetoh, M. Shibayama. (2000). Effects of non-uniform solvation on thermal response in poly(*N*-isopropylacrylamide) gels. *Polymer*, 41, 505–510.

T. Tokuhiro. (2001). Temperature dependence of density of polymer gels: 2. Poly[*N*-(1,3-dioxolan-2-ylmethyl)-*N*-methyl-acrylamide] networks -water or -alcohol system. *Journal of Physical Chemistry B*, 105, 11955–11960.

A.I. Triftaridou, S.C. Hadjiyannakou, M. Vamvakaki, C.S. Patrickios. (2002). Synthesis, characterization, and modelling of cationic amphiphilic model hydrogels: Effects of polymer composition and architecture. *Macromolecules*, 35, 2506–2513.

J. Valencia, I.F. Pierola. (2001). Equilibrium swelling properties of poly(*N*-vinylimidazole-co-sodium styrenesulfonate) hydrogels. *European Polymer Journal*, 37, 2345–2352.

J. Valencia, I.F. Pierola. (2002). Swelling kinetics of poly(*N*-vinylimidazole-co-sodium styrenesulfonate) hydrogels. *Journal of Applied Polymer Science*, 83, 191–200.

A.J.M. Valente, A.Y. Polishchuk, V.M.M. Lobo, G. Geuskens. (2002). Diffusion coefficients of lithium chloride and potassium chloride in hydrogel membranes derived from acrylamide. *European Polymer Journal*, 38, 13–18.

S. Varghese, A.K. Lele, R.A. Masjelkar. (2000). Designing new thermoreversible gels by molecular tailoring of hydrophilic-hydrophobic interactions. *Journal of Physical Chemistry*, 112, 3063–3070.

E.V. Vashuk, E.V. Vorobieva, I.I. Basalyga, N.P. Krutko. (2001). Water-absorbing properties of hydrogels based on polymeric complexes. *Materials Research Innovations*, 4, 350–352.

M.S. Vicente, J.C.Y. Gottifredi. (2000). Effect of volume changes due to absorption in polymer membranes. *Journal of Membrane Science*, 169, 249–254.

J.S. Vrentas, C.M. Vrentas. (1994). Solvent self-diffusion in rubbery polymer-solvent systems. *Macromolecules*, 27, 4684–4690.

T. Wallmersperger, B. Kroplin, J. Holdenried, W. Gulch. (2001). A Coupled Multi-Field-Formulation for Ionic Polymer Gels in Electric Fields. In: *Smart Structures and Materials 2001: Electroactive Polymer Actuators and Devices*, Y. Bar-Cohen (Ed.) SPIE Press, 4329, pp. 264–275.

B. Wang, T. Ymaguchi, S. Nakao. (2000). Solvent diffusion in amorphous glassy polymers. *Journal of Polymer Science Part B: Polymer Physics*, 38, 846–856.

J.A. White, W.M. Deen. (2002). Agarose-dextran gels as synthetic analogs of glomerular basement membrane water permeability. *Biophysical Journal*, 82, 2081–2089.

C.J. Whiting, A.M. Voice, P.D. Olmsted, T.C.B. Mcleish. (2001). Shear modulus of polyelectrolyte gels under electric field. *Journal of Physics: Condensed Matter*, 13, 1381–1393.

W. Xue, S. Champ, M.B. Huglin. (2001a). Thermoreversible swelling behaviour of hydrogels based on *N*-isopropylacrylamide with a zwitterionic comonomer. *European Polymer Journal*, 37, 869–875.

W. Xue, S. Champ, M.B. Huglin. (2001b). Network and swelling parameters of chemically crosslinked thermoreversible hydrogels. *Polymer*, 42, 3665–3669.

W. Xue, I.W. Hamley. (2002). Thermoreversible swelling behaviour of hydrogels based on *N*-isopropylacrylamide with a hydrophobic comonomer. *Polymer*, 43, 3069–3077.

L. Yao, S. Krause. (2003). Electromechanical responses of strong acid polymer gels in DC electric fields. *Macromolecules*, 36, 2055–2065.

H. Yasunaga, Y. Shirakawa, H. Urakawa, K. Kajiwara. (2002). Dynamic behaviour of water in hydrogel containing hydrophobic side chains as studied by pulse ^{1}H NMR. *Journal of Molecular Structure*, 602–603, 399–404.

B. Yildiz, B. Isik, M. Kis. (2002). Thermoresponsive poly(*N*-isopropylacrylamide-*co*-acrylamide-*co*-2-hydroxyethyl methacrylate) hydrogels. *Reactive and Functional Polymers*, 52, 3–10.

M.K. Yoo, Y.K. Sung, Y.M. Lee, C.S. Cho. (2000). Effect of polyelectrolyte on the lower critical solution temperature of poly(*N*-isopropyl acrylamide) in the poly(NIPAAm-*co*-acrylic acid) hydrogel. *Polymer*, 41, 5713–5719.

L. Zha, J. Hu, C. Wang, S. Fu, M. Luo. (2002). The effect of electrolyte on the colloidal properties of poly(*N*-isopropylacrylamide-*co*-dimethylaminoethylmethacrylate) microgel latexes. *Colloid and Polymer Science*, 280, 1116–1121.

J. Zhang, N.A. Peppas. (2000). Synthesis and characterization of pH- and temperature-sensitive poly(methacrylic acid)/poly(*N*-isopropylacrylamide) interpenetrating polymeric networks. *Macromolecules*, 33, 102–107.

X.Z. Zhang, D.Q. Wu, C.C. Chu. (2003). Effect of the crosslinking level on the properties of temperature-sensitive poly(*N*-isopropylacrylamide) hydrogels. *Journal of Polymer Science Part B: Polymer Physics*, 41, 582–593.

X.Z. Zhang, R.X. Zhuo. (2000a). Preparation of fast responsive, thermally sensitive poly(*N*-isopropylacrylamide) gel. *European Polymer Journal*, 36, 2301–2303.

X.Z. Zhang, R.X. Zhuo. (2000b). Novel synthesis of temperature-sensitive poly(*N*-isopropylacrylamide) hydrogel with fast deswelling rate. *European Polymer Journal*, 36, 643–645.

X.Z. Zhang, R.X. Zhuo. (2000c). Synthesis of temperature-sensitive poly(*N*-isopropylacrylamide) hydrogel with improved Surface Property. *Journal of Colloid and Interface Science*, 223, 311–313.

X.Z. Zhang, R.X. Zhuo. (2002). Synthesis, properties of thermosensitive poly(*N*-isopropylacrylamide-*co*-methyl methacrylate) hydrogel with rapid response. *Materials Letters*, 52, 5–9.

B. Zhao, J.S. Moore. (2001). Fast pH- and ionic strength-responsive hydrogels in microchannels. *Langmuir*, 17, 4758–4763.

X. Zhou, Y.C. Hon, S. Sun, A.F.T. Mak. (2002). Numerical simulation of the steady-state deformation of a smart hydrogel under an external electric field. *Smart Materials and Structures*, 11, 459–467.

Chapter 2
Multi-Effect-Coupling pH-Stimulus (MECpH) Model for pH-Sensitive Hydrogel

2.1 Introduction

In general, the degree of swelling/shrinking of a smart hydrogel is dependent upon many effects, such as the ionizable group and polymeric network structure of the hydrogel and the characteristics of environmental solutions including the composition, pH and temperature, in which there are different interactions between mechanical, chemical and electrical fields.

In this chapter, a multiphysics model is developed for simulation of the swelling/deswelling behaviour of the hydrogel responsive to surrounding pH when the hydrogel is immersed in the buffered solution, called the multi-effect-coupling pH- stimulus (MECpH) model. The model is based on Poisson–Nernst–Planck (PNP) formulation and derived in combination with the species diffusion, pendent ionizable group dissociation reaction and electric potential effect for the distributive profiles of ionic concentrations and electric potential within both the hydrogel and bathing solution domains in response to the change in pH of surrounding solution. The PNP equations are coupled with nonlinear mechanical equilibrium governing equation for analysis of large deformation of the pH-sensitive hydrogel. The MECpH model is validated by comparison between the simulation results and published experiment data available in the literature. It is followed by comprehensive parameter studies for the influences of hydrogel material properties and environmental solutions on the responsive performance of the pH-sensitive hydrogel when placed in the buffered solution.

2.2 Development of the MECpH Model

Extensive search of literature reveals that most of the theoretical models are either oversimplified to limit applicability or too complex to account flexibly for significant parameters requested by experiments. Therefore, the objective of this section is to formulate a theoretical model with clear fundamental, very robustness and possible extension for a wide range of applications. This is a multiphysics model and termed the multi-effect-coupling pH-stimulus (MECpH) model. It is developed

H. Li, *Smart Hydrogel Modelling*, DOI 10.1007/978-3-642-02368-2_2,

mathematically to simulate the behaviour of the pH-sensitive hydrogels in response to infinitesimal changes in environmental conditions.

Basically the Nernst–Planck flux system is employed to describe the mechanism of transports of diffusive species in a membrane. It is known that the Nernst–Planck system is insufficient as it only includes the effects of the gradients of species concentration and electrical potential. Over the past decades, many researchers chose the computational models simplified by assuming the electroneutrality condition or constant field. These are the two widely used assumptions and also the two limited cases of certain dimensionless parameters associated with the ratio of the Debye length to membrane thickness. Therefore, the Nernst–Planck system is not sufficiently accurate for some electrolytic solutions, especially for the presently considered hydrogels with fixed charge groups, which are often used for description of biological systems like thin membranes. A more rigorous model is necessarily developed to include the variation of the electric potential based on the spatial distribution of the electric charges, where the relationship between the electrical potential and the various ionic fluxes is required by coupling with the Poisson equation, which forms the PNP system. According to this system, the drift of an ionic species strongly influences those of all other ions dissolved in the electrolytic solutions. Further, the model couples the mechanical equilibrium equation by a finite deformation formulation with the PNP equations to simulate the large deformation of hydrogels.

One of the important contributions in development of the MECpH model is to incorporate a relation between the diffusive hydrogen ions and the fixed charge attached onto the polymeric chain network of the hydrogels, which is based on the Langmuir adsorption isotherm with consideration of the hydrogen ion bound by the fixed charge groups.

The MECpH model simulating the pH-responsive hydrogels is able to provide the concentration profiles of all diffusive ionic species, the electric potential distribution in both the hydrogel and bathing solution domains as well as the mechanical deformation of the hydrogels swollen in the surrounding solutions with various environmental conditions.

2.2.1 *Electrochemical Formulation*

As well known, there are several possible approaches to simulating the ion permeation at atomistic level, including all-atom molecular dynamics, Brownian dynamics and Monte Carlo simulations. However, the continuum formulations are still very valuable for highlighting fundamental principles in a particularly clear fashion, even though they often overlook the fine details of atomic-level reality that becomes significant at microscopic level. Further, when we deal with millions of ions interacting each other and with polymeric chains, it is much more practical in a sense to investigate the ion transport within the hydrogel in a continuum manner. In particular, the macroscopic continuum electrostatic simulation is based on the capability of choosing infinitesimal element volume of interest but large enough to cover sufficient number of charge groups, such that meaningful volume charge density can be characterized in a continuous sense and thus it can illustrate fundamental principles in a comprehensible way (Woodson and Melcher, 1968; Grodzinsky, 1974).

For the continuum theories developed for electrodiffusion modelling, Poisson–Nernst–Planck (PNP) equation has been applied successfully for describing the ion transportation phenomena in the polyelectrolyte gel (Gulch et al., 2000; Wallmersperger et al., 2001), the ion exchange or biological membrane (Helfferich, 1962; Carnay and Tasaki, 1971; Sjodin, 1971; Rubinstein, 1990), the biological ion channel (Kurnikova et al., 1999; Syganow and von Kitzin, 1999; Gillespie and Eisenberg, 2001, 2002; Roux et al., 2004), semiconductor (Selberherr, 1984), soil or clay (Samson et al., 1999; Samson and Marchand, 1999) and other porous media (MacGillivray, 1968; MacGillivray and Hare, 1969; Kato, 1995).

In the PNP system, the electric field is calculated self-consistently by the average ionic charge density. The ion–ion interactions are thus incorporated approximately at a mean field level (Eisenberg, 1999), namely it is the possible best way to find out the chemical properties of porous media by making them as small as possible for further understanding with mean electrostatic field. If the phenomena observed cannot be described well by the mean field, we then turn to chemically specific explanations for seeking out appropriate tools, for example, Langevin or molecular dynamics. This strategy will be followed up in the following sections for analysis of electrodiffusion within the hydrogel.

2.2.1.1 Ionic Flux

Hydrogel is generally composed of crosslinked polymer network matrix binding the fixed charge groups, where there exist mobile co-ions and counterions surrounding the mesh network within the interstitial fluid. The flow of fluxes rises generally due to the gradients of ionic concentration, electrical potential, chemical potential or pressure. As well known, the Nernst–Planck equation can characterize the ionic fluxes in the hydrogel in terms of the gradients of the ionic concentration, electrical potential and pressure. By the law of mass conservation, the ionic fluxes and species concentrations can be formulated in the form of macroscopic continuum. Therefore, the Nernst–Planck equations for ion transportation phenomena in the system can be written as follows for the flux of ionic species k (Teorell, 1953; Dresner, 1972):

$$\begin{aligned}\mathbf{J}_k &= -[[D_k]_i\mathbf{grad}(c_k) + F\mu_k z_k c_k \mathbf{grad}(\psi) + [D_k]_i c_k \mathbf{grad}(\ln \gamma_k)] + c_k\mathbf{U}_i \\ &= -[D_k]_i[\mathbf{grad}(c_k) + \tfrac{F}{RT} z_k c_k \mathbf{grad}(\psi) + c_k \mathbf{grad}(\ln \gamma_k)] + c_k\mathbf{U}_i \qquad (2.1)\\ &\qquad (k = 1,2,3,\ldots,N_{\text{ion}})\end{aligned}$$

where $\boldsymbol{J}_k$ (mM/s) is the flux of the kth species and N_{ion} is the number of total species in the system. D_k (m^2/s) is the diffusivity tensor of the kth species, i is the direction of flux flow. c_k (mM) and z_k are the concentration and valence number of the kth diffusive ionic species, respectively. μ_k (m.mol/s) is the mobility of the kth ion species, ψ (V) is the electrostatic potential, γ_k is the chemical activity coefficient of the kth species and $\boldsymbol{U}_i$ (m/s) is the fluid velocity relative to the polymer matrix network. F, R and T are the Faraday's constant (9.6487×10^4 C/mol), universal gas constant (8.314 J/mol·K) and absolute temperature (K), respectively.

The first term on the right-hand side of Eq. (2.1) represents the diffusive flux due to the gradient of concentration in the domain. This term is identical with the Fick's first law of diffusion equation. The spatial distribution of concentration $c_k(\mathbf{x})$ could be a linearly continuous gradient or becomes more complicated with irregular concentration pattern (Katchalsky and Curran, 1965).

The second term represents the migration flux arising from the gradient of the electrical potential. It is applicable when electrostatic force exists with or without the externally applied electric voltage. The electric potential could vary linearly across the domain that is so-called constant field condition. The distribution of the electric potential could also be governed by the Poisson equation. Therefore, the profiles of species concentration are influenced by the bulk concentration and the distributive electric field (Helfferich, 1962).

The third term is associated with the chemical activity coefficient of ion species in non-ideal electrolyte solution. There are several semi-empirical formulae developed for computing the chemical activity coefficient. However, the Debye–Huckel model may be one of the most popular mathematical descriptions to determine the activity coefficient. In addition, it should be noted that the rate of the ionic diffusion is much faster than the kinetics of chemical activity if an external electric field is applied. Therefore, the contribution of the chemical activity coefficient is very small relatively and thus negligible. More studies of the effect of the chemical activity coefficient can be found (Bockris and Reddy-Amulya, 1998; Samson et al., 1999).

The fourth term refers to the convective flux resulting from the fluid velocity due to the electro-osmosis solvent flow. The convection flux is described relatively to some reference velocities, which may be the average velocities of mass, molar, volume or solvent (Cussler, 1997). However, in the present analysis for the case of an unstirred solution in vibration-free experimental device, this term can be neglected for simplicity because the bulk fluid flow or hydrodynamic velocity vanishes and subsequently the convective flux becomes negligible (Marchiano and Arvia, 1983).

According to the law of mass conservation, the change in the amount of the kth species contained in the volume with respect to time t is characterized by the difference between the fluxes entering and leaving the reference volume (Yeager et al., 1983). Therefore, the continuity equation of the kth diffusive species is derived as follows:

$$\frac{\partial c_k}{\partial t} + \mathrm{div}(\mathbf{J}_k) = 0$$

or

$$\frac{\partial c_k}{\partial t} + \mathrm{div}\{-[[D_k]_i \mathbf{grad}(c_k) + F\mu_k z_k c_k \mathbf{grad}(\psi) + [D_k]_i c_k \mathbf{grad}(\ln \gamma_k)] + c_k \mathbf{U}_i\} = 0$$

or

$$\frac{\partial c_k}{\partial t} + \mathrm{div}\left\{ - [D_k]_i \left[\mathbf{grad}(c_k) + \frac{F}{RT} z_k c_k \mathbf{grad}(\psi) + c_k \mathbf{grad}(\ln \gamma_k) \right] + c_k \mathbf{U}_i \right\} = 0$$

$$(k = 1,2,3,\ldots,N_{\mathrm{ion}}) \tag{2.2}$$

The above equation, often known as the Nernst–Planck equation, involves the unknown mobility μ_k which can be determined by the Nernst–Einstein relationship as follows:

$$\mu_k = \frac{D_k}{RT} \tag{2.3}$$

Usually the diffusion coefficient of solute in hydrogels is dependent on many effects, including the size of the solute molecule relative to the structure and pore size of the polymeric gel, the polymer chain mobility and the water content. Peppeas et al. (2000) have put these effects into a general form as follows:

$$\frac{D_k}{D_0} = f(r_k, \phi_s, \xi) \tag{2.4}$$

where D_k is the effective diffusion coefficient and D_0 is the corresponding diffusion coefficient of solute in pure solvent. r_k is the radius of the diffusive molecule, ϕ_s is the volume fraction of polymer network in the hydrogel and ξ is network mesh size. The reviews of various theoretical models were made by Muhr and Blanshard (1982) and Amsden (1998). However, it is also shown from experimentally measured data that the diffusion coefficient is almost constant with swelling of the hydrogel but it is time dependent (Gehrke and Cussler, 1989).

2.2.1.2 Electrical Potential

For the distributive pattern of electrical potential, there are a few possible ways to describe its distribution, such as the electroneutrality or null current assumption (Hwang, and Helfferich, 1987; Doi et al., 1992; Samson and Marchand, 1999) or the constant field assumption (Malmivuo and Plonsey, 1995; Gillespie and Eisenberg, 2002). However, the Poisson equation is a more rigorous formulation (Helfferich, 1962; MacGillivray, 1968; MacGillivray and Hare, 1969). Each of them will be evaluated, and it will be shown that the constant field and electroneutrality assumptions are indeed the two limited cases of the Poisson equation.

Constant Field

The assumption of constant field in fact implies that the electric potential varies linearly across the system. In other words, the gradient of electric potential in the hydrogel is constant. This assumption was introduced first for analysis of ion transport through biological membrane (Goldman, 1943; Hodgkin and Katz, 1949), where the membrane is assumed to be uniform, planar and infinite in its lateral extent. Hence, the potential field ψ and ionic concentration c within the membrane are functions of x only. One can thus have

$$\frac{\partial^2 \psi}{\partial x^2} = 0 \tag{2.5}$$

If the membrane has a thickness of h

$$\frac{d\psi}{dx} \approx \frac{\psi(h) - \psi(0)}{h} = \frac{V_m}{h} \tag{2.6}$$

where V_m is the transmembrane voltage.

Constant Current

Following the assumptions of the electroneutrality conserved everywhere and the global flow of all ions across the boundary yielding null current, the additional conditions to support the Nernst–Planck flux equation (2.2) are summarized as follows:

Electroneutrality in the interior hydrogel:

$$\sum_k z_k c_k + z_f c_f = 0 \tag{2.7}$$

Electroneutrality in the exterior bathing solution:

$$\sum_k z_k c_k = 0 \tag{2.8}$$

Null current:

$$\sum_k z_k \boldsymbol{J}_k = 0 \tag{2.9}$$

where c_k is the concentration of the kth ion species either inside or surrounding the hydrogel. c_f is the density of the fixed charge group within the hydrogel.

Poisson Equation

The Poisson equation is a more rigorous approach to characterize the spatial distribution of the electric potential in the domain. In the studies of membrane immersed in physiological electrolytic environment, it is commonly accepted that the electric field near or in the membrane is of primary importance and the magnetic and electromagnetic phenomena inherently play a second role (Goldman, 1971). Therefore, the formulation can be limited to the case of an electrostatic field for deriving the Poisson equation (Panofsky and Phillips, 1964). The basic equations describing the electrostatic field are given as follows:

$$\nabla \times \mathbf{E}_{el} = 0 \tag{2.10}$$

$$\nabla \cdot \mathbf{E}_{el} = \frac{\rho_{el}}{\varepsilon_0} \tag{2.11}$$

$$\mathbf{E}_{el} \equiv -\nabla \psi \tag{2.12}$$

where $\mathbf{E}_{el}$ is the macroscopic average electric field acting on the charges within the medium, ψ is the electric potential, ρ_{el} (C/cm^3) is the charge density in an average volume and ϵ_0 the vacuum permittivity of free space or dielectric constant ($8.85418 \times 10^{-12} C^2/Nm^2$). The above three equations are termed Maxwell's equations and Eq. (2.12) results from Eq. (2.11), which means that the electrostatic field is irrotational (Panofsky and Phillips, 1964).

Let us define the divergence of tensor $\mathbf{E}_{el}$ in terms of a scalar potential ψ as follows:

$$\nabla \cdot \mathbf{E}_{el} = \nabla \cdot (-\nabla \psi) = -\nabla^2 \psi \tag{2.13}$$

By Maxwell's equation (2.11), Gauss's law proves

$$\nabla^2 \psi = -\frac{\rho_{el}}{\varepsilon_0} \tag{2.14}$$

Equation (2.14) is known as the Poisson equation. If the condition of zero charge, $\rho_{el} = 0$, is imposed, the Poisson equation is reduced to Laplace equation as follows:

$$\nabla^2 \psi = 0 \tag{2.15}$$

Equation (2.15) is equivalent to Eq. (2.5), which results from the constant field assumption.

From Maxwell's equation (2.10)

$$\nabla \times \mathbf{E}_{el} = \nabla \times (-\nabla \psi) = 0 \tag{2.16}$$

The curl law of Eq. (2.10) makes sure that $\mathbf{E}_{el}$ could be represented by the gradient of a scalar. Hence, $\nabla \times \mathbf{E}_{el} = 0$ permits $\mathbf{E}_{el} \equiv -\nabla \psi$; in return, $\mathbf{E}_{el} \equiv -\nabla \psi$

guarantees $\nabla \times \mathbf{E}_{el} = 0$. In addition, it is known that ψ can be determined just by a differential equation, i.e. Poisson equation, because ψ is a scalar. However, the determination of tensor $\mathbf{E}_{el}$ requires the presence of both the divergence and curl conditions.

In fact, the source of electric field is often separated into two types, the truly free charge ρ_{el} and the polarization or bound charge ρ_P, because of the material medium. Therefore, the Poisson equation (2.14) becomes

$$\nabla^2\psi = -\nabla \cdot \mathbf{E}_{el} = -\left(\frac{\rho_{el} + \rho_P}{\varepsilon_0}\right) \tag{2.17}$$

Further, it is more convenient to express ρ_P in terms of the divergence of the polarization $\mathbf{P}_{el}$:

$$-\nabla \cdot \left(\mathbf{E}_{el} + \frac{\mathbf{P}_{el}}{\varepsilon_0}\right) = -\frac{\rho_{el}}{\varepsilon_0} \tag{2.18}$$

Therefore, we can define the electric displacement $\mathbf{D}_{el}$ (C/m^2) as

$$\mathbf{D}_{el} = \varepsilon_0\mathbf{E}_{el} + \mathbf{P}_{el} \tag{2.19}$$

and subsequently Gauss's law in terms of $\mathbf{D}_{el}$ is given as

$$\nabla \cdot \mathbf{D}_{el} = \rho_{el} \tag{2.20}$$

For the medium with linear dielectrics, the polarization is $\mathbf{P}_{el} = \varepsilon_0(\varepsilon - 1)\mathbf{E}_{el}$ and Eq. (2.19) becomes

$$\mathbf{D}_{el} = \varepsilon\varepsilon_0\mathbf{E}_{el} \tag{2.21}$$

where ε is the relative dielectric constant of the surrounding medium. After rearrangement, the general Poisson equation is obtained as

$$\nabla^2\psi = -\frac{\rho_{el}}{\varepsilon\varepsilon_0} \tag{2.22}$$

The truly free charge ρ_{el} is a function of all ionic concentrations in solution and it is calculated by

$$\rho_{el} = F\left(\sum_k z_k c_k + z_f c_f\right) \tag{2.23}$$

Therefore, the Poisson equation can be finally written in the following form to characterize the spatial distribution of the electric potential in domain:

$$\nabla^2\psi = -\frac{F}{\varepsilon\varepsilon_0}\left(\sum_k z_k c_k + z_f c_f\right) \tag{2.24}$$

The Poisson equation embodies a mean field of electric potential ψ, which approximates the interactions of ion–ion and ion with the fixed charge. It is noted that ψ is generally the sum of the externally applied potential and diffusion potential. This is a mean field potential, instead of instantaneous potential, and it serves as the mean electrical driving force on the ions, about which the instantaneous potential fluctuates (Cooper et al., 1985). It is observed that the constant field and electroneutrality assumptions are in fact the special cases of the Poisson equation. The constant field assumption is valid for low ionic concentrations, while the electroneutrality with null current assumption is applicable only if the ionic concentrations are high (MacGillivray, 1968; MacGillivray and Hare, 1969).

If one needs to add the screening condition at microscopic level, the Poisson equation (2.22) can be extended conveniently to Poisson–Boltzmann equation, which is a mean field approximation approach including the electrostatic system through the Poisson equation and the effect of entropy, because of the mobility of the counterions through the Boltzmann distribution of statistical mechanics. Then, if $\rho(\mathbf{x}) = \rho_{\text{fixed}}(\mathbf{x}) + \rho_{\text{mobile}}(\mathbf{x})$ is the charge density as the function of coordinate position, the Poisson equation becomes

$$\nabla^2\psi = -\frac{\rho_{\text{fixed}}(\mathbf{x}) + \rho_{\text{mobile}}(\mathbf{x})}{\varepsilon\varepsilon_0} \tag{2.25}$$

According to the statistical mechanics, if only an ion species with charge q is mobile, the relative probability of finding an ion at position $\mathbf{x}$ is given by the Boltzmann expression, $\exp(-q\psi(\mathbf{x})/k_BT)$, where k_B is Boltzmann's constant (1.3807×10^{-23} J/K) and T is the absolute temperature. The profile of charge density is then expressed by

$$\rho_{\text{mobile}}(\mathbf{x}) = qc_0 \exp(-q\psi(\mathbf{x})/k_BT) \tag{2.26}$$

where c_0 is the ion density for the ions per volume at a point where the electrostatic potential vanishes. The Poisson equation (2.22) is thus extended into the Poisson–Boltzmann equation as follows:

$$\nabla^2\psi(\mathbf{x}) = -\rho_{\text{fixed}}(\mathbf{x})/\varepsilon - qc_0 \exp(-q\psi(\mathbf{x})/k_BT)/\varepsilon \tag{2.27}$$

One of the drawbacks of the Poisson–Boltzmann model is that, at short distance, e.g. within a few tenths of a nanometre from the membrane surface, the approximation of continuous charge density breaks down because of the atomistic nature of the system (Guldbrand et al., 1984; Redondo and Laser, 2004).

2.2.1.3 Fixed Charge Group

The capability of achieving large volume transition for the hydrogel is facilitated by the weakly acidic or weakly basic groups bound onto the polymer network chains, which is strongly dependent on the dissociation constant. The groups are readily ionizable and sensitive to the environment pH surrounding the hydrogel

(Katchalsky, 1949; Fragala et al., 1972; Tanaka et al., 1980; De Rossi et al., 1985). For example, the weakly acidic carboxyl groups exist in form of R–COOH when the hydrogen ion H^+ concentration is higher than the dissociation constant K_a, whereas the pendent groups are charged and become R–COO$^-$ if the H^+ concentration of medium is lower than K_a, which is characterized by

$$\text{R–COOH} \rightleftharpoons \text{R–COO}^- + \text{H}^+ \tag{2.28}$$

Therefore, the hydrogen ion H^+ is an important controller for electrochemical modulation of swelling of the pH-sensitive hydrogel. A relation between the fixed charge groups and the diffusive hydrogen ions was developed according to the Langmuir adsorption isotherm (Grimshaw et al., 1990). Based on the mechanism of single site binding single ion, the equilibrium constant is defined as follows:

$$K_a = \frac{[\text{R–COO}^-][\text{H}^+]}{[\text{R–COOH}]} \tag{2.29}$$

The density of total ionizable fixed charge groups within the dry gel is determined by titration process per volume of solid polymer (Grimshaw, 1989) and given as

$$c_{m0}^s = \frac{\text{moles of ionizable group}}{\text{volume of solid polymer}} = \frac{n}{V^s} \tag{2.30}$$

Since the ionic concentrations in Eqs. (2.1) and (2.2) are averaged over the interstitial fluid volume, we need to convert the density of total ionizable groups c_{m0}^s in a single unit of fluid volume:

$$c_{m0} = \frac{n}{V^f} = c_{m0}^s \frac{V^s}{V^f} = \frac{c_{m0}^s}{H} \tag{2.31}$$

where H is the hydration of the hydrogel and defined as the ratio of the interstitial fluid volume V^f to the solid polymer volume V^s as follows:

$$H = \frac{\text{volume of interstitial fluid}}{\text{volume of solid polymer}} = \frac{V^f}{V^s} \tag{2.32}$$

The reaction isotherm is described by

$$K_a = \frac{c_f c_{\text{H}^+}}{c_{m0} - c_f} \tag{2.33}$$

After rearrangement, the concentration of the fixed charge groups bound on the polymer chains, c, is finally written in the following form for an anionic hydrogel:

$$c_f = \frac{K_a c_{m0}}{K_a + c_{\text{H}^+}} = \frac{c_{m0}^s}{H} \frac{K_a}{K_a + c_{\text{H}^+}} \tag{2.34}$$

The above relation between the fixed charge density and the diffusive hydrogen ion concentration can be employed for computing the concentration of the fixed charge groups, as a function of the concentration of total ionizable groups in the dry gel, c^s_{m0}, and the concentration of free hydrogen ion c_{H^+} (Siegel et al., 1991; Siegel, 1990).

For a cationic hydrogel, the fixed charge density c_f based on the Langmuir isotherm relation is derived as

$$c_f = \frac{c^s_{m0}}{H} \frac{c_{H^+}}{K_a + c_{H^+}} \tag{2.35}$$

It should be noted that the fixed charge groups are bound onto the polymeric network chain and thus become immobile. In general, the profiles of the fixed charge groups are modified only by chemical reaction (Shibayama and Tanaka, 1993; Shiga et al., 1992a, b).

2.2.2 Mechanical Formulation

As mentioned before, the smart hydrogels are able to absorb or exude the fluid where they are immersed and thus swell or deswell until equilibrium is attained. At equilibrium state, the swelling force is balanced by the elastic retractive force exerted by crosslinked polymer solid matrix network of the hydrogel in order to maintain the current hydration state. The total swelling force could arise from the stretching of electrostatic effect, the polymer–solvent and polymer–solute interactions and entropic effect (Flory, 1962). However, it is known that the swelling force arising from the entropic effect, e.g. thermal motion or solvent interactions, reaches steady state faster than the ionic diffusion or water flow. Thus the swelling stress could be a state function of ionic environmental condition, and the hydration and fixed charge of the hydrogels. In this section, the mechanical equilibrium governing equations are formulated for swelling/deswelling deformation of the charged hydrogel.

As well known, usually the pH-sensitive hydrogel undergoes large deformation due to the effects of chemo-electro-mechanical multi-energy coupled fields, especially at higher pH level of surrounding solution. Then the difference between the initial and deformed configurations cannot be neglected as it is done for analysis of linear elasticity. The deformation gradient tensor $\boldsymbol{F}$ is thus defined as

$$\mathbf{F} = F_{ij} = \frac{\partial x_i^{\text{Deformed Configuration}}}{\partial X_j^{\text{Initial Configuration}}} = \frac{\partial (X_i + u_i)}{\partial X_j} = \delta_{ij} + \frac{\partial u_i}{\partial X_j} = \mathbf{I} + \nabla \cdot \mathbf{u} \tag{2.36}$$

For analysis of geometrically nonlinear problems, the mechanical governing equation of large deformation based on a total Lagrangian description is given as

$$\nabla \cdot \boldsymbol{P} + \boldsymbol{b} - \rho \dot{\boldsymbol{U}} = \boldsymbol{f} \tag{2.37}$$

where $\boldsymbol{P}$ is the first Piola–Kirchhoff stress tensor, ρ is membrane density, $\boldsymbol{b}$ is body force, $\boldsymbol{f}$ is external force, $\dot{\boldsymbol{U}}$is acceleration and $\rho\dot{\boldsymbol{U}}$ is inertial force. In the present mechanical analysis, no external force is considered so that $\boldsymbol{f}$=0, and the effects of $\boldsymbol{b}$ and $\rho\dot{\boldsymbol{U}}$ are neglected. Thus,

$$\nabla \cdot \boldsymbol{P} = 0 \text{ in } \Omega \tag{2.38}$$

$$\boldsymbol{u} = \boldsymbol{u}^* \text{ in } \Gamma_{\boldsymbol{u}^*} \tag{2.39}$$

$$\boldsymbol{P} \cdot \boldsymbol{n} = \boldsymbol{s}^* \text{ in } \Gamma_{\boldsymbol{s}^*} \tag{2.40}$$

where $\boldsymbol{u}^*$ is the specified displacement vector on the boundary portion $\Gamma_{\boldsymbol{u}*}$, $\boldsymbol{s}^*$ is the surface traction vector on the boundary $\Gamma_{\boldsymbol{s}*}$ and $\boldsymbol{n}$ is the unit outward normal vector. $\boldsymbol{u}$ is the displacement vector from the initial configuration $\boldsymbol{X}$ to the deformed configuration $\boldsymbol{x}$ where $\boldsymbol{x}= \boldsymbol{X}+u$. P as the first Piola–Kirchhoff stress tensor is a kind of expatriate, living partially in the deformed configuration $\boldsymbol{x}$ and partially in the reference configuration $\boldsymbol{X}$. Usually the second Piola–Kirchhoff stress $\boldsymbol{S}$ and the Green–Lagrangian strain $\boldsymbol{E}$ are used as the stress and strain measurements, respectively.

As $\boldsymbol{P}$ is unmeasurable and asymmetrical, the second Piola–Kirchhoff stress tensor $\boldsymbol{S}$ is required because $\boldsymbol{S}$ is symmetric and it is often used as the stress measurement (Malvern, 1969). For continuous solid materials, the relation between the first Piola–Kirchhoff stress tensor $\boldsymbol{P}$ and the second Piola–Kirchhoff stress tensor $\boldsymbol{S}$ is given as

$$\boldsymbol{P} = \boldsymbol{S}\boldsymbol{F}^{\mathrm{T}} \tag{2.41}$$

For the present hydrogel as a porous mixture, the above relation could be modified as follows:

$$\boldsymbol{P} = -J\boldsymbol{F}^{-1} p_{\text{osmotic}} \boldsymbol{I} + \boldsymbol{S}\boldsymbol{F}^{\boldsymbol{T}} \tag{2.42}$$

where $J = \det(\mathbf{F})$ is the determinant of deformation gradient tensor $\boldsymbol{F}$, p_{osmotic} is the osmotic pressure and $\boldsymbol{I}$ is identity tensor. One can also have

$$\boldsymbol{S} = \boldsymbol{C}{:}\boldsymbol{E} \text{ or } S_{ij} = C_{ijkl}E_{kl} \tag{2.43}$$

where C is the material moduli tensor and $\boldsymbol{E}$ is the Green–Lagrangian strain tensor used as strain measurement (here the symbol (:) in $\boldsymbol{A}{:}\boldsymbol{B}$ is the double contraction of inner indices, namely $\boldsymbol{A}{:}\boldsymbol{B}$ is given by $\boldsymbol{A}_{ij}\boldsymbol{B}_{ij}$. If $\boldsymbol{A}$ or $\boldsymbol{B}$ is symmetric, $\boldsymbol{A}{:}\boldsymbol{B}= \boldsymbol{A}_{ij}\boldsymbol{B}_{ji}$).

If the material is elastically isotropic

$$S_{ij} = [\lambda\delta_{ij}\delta_{kl} + \mu(\delta_{ik}\delta_{jl} + \delta_{il}\delta_{jk})]E_{kl} \tag{2.44}$$

Specially in one-dimensional domain

$$S_{11} = (\lambda + 2\mu)\left[\frac{1}{2}\left(2\frac{du}{dX} + \left(\frac{du}{dX}\right)^2\right)\right] = (\lambda + 2\mu)\left[\frac{du}{dX} + \frac{1}{2}\left(\frac{du}{dX}\right)^2\right] \tag{2.45}$$

For isotropic elastic materials, the material moduli tensor can be written as

$$C = \lambda I \otimes I + 2\mu I \quad \text{or} \quad C_{ijkl} = \lambda\delta_{ij}\delta_{kl} + \mu\left(\delta_{ik}\delta_{jl} + \delta_{il}\delta_{jk}\right) \tag{2.46}$$

where λ and μ are the Lamé coefficients of solid phase (here the symbol $\otimes$ in $a \otimes b$ indicates the vector product. In indicial notation, $a \otimes b \to a_i b_j$. In matrix notation, $a \otimes b \to \{a\}\{b\}^{\mathrm{T}}$). For example, in two-dimensional domain, there are two kinds of typical problems, the plane strain and plane stress problems.

In the plane strain problem, the thickness of solids in the z-direction is very large, compared with dimensions in the x- and y-directions. External loading is applied uniformly along the z-axis, and the movement in the z-direction at any point is constrained. The strain components in z-direction (ε_{zz}, ε_{xz}, ε_{yz}) are all zero, and there are only three in-plane strains (ε_{xx}, ε_{yy}, ε_{xy}) to deal with. Then

$$\mathbf{C} = \frac{E}{(1+\nu)(1-2\nu)}\begin{bmatrix} 1-\nu & \nu & 0 \\ \nu & 1-\nu & 0 \\ 0 & 0 & (1-2\nu)/2 \end{bmatrix} \text{or } \mathbf{C} = \begin{bmatrix} \lambda+2\mu & \lambda & 0 \\ \lambda & \lambda+2\mu & 0 \\ 0 & 0 & \mu \end{bmatrix} \tag{2.47}$$

In the plane stress problem, however, the thickness of solids in the z-direction is very small, compared with dimensions in the x- and y-directions. As external forces are applied only within the x–y plane and the stress components in z-direction (σ_{zz}, σ_{xz}, σ_{yz}) are all zero, there are only three in-plane stresses (σ_{xx}, σ_{yy}, σ_{xy}) to deal with. Then

$$C = \frac{E}{(1-\nu^2)}\begin{bmatrix} 1 & \nu & 0 \\ \nu & 1 & 0 \\ 0 & 0 & (1-\nu)/2 \end{bmatrix} \tag{2.48}$$

where the two Lamé elastic constants, λ and μ, are associated with the shear modulus G, Young's modulus E and Poisson's ratio v as follows:

$$\lambda = \frac{\nu E}{(1+\nu)(1-2\nu)} \quad \text{and} \quad \mu = G = \frac{E}{2(1+\nu)} \tag{2.49}$$

The Green–Lagrangian strain tensor $\boldsymbol{E}$ is given by

$$\boldsymbol{E} = \frac{1}{2}(\boldsymbol{F}^T \cdot \boldsymbol{F} - \boldsymbol{I}) \text{ or } E_{ij} = \frac{1}{2}(F_{ik}^T F_{kj} - \delta_{ij}) = \frac{1}{2}\left(\frac{\partial u_i}{\partial X_j} + \frac{\partial u_j}{\partial X_i} + \frac{\partial u_k}{\partial X_i}\frac{\partial u_k}{\partial X_j}\right) \tag{2.50}$$

For example, in two-dimensional domain

$$\boldsymbol{E} = E_{ij} = \frac{1}{2}\begin{bmatrix} 2\frac{\partial u_1}{\partial X_1} + \left(\frac{\partial u_1}{\partial X_1}\right)^2 + \left(\frac{\partial u_2}{\partial X_1}\right)^2 & \frac{\partial u_1}{\partial X_2} + \frac{\partial u_2}{\partial X_1} + \frac{\partial u_1}{\partial X_1}\frac{\partial u_1}{\partial X_2} + \frac{\partial u_2}{\partial X_1}\frac{\partial u_2}{\partial X_2} \\ \frac{\partial u_1}{\partial X_2} + \frac{\partial u_2}{\partial X_1} + \frac{\partial u_1}{\partial X_1}\frac{\partial u_1}{\partial X_2} + \frac{\partial u_2}{\partial X_1}\frac{\partial u_2}{\partial X_2} & 2\frac{\partial u_2}{\partial X_2} + \left(\frac{\partial u_2}{\partial X_2}\right)^2 + \left(\frac{\partial u_1}{\partial X_2}\right)^2 \end{bmatrix} \quad (2.51)$$

Therefore, the momentum equilibrium equation for the hydrogel mixture can be finally written as

$$\nabla \cdot (-\boldsymbol{J}\boldsymbol{F}^{-1} p_{\text{osmotic}} \boldsymbol{I} + \boldsymbol{S}\boldsymbol{F}^{\boldsymbol{T}}) = 0 \quad (2.52)$$

For example, in two-dimensional domain

$$\begin{aligned} \boldsymbol{P} &= -\boldsymbol{J}\boldsymbol{F}^{-1} p_{\text{osmotic}} \boldsymbol{I} + \boldsymbol{S}\boldsymbol{F}^{T} \\ &= -p_{\text{osmotic}} \begin{bmatrix} 1 + \frac{\partial u_2}{\partial X_2} & -\frac{\partial u_1}{\partial X_2} \\ -\frac{\partial u_2}{\partial X_1} & 1 + \frac{\partial u_1}{\partial X_1} \end{bmatrix} + \begin{bmatrix} S_{11} & S_{12} \\ S_{21} & S_{22} \end{bmatrix} \begin{bmatrix} 1 + \frac{\partial u_1}{\partial X_1} & \frac{\partial u_2}{\partial X_1} \\ \frac{\partial u_1}{\partial X_2} & 1 + \frac{\partial u_2}{\partial X_2} \end{bmatrix} \\ &= -p_{\text{osmotic}} \begin{bmatrix} 1 + \frac{\partial u_2}{\partial X_2} & -\frac{\partial u_1}{\partial X_2} \\ -\frac{\partial u_2}{\partial X_1} & 1 + \frac{\partial u_1}{\partial X_1} \end{bmatrix} \\ &\quad + \begin{bmatrix} S_{11}\left(1 + \frac{\partial u_1}{\partial X_1}\right) + S_{12}\frac{\partial u_1}{\partial X_2} & S_{11}\frac{\partial u_2}{\partial X_1} + S_{12}\left(1 + \frac{\partial u_2}{\partial X_2}\right) \\ S_{21}\left(1 + \frac{\partial u_1}{\partial X_1}\right) + S_{22}\frac{\partial u_1}{\partial X_2} & S_{21}\frac{\partial u_2}{\partial X_1} + S_{22}\left(1 + \frac{\partial u_2}{\partial X_2}\right) \end{bmatrix} \\ &= \begin{bmatrix} S_{11}\left(1 + \frac{\partial u_1}{\partial X_1}\right) + S_{12}\frac{\partial u_1}{\partial X_2} - p_{\text{osmotic}}\left(1 + \frac{\partial u_2}{\partial X_2}\right) \\ S_{11}\frac{\partial u_2}{\partial X_1} + S_{12}\left(1 + \frac{\partial u_2}{\partial X_2}\right) + p_{\text{osmotic}}\frac{\partial u_1}{\partial X_2} \\ S_{21}\left(1 + \frac{\partial u_1}{\partial X_1}\right) + S_{22}\frac{\partial u_1}{\partial X_2} + p_{\text{osmotic}}\frac{\partial u_2}{\partial X_1} \\ S_{21}\frac{\partial u_2}{\partial X_1} + S_{22}\left(1 + \frac{\partial u_2}{\partial X_2}\right) - p_{\text{osmotic}}\left(1 + \frac{\partial u_1}{\partial X_1}\right) \end{bmatrix} \end{aligned} \quad (2.53)$$

$$\nabla\cdot\boldsymbol{P}=0\Rightarrow\left[\begin{array}{l}\frac{\partial}{\partial X_1}\left[S_{11}\left(1+\frac{\partial u_1}{\partial X_1}\right)+S_{12}\frac{\partial u_1}{\partial X_2}-p_{\text{osmotic}}\left(1+\frac{\partial u_2}{\partial X_2}\right)\right]\\ \quad+\frac{\partial}{\partial X_2}\left[S_{21}\left(1+\frac{\partial u_1}{\partial X_1}\right)+S_{22}\frac{\partial u_1}{\partial X_2}+p_{\text{osmotic}}\frac{\partial u_2}{\partial X_1}\right]\\ \frac{\partial}{\partial X_1}\left[S_{11}\frac{\partial u_2}{\partial X_1}+S_{12}\left(1+\frac{\partial u_2}{\partial X_2}\right)+p_{\text{osmotic}}\frac{\partial u_1}{\partial X_2}\right]\\ \quad+\frac{\partial}{\partial X_2}\left[S_{21}\frac{\partial u_2}{\partial X_1}+S_{22}\left(1+\frac{\partial u_2}{\partial X_2}\right)-p_{\text{osmotic}}\left(1+\frac{\partial u_1}{\partial X_1}\right)\right]\end{array}\right]=0 \tag{2.54}$$

Specially for one-dimensional analysis

$$\nabla\cdot\boldsymbol{P}=\nabla\cdot(-J\boldsymbol{F}^{-1}p_{\text{osmotic}}\boldsymbol{I}+\boldsymbol{S}\boldsymbol{F}^T)=\frac{d}{dX}\left[-p_{\text{osmotic}}+S_{11}\left(1+\frac{du}{dX}\right)\right]=0 \tag{2.55}$$

that is

$$\begin{gathered}\frac{d}{dX}\left[-p_{\text{osmotic}}+(\lambda+2\mu)\left[\frac{du}{dX}+\frac{1}{2}\left(\frac{du}{dX}\right)^2\right]\left(1+\frac{du}{dX}\right)\right]=0\\ (\lambda+2\mu)\left[\frac{d^2u}{dX^2}+3\frac{du}{dX}\frac{d^2u}{dX^2}+\frac{3}{2}\left(\frac{du}{dX}\right)^2\frac{d^2u}{dX^2}\right]-\frac{dp_{\text{osmotic}}}{dX}=0\end{gathered} \tag{2.56}$$

The multiphase hydrogels may behave small deformation if the pH value of bathing solution is low. Then the linear elastic theory may provide sufficiently accurate simulation, and thus Eq. (2.52) is simplified to (Lai et al., 1991)

$$\nabla\cdot\sigma=\nabla\cdot[\lambda(\text{tr}(\boldsymbol{E}))\boldsymbol{I}+2\mu\boldsymbol{E}-p_{\text{osmotic}}\boldsymbol{I}]=0 \tag{2.57}$$

where $\boldsymbol{\sigma}$ is the Cauchy stress tensor. p_{osmotic} is the osmotic pressure contributed by the tendency of the hydrogel to absorb additional solvent, which is formulated by the difference of the concentrations between the interior hydrogel and the external medium (Helfferich, 1962; Katchalsky and Curran, 1965). Hence the osmotic pressure can be calculated according to

$$p_{\text{osmotic}}=RT\sum_k(c_k-\bar{c}_k) \tag{2.58}$$

where $\bar{c}_k$ is the concentration of the kth ion species in external solution and c_k the concentration of the kth ion species within the hydrogel.

2.3 Computational Domain, Boundary Condition and Numerical Implementation

In this section, the appropriate electrochemical and mechanical boundary conditions are specified to implement the MECpH model, which are associated with the experimental work conducted by Beebe and his team (Beebe et al., 2000a; De et al., 2002). In their experiment, the hydrogels were fabricated in a microchannel with upper and lower surfaces covered with glasses so that the deformation of hydrogels in axial direction was confined and the swelling occurs only in the radial direction. Due to the constraints, the equilibrium swelling/shrinking of a cylindrical hydrogel can be modelled as a one-dimensional problem along the diameter of the hydrogel. If the constant concentrations of surrounding bath solution and the macroscopic homogenous properties of the hydrogel are assumed, the one-dimensional problem can be further simplified to a symmetrical problem about the axis of the cylindrical hydrogel. Because of the axisymmetry, only half of the diameter is required in simulation, as shown in Fig. 2.1. In summary, the one-dimensional computational domain is thus composed of three subdomains, namely the radius of the hydrogel which represents the interior cylindrical hydrogel, the surrounding solution along the radius direction which represents the exterior bathing medium and the boundary effect domain that refers to the hydrogel–solution interface. For a system of N

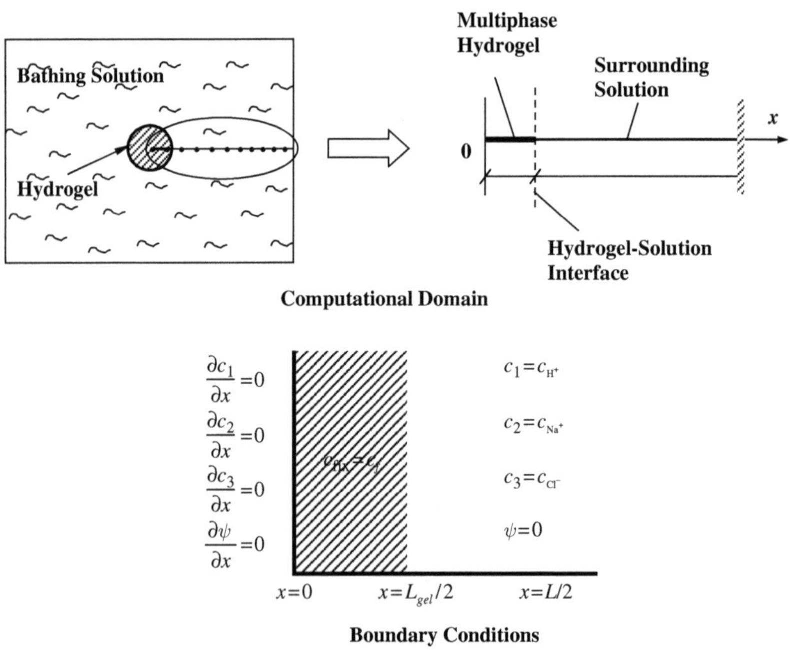

Fig. 2.1 Computational domain and boundary conditions for the numerical simulations, where the *shaded areas* represent the cylindrical pH-responsive hydrogel

diffusive species, the MECpH model generates the N Nernst–Planck equations (2.2) coupled with the Poisson equation (2.24). As usual, all the simulations are subjected to ambient temperature or stated otherwise.

Since this is an axisymmetric problem, the Neumann type of electrochemical boundary conditions is imposed within the hydrogel for continuity and the Dirichlet boundary conditions are applied at solution edges as prescribed by the experiment. Furthermore, the boundary conditions for mechanical equilibrium are imposed at the hydrogel–solution interface. These boundary conditions are illustrated in Fig. 2.1.

It should be pointed out that the MECpH model can also be employed for simulation of the pH and electric coupled driven hydrogel, for example, when a hydrogel strip is immersed in pH buffered solution and centred between a pair of electrodes and aligned in parallel with them, where a DC electric field is applied across the domain. Then we can stimulate the degree of swelling/deswelling of the charged hydrogels subject to the coupled stimuli of chemical pH milieu and DC electric field. Due to the electric voltage applied across the system, the symmetric conditions are no longer applicable. The computational domain has to cover whole domain of interest, from the cathode on the left region to the anode on the right, where the boundary conditions of ionic concentrations are identical for each species in both the anode and cathode regions, i.e. $c_k^*=c_k^{\text{Left}}=c_k^{\text{Right}}$. The electric boundary conditions are given as prescribed by the applied electric voltagc.

The MECpH model consisting of a set of nonlinear coupled electrochemical and mechanical equations has been developed for simulation of the swelling/deswelling equilibrium of the pH-sensitive hydrogel. Modelling of equilibrium behaviour of the pH-responsive hydrogel requires a good understanding of the diffusion of hydrogen ions inside and outside the hydrogel. This requirement also takes into account the chemical reactions of the hydrogen ions with the fixed charge groups and the buffering effect on the diffusion of hydrogen ion. Equations (2.2) coupled with Eq. (2.24) form a famous formulation known as Poisson–Nernst–Planck (PNP) system, expressed by a set of nonlinear partial differential equations. The PNP theory of continuum diffusion attempts to characterize the average ion fluxes in terms of the gradients of concentrations and/or electric potential. Based on the difference in the ionic concentrations and electric potential across the hydrogel, the degree of equilibrium swelling/shrinking can be computed by the finite deformation mechanical governing equation (2.52) or (2.57).

It is definitely impossible for the MECpH model to have any closed-form analytical solutions, which is composed of the PNP system (2.2) and (2.24) coupled with the mechanical equation (2.52) or (2.57). We have to employ appropriate numerical approaches for approximate solutions of the MECpH model. A strong-form meshless numerical technique termed the Hermite-cloud method is thus used (Li et al., 2003), which faces several computing challenges of the MECpH model, such as the remeshing of computational domain due to moving boundaries at the hydrogel–solution interfaces and the localized high gradient over the hydrogel–solution interfaces. For simplification of numerical computation, a set of non-dimensional parameters is defined as follows:

$$\tilde{x} = \frac{x}{L_{\text{ref}}}, \tilde{u} = \frac{u}{L_{\text{ref}}}, \tilde{c}_k = \frac{c_k}{c_{\text{ref}}}, \tilde{c}_f = \frac{c_f}{c_{\text{ref}}}, \tilde{\psi} = \frac{\psi}{\psi_{\text{ref}}} = \frac{F\psi}{\eta RT} \tag{2.59}$$

where $\tilde{x}$, $\tilde{u}$, $\tilde{c}_k$, $\tilde{c}_f$, $\tilde{\psi}$ are the dimensionless variables of coordinate, displacement, diffusive ionic concentration, fixed charge density and electric potential, respectively. L_{ref}, c_{ref}, ψ_{ref} and η are the characteristic length, concentration, electric potential and weighted coefficient, respectively. As a result, the non-dimensional form of the MECpH governing equations can be written to simplify numerical simulation. Another reason for the non-dimensional treatment is to overcome the difficulty of the unknown variables with different scales and units.

Computational flowchart is illustrated in Fig. 2.2. By following the flowchart, the fixed charge density c_f is computed first by Eq. (2.34) or (2.35), followed by solving the Nernst–Planck equations (2.2) and Poisson equation (2.24) with a Newton iterative technique and a relaxation approach to the self-consistent PNP system for convergences of the mobile ionic concentrations c_k and the electric potential ψ. Both the converged concentrations c_k and potential ψ are in turn substituted into the mechanical equilibrium equation (2.52) or (2.57) for determining the corresponding displacement u of the hydrogel. Because of the deformation u, the fixed charge density c_f is redistributed within the hydrogel and thus this requests new computation again. This computational loop is carried out until all the independent variables converge, including the diffusive ionic concentrations $c_k (k = 1,2,3,...,N_{\text{ion}})$ and the electric potential ψ as well as the hydrogel displacement u.

The swelling/deswelling of the pH-sensitive hydrogel in equilibrium can be predicted by the steady-state simulation based on the Nernst–Planck equations, the Poisson equation and the mechanical equilibrium governing equation, collectively known as MECpH model. The set of equations is amenable by numerical solution. Nevertheless, appropriate approximations can greatly reduce the computing time necessarily for the solution with desired accuracy.

First, the diffusion coefficients are eliminated in the steady-state simulation as they affect the diffusing rate only but not the final equilibrium state. In other words, the time derivative term, $\partial c_k/\partial t$, in Eq. (2.2) is removed for the steady-state simulation. This significantly reduces the computational cost when solving the system of coupled nonlinear partial differential equations, as the computing time is saved for transient solutions of the N continuity equations (2.2). Moreover, it is observed in many experiments that the diffusion coefficient varies slightly with change in the degree of swelling (Gehrke and Cussler, 1989). Second, the convection term in the Nernst–Planck flux equations (2.2) is negligible as the fluid pressure remains constant across the hydrogel and the fluid velocity is unchanged with swelling of the hydrogel. This in fact assumes that the change in the concentration profiles due to convective flow of ions is much smaller than the concentrations resulting from the fluxes of diffusion and migration (Grimshaw, 1989). Third, it is also assumed

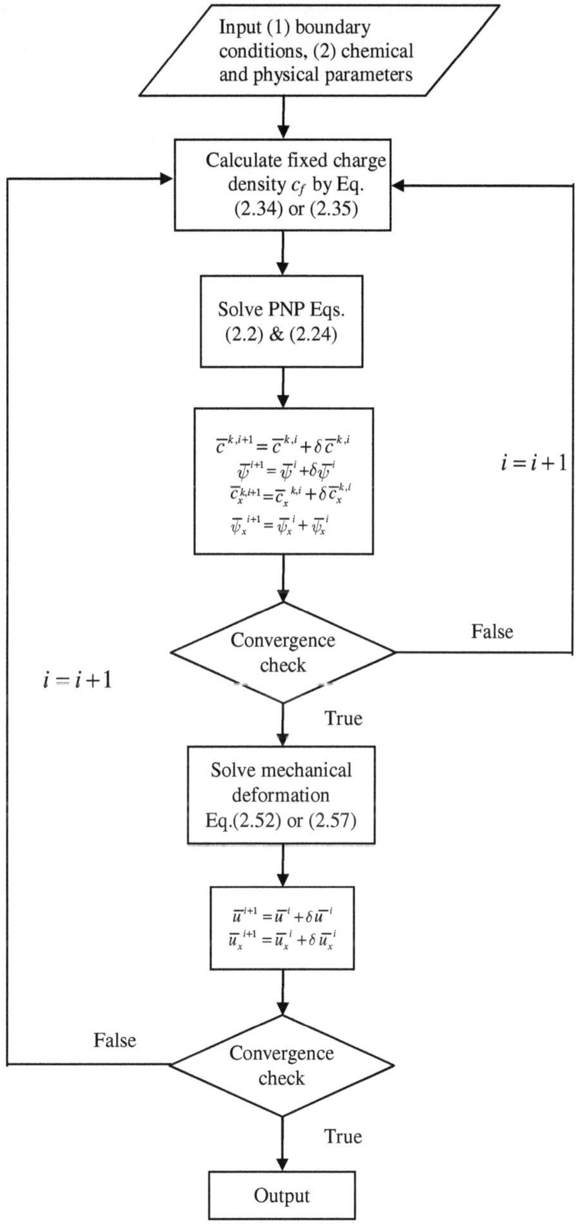

Fig. 2.2 Computational flowchart of relaxation approach for the self-consistent MECpH model

that the flux due to the chemical reaction of ions is much smaller than the net diffusion and migration fluxes. Therefore, the contribution of the chemical activity coefficient to the flux becomes negligible. In addition, the chemical activity coefficients of ionic species are always equal to unity in the dilute condition (Bockris and Reddy-Amulya, 1998). Lastly, the properties of the hydrogel considered are assumed to be homogeneous so that only the isotropic equilibrium swelling occurs. Although the actual dielectric constants of ions depend on the species and applied electric current, they are chosen equally each other for simplification. Besides that, electrolytic process due to the externally applied electric field can be ignored by limiting to low-range voltages in simulations.

In summary, for implementation of the MECpH model, the mechanical equilibrium governing equation (2.52) or (2.57) is coupled with the PNP equations (2.2) and (2.24) and the fixed charge dissociation equation (2.34) or (2.35) through the local hydration H. These equations complete the MECpH model. Therefore, there are N_{ion} Nernst–Planck equations (2.2) coupled with the Poisson equation (2.24) for the (N_{ion}+1) unknown variables $c_1, c_2, \ldots, c_N, \psi$. Apart from that, the mechanical equilibrium governing equation (2.52) or (2.57) is solved for the (N_{ion}+2)th unknown variable, i.e. the deformation u of the hydrogel. If appropriate boundary conditions are specified, these equations construct a complete partial differential boundary value (PDBV) problem with (N_{ion}+2) unknown variables. In principle, these equations can be solved approximately for the concentrations of diffusive species and fixed charge group, the electric potential and displacement. However, these equations are coupled nonlinearly which complicates the computation of the MECpH model.

2.4 Model Validation with Experiment

In order to examine the MECpH model, one-dimensional steady-state simulations are conducted and compared with the experimental data obtained by Beebe and his group (Beebe et al., 2000, 2002; De et al., 2002). For illustration of the single dimensionality of the present problem, an insight is given into the experimental procedure. With reference to the experimental work by Beebe et al. (2000a), the polymerizable mixture, composed of acrylic acid and 2-hydroxyethyl methacrylate (HEMA) with the photoinitiator (3 wt%) and ethylene glycol dimethacrylate (1 wt%) as the crosslinker, was filled up in the microchannel where the experiment was carried out. The microchannel was covered at top and bottom with two pieces of glasses. When the mixture reached a quiescent state, it was exposed to UV light through a circular photomask placed on the top of the microchannel. The channel was then flushed with water to remove the unpolymerized monomers mixture. The axisymmetric cylindrical hydrogel was formed in the microchannel and constrained from axial displacement by the cover glasses placed on the top and bottom. Thus it is quite reasonable to assume that the cylindrical hydrogel undergoes displacement mainly in the radial direction. The cylindrical hydrogel with diameter of 400 μm at dry state was placed in a bathing solution with ionic strength of 300 mM. It was observed that the hydrogel swelled to a certain degree after submerged in the solution. The instantaneous swelling of the hydrogel is herein referred to as initial hydration state. The hydrogel was presumed to attain an equilibrium swelling state before it was

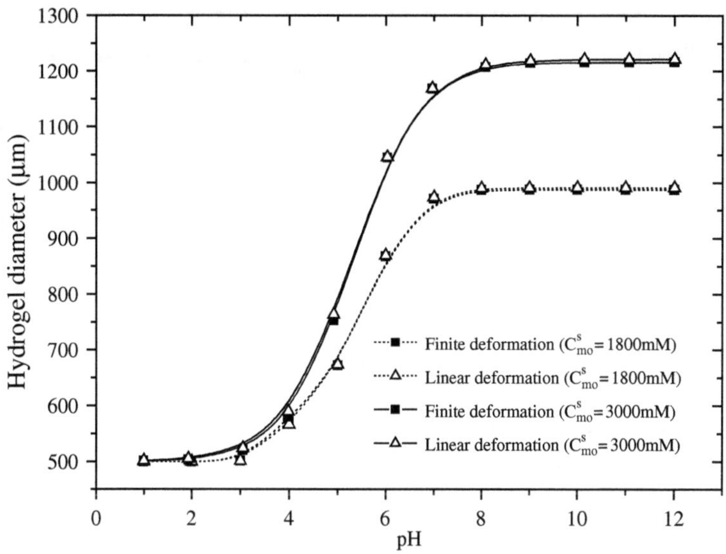

Fig. 2.3 Comparison of finite and linear deformation theories

subjected to step changes in pH of the surrounding solution. The volume changes of the hydrogel were observed and the diameter of the cylindrical hydrogel was measured after reaching equilibrium at various environmental pH values (circular markers in Fig. 2.4). Using the Hermite-cloud meshless technique (Li et al., 2003), the MECpH model, with the effects of chemo-electro-mechanical multi-energy fields expressed by fully coupled nonlinear partial differential equations, is solved numerically for simulation of the response performance of the given pH stimulus-responsive hydrogel. The simulation domain and corresponding boundary conditions are prescribed in Fig. 2.1.

Computations are conducted by both the finite and linear deformation theories for comparison. As shown in Fig. 2.3, the comparison between the two theories shows almost identical results. As the swelling of the hydrogel becomes larger, for example, when the ionizable fixed charge concentration increases, the linear deformation theory gives a slightly greater degree of swelling than that by the finite deformation theory.

The square markers in Fig. 2.4 indicate the simulation results obtained, where the solid line is plotted to assist visualization. It is apparent that they compare very well with the experimental data qualitatively and quantitatively. As well known, the change of environmental pH alters the degree of ionization of the fixed charge groups and the state of equilibrium swelling concurrently. The figure demonstrates that the hydrogel volume changes abruptly in the range from pH 4 to 7. It is also observed that the hydrogel remains almost unchanged from the initial hydrated state in the lower pH solution and starts to expand at about pH 4. As the pH of bathing solution increases, the ionization of the pendent poly(HEMA) carboxylic

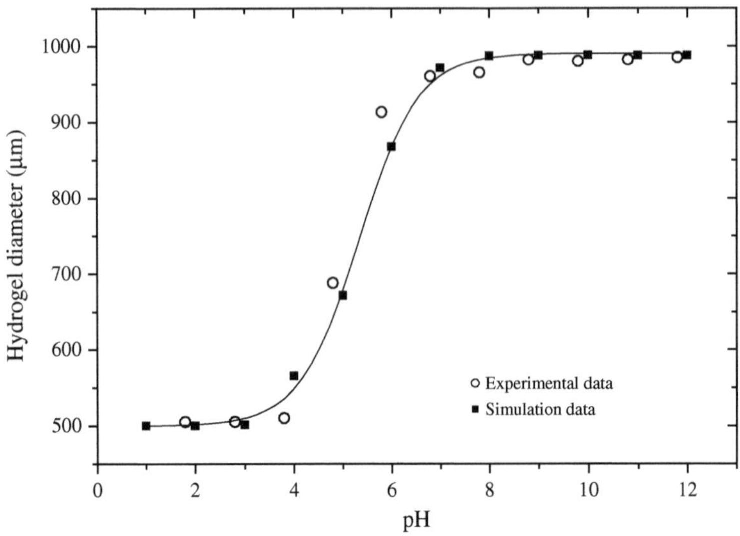

Fig. 2.4 Comparison between the experimental data (Beebe et al., 2000, 2002) and computational results by MECpH model for equilibrium swelling of the PHEMA hydrogel as function of pH

acid groups onto the backbone of the network increases, where the acid–base equilibrium process is described by the formula (2.28).

As a result, the surplus charge within the hydrogel increases. In order to maintain the neutrality within the hydrogel, more mobile ions with opposite sign to the fixed charge groups diffuse into the hydrogel. The concomitant increase of the ionic concentrations in the interior hydrogel generates higher osmotic pressure, which in turn causes the observable increase of swelling. As the ionization process of the carboxylic acid group approaches the saturation state at pH 7, the further increase of environmental pH after pH 7 does not enlarges significantly the degree of swelling. In summary, based on the above comparison and discussion of Fig. 2.4, it is concluded that the present MECpH model is accurate and robust to provide the prediction of the responsive behaviour of the pH-sensitive hydrogel with large deformation.

2.5 Parameter Studies by Steady-State Simulation for Equilibrium of Hydrogel

As mentioned before, the pH-sensitive hydrogel is the polymer network containing pendent ionizable groups which are usually weakly acidic or weakly basic. Ionization occurs when the environment pH changes, and then the pendent groups become charged. The responsive performance of the crosslinked charged hydrogel is mainly pH dependent. However, many effects also have to be considered, such as the crosslinking density, the chemical nature of the fixed charge groups, the ionic strength and composition of bathing solution. They greatly affect the equilibrium swelling of the hydrogel (Brondsted and Kopecek, 1992).

Due to the intrinsic swelling property, the pH-sensitive hydrogel is explored for development of novel chemomechanical sensing system, which would convert chemical energy directly to mechanical work, for example, chemically driven functional actuator or sensor (Osada and Gong, 1993; Beebe et al., 2000a, b). The pH-sensitive hydrogel has significant advantages over conventional microfluidic actuators owing to the capability of undergoing the abrupt volume changes in response to surrounding environmental pH without requirement of external power source. In addition, the hydrogel can perform well for sensing, actuating and regulating functions that are usually performed by discrete components such as valve and sensor in traditional system.

Currently the applications of the smart hydrogels in medical and pharmaceutical fields are increasingly becoming one of the most active research areas. The pH-sensitive hydrogels are found to be appropriate carriers as swelling-controlled drug release devices. The ability to dynamically control the swelling of the hydrogels subject to the changes in the pH and ionic strength of the external medium provides various opportunities for the pH-sensitive hydrogels. For example, the pH of fluid varies considerably along the length of the gastrointestinal tract (1–3 in the stomach, 5–8 in the small intestine), a weakly acidic hydrogel is a good candidate as smart device to deliver drug into the small intestines while avoiding release in the stomach. Various drug delivery systems based on the pH-sensitive hydrogels are developed or reviewed by many researchers (Siegel, 1990; Lowman and Peppas, 1999; Peppas et al., 2000). Design and optimization of these systems greatly demand insight into the underlying mechanism of the equilibrium swelling of the smart hydrogels.

In this section, we pay our attention to parameter studies by steady-state simulation for equilibrium of the pH-responsive hydrogels. The emphasis is placed on the influences of various hydrogel material properties and surrounding environmental conditions on the equilibrium swelling responses of the pH-responsive hydrogels. They include the environmental pH, the physical properties of the hydrogel, the chemical nature of the fixed charge groups and the ionic strength and compositions of surrounding solution.

Simulations are worked out to discuss the influences of the hydrogel properties and bath conditions on the distributive profiles of diffusive ionic concentrations and electrical potential as well as mechanical displacement. The input data used for implementation of numerical simulations are tabulated in Table 2.1, which are obtained from the experimental works by Beebe's research group (Beebe et al., 2000a, b; Johnson et al., 2002, 2004a, b; Yu et al., 2001; Zhao and Moore, 2001) and the handbook (Lide, 2002).

As shown in Fig. 2.1, a cylindrical pH-sensitive hydrogel with circular cross-sectional area is immersed in a bath solution consisting of sodium chloride (NaCl) and hydrochloric acid (HCl), which is referred as ideal solution here. In order to investigate the influences of various buffer contents on the swelling equilibrium of the pH-sensitive hydrogel, two buffer solution systems, namely the phosphate buffer and the Britton–Robinson buffer, are considered. Advantage of using pH buffers in

Table 2.1 Essential chemical and physical parameters used as input data for implementation of the numerical simulations

K	$10^{-2.0}$mM
ν	0.45
ϵ	80
ϵ_0	8.8542 × 10^{-12}C^2/N·m^2
F	9.6487 × 10^4C/mol
T	298 K
R	8.314 J/mol·K

experiment is to stabilize the pH solution and swelling response rate, where the buffer may resist the change in the solution pH when small amount of acid or base is added. In general, a buffer solution is a mixture of weak acid or weak base and corresponding salt. For investigation of the influences of buffer content and hydrogel properties on the swelling equilibrium, the cylindrical hydrogels are placed into the two systems of the buffered solutions, namely the phosphate buffer and the Britton–Robinson buffer (Townshend, 1995). The phosphate buffer is prepared with the ionic strength specified by controlling NaCl, and the pH of the solution can be regulated by adding certain amount of NaH_2PO_4, Na_2HPO_4, HCl or NaOH. The Britton–Robinson buffer solution is prepared by mixing the specified amount of strong base NaOH with the solution containing weak acid CH_3COOH, H_3PO_4 and H_3BO_3 in a proportional amount until a desired pH value is attained. Usually the ionic strength and valence numbers of the diffusive ionic species have important effects that discern the difference between the phosphate and Briton–Robinson buffers. These effects are reflected clearly in the PNP equations (2.2) and (2.24) of the present MECpH model.

As mentioned before, the present problem domain is considered as axisymmetry so that the problem can be simplified to one-dimensional computation, where the swelling takes place along the radial direction of the cylindrical hydrogel, as shown in Fig. 2.1. Before swelling, the dry gel has diameter of 400 μm. In general, the diameter of the dry gel is determined by the diameter of the circular photomask. However, it is an inevitable consequence that the hydrogel dimensions vary from the size of the photomask after the microchannel is flushed to remove the unpolymerized monomers. Light scattering and reflection are identified by Beebe et al. (2000b) as the possible reasons behind the shift in the hydrogel dimension. Therefore, a boundary-effect domain is necessarily defined as describing the hydrogel–solution interface. Usually very short range is given as the length scale of the interface, for example, about 1/20 of the total length of the computational domain. As an example, at the state before swelling, $x(\mu m) \in [0, 200]$ is the interior dry gel domain and $x(\mu m) \in [300, 2000]$ is the solution domain. The hydrogel–solution interface, x $(\mu m) \in [200, 300]$, is referred to as the boundary-effect domain. A hyperbolic tangent profile is employed to polish the distribution of the fixed charge density, ranging from c_f within the hydrogel to zero in the bathing solution. The fixed charge concentration inside the hydrogel is calculated by Eq. (2.34) or (2.35) with $c^s_{m0} = 1800$ mM and $K = 10^{-2.0}$ mM, respectively. According to the experimental data obtained by

Beebe's research group (Johnson et al., 2002), it is well known that Young's modulus of the hydrogel varies with the environmental pH value. In similar manner to the fixed charge concentration, a hyperbolic tangent profile is used to polish the variation of Young's modulus as function of solution pH. The distributive profile of Young's modulus consists of three segments, namely constant Young's moduli of 0.29 MPa if pH≤5.5 and 0.21 MPa if pH≥7.5. In the third segment for 5.5<pH<7.5, Young's modulus is polished to vary linearly from 0.29 to 0.21 MPa.

Despite the complexity of the swelling/deswelling mechanism of the smart hydrogel at various levels of environmental pH, much insight can be obtained via independent investigations of the influences of various material properties of the hydrogel and the environmental conditions of the bath solution. The nonlinear coupled partial differential equations of the MECpH model are solved numerically with appropriate parameters to further understand the fundamental mechanism of the swelling or deswelling of the pH-sensitive hydrogels.

2.5.1 *Influence of Initially Fixed Charge Density of Hydrogel*

Figures 2.5, 2.6 and 2.7 are plotted for analysis of the distributive profiles of ionic concentrations, electric potential and mechanical displacement as functions of ionizable fixed charge density within the hydrogel.

Figure 2.5a–c shows the distributive patterns of the diffusive ionic concentrations of the hydrogen (H^+), sodium (Na^+) and chlorine (Cl^-) ions, when the hydrogel is submerged in the acidic solution of pH 3.0. The results simulated are intuitively correct in the sense that the PHEMA gel is an acidic hydrogel. The distributions of H^+ and Na^+ions are high in the hydrogel domain, but decreases in the bath solution; while the opposite characteristics are true for the Cl^- ion. The concentration of the H^+ ion is much lower than those of Na^+ and Cl^- ions. The electroneutrality is conserved at every point in the bath solution, which is proved straightforward from the summation of the concentration profiles in Fig. 2.5. In the interior hydrogel, there is a difference between the concentrations of the Na^+ and Cl^- ions, and the Na^+ ion has higher concentration. This concentration difference between the sodium and chlorine ions corresponds closely to the fixed charge concentration as shown in Fig. 2.5d. Figure 2.5e illustrates the distribution of electric potential. The net difference of drift forces among all ionic concentrations in the interior hydrogel results in a constant electric potential within the hydrogel, but it is usually very small, e.g. in millivolts. In the bath solution, however, the distributive electric potential is zero due to the electroneutrality resulting from the net resultant concentration of the all-ionic species. Figure 2.5f demonstrates the mechanical displacement of the cylindrical hydrogel in radial direction. It is seen that the degree of swelling of the hydrogel decreases by following the sequence of c^s_{m0} = 2400, 1800 and 1200 mM. It is predicted by the MECpH model that the PHEMA hydrogels with greater amount of initially fixed charge groups may exhibit greater swelling, even though the swelling

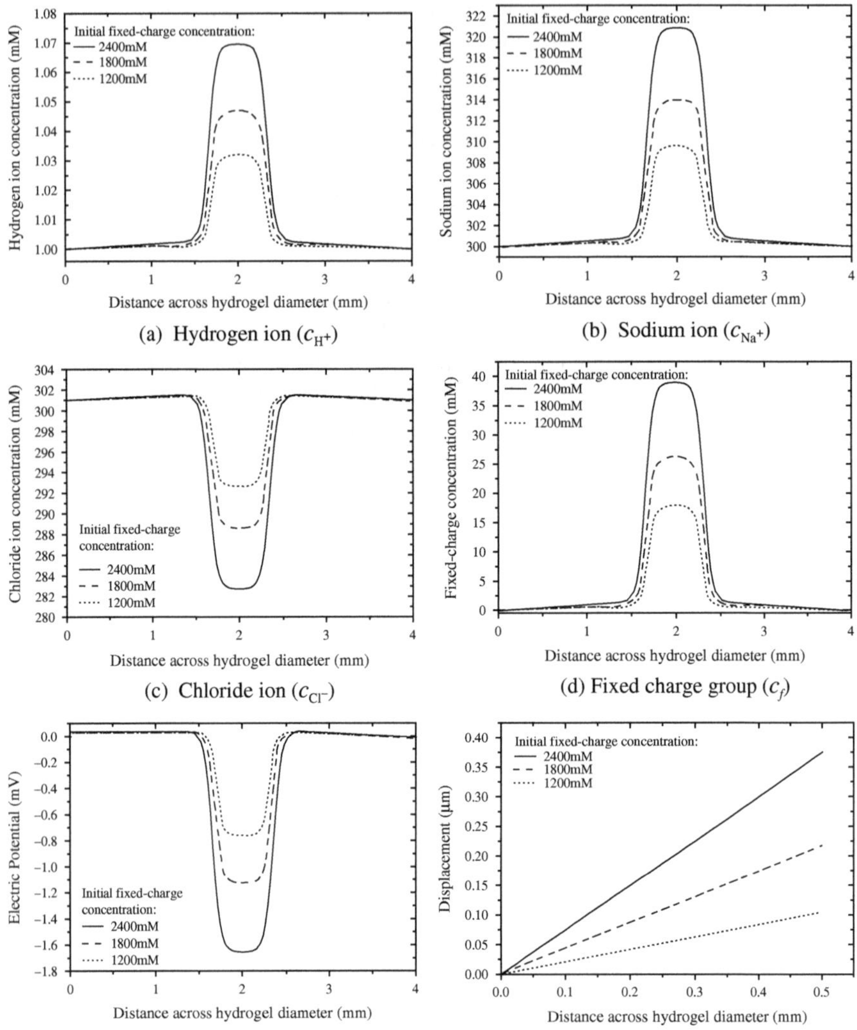

(a) Hydrogen ion (c_{H^+}) (b) Sodium ion (c_{Na^+})

(c) Chloride ion (c_{Cl^-}) (d) Fixed charge group (c_f)

(e) Electric potential (ψ) (f) Displacement (u)

Fig. 2.5 Distributive profiles of c_{H^+}, c_{Na^+}, c_{Cl^-}, c_f, ψ and u as the function of ionizable fixed charge concentration c^s_{mo}, where the PHEMA hydrogel is equilibrated in acidic medium of pH 3 with NaCl added to control the ionic strength

is very small in acidic solution as illustrated in Fig. 2.5, compared with that in basic solution as illustrated in Fig. 2.7.

Figures 2.6 and 2.7 demonstrate the distributive profiles of the diffusive ionic concentrations, electric potential and mechanical displacement of the hydrogel with the initial diameter of 400 μm in the neutral solution and the alkaline solution of pH 12 respectively, for the fixed charge groups at various initial concentrations. It is observed that the hydrogel behaves similar characteristics to those in

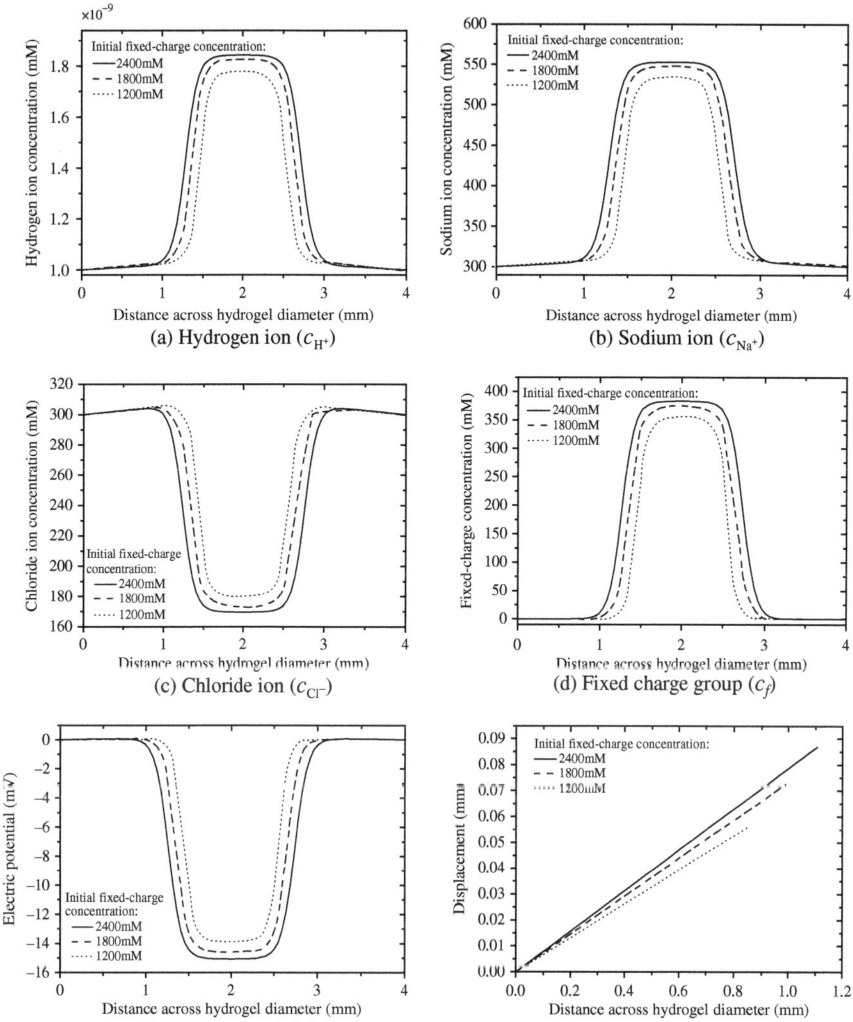

Fig. 2.6 Distributive profiles of $c_{H^+}, c_{Na^+}, c_{Cl^-}, c_f, \psi$ and u as the function of ionizable fixed charge concentration c^s_{mo}, where the PHEMA hydrogel is equilibrated in a neutral medium with NaCl added to control the ionic strength

the corresponding acidic bathing solution as shown in Fig. 2.5. However, there are dissimilarities quantitatively in the degree of swelling between the hydrogels when immersed in acidic and alkaline solutions. The degree of swelling of the PHEMA hydrogels is insignificant in acidic solution. In alkaline solution, however, they expand much more than the double size at dry state. On top of that, the differences in the degrees of swelling are also large among the hydrogels with different initially fixed charge densities.

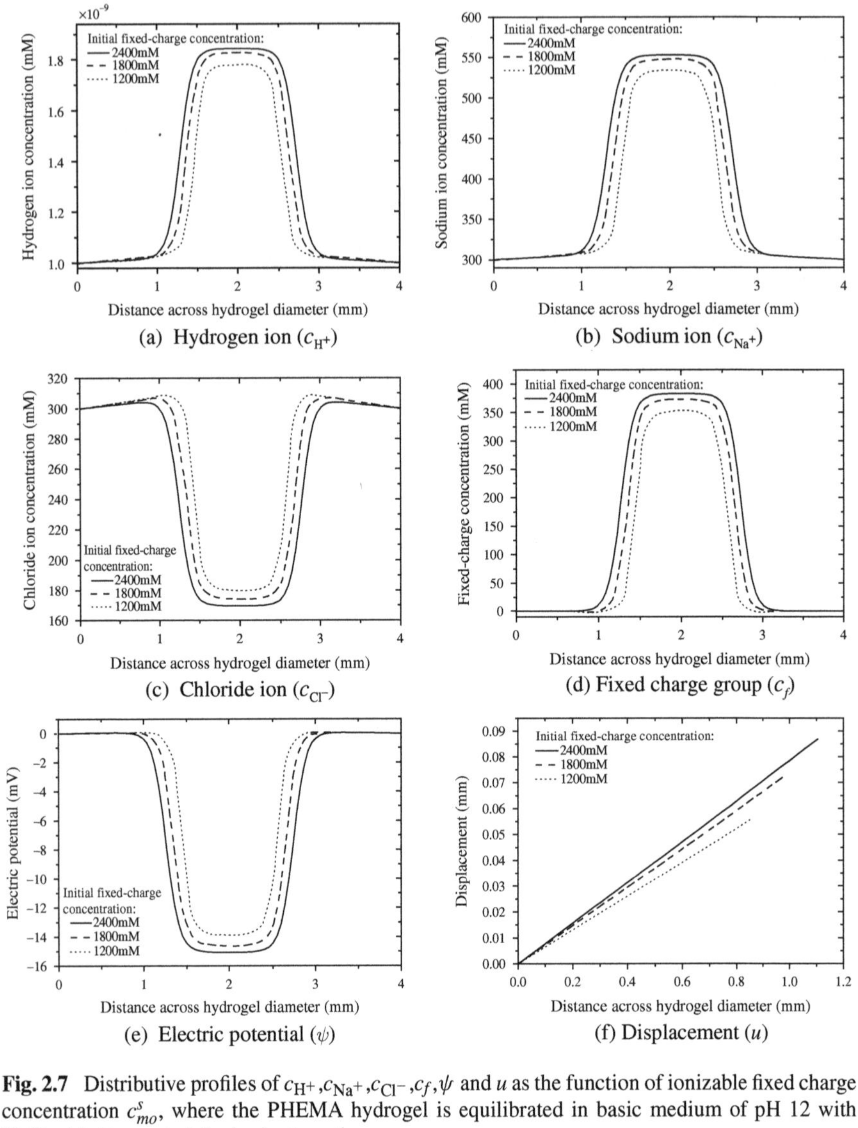

Fig. 2.7 Distributive profiles of c_{H^+},c_{Na^+},c_{Cl^-},c_f,ψ and u as the function of ionizable fixed charge concentration c^s_{mo}, where the PHEMA hydrogel is equilibrated in basic medium of pH 12 with NaCl added to control the ionic strength

The fixed charge groups attached onto the backbone of the PHEMA network, the carboxyl groups, exist in the form of R–COO$^-$ in basic solution or in the form of R–COOH in acidic medium, which are characterized by Eq. (2.28). The dissociation constant K describes the ionization of the pendent fixed charge groups, like the acidic or basic groups of monobasics or monobases. If the concentration of the hydrogen ion H$^+$ is smaller than the dissociation constant K, the carboxyl groups try to release more protons. In order to maintain the electroneutrality, more mobile counterions, e.g. Na^+in the present case, diffuse into the interior hydrogel to compensate the surplus charge. In contrast, the mobile anions are repulsed from entering the interior hydrogel. It is thus evident that the concentration of the Na^+ ion is higher whereas that of Cl^- ion is lower in the case of the alkaline solution, compared with the acidic solution. As a consequence, the concentration differences between the interior hydrogel and the exterior solution increase tremendously, leading to higher osmotic pressure which effectively drives higher degree of swelling.

Figure 2.8a, b shows the theoretically predicted dependence of the swelling of the pH-sensitive hydrogel response to the changes in the initially fixed charge density for an ideal solution at different pH levels. It is obvious that the change of the fixed charge concentration c_{m0}^s at dry state strongly influences the equilibrium swelling of the hydrogels at high pH values, whereby the decrease of c_{m0}^s dramatically reduces the degree of swelling at high pH values. The initial concentration of fixed charge c_{m0}^s is a function of molar ratio of the comonomers during preparation (Chu et al., 1995). As the molar ratio of carboxylic acid to 2-hydroxyethyl methacrylate decreases, the initially fixed charge density decreases dramatically. The concentration difference thus decreases between the interior hydrogel and the exterior solution. As a result, this in turn mitigates the osmotic pressure and generates smaller degree of hydrogel swelling. These observations are in agreement with the experimental trend reported by Siegel (1990). Figure 2.8b characterizes well the experimental phenomena, where a monotonic swelling is predicted with increasing the total molar concentration of ionizable groups per volume of solid network polymer.

Figure 2.9 exhibits the relation between the equilibrium swelling of the hydrogel and the concentration of fixed charge group c_{m0}^s for three different buffer solutions, the ideal solution, the phosphate buffer and the Briton–Robinson buffer. The larger c_{m0}^s is, the greater degree of swelling the hydrogel performs for both the buffer solutions at higher pH. At low pH level, however, the degree of swelling keeps almost constant for both the buffer systems. The figure evidently shows that the swelling equilibrium achievable in the Britton–Robinson system is always higher than that in the ideal solution and the phosphate buffer, especially as the initial concentration of fixed charge group c_{m0}^s increases highly. Swelling of the hydrogel is almost the same if bathed with either the phosphate buffer or the ideal solution, e.g. only NaCl and/or HCl in solution. However, the phosphate buffer shows greater degree of swelling with the increase of initially fixed charge group concentration c_{m0}^s.

2.5.2 *Influence of Young's Modulus of Hydrogel*

Figures 2.10, 2.11 and 2.12 are plotted for the distributions of ionic concentrations and electric potential as well as mechanical displacement as function of the normalized Young's modulus of the hydrogel, if placed in an ideal bathing solution composed of NaCl and HCl electrolytes. The characteristic profiles of the ionic concentrations and electric potential are similar to those in Figs. 2.5, 2.6 and 2.7 for environmental conditions of the acidic solution (pH 3), neutral solution (pH 7)

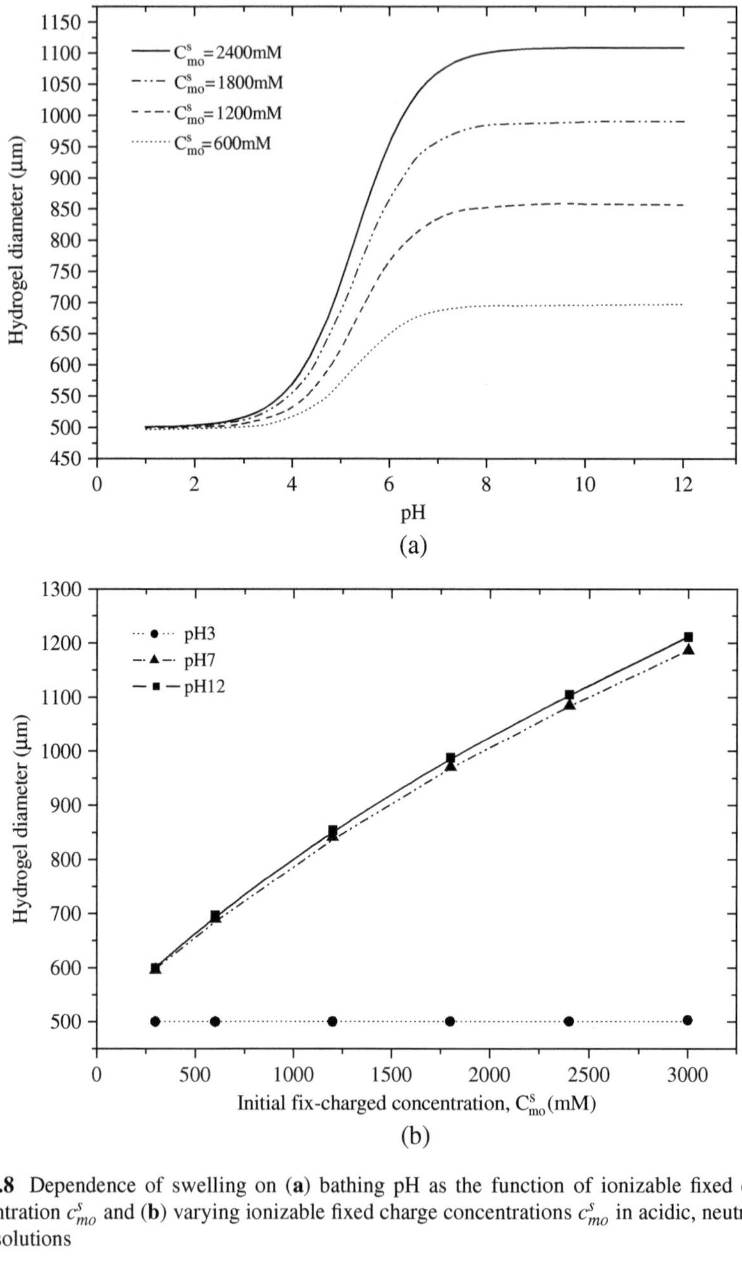

Fig. 2.8 Dependence of swelling on (**a**) bathing pH as the function of ionizable fixed charge concentration c^s_{mo} and (**b**) varying ionizable fixed charge concentrations c^s_{mo} in acidic, neutral and basic solutions

and basic solution (pH 12). The cationic concentrations, e.g. H^+ and Na^+ within the hydrogel are higher than those in the bath solution. In contrast, the anion concentration in the interior hydrogel is at a lower level than that in the external solution. Electroneutrality is conserved everywhere.

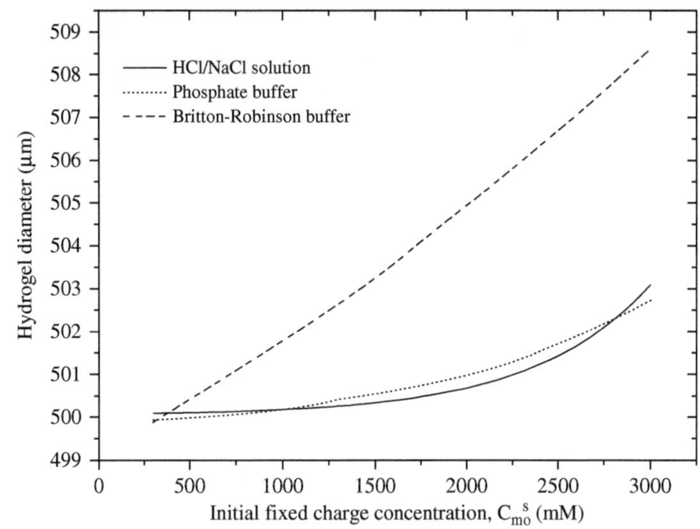

(a)

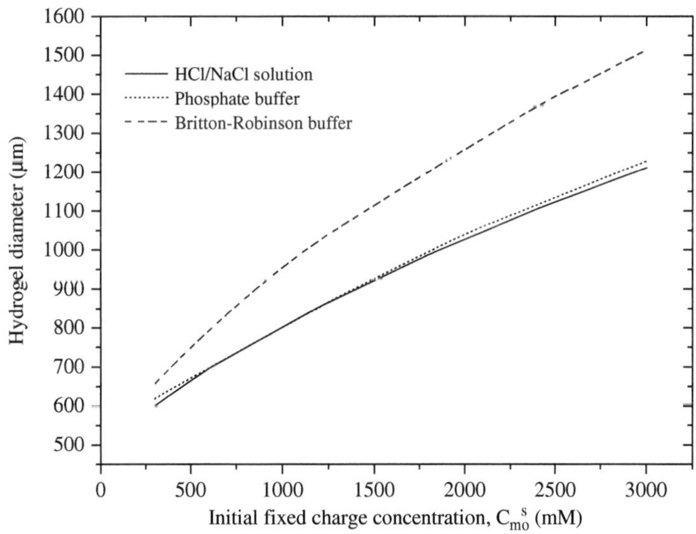

(b)

Fig. 2.9 Influences of buffer systems on swelling equilibrium as the function of ionizable fixed charge concentration in (**a**) acidic medium of pH 3 and (**b**) basic medium of pH 9

The dissimilarity of the swelling response at lower and higher pH levels displays two different conditions. As discerned from Fig. 2.10, the changes in Young's modulus values of the hydrogel seem to have no significant effect on the degree of swelling at low pH. Probably as the hydrogel is still in compact state at low pH, the effect of changing Young's modulus is very tiny on the swelling equilibrium. In contrast, the degree of swelling is controlled greatly by changing Young's modulus

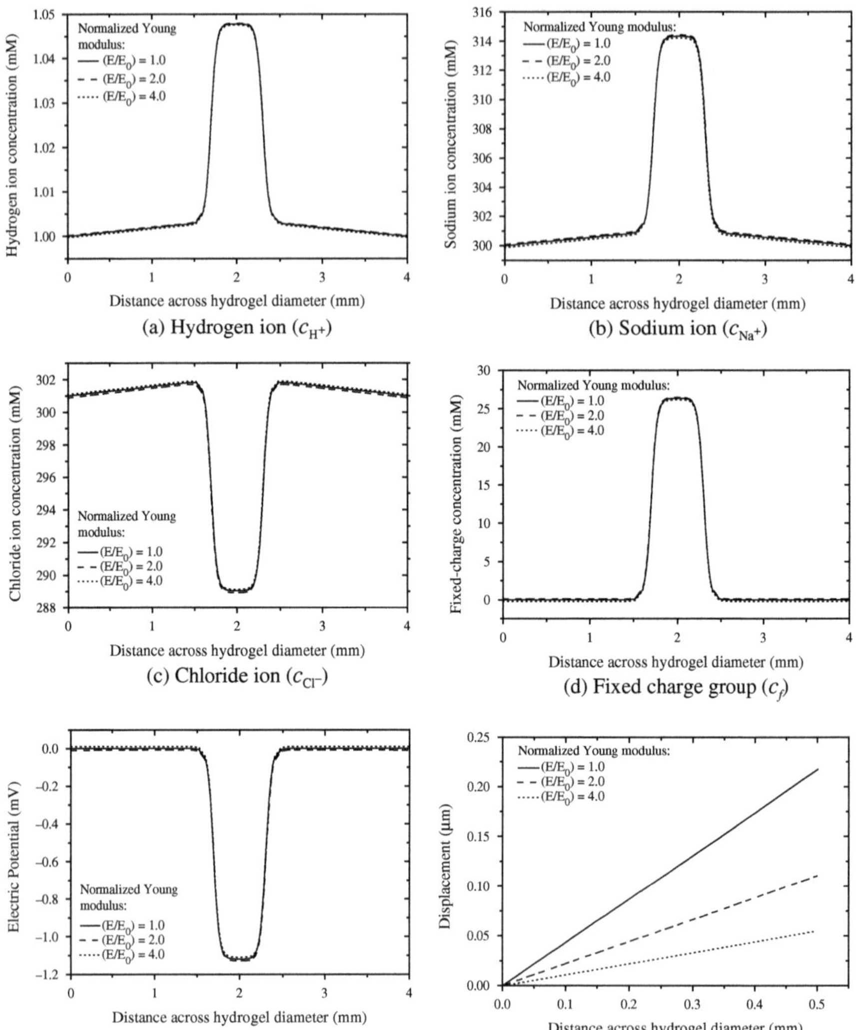

(a) Hydrogen ion (c_{H^+}) (b) Sodium ion (c_{Na^+}) (c) Chloride ion (c_{Cl^-}) (d) Fixed charge group (c_f) (e) Electric potential (ψ) (f) Displacement (u)

Fig. 2.10 Distributive profiles of $c_{H^+}, c_{Na^+}, c_{Cl^-}, c_f, \psi$ and u as the function of normalized Young's modulus (E/E_0), where the PHEMA hydrogel is equilibrated in acidic medium of pH 3 with NaCl added to control the ionic strength

if the environmental pH level is high, as observed from Figs. 2.11 and 2.12. The phenomena mentioned occur owing to the fact that the more fixed charge groups are ionized as pH increases and thus the degree of swelling increases. However, the swelling is constrained as Young's modulus increases. The interaction between expanding and retracting forces lasts until new equilibrium is reached.

Figure 2.13a shows the dependence of swelling of the hydrogel on the changes of environmental pH as function of Young's modulus of the pH-sensitive hydrogel.

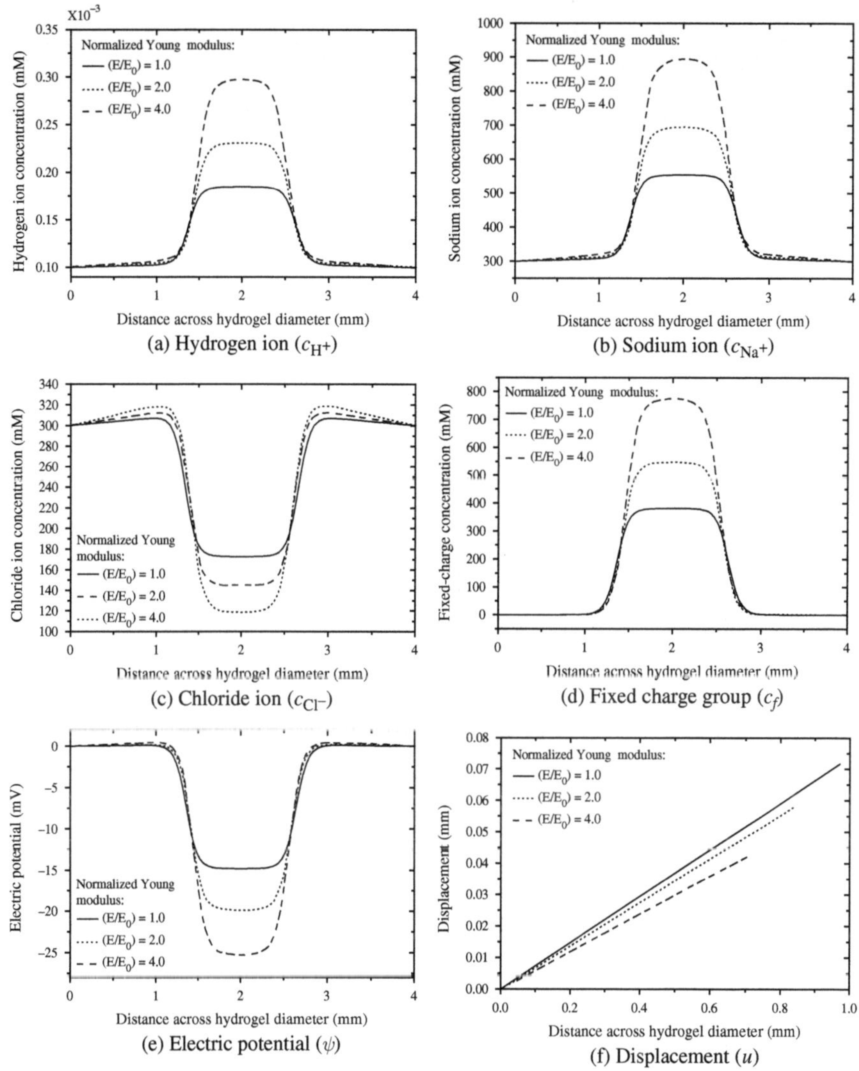

Fig. 2.11 Distributive profiles of $c_{\mathrm{H}}^{+}, c_{\mathrm{Na}}^{+}, c_{\mathrm{Cl}}^{-}, c_f, \psi$ and u as the function of normalized Young's modulus (E/E_0), where the PHEMA hydrogel is equilibrated in neutral medium with NaCl added to control the ionic strength

The MECpH model theoretically predicts that, for the hydrogels with larger Young's modulus, the degree of swelling decreases at higher solution pH. The characteristics become more visible in Fig. 2.13b when the normalized Young's modulus is plotted against the diameters of hydrogels at equilibrium state. The magnitude of swelling reduces exponentially with the increase of Young's modulus. Usually it is known that Young's modulus of the hydrogel is strongly dependent on preparation process, where the modulus is primarily determined by the volume per molar

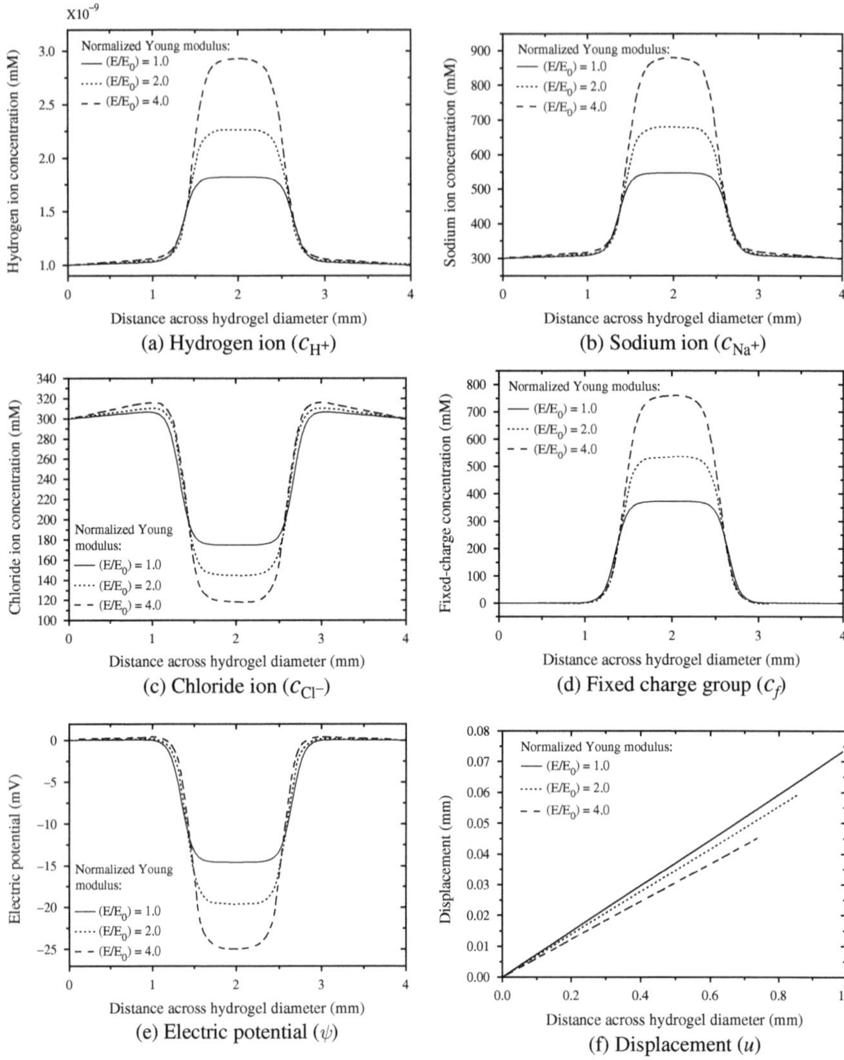

Fig. 2.12 Distributive profiles of c_{H^+},c_{Na^+},c_{Cl^-},c_f,ψ and u as the function of normalized Young's modulus (E/E_0), where the PHEMA hydrogel is equilibrated in basic medium of pH 12 with NaCl added to control the ionic strength

ratio of copolymer mixture which directly quantifies the density of entanglement strands or crosslinking ratio. As the crosslinking content increases in the polymer network, the hydrogel enhances larger retraction force and thus develops higher Young's modulus. The phenomenon always exists regardless of buffer contents as depicted in Fig. 2.14. The diameters of the swollen hydrogels are plotted against the normalized Young's modulus for the buffer solutions of pH 3 and 9. The increase of Young's modulus reduces exponentially the swelling of the hydrogels for the three

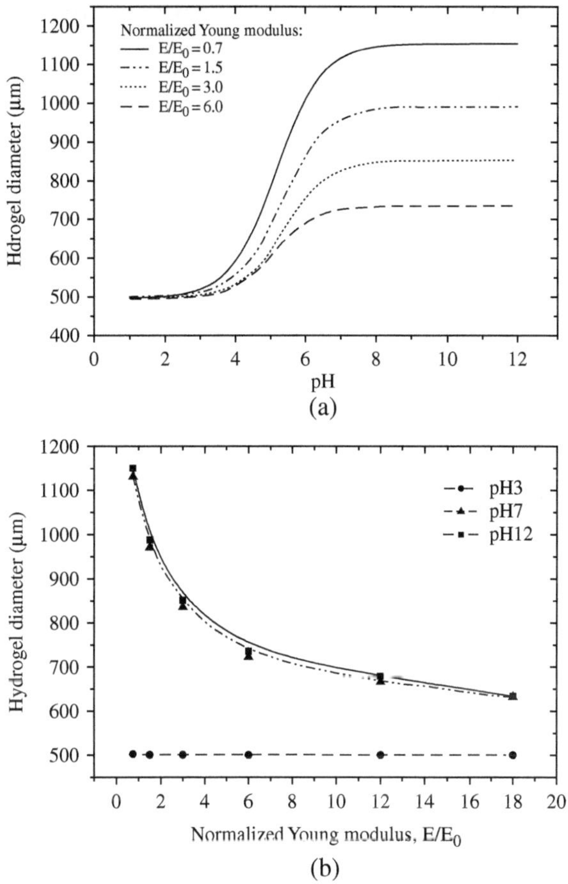

Fig. 2.13 Dependence of swelling on (**a**) bathing pH as the function of normalized Young's modulus (E/E_0) and (**b**) varying normalized Young's modulus (E/E_0) in acidic, neutral and basic solutions

different buffer systems as illustrated in Fig. 2.14. Influence of buffer contents on the swelling equilibrium at higher pH is more significant than that at lower pH. When the pH of buffer solutions is low, the degree of swelling of the hydrogel is almost insignificant even in different buffer systems and Young's moduli. If the pH is high, the obvious differences in the degree of swelling are observed for different buffer systems. Further, the Britton–Robinson is the unchanging leader for providing the better buffer solution when large swelling scale is required. It should be pointed out that the influence of the buffer contents vanishes if Young's modulus is high enough.

2.5.3 *Influence of Initial Geometry of Hydrogel*

Figures 2.15, 2.16 and 2.17 are plotted for the distributive profiles of the diffusive ionic concentrations and the electric potential as well as the mechanical displacement as function of the initial diameter of the hydrogel at the state before it is

Fig. 2.14 Influences of buffer systems on swelling equilibrium as the function of normalized Young's modulus (E/E_0) in (**a**) acidic medium of pH 3 and (**b**) basic medium of pH 9

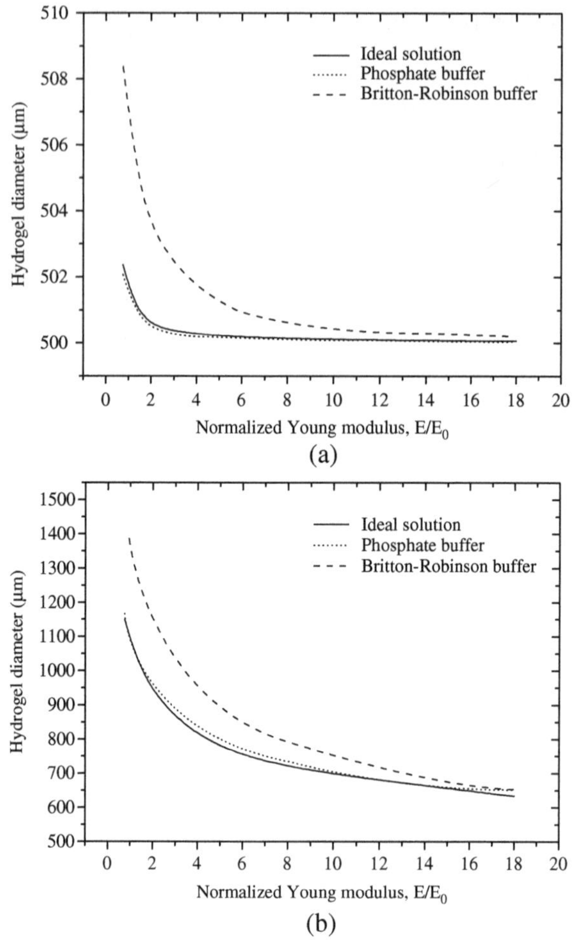

immersed in solution. The bathing solution is the electrolyte which is composed of NaCl and/or HCl. For the three pH buffered conditions, pH 3, 7 and 12, the concentrations of cations inside the hydrogel are higher, whereas those of anions are lower in comparison with those in the external solution. These are the performance expected because of the negative sign of the ionized charge groups dangling on the network matrix of the PHEMA hydrogel. Similarly, the electric potential exists as long as the migrations of ions occur between the interior hydrogel and the exterior solution.

Since both the hydration and fixed charge concentration kept almost constant against different initial geometries of the hydrogel as shown in Figs. 2.17 and 2.18, the swelling of the hydrogels behaves negligibly when they are soaked in electrolyte solution. The characteristics are found more clearly in Fig. 2.18, where the hydrations of the swollen hydrogels are plotted against the different pH levels

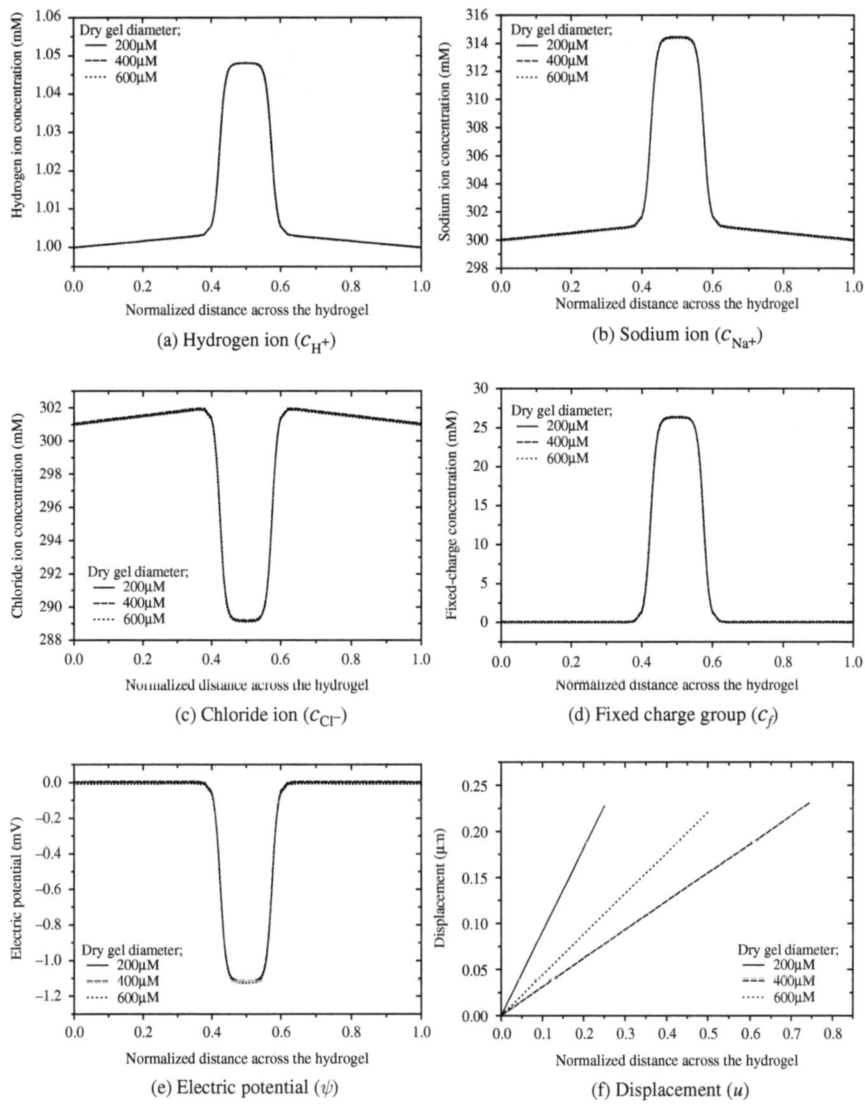

Fig. 2.15 Distributive profiles of c_{H^+},c_{Na^+},c_{Cl^-},c_f,ψ and u as the function of initial diameter of the dry gel, where the PHEMA hydrogel is equilibrated in an acidic medium of pH 3 with NaCl added to control the ionic strength

and dry gel diameters. As mentioned before, the diameters of dry gel are determined by the diameters of the nominal photomasks and the hydration is calculated by $H = V^f/V^s$.

In order to understand the responsive characteristics of the hydrogel bathed in different buffer solution systems, several simulations are carried out to predict the

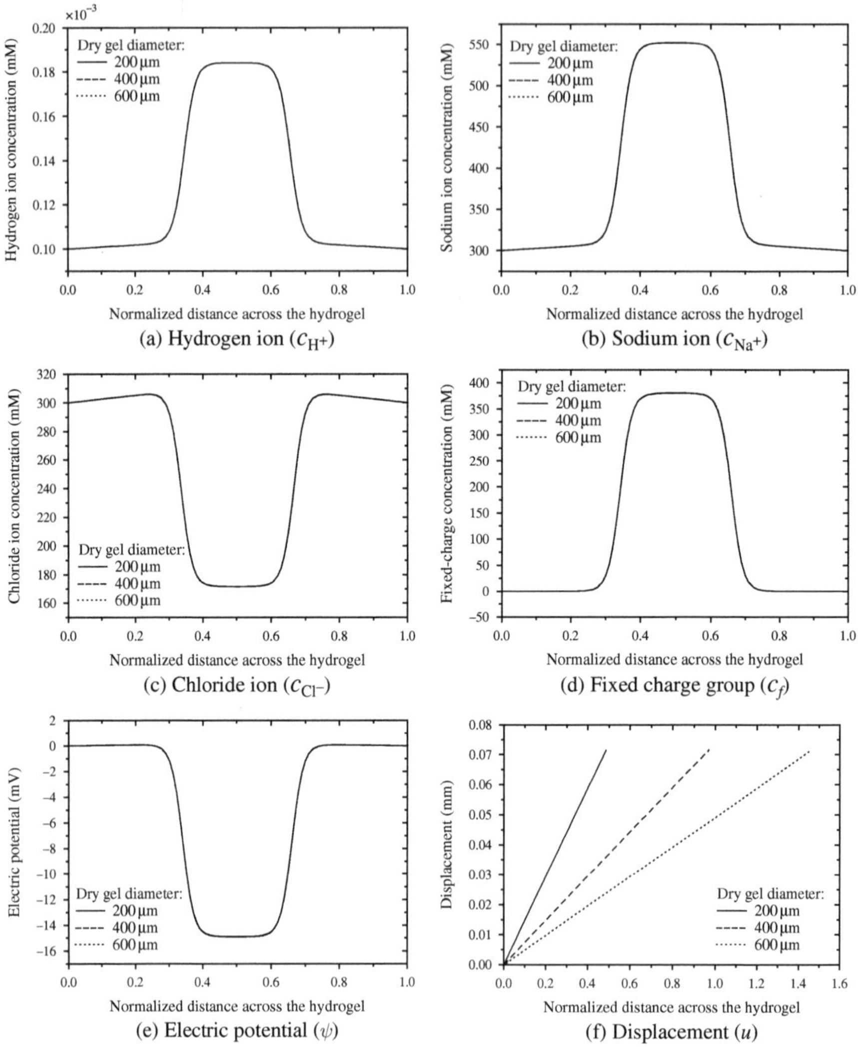

Fig. 2.16 Distributive profiles of $c_{H^+}, c_{Na^+}, c_{Cl^-}, c_f, \psi$ and u as the function of initial diameter of the dry gel, where the PHEMA hydrogel is equilibrated in a neutral medium with NaCl added to control the ionic strength

dilation of the hydrogels soaking in the phosphate buffer and Britton–Robinson buffer solutions. Figure 2.19 provides the comparison of swelling degree of the hydrated hydrogel for the three buffer solutions at pH 3 and 9 levels. The figure shows the significant independency of swelling on the initial diameters of the hydrogels. Similar to the previous discussions, the Britton–Robinson buffer always demonstrates a potentiality as buffer environment solution to meet the need of voluminous swelling.

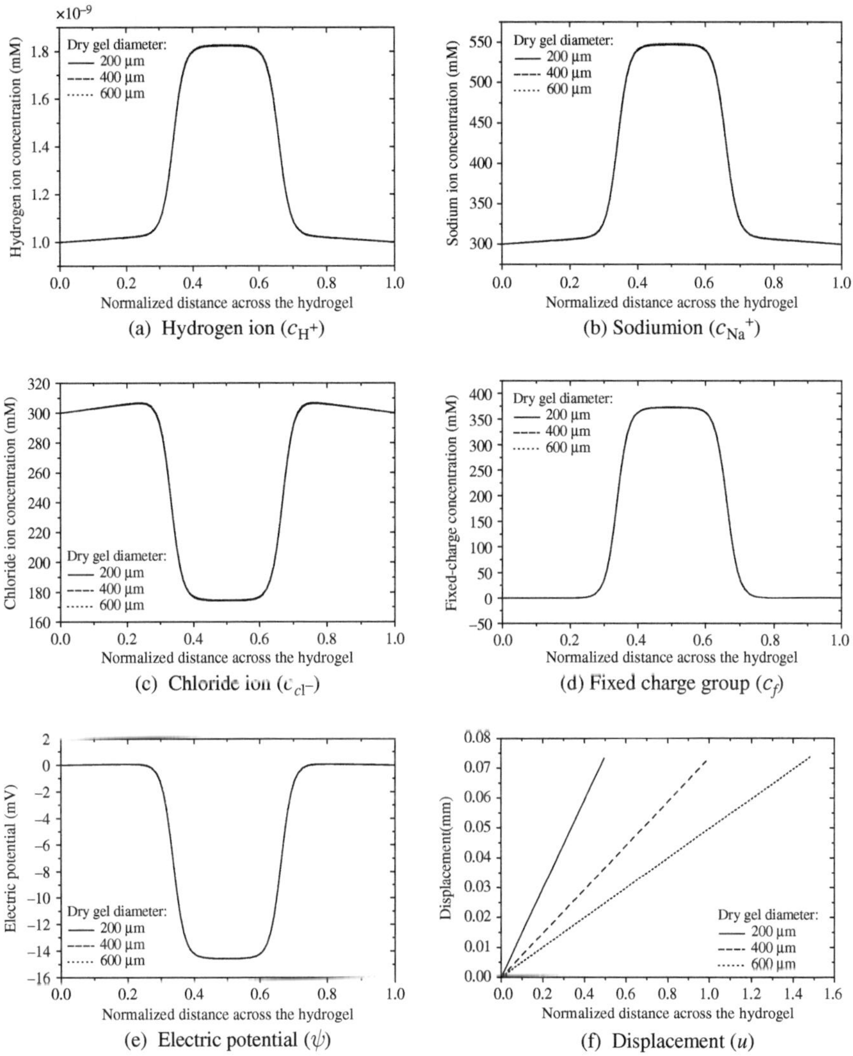

Fig. 2.17 Distributive profiles of $c_{H^+}, c_{Na^+}, c_{Cl^-}, c_f, \psi$ and u as the function of initial diameter of the dry gel, where the PHEMA hydrogel is equilibrated in a basic medium of pH 12 with NaCl added to control the ionic strength

2.5.4 Influence of Ionic Strength of Bath Solution

Figures 2.20, 2.21 and 2.22 are plotted to investigate the characteristics of diffusive ionic species, electrical potential and mechanical deformation of the pH-responsive hydrogel as function of environmental ionic strength conditioning.

It is widely accepted that the osmotic pressure arises from the concentration difference between the interior hydrogel and the external solution, and the swelling of

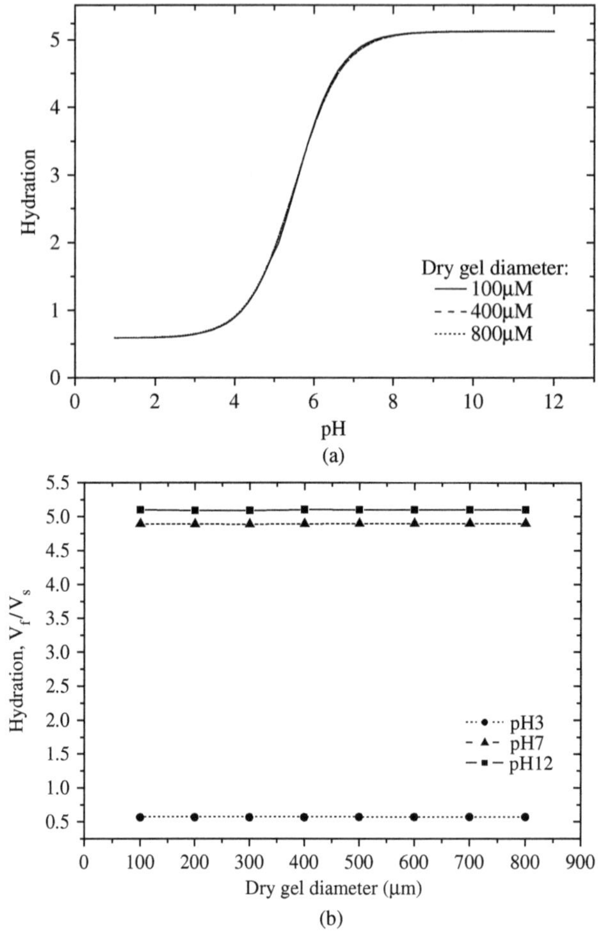

Fig. 2.18 Dependence of hydration on (**a**) bathing pH as the function of initial diameter of the dry gel and (**b**) the diameter of the dry gel in acidic, neutral and basic solutions

the hydrogel in equilibrium can be approximated by examination of the concentration ratio between the interior and exterior of the hydrogel, which is coherent with the definition of Donnan equilibrium (Flory, 1962; Helfferich, 1962). The osmotic pressure is usually regarded as outward pressure in excess of the pressure of surrounding solution, which would result in expanding of the polymer network. In order to discuss the problems with different bath solution concentrations, it is convenient to introduce the density ratio or better known as Donnan ratio λ_D as follows (Ricka and Tanaka, 1984; Siegel, 1990; Homma et al., 2000):

$$\lambda_D = \left(\frac{c_k}{\bar{c}_k}\right)^{1/z_k} \tag{2.60}$$

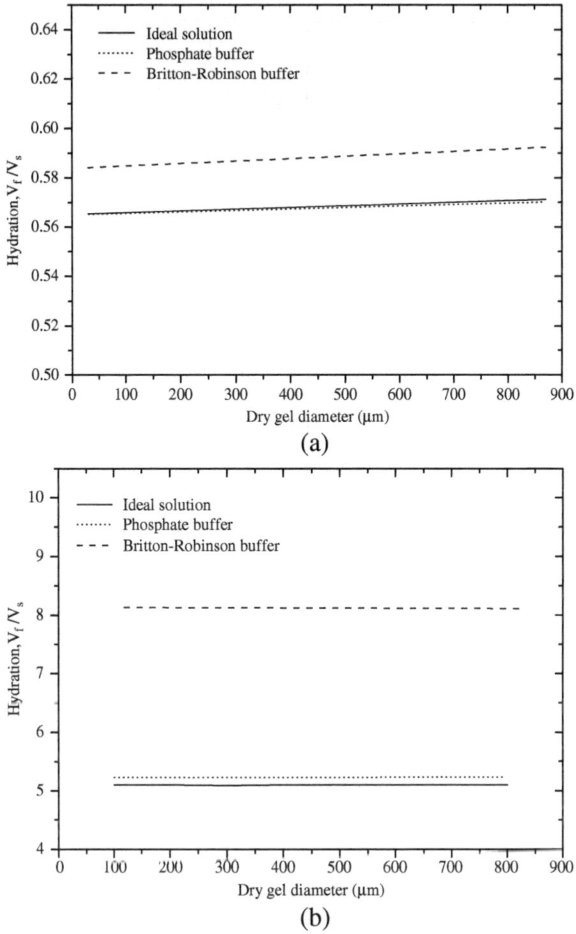

Fig. 2.19 Influences of buffer systems on hydration as the function of initial diameter of the dry gel in (**a**) acidic medium of pH 3 and (**b**) basic medium of pH 9

where z_k, c_k and $\bar{c}_k$ ($k = 1,2,3,..., N_{\text{ion}}$) are the valence number and the concentrations in the hydrogel and the external solution for the kth diffusive ion species, respectively. As the concentrations of the external solution keep constant everywhere in equilibrium sate, $\bar{c}_k$ are allowed to take the concentration boundary values of corresponding ionic species.

The distributions of the concentrations of diffusive Na^+ and Cl^- species are presented in the form of Donnan ratio and shown in Fig. 2.20. It is evident that the mobile cations, hydrogen (H^+) and sodium (Na^+) ions, have higher concentrations within the interior hydrogel than those in the exterior solution, while the mobile anions, chloride (Cl^-) ion, show a contrary pattern. However, the requirement of electroneutrality condition is always met everywhere in the domain of surrounding solution.

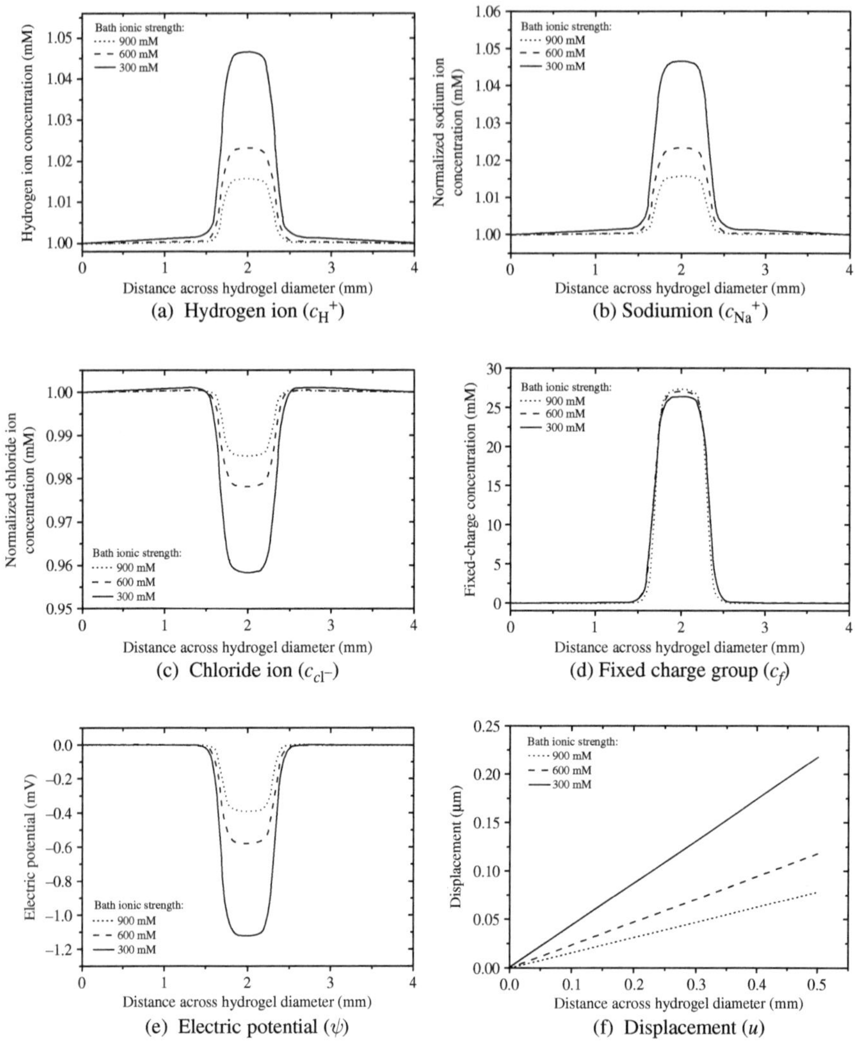

Fig. 2.20 Distributive profiles of c_{H^+},c_{Na^+},c_{Cl^-},c_f,ψ and u as the function of the ionic strength of medium, where the PHEMA hydrogel is equilibrated in acidic medium of pH 3 with NaCl added to control the ionic strength

As the ionic strength of bath solution increases, the ratio of the ionic concentrations decreases between the interior hydrogel and the exterior bath solution (Siegel, 1990). On this account, the osmotic pressure becomes less and the reduction of the hydration is expected. The fixed charge density is a function of the hydration of the swollen hydrogel, the total ionizable groups per volume of network polymer and the concentration of diffusive H^+ ion provided by the outer solution. Since the increases of the ionic strength is controlled by NaCl, the totally resultant concentration of

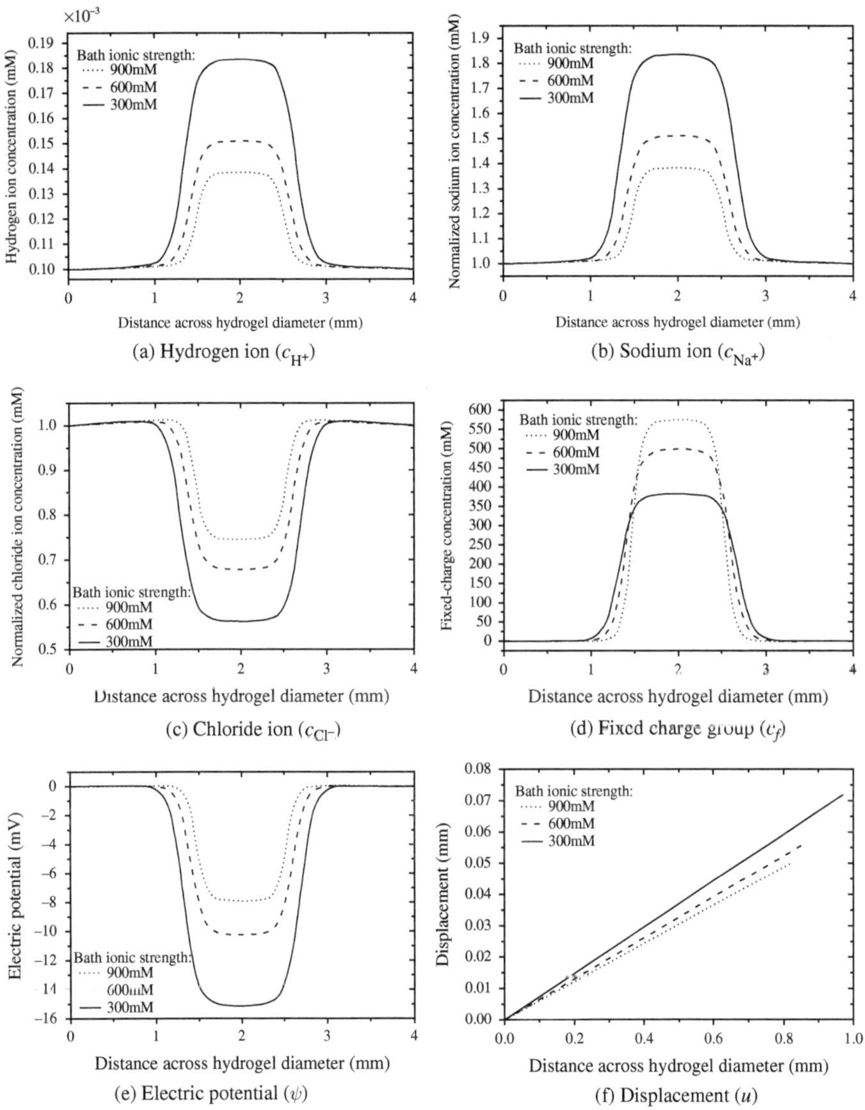

Fig. 2.21 Distributive profiles of c_{H^+},c_{Na^+},c_{Cl^-},c_f,ψ and u as the function of the ionic strength of medium, where the PHEMA hydrogel is equilibrated in neutral medium with NaCl added to control the ionic strength

H^+ ion and ionizable groups remains constant. Therefore, it is foreseeable that the redistribution of the fixed charge concentration is controlled mainly by the hydration. Due to the change of the fixed charge concentrations, there are the concurrent changes in the concentration profiles of the diffusive cations and anions. The interaction continues until new equilibrium state is achieved. The phenomena become more obvious at higher solution pH, as seen in Figs. 2.21 and 2.22.

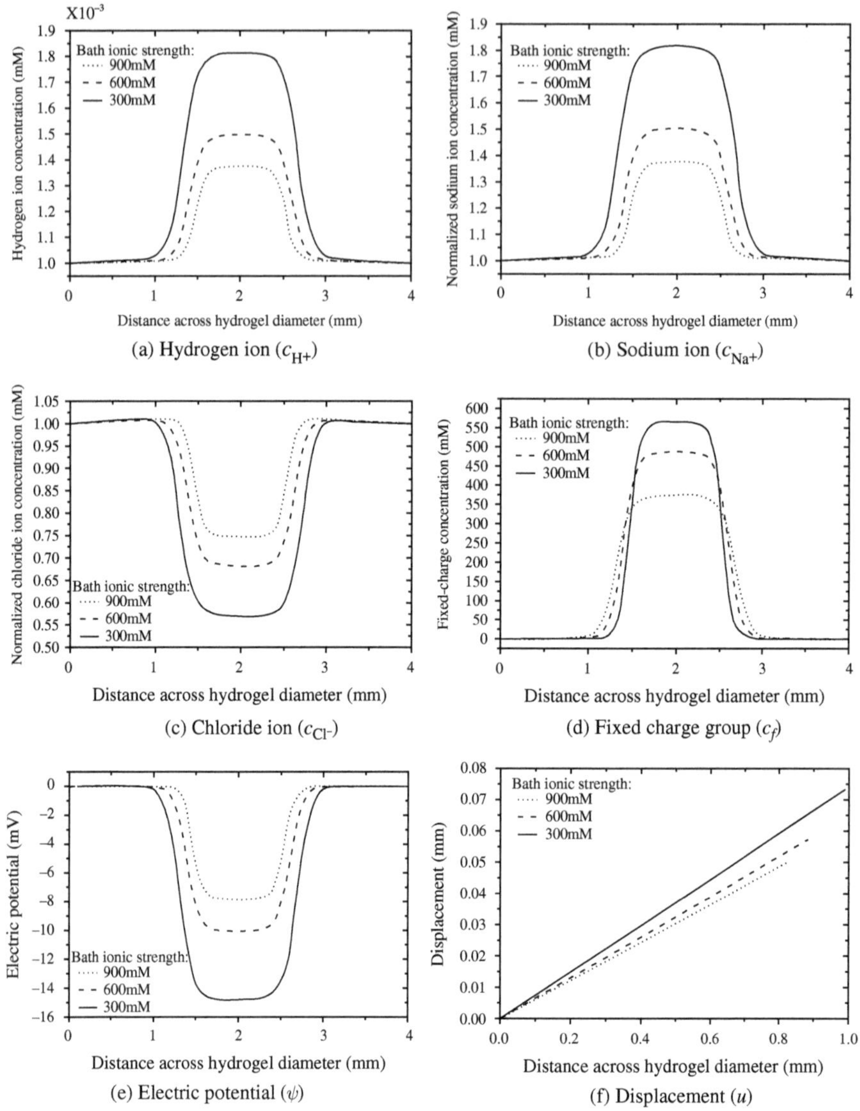

Fig. 2.22 Distributive profiles of $c_{H^+}, c_{Na^+}, c_{Cl^-}, c_f, \psi$ and u as the function of the ionic strength of medium, where the PHEMA hydrogel is equilibrated in basic medium of pH 12 with NaCl added to control the ionic strength

Figure 2.23a, b theoretically demonstrates the influences of the ionic strength of bathing solution on the equilibrium swelling of the HEMA hydrogel with the identical fixed charge density and Young's modulus. As predicted, the hydrogel behaves like a hydrophobic polymer network at low pH. After pH 4, however, the fluid phase content within the hydrogel increases abruptly and thus results in highly

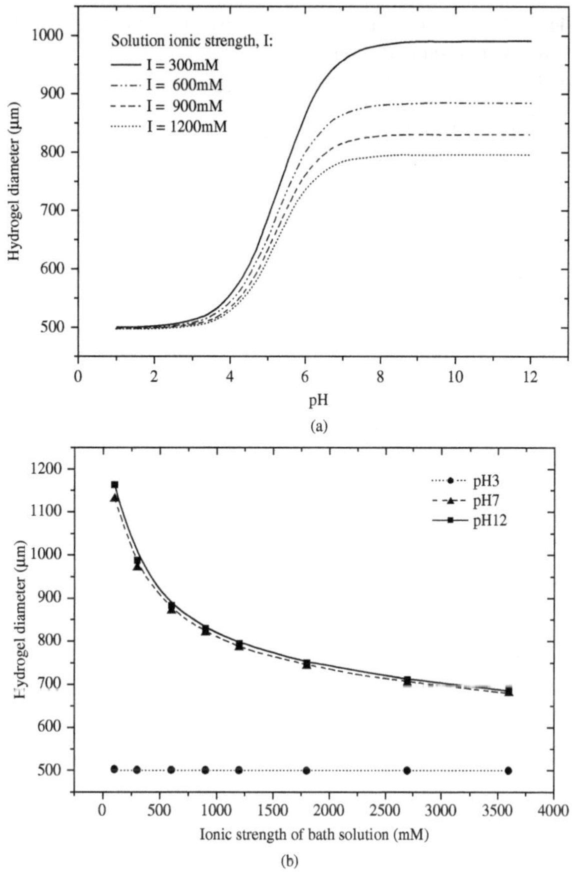

Fig. 2.23 Dependence of swelling on (**a**) bathing pH as the function of ionic strength of bath medium and (**b**) varying ionic strengths in acidic, neutral and basic mediums

swollen hydrogel. Apart from that, the highest curve is associated with the solution for the ionic strength of 300 mM, while subsequently lower curves correspond to the bath solutions with higher ionic strength, increasing regularly from 600 to 1200 mM. In the surrounding solutions with very low ionic strengths, the hydrogen (H^+) ions play an essential role in association or dissociation process. This implies that the contributions of other mobile ion species to the osmotic pressure are minimized. However, as the ionic strength of bathing solution increases, the degree of swelling decreases for high ambient pH. This phenomenon is in accordance with the experimental observation by Siegel and his team (1988, 1990, 1991).

Figure 2.24 focuses on the dependence of equilibrium swelling on the ionic strength of certain buffer systems, where two diverse solutions are considered for comparison and they are the NaCl/HCl solution and the phosphate buffer solution with a calculated amount of NaCl added to adjust the ionic strength at a desired level. The responsive characteristics of the pH-sensitive hydrogel perform exponentially the decrease of the degree of swelling as the environmental ionic strength

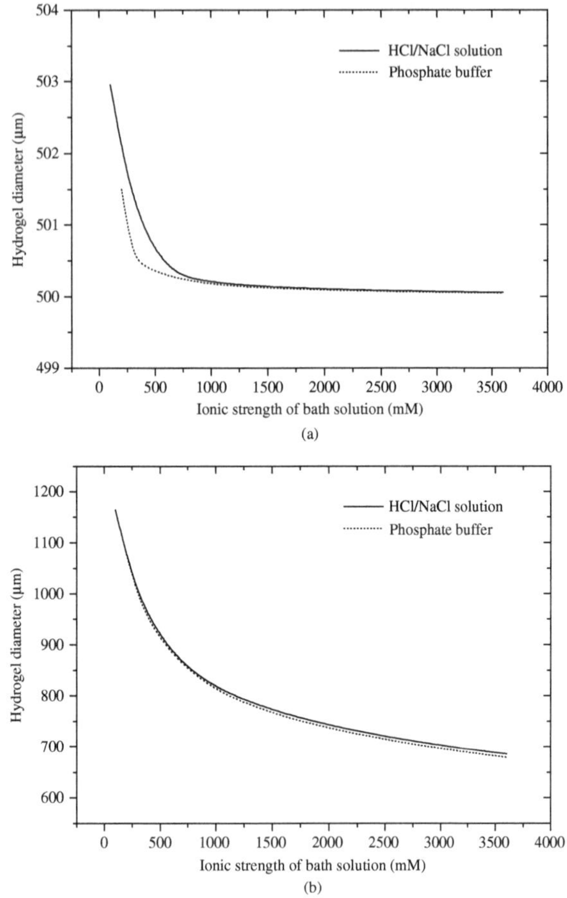

Fig. 2.24 Influences of buffer systems on hydration as the function of ionic strength of bath medium in (**a**) acidic medium of pH 3 and (**b**) basic medium of pH 9

increases for both the bathing solutions. The observations are comparable with the experimental work by Brannon-Peppas and Peppas (1991). Insignificant difference of swelling is found for both the solutions. In all likelihood, both the solutions do not make enormous difference as the dominant counterions are essentially the univalent ions which are mainly the sodium ions (Siegel et al., 1991).

2.5.5 Influence of Multivalent Ionic Composition of Bath Solution

In order to investigate the influences of multivalent ionic compositions of bath solutions on the equilibrium swelling, the multivalent polyelectrolyte solution is considered in this section. Influences of bath compositions with multivalent ions on the characteristics of pH-dependent equilibrium swelling are shown in

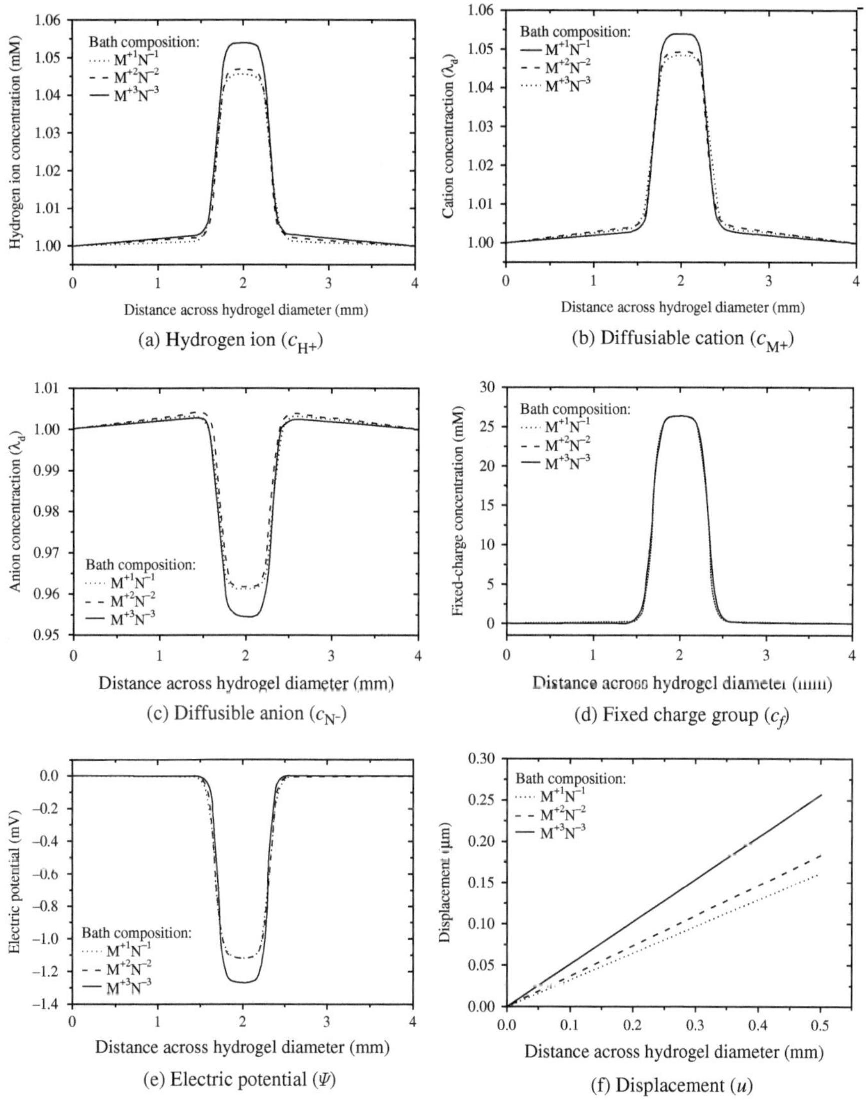

Fig. 2.25 Distributive profiles of c_{H^+},c_{M^+},c_{N^-},c_f,ψ and u for specified solvent composition (monovalent, divalent and trivalent), where the PHEMA hydrogel is equilibrated in acidic medium of pH 3

Figs. 2.25, 2.26, 2.27, 2.28 and 2.29, where the relevant conditions are maintained at the temperature of 25°C and the ionic strength of 300 mM. For analysis of the characteristics of the hydrogel swelling, the bathing solution is assumed reasonably to be primarily composed of symmetrical salt (z:z) with varying ionic valences, namely $M^{+1}N^{-1}$, $M^{+2}N^{-2}$ and $M^{+3}N^{-3}$.

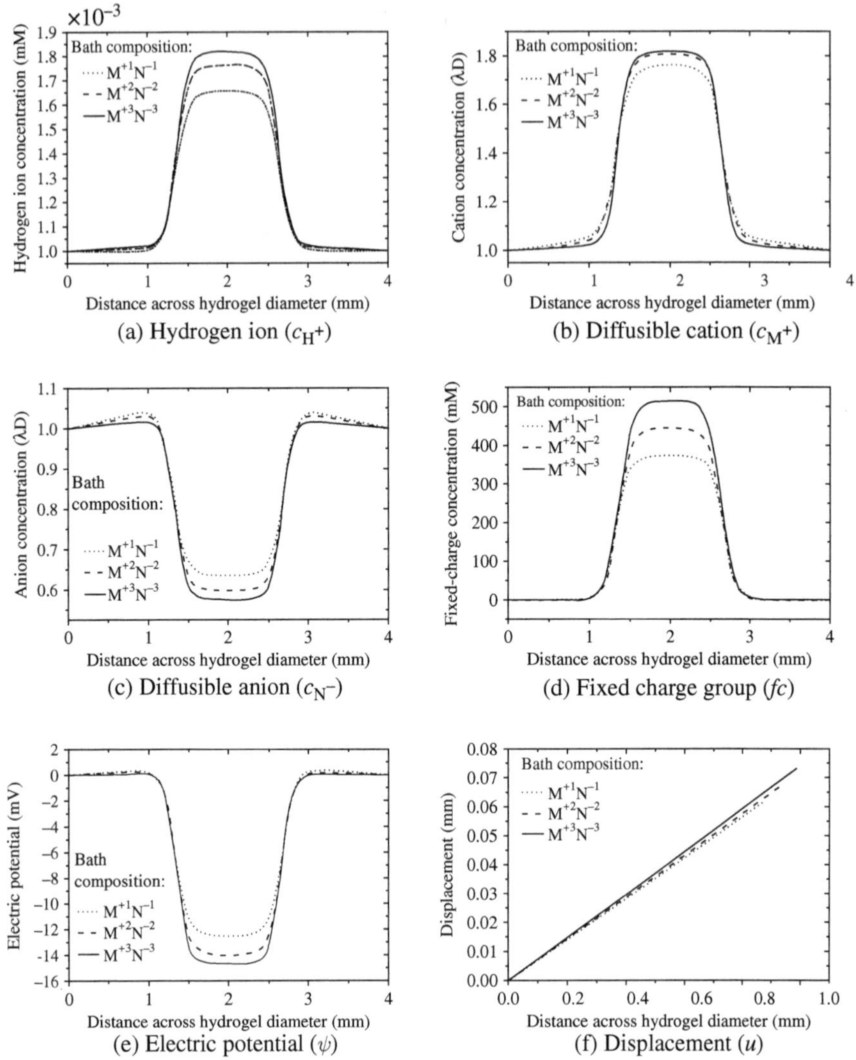

Fig. 2.26 Distributive profiles of $c_{H^+}, c_{M^+}, c_{N^-}, c_f, \psi$ and u for specified solvent composition (monovalent, divalent and trivalent), where the PHEMA hydrogel is equilibrated in a neutral medium

Figures 2.25, 2.26 and 2.27 demonstrate the concentration profiles of mobile ion species and fixed charge groups, and the profiles of electric potential and mechanical displacement predicted in equilibrium. The Donnan concentrations of the sodium and chloride ions are presented for the sake of apprehensible comparison of the osmotic pressure. When the profiles of the ionic concentrations are determined, Donnan ratio λ_D becomes an elegant tool for estimating quantitatively the

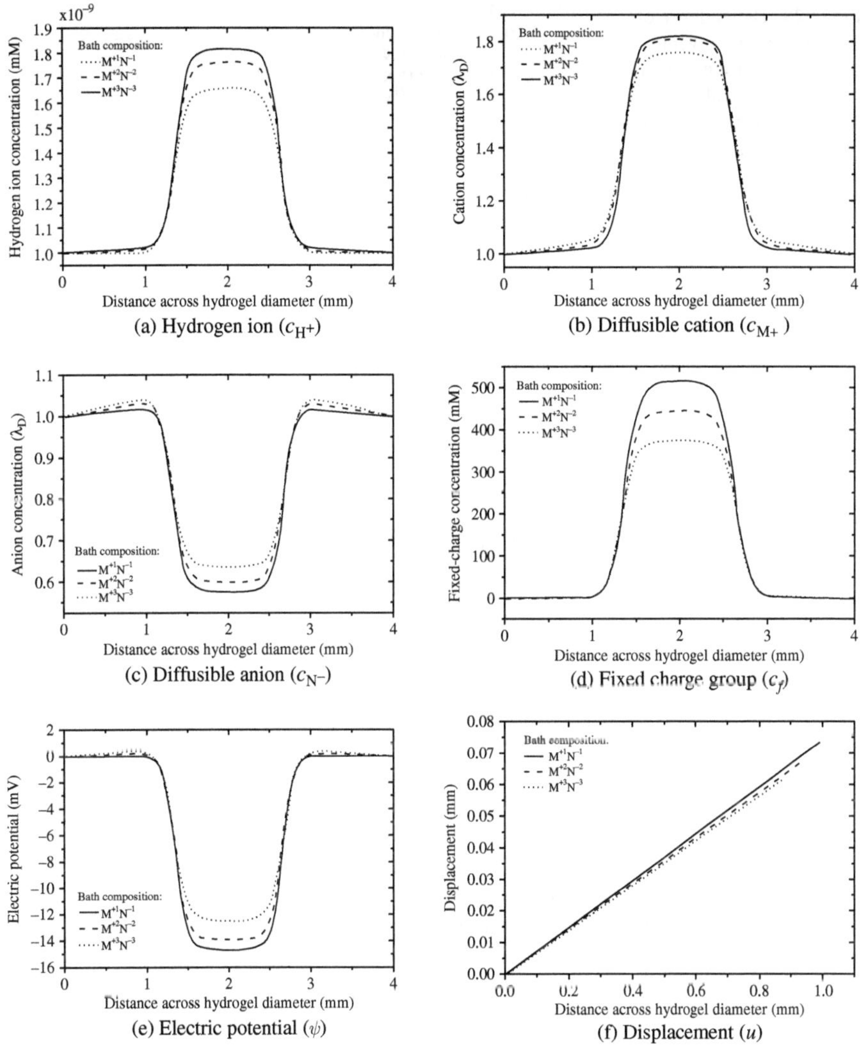

Fig. 2.27 Distributive profiles of $c_{H^+}, c_{M^+}, c_{N^-}, c_f, \psi$ and u for specified solvent composition (monovalent, divalent and trivalent), where the PHEMA hydrogel is equilibrated in basic medium of pH 12

swelling of the hydrogel, if the influence of bath composition is studied with different ion valences and ionic strengths. Discrepancies in the profiles and the degrees of swelling are insignificant for acidic solution, as observed from Fig. 2.25. However, they show the trend similar to those for the cases with neutral and base solutions, as shown in Figs. 2.26 and 2.27. Specifically, the Donnan ratio λ_D or the concentrations of mobile ions within the hydrogel decreases as the ionic valence increases, and consequently the equilibrium swelling reduces as predicted. The distribution

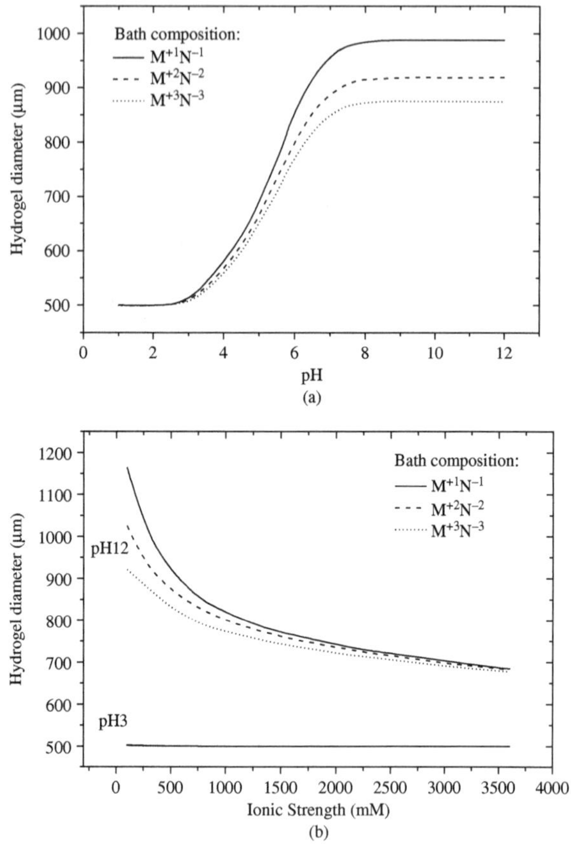

Fig. 2.28 Dependence of swelling on (**a**) solvent composition (monovalent, divalent and trivalent) as the function of bathing pH and (**b**) ionic strength of bath medium with different ionic valences

of the fixed charge concentration in Fig. 2.25d seems to be the same for all conditions, but actually the values of the concentrations are different from each other, namely 26.41631 mM for $M^{+3}N^{-3}$, 26.39451 mM for $M^{+2}N^{-2}$ and 26.38594 mM for $M^{+1}N^{-1}$. Therefore, the ionized groups attached on the hydrogel network tend to increase as the ionic valences of the solution increase due to the reduction of equilibrium swelling.

Figure 2.28a, b exhibits the influences of the ionic valences on the responsive equilibrium swelling of an acidic hydrogel at certain pH levels. It is seen that the solvent composition strongly influences the characteristics of swelling. This is consistent with the published experimental works (Siegel and Firestone, 1988; Siegel, 1990), in which as the salt (solvent) valence of the solution increases, the ion osmotic effect is expected to decrease significantly, because less amount of counterions diffuse into the hydrogel for neutralization of the charged groups. All the present discussions make the simulations in agreement with the experiment and they are also found in Fig. 2.28a, b, from which it is clearly known that the response of the hydrogel immersed in the solution with the larger ionic valence behaves the smaller deformation of the hydrogel. The effects of multivalent ionic composition of bath solution gradually fade out as the ionic concentration increases.

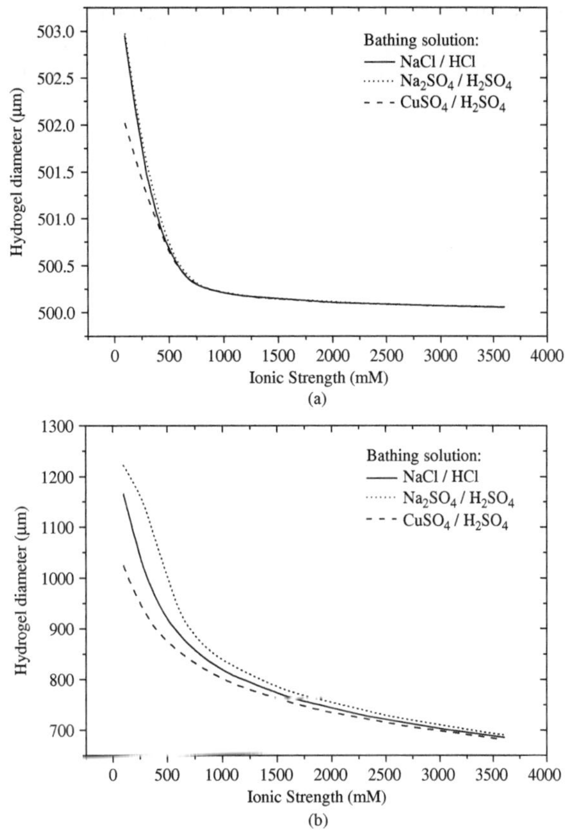

Fig. 2.29 Effects of ionic valences of bathing solution on swelling equilibrium as the function of ionic strength in (**a**) acidic medium of pH 3 and (**b**) basic medium of pH 9

A further comparison study is made between an asymmetrical salt (Na_2SO_4) solution and a symmetrical salt (NaCl and $CuSO_4$) solution, and demonstrated in Fig. 2.29 for both the acidic pH 3 and basic pH 9 mediums. The degree of changes in equilibrium swelling of the hydrogel in Na_2SO_4 solution is higher than that in NaCl for base environment, even though the counterion species is identical. This may be explained by the difference of the sodium ion (counterion) concentrations in the two bathing solutions. In order to maintain the electroneutrality and the ionic strength, the sodium ion concentration in Na_2SO_4 electrolyte solution is at a lower level. As a result, the pressure gradient between the interior hydrogel and the exterior solution tends to increase, leading to larger amount of swelling. On the contrary, the $CuSO_4$ solution shows the lowest degree of swelling. Since fewer divalent and monovalent counterions are required to neutralize the carboxylic acid groups, this causes relatively low concentration gradient and thus reduces the equilibrium swelling. On the other hand, the differences in the degree of swelling are insignificant until the ionic strength decreases below 600 mM as shown Fig. 2.29a. Therefore, the ionic valences still play some part in the mechanism of swelling but a secondary role as compared with the pH condition and ionic strength of the bathing solution (Siegel and

Firestone, 1988). The extent of swelling increases exponentially with the decrease of the ionic strength regardless of the bath contents at high pH. However, the asymmetrical electrolyte solution shows the inclination of swelling at low ionic strength with high pH.

2.6 Remarks

The modulating capability of absorbing or exuding the fluid for the smart hydrogel stimulated by surrounding environmental pH enables us to dynamically control the swelling/deswelling, and thereby achieve the effective diffusibility and permeability of the solutes and mechanical energy in the hydrogel. In addition, the presence of the electrostatic potential that is locally induced in the electrolytic solution by movement of all diffusive ionic species is an important phenomenon occurring in an ionic diffusion, but not in non-electrolyte species diffusion. In the ionic solution, the local electroneutrality is conserved everywhere. During the diffusion, all ions do not move at the same speed because different ionic species tend to diffuse at different rates. However, excessive charges contributed by the faster ions build up a local electric field, also called the diffusion potential, which slows down the faster ions and reciprocally accelerates the slower ionic species. Further, the diffusion potential should also be considered even if an external electrical field is applied to the system, by superimposing the diffusion potential upon the externally applied electrical field, as shown in Fig. 2.30.

The interactions between the hydrogel elastic polymer network and chemical medium strongly influence the degree of responsive swelling/deswelling of the smart hydrogel. The ionizable polymer fractions in the hydrogel are capable of associating and dissociating themselves, which characterizes the physicochemical properties of the hydrogels. When the hydrogel is immersed in a buffered solution, the electrolytic composition of the surrounding solution diffuses into the hydrogel and this determines the dissociation or association of the polyelectrolyte fraction of the hydrogels. Chemical reactions occur as a result of the reversible process of dissociation/association between the diffusive mobile ions and the ionizable groups attached on the hydrogel network, and subsequently redistribute the ionic concentrations within the hydrogel. The redistributions of ionic concentrations within the hydrogel generate both the electrostatic field and the osmotic pressure due to the difference of ionic concentrations between the hydrogel and surrounding environment. The osmotic pressure drives the expansion or contraction of the hydrogel. The swelling or shrinking subsequently redistributes the ion concentration of the interior hydrogel. The loop continues until equilibrium is achieved.

In this chapter, the formulated electrochemical and mechanical equations that are coupled together through the hydration are known collectively as the multi-effect-coupling pH-stimulus (MECpH) model for simulation of the responsive characteristics of the multiphase pH-sensitive hydrogel and surrounding solution. The MECpH model does have the following advantages that make this model more attractive than other models in certain aspects.

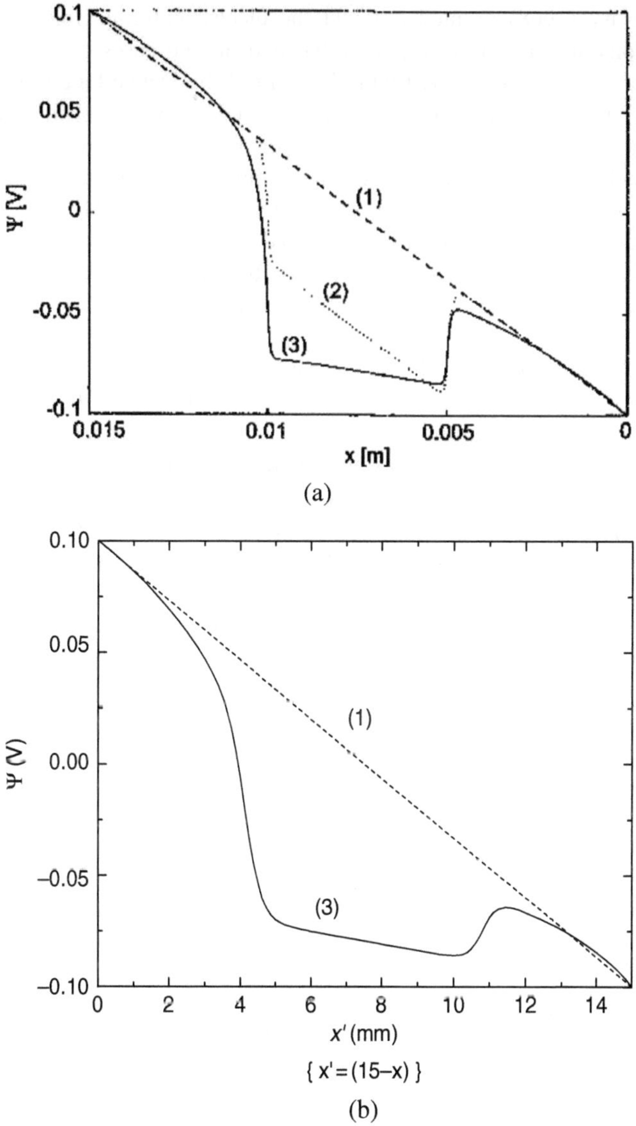

Fig. 2.30 Comparison of electrical potential in the hydrogel and bathing solution due to externally applied electric field between (**a**) stabilized space–time FEM (Wallmersperger, 2001a) and (**b**) Hermite-cloud meshless methods (Li et al., 2003)

- Computational domains of interest cover both the interior hydrogel and the exterior bathing solution. The model is thus able to predict the distributions of concentrations of all diffusive species, electric potential in both the hydrogel and surrounding solution simultaneously.

- MacGillivary (1968) pointed out that the electroneutrality and constant field assumptions are in fact nothing new but two limited cases of a dimensionless parameter, which is associated with the ratio of the Debye length to membrane thickness. The Poisson equation provides a more robust approach to account for the effect of electrical potential coupled with various ionic fluxes.
- Theoretically, the distribution of the fixed charge density is a function of the material properties and hydration of the hydrogel. In the MECpH model, the Langmuir adsorption isotherm is introduced to derive the fixed charge concentration, expressed by a function of the hydration, the diffusive hydrogen ion H^+ concentration and the concentration of fixed charge groups per polymer network volume.
- The MECpH model can easily handle the large deformation of the pH-sensitive hydrogels, based on the geometrically nonlinear finite deformation theory. This is another advantage of the model since the pH-sensitive hydrogels usually undergo very large displacement due to the chemo-electro-mechanical multi-energy coupled effects, especially at higher pH levels of environmental solutions.
- The MECpH model can also easily incorporate multiple ionic species, unlike other models where only two monovalent ion species, an anion and a cation, are considered.
- The MECpH model is formulated in very elegant form, in which several key effects on the responsive behaviours of the pH-sensitive hydrogel and surrounding solution are expressed in a straightforward manner. The model is readily applicable for numerical implementation as compared with other models.

We believe that the pH-sensitive hydrogel, like ionic exchanger found in natural charged membranes, contains substantial ionizable groups capable of dissociating and subsequently achieving net charge due to buffered medium. These groups are capable of ionizing as a function of the electrolyte pH and the ionic strength and thereby producing positive or negative charges fixed onto the polymeric network chains.

The pendent charges fixed on the backbone of the polymer network of the hydrogels, e.g. carboxylic group, exist in the form of R–COO^- in basic solution and in the form of R–COOH in acid medium as mentioned above. When the pH of surrounding medium is higher than the pK_a of the weakly acid group bound to the polymeric network, the chemical reactions proceed to the forward direction. As a result, the hydrogel achieves higher fixed charge density. In order to maintain the electroneutrality within the hydrogel, more mobile counterions (e.g. Na^+if the sodium chloride electrolyte solution is added to adjust the ionic strength) diffuse into the interior hydrogel to compensate the surplus charges. In vice versa, the mobile ions with the same sign of the fixed charge groups are repulsed from entering the interior hydrogel. However, besides those compensating counterions, there are also the absorbed counterions which are accompanied by equivalent amount of co-ions (Helfferich, 1962). By understanding of the mechanism of the distributive ionic concentrations in the interior hydrogel, it is clearly known that the concentration of the Na^+ ion is higher whereas that of Cl^- ion is lower for alkaline solution if compared with acidic solution.

The uptake or sorption of mobile ions essentially redistributes the mobile ions between the two liquid states, the interstitial liquid phase within the hydrogel and the bathing solution, until equilibrium is attained. As a result, the concentration difference increases tremendously between the interior hydrogel and the exterior solution, leading to higher osmotic pressure which drives larger degree of swelling. In the meantime, the elastic retractive force of the polymer network balances with the expanding network. The interacting process carries on forward and backward until equilibrium state is achieved.

Theoretical simulation reveals that the swelling behaviour of the HEMA hydrogels may be divided into three states: (1) the insignificant swelling because of the compact and hydrophobic states at pH values lower than 4 regardless of monomer composition, (2) the ionization takes place actively if the pH ranges from 4 to 7 where the fluid content within the hydrogels increases abruptly and gives rise to highly swollen hydrogel and (3) the binding sites of charge groups are saturated at environmental pH higher than pH 7 and the hydrogels do not further expand significantly even the surrounding pH increases continuously.

There are also secondary parameters influencing the expansion and contraction of the charged crosslinked hydrogels. In addition to the pH sensitivity, swelling and deswelling are also dependent on physical and chemical properties of the smart hydrogel as well as the ionic strength and composition of surrounding medium. These observations are all consistent with experimental phenomena reported in open literature.

Similar phenomena are observed for analysis of response characteristics of ionic species and electrical potential when the smart hydrogel is exposed to the phosphate and Britton–Robinson buffer systems, as compared with HCl/NaCl solution. Quantitatively, the degree of swelling is almost the same between the phosphate buffer and HCl/NaCl solution, where the pH-sensitive hydrogel is placed. In another study, the equilibrium swelling of the smart hydrogel in the Britton–Robinson buffer solution is predicted to be always larger than that bathed in the phosphate buffer system and HCl/NaCl solution. Probably one of the reasons is that the Britton–Robinson buffer system has lower ionic strength than those of the phosphate buffer system and HCl/NaCl solution. However, the identity of buffer ions may also play an important role.

References

D.J. Beebe, J.S. Moore, J.M. Bauer, Q. Yu, R.H. Liu, C. Devadoss, B.H. Jo. (2000a). Functional hydrogel structures for autonomous flow control inside microfluidic channels. *Nature*, 404, 588–590.

D.J. Beebe, J.S. Moore, Q. Yu, H. Liu, M.L. Kraft, B.H. Jo, C. Devadoss. (2000b). Microfluidic tectonics: A comprehensive construction platform for microfluidic systems. *Proceedings of the National Academy of Sciences of the United States of America*, 97, 13488–13493.

J.O'M. Bockris, B.E. Conway, E. Yeager (Eds.) (1983). *Comprehensive Treatise of Electrochemistry*, Vol. 6, *Electrodics: Transport*, New York: Plenum Press.

J.O.M. Bockris, K.N. Reddy-Amulya. (1998). *Modern Electrochemistry: Ionics*, 2nd edn. New York: Plenum Press.

L. Brannon-Peppas, N.L. Peppas, (1991). Equilibrium swelling behavior of pH-sensitive hydrogels. *Chemical Engineering Science*, 46, 715–722.

H. Brondsted, J. Kopecek. (1992). pH-Sensitive Hydrogels: Characteristics and Potential in Drug Delivery. In: *Polyelectrolyte Gels: Properties Preparation and Applications*, ACS Symposium Series 480, R.S. Harland, R.K. Prud'homme (Eds.) Washington DC: American Chemical Society, pp. 285–304.

L.D. Carnay, I. Tasaki. (1971). Ion Exchange Properties and Excitability of the Squid Giant Axon. In: *Biophysics and Physiology of Excitable Membranes*, W.J. Adelman Jr. (Ed.) New York: Van Nostrand Reinhold Co, pp. 379–422.

Y. Chu, P.P. Varanasi, M.J. McGlade, S. Varanasi. (1995). pH-induced swelling kinetics of polyelectrolyte hydrogels. *Journal of Applied Polymer Science*, 58, 2161–2176.

K. Cooper, E. Jakobsson, P. Wolynes. (1985). The Theory of ion transport through membrane channels. *Progress in Biophysics and Molecular Biology*, 46, 51–96.

E.L. Cussler. (1997). *Diffusion Mass Transfer in Fluid System*, 2nd edn. Cambridge: Cambridge University Press.

D. De Rossi, P. Parrini, P. Chiarelli, G. Buzzigoli. (1985). Electrically induced contractile phenomena in charged polymer networks: Preliminary study on the feasibility of muscle-like structures. *Transactions of the American Society for Artificial Internal Organs*, XXXI, 60–65.

M. Doi, M. Matsumoto, Y. Hirose. (1992). Deformation of ionic polymer gels by electric fields. *Macromolecules*, 25, 5504–5511.

L. Dresner. (1972). Some remarks on the integration of extended Nernst–Planck equations in the hyperfiltration of multicomponent solution. *Desalination*, 10, 27–46.

R.S. Eisenberg. (1999). From structure to function in open ionic channel. *Journal of Membrane Biology*, 171, 1–24.

P.J. Flory. (1953). *Principles of Polymer Chemistry*, Ithaca, New York: Cornell University Press.

A. Fragala, J. Enos, A. LaConti, J. Boyack. (1972). Electrochemical activation of a synthetic artificial muscle membrane. *Electrochimica Acta*, 17, 1507–1522.

S.H. Gehrke, E.L. Cussler. (1989). Mass transfer in pH-sensitive hydrogels. *Chemical Engineering Science*, 44, 559–566.

D. Gillespie, R.S. Eisenberg. (2001). Modified Donnan potentials for ion transport through biological ion channels. *Physical Review E*, 63, 061902.

D. Gillespie, R.S. Eisenberg. (2002). Physical descriptions of experimental selectivity measurements in ion channels. *European Biophysics Journal*, 31, 454–466.

D.E. Goldman. (1943). Potential, impedance and rectification in membranes. *Journal of General Physiology*, 27, 37–60.

D.E. Goldman. (1971). Excitability Models. In: *Biophysics and Physiology of Excitable Membranes*, W.J. Adelman Jr. (Ed.) New York: Van Nostrand Reinhold Co, pp. 337–358.

P.E. Grimshaw. (1989). Electrical control of solute transport across polyelectrolyte membranes. Ph.D Thesis, Massachusetts Institute of Technology.

P.E. Grimshaw, J.H. Nussbaum, A.J. Grodzinsky. (1990). Kinetics of electricity and chemically induced swelling in polyelectrolyte gels. *Journal of Chemical Physics*, 93, 4462–4472.

A.J. Grodzinsky. (1974). Electromechanics of deformable polyelectrolyte membranes. Sc.D Thesis, Massachusetts Institute of Technology.

R.W. Gulch, J. Holdenried, A. Weible, T. Wallmersperger, B. Kroplin. (2000). Polyelectrolyte Gels in Electric Fields: A Theoretical and Experimental Approach. In: *Smart Structures and Materials 2000: Electroactive Polymer Actuators and Devices, Proceedings of the SPIE 3987*, Y. Bar-Cohen (Ed.) Bellingham, Washington: SPIE Press, pp. 193–202.

L. Guldbrand, B. Jonsson, H. Wennerstrom, P. Linse. (1984). Electrical double layer forces: A Monte Carlo study. *Journal of Chemical Physics*, 80, 2221–2228.

F. Helfferich. (1962). *Ion Exchange*, New York: McGraw-Hill.

A.L. Hodgkin, B. Katz. (1949). The effect of sodium ions on the electrical activity of the giant axon of the Squid. *The Journal of Physiology*, 108, 37–77.

M. Homma, Y. Seida, Y. Nakano. (2000). Evaluation of optimum condition for designing high-performance electro-driven polymer hydrogel systems. *Journal of Applied Polymer Science*, 75, 111–118.

Y. Hwang, F. Helfferich. (1987). Generalized model for multispecies ion-exchange kinetics including fast reversible reactions. *Reactive and Functional Polymers*, 5, 237–253.

B.D. Johnson, J.M. Bauer, D.J. Niedermaier, W.C. Crone, D.J. Beebe. (2004a). Experimental techniques for mechanical characterization of hydrogels at the microscale. *Experimental Mechanics*, 44, 21–28.

B.D. Johnson, D.J. Beebe, W.C. Crone. (2004b). Effects of swelling on the mechanical properties of a pH-sensitive hydrogel for use in microfluidic devices. *Materials Science and Engineering C: Biomimetic and Supramolecular Systems*, 24, 575–581.

B.D. Johnson, D.J. Niedermaier, W.C. Crone, J. Moorthy, D.J. Beebe. (2002). Mechanical properties of a pH sensitive hydrogel, *Proceedings of the 2002 Annual Conference of Society for Experimental Mechanics*, Milwaukee, Wisconsin.

A. Katchalsky. (1949). Rapid swelling and deswelling of reversible gels of polymeric acids by ionization. *Experientia*, 5, 319–320.

A. Katchalsky, P.F. Curran. (1965). *Nonequilibrium Thermodynamics in Biophysics*, Massachusetts: Harvard University Press.

M. Kato. (1995). Numerical analysis of the Nernst–Planck–Poisson system. *Journal of Theoretical Biology*, 177, 299–304.

M.G. Kurnikova, R.D. Coalson, P. Graft, A. Nitzan. (1999). A lattice relaxation algorithm for three-dimensional Poisson–Nernst–Planck theory with application to ion transport through the gramicidin a channel. *Biophysical Journal*, 76, 642–656.

W.M. Lai, J.S. Hou, V.C. Mow. (1991). A triphasic theory for the swelling and deformation behaviors of articular cartilage. *ASME Journal of Biomechanical Engineering*, 113, 245–258.

H. Li, T.Y. Ng, J.Q. Cheng, K.Y. Lam. (2003). Hermite-cloud: A novel true meshless method. *Computational Mechanics*, 33, 30–41.

D.R. Lide. (Ed.) (2002). *CRC Handbook of Chemistry and Physics*, 83rd edn. Boca Raton: CRC Press.

A.M. Lowman, N.A. Peppas. (1999). Hydrogels. In: *Encyclopedia of Controlled Drug Delivery*, E. Mathiowitz (Ed.) New York: Wiley, pp. 397–418.

A.D. MacGillivray. (1968). Nernst–Planck equation and the electroneutrality and Donnan equilibrium assumptions. *Journal of Chemical Physics*, 48, 2903–2907.

A.D. MacGillivray, D. Hare. (1969). Applicability of goldman's constant field assumption to biological systems. *Journal of Theoretical Biology*, 25, 113–126.

J. Malmivuo, R. Plonsey. (1995). *Bioelectromagnetism: Principles and Applications of Bioelectric and Biomagnetic Fields*, New York: Oxford University Press.

L.E. Malvern. (1969). *Introduction to the Mechanics of A Continuum Medium*, Englewood Cliffs, New Jersey: Prentice-Hall.

Y. Osada, J.P. Gong. (1993). Stimuli-responsive polymer gels and their application to chemomechanical systems. *Progress in Polymer Science*, 18, 187–226.

W.K. Panofsky, M. Phillips. (1964). *Classical Electricity and Magnetism*, 2nd edn. Reading, Massachusetts: Addison-Wesley.

N.A. Peppas, P. Bures, W. Leobandung, H. Ichikawa. (2000). Hydrogels in pharmaceutical formulations. *European Journal of Pharmaceutics and Biopharmaceutics*, 50, 27–46.

A. Redondo, R. LeSar. (2004). Modelling and simulation of biomaterial. *Annual Review of Materials Research*, 34, 279–314.

J. Ricka, T. Tanaka. (1984). Swelling of ionic gels: Quantitative performance of the Donnan theory. *Macromolecules*, 17, 2916–2921.

B. Roux, T. Allen, S. Berneche, W. Im. (2004). Theoretical and computational models of biological ion channels. *Quarterly Reviews of Biophysics*, 37, 15–103.

I. Rubinstein. (1990). *Electro-Diffusion of Ions SIAM Studies in Applied Mathematics*, Philadelphia: SIAM.
E. Samson, J. Marchand. (1999). Numerical solution of the extended Nernst–Planck model. *Journal of Colloid and Interface Science*, 215, 1–8.
E. Samson, J. Marchand, J.L. Robert, J.P. Bournazel. (1999). Modelling ion diffusion mechanisms in porous media. *International Journal for Numerical Methods in Engineering*, 46, 2043–2060.
S. Selberherr. (1984). *Analysis and Simulation of Semiconductor Devices*, New York: Springer.
M. Shibayama, T. Tanaka. (1993). Volume Phase Transition and Related Phenomena of Polymer Gels. In: *Responsive Gels: Volume Transitions I, Advances in Polymer Science* Vol. 109, K. Dusek (Ed.) Berlin: Springer-Verlag, pp. 1–62.
T. Shiga, Y. Hirose, A. Okada, T. Kurauchi. (1992a). Bending of poly(vinyl alcohol)-poly(sodium acrylate) composite hydrogel in electric fields. *Journal of Applied Polymer Science*, 44, 249–253.
T. Shiga, Y. Hirose, A. Okada, T. Kurauchi. (1992b). Electric field-associated deformation of polyelectrolyte gel near a phrase transition point. *Journal of Applied Polymer Science*, 46, 635–640.
R.A. Siegel. (1990). pH Sensitive Gels: Swelling Equilibria, Kinetics and Applications for Drug Delivery. In: *Pulse and Self-Regulated Drug Delivery*, J. Kost (Ed.) Boca Raton: CRC Press, pp. 129–155.
R.A. Siegel, B.A. Firestone. (1988). pH-dependent equilibrium swelling properties of hydrophobic polyelectrolyte copolymer gels. *Macromolecules*, 21, 3254–3259.
R.A. Siegel, B.A. Firestone, J. Cornejo-Bravo, B. Schwarz. (1991). Hydrophobic Weak Polybasic Gels: Factors Controlling Swelling Equilibrium. In: *Polymer Gels: Fundamental and Biomedical Applications*, D. DeRossi, K. Kajiwara, Y. Osada, A. Yamauchi (Eds.) New York: Plenum Press, pp. 309–317.
R.A. Sjodin. (1971). Ion Transport across Excitable Cell Membranes. In: *Biophysics and Physiology of Excitable Membranes*, W.J. Adelman Jr. (Ed.) New York: Van Nostrand Reinhold Co, pp. 96–124.
A. Syganow, E. von Kitzing. (1999). The drift approximation solves the Poisson, Nernst–Planck, and continuum equation in the limit of large external voltages. *European Biophysics Journal*, 28, 393–414.
T. Tanaka, D. Fillmore, S.T. Sun, I. Nishio, G. Swislow, A. Shah. (1980). Phase transition in ionic gels. *Physical Review Letters*, 45, 1636–1639.
T. Teorell. (1953). Transport processes and electrical phenomena in ionic membranes. *Progress in Biophysics & Molecular Biology*, 3, 305–369.
A. Townshend. Ed. (1995). *Encyclopedia of Analytical Science*, Vol. 1 (A-Che), London: Academic Press.
T. Wallmersperger, B. Kroeplin. (2001). Modelling and Analysis of the Chemistry and Electromechanics. In: *Electroactive Polymer Actuators as Artificial Muscles*, Y. Bar-Cohen (Ed.) SPIE Press, pp. 285–307.
H.H. Woodson, J.R. Melcher. (1968). *Electromechanical Dynamics Part I: Discrete Systems*, New York: John Wiley and Sons.
Q. Yu, J.M. Bauer, J.S. Moore, D.J. Beebe. (2001). Responsive biomimetic hydrogel valve for microfluidics. *Applied Physics Letters*, 78, 2589–2591.
B. Zhao, J.S. Moore. (2001). Fast pH- and ionic strength-responsive hydrogels in microchannels. *Langmuir*, 17, 4758–4763.

Chapter 3
Multi-Effect-Coupling Electric-Stimulus (MECe) Model for Electric-Sensitive Hydrogel

3.1 Introduction

In this chapter, the two models previously published are reviewed first for the responsive hydrogels, which providing the basis for the present multi-effect-coupling electric-stimulus (MECe) model. It is followed by development of the MECe model, in which four important governing equations are formulated to characterize the diffusive ion concentrations, the electric potential, the interstitial fluid pressure and the deformation of hydrogel, respectively. The non-dimensional form of the MECe governing equations is then presented and the boundary and initial conditions are proposed accordingly. After validation of the MECe model by comparison with the experimental data published in open literature, the steady-state simulation is conducted for equilibrium analysis of the electric stimulus-responsive hydrogel and the transient simulation for kinetics analysis of the smart hydrogel.

3.2 Development of the MECe Model

As well known, the fixed charge groups attached onto the polymer network chains of the hydrogel attract the diffusive electro-opposite ions from the surrounding solution to maintain the electroneutrality, when the hydrogel is immersed into a bath solution subject to an externally applied electric field. Meanwhile, the external electric field also drives the mobile ions moving towards electro-opposite electrodes. These two effects result in the difference of the diffusive ionic concentrations between the interior hydrogel and the exterior solution, and thus induce the fluid pressure. As the main driving source, the fluid pressure makes the hydrogel deformed, and then the deformation of the hydrogel causes the redistribution of the fixed charge groups. The mobile ions in the solution will diffuse further and redistribute again due to the change of the distributive fixed charge density. This cycle process continues until the hydrogel mixture reaches the equilibrium.

For simulation of the performance of the hydrogels responding to the applied electric voltage, several models were published. Two of them are important and reviewed here because they provide the basis for development of the present MECe model. The first model called the triphase model was proposed by Hon et al. (1999),

H. Li, *Smart Hydrogel Modelling*, DOI 10.1007/978-3-642-02368-2_3,

Zhou et al. (2002). It is based on the generalized law of thermodynamics for an irreversible thermodynamic system and the triphase mixture theory of Lai et al. (1991), composed of coupled governing equations, in which several disadvantages are found. For example, the computational domain of this model is limited to the region of the hydrogel only, excluding the surrounding solution, since the governing equations are derived from the classical thermodynamics. It is thus unable to provide the complete information of diffusive ionic concentration distributions. In addition, the electroneutrality condition is required and the electric potential is not a field variable in the governing equation so that no electric potential distributing over the exterior solution is simulated. The second model is a multi-field formulation proposed by Wallmersperger et al. (2004), in which the diffusion–convection equations, Poisson equation and the motion equation are coupled together to describe the chemical, electric and mechanical fields. However, the effect of the fixed charge density is not considered, although the computational domain covers both the hydrogel and external bath solution. Additionally, Newton's second law is employed directly, which is far away from the complicated mechanical behaviours of the multiphase hydrogels.

3.2.1 Formulation of the MECe Governing Equations

In order to overcome the drawbacks of the above models, a novel multiphysics model is developed mathematically in this chapter, called the multi-effect-coupling electric-stimulus (MECe) model, which is based on the work done by Hon et al. (1999). Compared with the triphase model of Hon et al. (1999), the presently developed MECe model defines a computational domain as covering both the hydrogel and the surrounding solution. The electric potential is also considered in the nonlinear coupled partial differential governing equations. By the multi-field formulation, the MECe model characterizes the mechanical deformation of the hydrogels more accurately and explicitly, making it more convenient for steady and transient simulations. As such, the multiphysics MECe model is a more precise mathematical formulation with capability of providing more reliable simulation of the multiphase smart hydrogel responding to externally applied electric voltage.

It is assumed in the MECe model that the mixture of the hydrogel is composed of the three phases, the polymeric network matrix solid phase denoted by superscript s, the interstitial water/fluid phase by w and the mobile ion phase by k ($k = 1,2,\ldots,N_{\text{ion}}$) where N_{ion} is the number of total mobile ion species. If ϕ^{α} ($\alpha = s, w, k$) is defined as the volume fraction of the phase α

$$\phi^{\alpha} = \frac{dV^{\alpha}}{dV} \quad (\alpha = s, w, k) \tag{3.1}$$

where V^{α} is the true volume of the phase α and V is the volume of the hydrogel mixture; the saturation condition of the hydrogel mixture can be written as follows:

$$\sum_{\alpha=s,w,k} \phi^{\alpha} = 1 \tag{3.2}$$

If infinitesimal deformation of the mixture occurs, the apparent volume ratio of the polymeric network matrix solid phase is given by

$$J = \frac{dV_0}{dV} = \frac{1}{1 + tr(E)} \tag{3.3}$$

in which V_0 is the volume of hydrogel mixture at reference configuration and $\boldsymbol{E}$ the elastic strain vector of the polymeric solid phase. The volume fraction of the polymeric network matrix solid phase is thus derived as

$$\phi^s = \frac{dV^s}{dV} = \frac{dV^s}{dV_0} \cdot \frac{dV_0}{dV} = \phi_0^s J = \frac{\phi_0^s}{(1 + tr(E))} \tag{3.4}$$

where ϕ_0^s is the volume fraction of the polymeric solid phase at reference configuration. Due to the extremely small volume of the ion phase, it is reasonable to neglect ϕ^k when compared with ϕ^s and ϕ^w. The volume fraction of the water/fluid phase is expressed as

$$\phi^w \approx 1 - \phi^s = 1 - \frac{\phi_0^s}{(1 + tr(E))} \tag{3.5}$$

If chemical reaction is negligible, each phase should follow the law of mass conservation

$$\frac{\partial \rho^\alpha}{\partial t} + \nabla \cdot (\rho^\alpha v^\alpha) = 0 \ (\alpha = s, w, k) \tag{3.6}$$

where $\boldsymbol{v}^\alpha (\alpha = s,\ w,\ k)$ is the velocity of the phase α and $\rho^\alpha (\alpha = s,\ w,\ k)$ is the apparent mass density of the phase α. It is known that the apparent mass density ρ^α can be expressed by the respective true mass density ρ_T^α, namely $\rho^\alpha = \rho_T^\alpha \phi^\alpha (\alpha = s, w, k)$. Meanwhile, on the basis of incompressibility restriction, ρ_T^α may be assumed reasonably to be constant. Equation (3.6) is then rewritten as

$$\frac{\partial \phi^\alpha}{\partial t} + \nabla \cdot (\phi^\alpha v^\alpha) = 0 \ (\alpha = s, w, k) \tag{3.7}$$

Considering Eqs. (3.2) and (3.7), the continuity condition of the mixture of the hydrogel is obtained as

$$\nabla \cdot \left(\sum_{\alpha = s,w,k} \phi^\alpha v^\alpha \right) = 0 \tag{3.8}$$

By the tensor analysis, one can have

$$\nabla \cdot (\phi^\alpha v^\alpha) = \phi^\alpha \nabla \cdot v^\alpha + v^\alpha \cdot \nabla \phi^\alpha = \phi^\alpha \mathrm{I}{:}\nabla v^\alpha + v^\alpha \cdot \nabla \phi^\alpha \tag{3.9}$$

Equation (3.8) is thus rewritten as

$$\sum_{\alpha = s,w,k} (\phi^\alpha \mathrm{I}{:}\nabla v^\alpha + v^\alpha \cdot \nabla \phi^\alpha) = 0 \tag{3.10}$$

In addition, the rate of kinetic energy of the hydrogel mixture is given by

$$\dot{K} = \sum_{\alpha=s,w,k} \dot{K}^{\alpha} = \int_V \left(\sum_{\alpha=s,w,k} v^{\alpha} \cdot \dot{v}^{\alpha} \rho^{\alpha} \right) dV \tag{3.11}$$

If the internal energy of the hydrogel mixture U can be expressed by the Helmholtz energy function F as

$$U = F + TS \tag{3.12}$$

the rate of internal energy $\dot{U}$ is written as follows:

$$\dot{U} = \dot{F} + (\dot{T}S + T\dot{S}) = \int_V \sum_{\alpha=s,w,k} (\dot{\Gamma}^{\alpha} + \dot{T}\eta^{\alpha} + T\dot{\eta}^{\alpha})\rho^{\alpha} dV \tag{3.13}$$

where T is the absolute temperature, S the entropy of the system, Γ^{α} and η^{α} are the density of Helmholtz energy and the entropy per unit mass for the phase α, respectively.

It is noted that both the internal energy U and the Helmholtz energy F are the state functions depending on their state variables, such as the absolute temperature T, the elastic strain tensor $\boldsymbol{E}$, the apparent densities ρ^{α} and the fixed charge density c^f. Based on such constitutive consideration, the Helmholtz energy density is expressed by

$$\Gamma^{\alpha} = \Gamma^{\alpha}(T, E, \rho^s, \rho^w, \rho^k, c^f) \ (\alpha = s, w, k) \tag{3.14}$$

The rate of Helmholtz energy density is thus derived as

$$\dot{\Gamma}^{\alpha} = \frac{\partial \Gamma^{\alpha}}{\partial T}\dot{T} + \frac{\partial \Gamma^{\alpha}}{\partial E}\dot{E} + \sum_{\beta=s,w,k} \frac{\partial \Gamma^{\alpha}}{\partial \rho^{\beta}}\dot{\rho}^{\beta} + \frac{\partial \Gamma^{\alpha}}{\partial c^f}\dot{c}^f \ (\alpha = s, w, k) \tag{3.15}$$

From Eq. (3.9) one can know

$$\dot{\rho}^{\beta} \equiv \frac{D\rho^{\beta}}{Dt} = -\rho^{\beta} \mathrm{I}{:}\nabla v^{\beta} \tag{3.16}$$

and with the fact $\dot{J} = -J\nabla \cdot v^s$, we have

$$\dot{c}^f = -c_0^f J \mathrm{I}{:}\nabla v^s \tag{3.17}$$

Considering the relation $\dot{E} = (F^s)^T \cdot \nabla v^s \cdot F^s$, one can obtain

$$\frac{\partial \Gamma^{\alpha}}{\partial E}{:}\dot{E} = \frac{\partial \Gamma^{\alpha}}{\partial E}{:}((F^s)^T \cdot \nabla v^s \cdot F^s) = \left(F^s \cdot \frac{\partial \Gamma^{\alpha}}{\partial E} \cdot (F^s)^T \right){:}\nabla v^s \tag{3.18}$$

where $\boldsymbol{F}^s$ and $(\boldsymbol{F}^s)^T$ are the deformation gradient tensor and its transpose, respectively.

By substituting Eqs. (3.16), (3.17) and (3.18) into Eq. (3.15), we have

$$\dot{\Gamma}^\alpha = \frac{\partial \Gamma^\alpha}{\partial T}\dot{T} + \left(F^s \cdot \frac{\partial \Gamma^\alpha}{\partial E} \cdot (F^s)^T - c_0^f J \frac{\partial \Gamma^\alpha}{\partial c^f}\mathrm{I}\right) : \nabla v^s - \sum_{\beta=s,w,k} \rho^\beta \frac{\partial \Gamma^\alpha}{\partial \rho^\beta}\mathrm{I} : \nabla v^\beta \quad (3.19)$$

Substituting Eq. (3.19) into (3.13), we have

$$\dot{U} = \int_V \sum_{\alpha=s,w,k} \left(\frac{\partial \Gamma^\alpha}{\partial T}\dot{T} + \left(F^s \cdot \frac{\partial \Gamma^\alpha}{\partial E} \cdot (F^s)^T - c_0^f J \frac{\partial \Gamma^\alpha}{\partial c^f}\mathrm{I}\right) : \nabla v^s - \right.$$
$$\left. - \sum_{\beta=s,w,k} \rho^\beta \frac{\partial \Gamma^\alpha}{\partial \rho^\beta}\mathrm{I} : \nabla v^\beta + \dot{T}\eta^\alpha + T\dot{\eta}^\alpha \right) \rho^\alpha dV \quad (3.20)$$

Rate of total work $\dot{W}$ consists of two portions, the rate of work done by external forces $\dot{W}_e$ and the rate of work done by pressure $\dot{W}_p$

$$\dot{W} = \dot{W}_e + \dot{W}_p \quad (3.21)$$

The rate of work done by external forces $\dot{W}_e$ is defined as

$$\dot{W}_e = \sum_{\alpha=s,w,k} \left(\int_V \rho^\alpha f^\alpha \cdot v^\alpha dV + \int_S t^\alpha \cdot v^\alpha dS \right) \quad (3.22)$$

where $\mathbf{f}^\alpha$ is the body force per unit mass of the phase α, $t^\alpha = \sigma^\alpha \cdot v$ is the drag force applied on the surface, $\boldsymbol{\sigma}^\alpha$ the stress tensor of the phase α and $\mathbf{v}$ the external normal on the surface. Due to the symmetry of stress tensor and the Gaussian gradient formula, we have the following transformation:

$$\int_S t^\alpha \cdot v^\alpha dS = \int_S (\sigma^\alpha \cdot v^\alpha) \cdot v dS = \int_V \nabla \cdot (\sigma^\alpha \cdot v^\alpha) dV$$
$$= \int_V ((\nabla \cdot \sigma^\alpha) \cdot v^\alpha + \sigma^\alpha : \nabla v^\alpha) dV \quad (3.23)$$

Equation (3.22) is thus rewritten as

$$\dot{W}_e = \sum_{\alpha=s,w,k} \int_V ((\rho^\alpha f^\alpha + \nabla \cdot \sigma^\alpha) \cdot v^\alpha + \sigma^\alpha : \nabla v^\alpha) dV \quad (3.24)$$

The rate of work done by pressure $\dot{W}_p$ is defined as

$$\dot{W}_p = -p\dot{V} \quad (3.25)$$

Considering that the continuity equation (3.10) is adopted at the incompressible condition $\dot{V} = 0$, one can get

$$\dot{W}_p = \int_V -p \sum_{\alpha=s,w,k} (\phi^\alpha \mathrm{I}:\nabla v^\alpha + v^\alpha \cdot \nabla \phi^\alpha) dV \tag{3.26}$$

Substituting Eqs. (3.24) and (3.26) into Eq. (3.21), we obtain

$$\dot{W} = \sum_{\alpha=s,w,k} \int_V ((\rho^\alpha f^\alpha + \nabla \cdot \sigma^\alpha - p\nabla\phi^\alpha) \cdot v^\alpha + (\sigma^\alpha - p\phi^\alpha I):\nabla v^\alpha) dV \tag{3.27}$$

Rate of the heat transferred into the mixture is defined as

$$\dot{Q} = \sum_{\alpha=s,w,k} \left(\int_V \rho^\alpha \gamma^\alpha dV - \int_S q^\alpha \cdot v dS \right) \tag{3.28}$$

where γ^α is the rate of heat generation per unit mass of the phaseα and $\boldsymbol{q}^\alpha$ the heat flux vector. Similarly, by the Gaussian gradient formula we have

$$\dot{Q} = -\sum_{\alpha=s,w,k} \int_V (\nabla \cdot q^\alpha - \rho^\alpha \gamma^\alpha) dV \tag{3.29}$$

Rate of energy dissipation $\dot{D}$ is defined as

$$\dot{D} = \sum_{\alpha=s,w,k} \int_V \Pi^\alpha \cdot v^\alpha dV \tag{3.30}$$

where $\prod^\alpha$ is the diffusive momentum exchange between the phases, and it is a physical parameter indicating the diffusive resistance to the relative flow between the two phases. $\prod^\alpha$ can thus be expressed by their relative velocities as follows:

$$\Pi^\alpha = \sum_{\beta=s,w,k} f_{\alpha\beta}(v^\beta - v^\alpha) \tag{3.31}$$

where $f_{\alpha\beta}$ is the diffusive drag coefficient between the phases α and β (or constituents) and $f_{\alpha\beta}$=$f_{\beta\alpha}$. Obviously, $\prod^\alpha$ satisfies the following condition:

$$\sum_{\alpha=s,w,k} \Pi^\alpha = 0 \tag{3.32}$$

Based on the first law of thermodynamics, we have the following relation of energy conservation for the irreversible thermodynamic system:

$$\dot{K} + \dot{U} - \dot{D} = \dot{W} + \dot{Q} \tag{3.33}$$

With substitution of Eqs. (3.11), (3.20), (3.27), (3.29) and (3.30) into Eq. (3.33), we obtain

$$
\begin{aligned}
&-\int_V \sum_{\alpha=s,w,k} (\nabla \cdot \sigma^\alpha + \rho^\alpha f^\alpha - \rho^\alpha \dot{v}^\alpha + \Pi^\alpha + p\nabla\phi^\alpha) \cdot v^\alpha dV \\
&+\int_V \left\{ \left[F^s \cdot \left(\sum_{\alpha=s,w,k} \frac{\partial \Gamma^\alpha}{\partial E} \right) \cdot (F^s)^T - c_0^f J \left(\sum_{\alpha=s,w,k} \frac{\partial \Gamma^\alpha}{\partial c^f} \right) \mathrm{I} \right] : \nabla v^s \right. \\
&\left. + \sum_{\alpha=s,w,k} \left[-\sigma^\alpha - p\phi^\alpha \mathrm{I} - \left(\sum_{\beta=s,w,k} \rho^\alpha \frac{\partial \Gamma^\beta}{\partial \rho^\alpha} \right) \mathrm{I} \right] : \nabla v^\alpha \right\} dV \\
&+\int_V \sum_{\alpha=s,w,k} (\nabla \cdot q^\alpha + T\rho^\alpha \dot{\eta}^\alpha - \rho^\alpha \gamma^\alpha) dV + \int_V \sum_{\alpha=s,w,k} \left(\frac{\partial \Gamma^\alpha}{\partial T} + \rho^\alpha \eta^\alpha \right) \dot{T} dV = 0
\end{aligned}
\tag{3.34}
$$

If the chemical potential is defined as

$$\mu^\alpha = \frac{\partial \Gamma}{\partial \rho^\alpha} \tag{3.35}$$

the chemical term in Eq. (3.34) becomes

$$\sum_{\beta=s,w,k} \frac{\partial \Gamma^\beta}{\partial \rho^\alpha} \rho^\alpha = \rho^\alpha \frac{\partial (\sum_\beta \Gamma^\beta)}{\partial \rho^\alpha} = \rho^\alpha \frac{\partial \Gamma}{\partial \rho^\alpha} = \rho^\alpha \mu^\alpha \tag{3.36}$$

The mechanical term in Eq. (3.34) can be expressed by the second Piola–Kirchhoff stress tensor τ_E^s and the Cauchy stress tensor σ_E^s, which are defined as follows:

$$\tau_E^s = \frac{\partial \Gamma}{\partial E} = \sum_{\alpha=s,w,k} \frac{\partial \Gamma^\alpha}{\partial E} \tag{3.37}$$

$$\sigma_E^s = F^s \cdot \tau_E^s \cdot (F^s)^T \tag{3.38}$$

In order to simplify Eq. (3.34), a chemical expansion stress T_C is introduced to replace the complicated expression

$$T_C = c_0^f J \left(\sum_{\alpha=s,w,k} \rho^\alpha \frac{\partial \Gamma^\alpha}{\partial \rho^\alpha} \right) \tag{3.39}$$

By substituting Eqs. (3.36), (3.37), (3.38) and (3.39) into Eq. (3.34), a more complete formulation of Eq. (3.33) is obtained as

$$
\begin{aligned}
&-\int_V \sum_{\alpha=s,w,k} (\nabla \cdot \sigma^\alpha + \rho^\alpha f^\alpha - \rho^\alpha \dot{v}^\alpha + \Pi^\alpha + p\nabla\phi^\alpha) \cdot v^\alpha dV \\
&+\int_V \left\{ (\sigma_E^s - T_C \mathrm{I}):\nabla v^s + \sum_{\alpha=s,w,k} \left[-\sigma^\alpha - p\phi^\alpha \mathrm{I} - \left(\sum_{\beta=s,w,k} \rho^\alpha \frac{\partial \Gamma^\beta}{\partial \rho^\alpha} \right) \mathrm{I} \right] : \nabla v^\alpha \right\} dV \\
&+\int_V \sum_{\alpha=s,w,k} (\nabla \cdot q^\alpha + T\rho^\alpha \dot{\eta}^\alpha - \rho^\alpha \gamma^\alpha) dV + \int_V \sum_{\alpha=s,w,k} \left(\frac{\partial \Gamma^\alpha}{\partial T} + \rho^\alpha \eta^\alpha \right) \dot{T} dV = 0
\end{aligned}
\tag{3.40}
$$

Due to the independence of the variables $\boldsymbol{v}^\alpha$, $\nabla \boldsymbol{v}^\alpha$ and $\dot{T}$ satisfying Eq. (3.40), the following equations are achieved:

Momentum equations:

$$\nabla \cdot \sigma^\alpha + \rho^\alpha f^\alpha - \rho^\alpha \dot{v}^\alpha + \Pi^\alpha + p\nabla\phi^\alpha = 0 \ (\alpha = s, w, k) \tag{3.41}$$

Constitutive equations:

$$\sigma^s = -\phi^s p\mathrm{I} + \sigma_E^s - \rho^s \mu^s \mathrm{I} - T_C \mathrm{I} \tag{3.42}$$

$$\sigma^\alpha = -\phi^\alpha p\mathrm{I} - \rho^\alpha \mu^\alpha \mathrm{I} \ (\alpha = w, k) \tag{3.43}$$

Heat transfer equation:

$$\nabla \cdot q^\alpha + T\rho^\alpha \dot{\eta}^\alpha - \rho^\alpha \gamma^\alpha = 0 \ (\alpha = s, w, k) \tag{3.44}$$

By summation of Eq. (3.41), the momentum equation for the multiphase mixture of the hydrogel is written as

$$\nabla \cdot \sigma + \rho f - \rho \dot{v} = 0 \tag{3.45}$$

in which

$$\sigma = \sum_{\alpha=s,w,k} \sigma^\alpha, \ \mathrm{f} = \left(\sum_{\alpha=s,w,k} \rho^\alpha \mathrm{f}^\alpha \right) \Big/ \rho \ \text{and} \ v = \left(\sum_{\alpha=s,w,k} \rho^\alpha v^\alpha \right) / \rho$$

If the body force $\mathbf{f}$ and the inertial force $\rho\dot{v}$ are neglected, Eq. (3.45) is simplified to

$$\nabla \cdot \sigma = 0 \tag{3.46}$$

As well known, for an osmotic process at constant temperature, the relation between the chemical potential and the osmotic pressure p_{osm} can be derived by the Gibbs–Duhem equation

$$dp_{\text{osm}} = \sum_{\alpha=s,w,k} \rho^{\alpha} d\mu^{\alpha} \tag{3.47}$$

Integrating Eq. (3.47) we have

$$p_{\text{osm}} = \sum_{\alpha=s,w,k} \rho^{\alpha}(\mu^{\alpha} - \mu_0^{\alpha}) \tag{3.48}$$

By summation of Eqs. (3.42) and (3.43), the constitutive equation for the stress tensor of the hydrogel mixture is given as

$$\sigma = \sigma_E^s - (p + T_C)\mathrm{I} \tag{3.49}$$

where the total pressure p includes the osmotic pressure p_{osm}. If the chemical expansion stress T_C is neglected and the isotropic elastic material is assumed

$$\sigma_E^s = \lambda_s tr(E)\mathrm{I} + 2\mu_s E \tag{3.50}$$

in which λ_s and μ_s are Lamé coefficients of the polymer network solid matrix. Equation (3.49) is rewritten as

$$\sigma = -p\mathrm{I} + \lambda_s tr(E)\mathrm{I} + 2\mu_s E \tag{3.51}$$

By neglecting the body and inertial forces and considering Eqs. (3.41) and (3.43), the momentum equations of the water and ion phases in terms of their chemical potential are obtained as

$$\rho^{\alpha}\nabla\mu^{\alpha} - \Pi^{\alpha} = (\alpha = w, k) \tag{3.52}$$

Substituting Eq. (3.31) into (3.52), one can derive the momentum equations in terms of the chemical potential and the velocities as follows:

$$-\rho^w\nabla\mu^w + f_{ws}(v^s - v^w) + \sum_{k=1}^{N_{\text{ion}}} f_{wk}(v^k - v^w) = 0 \tag{3.53}$$

$$-\rho^k\nabla\mu^k + f_{ks}(v^s - v^k) + f_{kw}(v^w - v^k) + \sum_{j=1(j\neq k)}^{N_{\text{ion}}} f_{kj}(v^j - v^k) = 0 \tag{3.54}$$

Based on the work done by Lai et al. (1991), the following constitutive equations are obtained for the chemical potential of the water and ion phases:

$$\mu^w = \mu_0^w + \frac{1}{\rho_T^w}\left(p - RT\sum_{k=1}^{N_{\text{ion}}} \Phi^k c^k + B_w tr(E)\right) \tag{3.55}$$

$$\mu^k = \mu_0^k + \frac{RT}{M^k}\ln(\gamma_k c^k) + \frac{z^k F_c \psi}{M^k} \tag{3.56}$$

where μ_0^α ($\alpha = w, k$) is the chemical potential of the phase α at reference configuration. R is the universal gas constant, F_c Faraday constant, $\boldsymbol{B}_w$ the coupling coefficient, ψ the electric potential. Φ^k, c^k, γ_k, M^k and z^k are the osmotic coefficient, the concentration, the activity coefficient, the molar weight and the valance number for the kth ion species.

The works previously published by Hon et al. (1999) and Lai et al. (1991) have so far been summarized. On the basis of their work, the MECe model is developed as follows.

It is reasonably assumed that f_{sk} and f_{kj} are negligible in comparison with f_{ws} and f_{wk} for Eqs. (3.53) and (3.54). As such, the formulation for the momentum equation of the water/fluid phase is simplified to

$$f_{ws}(v^s - v^w) = \sum_{\alpha=w,k} \rho^\alpha \nabla \mu^\alpha \tag{3.57}$$

Substituting the constitutive relations of the hydrogel mixture and each phase expressed by Eqs. (3.51), (3.55) and (3.56) into the momentum equations (3.46) and (3.57), we have

$$\nabla \cdot (-p\mathrm{I} + \lambda_s tr(E)\mathrm{I} + 2\mu_s E) = 0 \tag{3.58}$$

$$f_{ws}(v^s - v^w) = \phi^w \left(\nabla p + RT\nabla \sum_k (1 - \Phi^k)c^k + F_c \nabla \psi \sum_k z^k c^k + B_w \nabla tr(E) \right) \tag{3.59}$$

With the assumption that ϕ^k is negligibly small, and by Eqs. (3.2) and (3.8), one can write

$$\nabla \cdot (\phi^w (v^s - v^w)) = \nabla \cdot v^s \tag{3.60}$$

Due to Eq. (3.60), Eq. (3.59) is rewritten as

$$\nabla \frac{\partial u}{\partial t} = \nabla \left[\frac{(\phi^w)^2}{f_{ws}} \left(\nabla p + RT\nabla \sum_k (1 - \Phi^k)c^k + F_c \sum_k z^k c^k \nabla \psi + B_w \nabla tr(E) \right) \right] \tag{3.61}$$

where u is the displacement of polymeric network matrix solid phase of the hydrogel.

As well known, the pressure results from the difference of the diffusive ionic concentrations between the hydrogel and the surrounding solution. In the MECe model, the ionic concentration is determined by the Nernst–Planck flux as follows:

$$J_i = -D_k c^k_{,i} - \frac{F_c}{RT} z^k D_k c^k \psi_{,i} + c^k v_i \ (k = 1, 2, \ldots, N_{\mathrm{ion}}) \tag{3.62}$$

where i denotes the spatial direction x_i, the subscript i after a comma indicates partial differentiation with respect to the variable x_i and D_k is the diffusive coefficient of the ion species k. The diffusion equation of the ionic species k is given as

$$\frac{\partial c^k}{\partial t} = -J_{i,i} + r_k \ (k = 1, 2, \ldots, N_{\text{ion}}) \tag{3.63}$$

in which r_k is the source term resulting from the chemical conversion of the molecules.

By Eqs. (3.62) and (3.63), the following diffusion–convection equations are obtained:

$$(D_k c^k_{,i})_{,i} + \frac{F_c z^k}{RT} (D_k c^k \psi_{,i})_{,i} = \frac{\partial c_k}{\partial t} + (c^k v_i)_{,i} \ (k = 1, 2, \ldots, N_{\text{ion}}) \tag{3.64}$$

where r_k is neglected due to the assumption of ideal solution.

For the MECe model, it should be noted that the externally applied electric field has important effect on the deformation of the hydrogels, and it is determined by the Poisson equation as

$$\nabla^2 \psi = -\frac{F_c}{\varepsilon \varepsilon_0} \left[\sum_{k=1}^{N_{\text{ion}}} z^k c^k + z^f c^f \right] \tag{3.65}$$

in which the fixed charge density c^f is given by

$$c^f = \frac{\phi_0^w c_0^f}{\phi^w (1 + tr(E))} = \frac{c_0^f}{(1 + tr(E)/\phi_0^w)} \tag{3.66}$$

where ε is the dielectric constant, ε_0 the permittivity of free space, z^f the valence of the fixed charge groups, c_0^f the fixed charge density at reference configuration and ϕ_0^w the volume fraction of the water phase at reference configuration.

The formulation of the MECe model has thus far been completed. It consists of the Nernst–Planck type of diffusion–convection equations (3.64) for the diffusive ionic concentrations, the Poisson equation (3.65) for the electric potential, the continuity equation of the hydrogel mixture (3.61) for the fluid pressure and the momentum equation of the hydrogel mixture (3.58) for the displacement of the hydrogel. They are coupled together to form a set of nonlinear partial differential equations and solved numerically by a hierarchical iteration technique. In the inner iteration process, the diffusive ionic concentrations c^k and the electric potential ψ are computed first by solving Eqs. (3.64) and (3.65) simultaneously. Then substituting the converged concentrations c^k and the electric potential ψ into the outer iteration, the fluid pressure p and the hydrogel displacement u are obtained by Eqs. (3.58) and (3.61), respectively. In addition, ϕ^w and c^f required for solving these equations are calculated by Eqs. (3.5) and (3.66). Following the computational procedure, the presently developed MECe model can be used for both the steady-state and transient simulations for analysis of equilibrium and kinetics of the electric-sensitive hydrogels.

For convenience of coding and computing, non-dimensional variables are defined as follows:

$$\bar{\zeta} = \frac{\zeta}{L_{\text{ref}}};\ \bar{u} = \frac{u}{L_{\text{ref}}};\ \bar{c}_k = \frac{c_k}{c_{\text{ref}}};\ \bar{c}_f = \frac{c_f}{c_{\text{ref}}};\ \bar{\psi} = \frac{\psi}{\psi_{\text{ref}}} = \frac{F\psi}{\alpha RT};\ \bar{p} = \frac{p}{p_{\text{ref}}} = \frac{p}{\beta c_{\text{ref}} RT} \tag{3.67}$$

where ζ denotes the spatial coordinate variable, α and β are non-dimensional adjustable parameters.

Therefore, the non-dimensional form of the partial differential governing equations of the MECe model can be derived as follows:

$$(D_k\bar{c}^k_{,i})_{,i} + \alpha z^k (D_k \bar{c}^k \bar{\psi}_{,i})_{,i} = L^2_{\text{ref}} \frac{\partial \bar{c}_k}{\partial t} + L_{\text{ref}}(\bar{c}^k v_i)_{,i} \tag{3.68}$$

$$\nabla^2 \bar{\psi} = -\frac{F_c^2 L^2_{\text{ref}} c_{\text{ref}}}{\varepsilon\varepsilon_0 RT\alpha}\left[\sum_{k=1}^{N_{\text{ion}}} z^k \bar{c}^k + z^f \bar{c}^f\right] \tag{3.69}$$

$$\beta RT c_{\text{ref}} \nabla \cdot (p\mathrm{I}) = \nabla(\lambda_s tr(E)\mathrm{I} + 2\mu_s E) \tag{3.70}$$

$$\frac{L^2_{\text{ref}}}{c_{\text{ref}} RT}\nabla\frac{\partial \bar{u}}{\partial t} = \nabla\left[\frac{(\phi^w)^2}{f_{ws}}\left(\beta\nabla\bar{p} + RT\nabla\sum_k (1-\Phi^k)\bar{c}^k + \alpha\sum_k z^k \bar{c}^k \nabla\bar{\psi}\right.\right.$$
$$\left.\left. + \frac{1}{c_{\text{ref}} RT} B_w \nabla tr(E)\right)\right] \tag{3.71}$$

3.2.2 Boundary and Initial Conditions

In this chapter, only one-dimensional simulations are conducted, and two kinds of boundary conditions are imposed at the solution ends (electrodes) and the hydrogel–solution interfaces, respectively. The first is the Dirichlet boundary conditions for the ionic concentrations and the electric potential applied at two ends of the solution:

$$c|_{\text{Anode}} = c|_{\text{Cathode}} = c^* \tag{3.72}$$

$$\psi|_{\text{Anode}} = 0.5V_e \text{ and } \psi|_{\text{Cathode}} = -0.5V_e \tag{3.73}$$

where c^* is the initial ion concentration of the bath solution and V_e the externally applied electric voltage. The second is to assign the boundary values of the fluid pressure and the hydrogel displacement at the hydrogel–solution interfaces. Based on the assumption that the chemical potentials of the water and ion phases within the hydrogels at equilibrium state should be equal to those outside the hydrogels, the boundary condition of the fluid pressure is imposed at the hydrogel–solution interfaces as follows:

$$p_{\text{interface}} = RT \sum_{k=1}^{N_{\text{ion}}} (c^k_{\text{in-interface}} - c^k_{\text{out-interface}}) - p_0 \tag{3.74}$$

where $c^k_{\text{in-interface}}$ is the ion concentration within the hydrogels near the interfaces and $c^k_{\text{out-interface}}$is the ion concentration within the exterior solution near the interfaces. p_0 denotes the fluid pressure at reference configuration. Due to the zero stress of the hydrogel mixture phase on the hydrogel–solution interface, the boundary condition of displacement of the hydrogels on the hydrogel–solution interface is written as

$$\lambda_s tr(E_{\text{interface}})\mathrm{I} + 2\mu_s E_{\text{interface}} = p_{\text{interface}}\mathrm{I} \tag{3.75}$$

The boundary conditions mentioned above are applicable for both the steady-state and transient simulations. However, for implementation of the transient computations for kinetics of the hydrogels, additional initial conditions are required. It is assumed here that the hydrogel is initially in the equilibrium state only subject to the effect of bath solution without the externally applied electric field $V_e = 0$. By taking this equilibrium state as the initial state for transient analysis, the steady-state simulations are carried out first and the computed steady-state results of all the variables are used as the initial values for the initial conditions of the transient analysis, namely

$$c^{\text{transient}}_{\text{initial}} = c^{\text{steady}}_{Ve=0};\ \psi^{\text{transient}}_{\text{initial}} = \psi^{\text{steady}}_{Ve=0};\ p^{\text{transient}}_{\text{initial}} = p^{\text{steady}}_{Ve=0};\ u^{\text{transient}}_{\text{initial}} = u^{\text{steady}}_{Ve=0} \tag{3.76}$$

where $c^{\text{steady}}_{Ve=0}$, $\psi^{\text{steady}}_{Ve=0}$, $p^{\text{steady}}_{Ve=0}$ and $u^{\text{steady}}_{Ve=0}$denote the steady-state results computed without the externally applied electric voltage ($V_e = 0$).

3.3 Steady-State Simulation for Equilibrium of Hydrogel

In this section, the MECe model is simplified for one-dimensional steady-state simulations of a hydrogel strip, in which the steady-state formulations of the governing equations and corresponding discretizations are reduced. After validation of the developed MECe model by comparison with the experiments published, several parameter studies are carried out in detail.

3.3.1 Numerical Implementation

As shown in Fig. 3.1, a hydrogel strip is immersed into a bath solution subject to an externally applied electric voltage. With the effects of the chemical and electric coupled fields, the hydrogel strip is expected to be bent towards either the cathode or anode direction, depending on the electric attribution of the fixed charges in the hydrogels. It is noted that only the one-dimensional studies on the hydrogel strips are made in this chapter for both the steady-state and transient simulations.

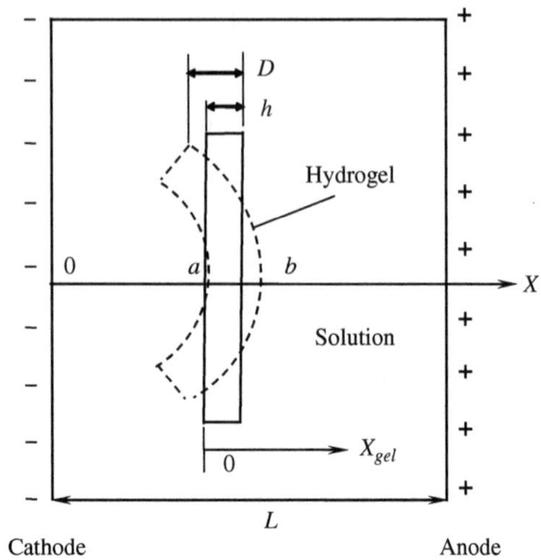

Fig. 3.1 Schematic diagram of a hydrogel strip immersed in bath solution subject to externally applied electric field, where the whole computational domain is defined as the *X*-coordinate system and the hydrogel domain as the *X*gel-coordinate system

Thus the deformation of the hydrogels along the *x*-direction is only concerned. It is also assumed that the middle point of hydrogel thickness *h* is fixed to eliminate rigid body motion, and thus zero-displacement condition is imposed at the middle point. Further, although the MECe model is applicable for the multivalent electrolyte bath solution, the present simulations and discussions focus on the simple ideal monovalent solution consisting of two ionic species, such as NaCl solution.

For the one-dimensional steady-state simulations, the non-dimensional partial differential governing equations (3.68), (3.69), (3.70) and (3.71) of the MECe model can be reduced to

Diffusion equations for the ion concentration c^k

$$\frac{\partial^2 \bar{c}^k}{\partial \bar{x}^2} + \alpha z^k \frac{\partial \bar{c}^k}{\partial \bar{x}} \frac{\partial \bar{\psi}}{\partial \bar{x}} + \alpha z^k \bar{c}^k \frac{\partial^2 \bar{\psi}}{\partial \bar{x}^2} = 0 \ (k = +, -) \tag{3.77}$$

Poisson equation for the electric potential ψ

$$\frac{\partial^2 \bar{\psi}}{\partial \bar{x}^2} + \frac{F_c^2 L_{\text{ref}}^2 c_{\text{ref}}}{\varepsilon \varepsilon_0 RT\alpha} \left(\sum_{k=+,-} z^k \bar{c}^k + z^f \bar{c}^f \right) = 0 \tag{3.78}$$

Continuity equation of the hydrogel mixture for the fluid pressure p

$$\begin{aligned} &\beta \left(\phi^w \frac{\partial^2 \bar{p}}{\partial \bar{x}^2} + 2 \frac{\partial \phi^w}{\partial \bar{x}} \frac{\partial \bar{p}}{\partial \bar{x}} \right) + \alpha \left[2 \frac{\partial \phi^w}{\partial \bar{x}} \frac{\partial \bar{\psi}}{\partial \bar{x}} \sum_{k=+,-} z^k \bar{c}^k \right. \\ &\left. + \phi^w \left(\frac{\partial \bar{\psi}}{\partial \bar{x}} \sum_{k=+,-} \left(z^k \frac{\partial \bar{c}^k}{\partial \bar{x}} \right) + \frac{\partial^2 \bar{\psi}}{\partial \bar{x}^2} \sum_{k=+,-} z^k \bar{c}^k \right) \right] = 0 \end{aligned} \tag{3.79}$$

where $\Phi^+ = \Phi^- = 1$ is assumed for an ideal solution without chemical reaction. $\boldsymbol{B}_w = 0$ is also assumed since it is negligibly small if $\boldsymbol{B}_w tr(\boldsymbol{E})$ is compared with the pressure p.

Momentum equation of the hydrogel mixture for the hydrogel displacement u

$$(3\lambda_s + 2\mu_s)\frac{\partial^2 \bar{u}}{\partial \bar{x}^2} - \beta RTc_{\text{ref}}\frac{\partial \bar{p}}{\partial \bar{x}} = 0 \tag{3.80}$$

where the isotropic strain is assumed for the hydrogel strip, namely $tr(E) = 3e_{11} = 3(\partial u/\partial x)$ where e_{11} is the x-component of the strain vector $\boldsymbol{E}$. With this assumption, the boundary condition (3.75) is simplified further to

$$(3\lambda_s + 2\mu_s)\frac{\partial \bar{u}_{\text{interface}}}{\partial \bar{x}} = \beta RTc_{\text{ref}}\bar{p}_{\text{interface}} \tag{3.81}$$

For discretization of the reduced governing equations via the Hermite-cloud method (Li et al., 2003), we construct the discrete variables as follows:

$$\bar{c}^k(\bar{x}_i) = \sum_{j=1}^{N_{\text{total}}} N_j(\bar{x}_i)\bar{c}_j^k - \sum_{m=1}^{N_{\text{total}}}\left(\bar{x}_i - \sum_{j=1}^{N_{\text{total}}} N_j(\bar{x}_i)\bar{x}_j\right) M_m(\bar{x}_i)\bar{c}_{xm} \tag{3.82}$$

$$\bar{\psi}(\bar{x}_i) - \sum_{j=1}^{N_{\text{total}}} N_j(\bar{x}_i)\bar{\psi}_j - \sum_{m=1}^{N_{\text{total}}}\left(\bar{x}_i - \sum_{j=1}^{N_{\text{total}}} N_j(\bar{x}_i)\bar{x}_j\right) M_m(\bar{x}_i)\bar{\psi}_{xm} \tag{3.83}$$

$$\bar{p}(\bar{x}_i) = \sum_{j=1}^{N_{\text{gel}}} N_j(\bar{x}_i)\bar{p}_j - \sum_{m=1}^{N_{\text{gel}}}\left(\bar{x}_i - \sum_{j=1}^{N_{\text{gel}}} N_j(\bar{x}_i)\bar{x}_j\right) M_m(\bar{x}_i)\bar{p}_{xm} \tag{3.84}$$

$$\bar{u}(\bar{x}_i) = \sum_{j=1}^{N_{\text{gel}}} N_j(\bar{x}_i)\bar{u}_j - \sum_{m=1}^{N_{\text{gel}}}\left(\bar{x}_i - \sum_{j=1}^{N_{\text{gel}}} N_j(\bar{x}_i)\bar{x}_j\right) M_m(\bar{x}_i)\bar{u}_{xm} \tag{3.85}$$

Using the Hermite-cloud method (Li et al., 2003), the one-dimensional steady-state governing equations of the MECe model and the auxiliary conditions are discretized as

$$\begin{aligned}
&\sum_{j=1}^{N_{\text{total}}} N_{xxj}(\bar{x}_i)\bar{c}_j^k + \alpha z^k \left[\sum_{m=1}^{N_{\text{total}}} M_m(\bar{x}_i)\bar{c}_{xm}^k\right]\left[\sum_{m=1}^{N_{\text{total}}} M_m(\bar{x}_i)\bar{\psi}_{xm}\right] \\
&+\alpha z^k \left[\sum_{j=1}^{N_{\text{total}}} N_j(\bar{x}_i)\bar{c}_j^k - \sum_{m=1}^{N_{\text{total}}}\left(\bar{x}_i - \sum_{j=1}^{N_{\text{total}}} N_j(\bar{x}_i)\bar{x}_j\right) M_m(\bar{x}_i)\bar{c}_{xm}\right] \\
&\left[\sum_{j=1}^{N_{\text{total}}} N_{xxj}(\bar{x}_i)\bar{\psi}_j\right] = 0
\end{aligned} \tag{3.86}$$

$$\sum_{j=1}^{N_{\text{total}}} N_{xxj}(\bar{x}_i)\bar{\psi}_j + \frac{F_c^2 L_{\text{ref}}^2 c_{\text{ref}}}{\varepsilon\varepsilon_0 RT\alpha}\left\{z^f \bar{c}^f + \sum_{k=+,-} z^k \left[\sum_{j=1}^{N_{\text{total}}} N_j(\bar{x}_i)\bar{c}_j^k - \sum_{m=1}^{N_{\text{total}}} (\bar{x}_i - \sum_{j=1}^{N_{\text{total}}} N_j(\bar{x}_i)\bar{x}_j) M_m(\bar{x}_i)\bar{c}_{xm}\right]\right\} = 0 \tag{3.87}$$

$$(3\lambda_s + 2\mu_s)\sum_{j=1}^{N_{\text{gel}}} N_{xxj}(\bar{x}_i)\bar{u}_j - \beta RT c_{\text{ref}} \sum_{m=1}^{N_{\text{gel}}} M_m(\bar{x}_i)\bar{p}_{xm} = 0 \tag{3.88}$$

$$\begin{aligned}
&\beta\left[\phi^w \sum_{j=1}^{N_{\text{gel}}} N_{xxj}(\bar{x}_i)\bar{p}_j + 2\frac{\partial \phi^w}{\partial \bar{x}} \sum_{m=1}^{N_{\text{gel}}} M_m(\bar{x}_i)\bar{p}_{xm}\right] \\
&\quad + \alpha\left\{2\frac{\partial \phi^w}{\partial \bar{x}} \sum_{m=1}^{N_{\text{gel}}} M_m(\bar{x}_i)\bar{\psi}_{xm}\left[\sum_{k=+,-} z^k \sum_{j=1}^{N_{\text{gel}}}\left(N_j(\bar{x}_i)\bar{c}_j^k \right.\right.\right. \\
&\quad \left.\left. - \sum_{m=1}^{N_{\text{gel}}}\left(\bar{x}_i - \sum_{j=1}^{N_{\text{gel}}} N_j(\bar{x}_i)\bar{x}_j\right) M_m(\bar{x}_i)\bar{c}_{xm}\right)\right] \\
&\quad + \phi^w\left[\sum_{m=1}^{N_{\text{gel}}} M_m(\bar{x}_i)\bar{\psi}_{xm}\left(\sum_{k=+,-} z^k \sum_{m=1}^{N_{\text{gel}}} M_m(\bar{x}_i)\bar{c}_{xm}^k\right)\right. \\
&\quad + \sum_{j=1}^{N_{\text{gel}}} N_{xxj}(\bar{x}_i)\bar{\psi}_j \sum_{k=+,-} z^k \sum_{j=1}^{N_{\text{gel}}}\left(N_j(\bar{x}_i)\bar{c}_j^k - \sum_{m=1}^{N_{\text{gel}}}\left(\bar{x}_i\right.\right. \\
&\quad \left.\left.\left.\left. - \sum_{j=1}^{N_{\text{gel}}} N_j(\bar{x}_i)\bar{x}_j\right) M_m(\bar{x}_i)\bar{c}_{xm}\right)\right]\right\} = 0
\end{aligned} \tag{3.89}$$

$$\sum_{j=1}^{N_{\text{total}}} N_{xj}(\bar{x}_i)\bar{c}_j^k - \left[\sum_{j=1}^{N_{\text{total}}} N_{xj}(\bar{x}_i)\bar{x}_j\right] \sum_{m=1}^{N_{\text{total}}} M_m(\bar{x}_i)\bar{c}_{xm}^k = 0 \tag{3.90}$$

$$\sum_{j=1}^{N_{\text{total}}} N_{xj}(\bar{x}_i)\bar{\psi}_j - \left[\sum_{j=1}^{N_{\text{total}}} N_{xj}(\bar{x}_i)\bar{x}_j\right] \sum_{m=1}^{N_{\text{total}}} M_m(\bar{x}_i)\bar{\psi}_{xm} = 0 \tag{3.91}$$

$$\sum_{j=1}^{N_{\text{gel}}} N_{xj}(\bar{x}_i)\bar{u}_j - \left[\sum_{j=1}^{N_{\text{gel}}} N_{xj}(\bar{x}_i)\bar{x}_j\right] \sum_{m=1}^{N_{\text{gel}}} M_m(\bar{x}_i)\bar{u}_{xm} = 0 \tag{3.92}$$

$$\sum_{j=1}^{N_{\text{gel}}} N_{xj}(\bar{x}_i)\bar{p}_j - \left[\sum_{j=1}^{N_{\text{gel}}} N_{xj}(\bar{x}_i)\bar{x}_j\right] \sum_{m=1}^{N_{\text{gel}}} M_m(\bar{x}_i)\bar{p}_{xm} = 0 \tag{3.93}$$

where N_{total} is the number of total points scattered over the whole domain covering both the hydrogel and surrounding solution and N_{gel} that within the hydrogel domain only.

3.3.2 Model Validation with Experiment

In order to validate the presently developed MECe model, a simulation is carried out and compared numerically with the experimental data published (Zhou et al., 2002) for a specified hydrogel strip with positive fixed charge groups ($z^f = +1$). The parameters used as input of the MECe simulation are given as follows: $T = 298$ (K), $R = 8.314$ (J/mol·K), $F = 9.648 \times 10^4$ (C/mol), $c_0^f = 20$ (mol/m^3), $c^* = 5.5$ (mol/m^3), $\varepsilon_0 = 8.854 \times 10^{-12}$ (C^2/Nm^2), $\varepsilon = 80$ $\phi_0^w = 0.8$, $3\lambda + 2\mu = 1.2 \times 10^5$ (Pa), the one-dimensional computational domain $L = 20$(mm), the thickness of hydrogel strip in the simulation direction $h = 1$ (mm). The comparison is presented in Fig. 3.2, where an average curvature measuring the extent of the hydrogel deformation, Ka, is defined as $Ka = 2(e_1 - e_2)/(h(2 + e_1 + e_2))$ at the middle point of hydrogel thickness, in which e_1 and e_2 are the strains of the hydrogel at the two ends of the hydrogel strip thickness. It is observed from Fig. 3.2 that the average curvature Ka increases almost linearly with increasing the applied electric voltage V_e. The computational results agree well with the experimental data (Zhou et al., 2002) when the applied electric voltage is lower than 5 V. However, they seem to have larger discrepancy as the applied electric voltage increases at higher than 5 V. Probably the reasons are that the bending deformation of the electric-sensitive hydrogels depends directly on many nonlinear effects, such as the voltage of applied electric field, fixed charge, ionic diffusion and/or convection, electrolyte composition, chemical

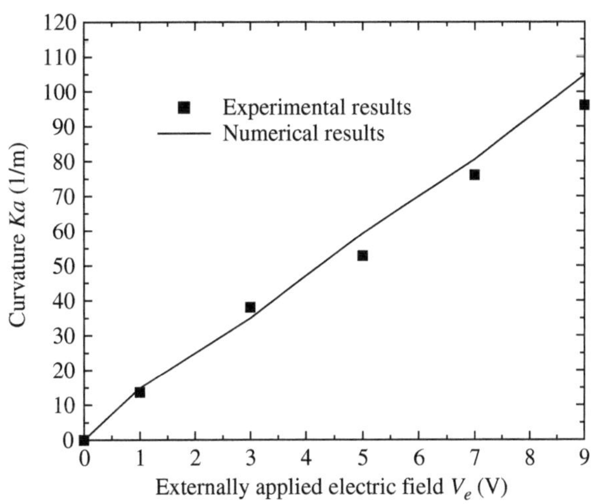

Fig. 3.2 Comparison between the steady-state simulations and the experimental data (Zhou et al., 2002)

reaction, temperature and heat conduction. As a preliminary work with isotropy assumption, the effects of chemical reaction, temperature and heat conduction have not been included in the presently developed MECe model. The simplifications are feasible for the situations subject to lower voltage of the applied electric field such as 5 V. However, with the increase of the applied electric voltage, these nonlinear effects become more and more significant and then they should be considered. In other words, these nonlinear effects may be the main reasons for the different trends between the computational numerical results and the experimental data, when the electric voltage applied is higher than 5 V. In brief, the present comparison achieves good agreement between the simulation and the experiment (Zhou et al., 2002).

3.3.3 Parameter Studies

For analysis of the influences of the hydrogel material properties and the environmental conditions on the responsive characteristics of the electric-sensitive hydrogels, simulations are conducted with the parameters used as input, $T = 298$ (K), $R = 8.314$ (J/mol·K), $F = 9.648 \times 10^4$ (C/mol), $\varepsilon_0 = 8.854 \times 10^{-12}$ (C^2/N·m^2), $\varepsilon = 80$, $\phi_0^w = 0.8$, $z^f = -1$, $L = 1.5 \times 10^{-2}$(m), $h = 5.0 \times 10^{-3}$(m) and $3\lambda + 2\mu = 1.2 \times 10^5$ (Pa). Figures 3.3, 3.4 and 3.5 demonstrate a preliminary study before the effects of the physical parameters are discussed, where the distributions of the diffusive ion concentrations and the electric potential in both the interior hydrogel and exterior solution and the displacement field of the hydrogel strip are presented with $c^* = 1$ (mol/m^3), $c_0^f = 10$ (mol/m^3) and $V_e = 0.2$(V). It is observed from Fig. 3.3 that the concentration of diffusive Na^+ ion in the surrounding solution is equal to that of the diffusive Cl^- ion. In the environmental bath solution, the ionic concentrations

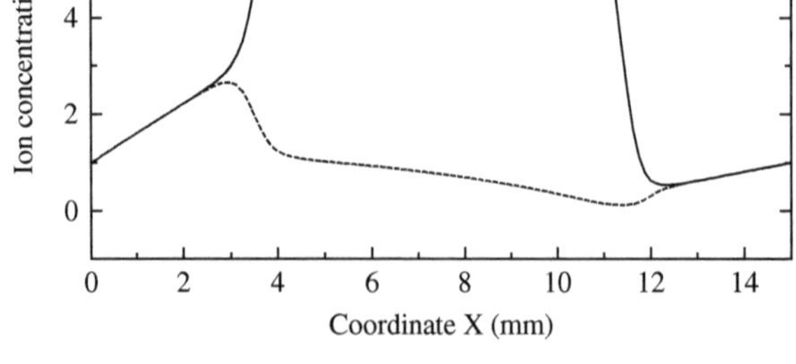

Fig. 3.3 Distribution of the diffusive ion concentrations

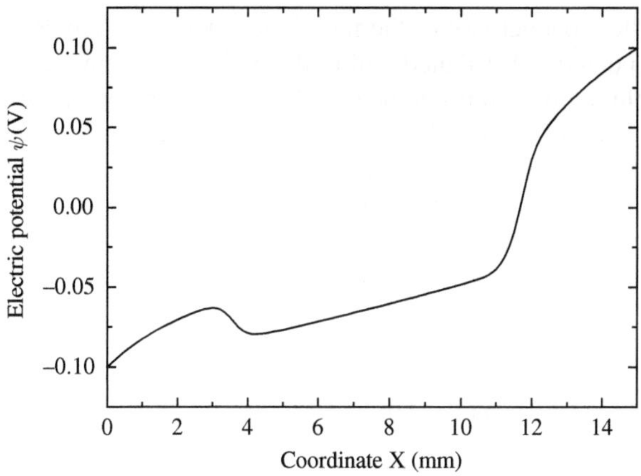

Fig. 3.4 Distribution of the electric potential

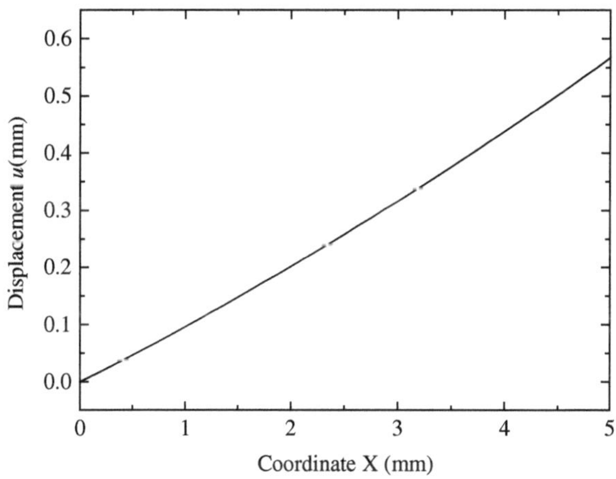

Fig. 3.5 Distribution of displacement of the hydrogel

near the cathode increases with the distance away from the cathode, while the ionic concentrations near the anode decreases with the distance away from the anode. In the hydrogel domain, however, a difference of the ionic concentrations exists between the diffusive Na^+ and Cl^- due to the effect of the fixed charge groups, and both the Na^+ and Cl^- ion concentrations decrease with the distance away from the cathode. It is also noted that there is an evident difference of the ionic concentrations over the hydrogel–solution interfaces between the interior hydrogels and surrounding solution, which results in the pressure to drive the hydrogels to deformed. It is seen from Fig. 3.4 that the electric potential is no longer distributed linearly over the whole computational domain because of the effect of the fixed charge groups. The distributive profile of the electric potential collapses within the hydrogel domain.

Due to the higher conductivity of the mobile ions within the hydrogels, the gradient of electric potential distributed within the hydrogels is smaller than that in the surrounding solution, which is compensated by a smaller step of the electric potential distributed over the hydrogel–solution interface near the cathode in comparison with that near the anode. These simulated phenomena are in good agreement with the FEM results completed by Wallmersperger et al. (2004). In Fig. 3.5, it is shown that the displacement of the hydrogel strip increases with the coordinate x, resulting from the non-uniform distributions of the diffusive ionic concentrations and electric potential.

3.3.3.1 Influence of Externally Applied Electric Voltage

In this section, the influence of externally applied electric field is discussed on the responsive behaviours of the electric-sensitive hydrogels. For given different levels of the electric voltages V_e, Figs. 3.6, 3.7, 3.8 and 3.9 show the distributions of the diffusive ionic concentrations and the electric potential as well as the hydrogel displacement and Figs. 3.10, 3.11 and 3.12 the variations of the distributive average curvature Ka with various physical parameters, including the fixed charge density c_0^f, the bath solution concentration c^* and the hydrogel strip thickness h.

Figures 3.6 and 3.7 are plotted at $c_0^f = 2\ (\text{mol/m}^3)$, $c^* = 1\ (\text{mol/m}^3)$ and $V_e = 0.02, 0.04, 0.08, 0.16$ (V), respectively, namely $\psi = +0.01, +0.02, +0.04$ and $+0.08$ (V) at anode $\psi = -0.01, -0.02, -0.04$ and -0.08 (V) at cathode, respectively. It

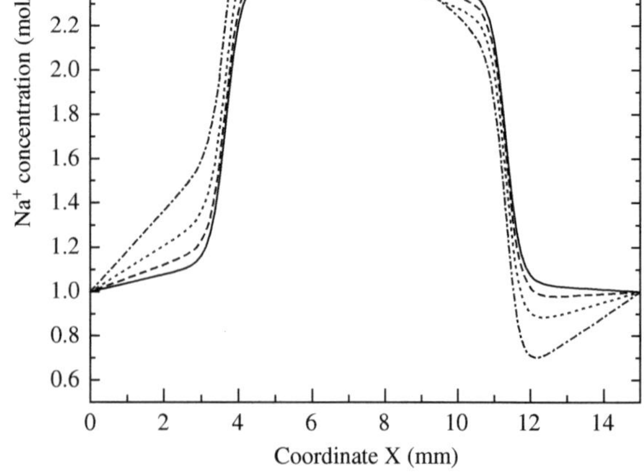

Fig. 3.6 Effect of the externally applied electric field on the variation of the diffusive Na^+ concentration distribution

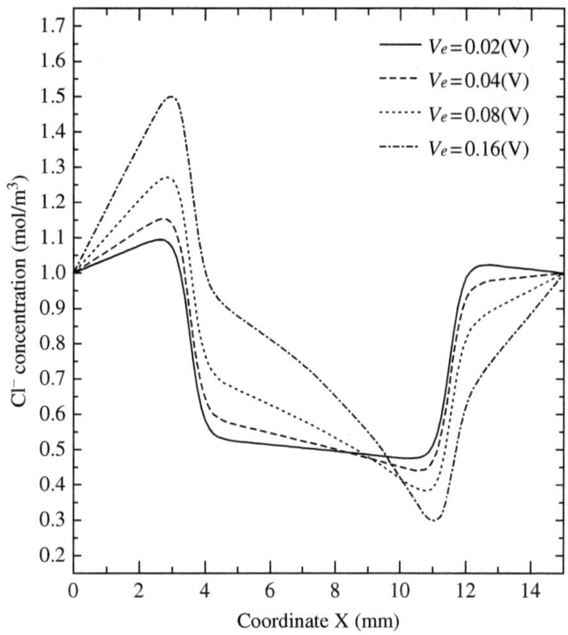

Fig. 3.7 Effect of the externally applied electric field on the variation of the diffusive Cl^- concentration distribution

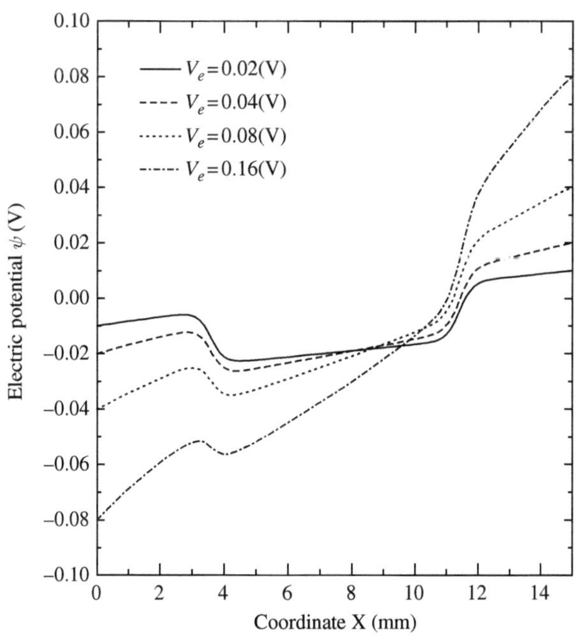

Fig. 3.8 Effect of the externally applied electric voltage on the variation of the resulting electrostatic potential distribution

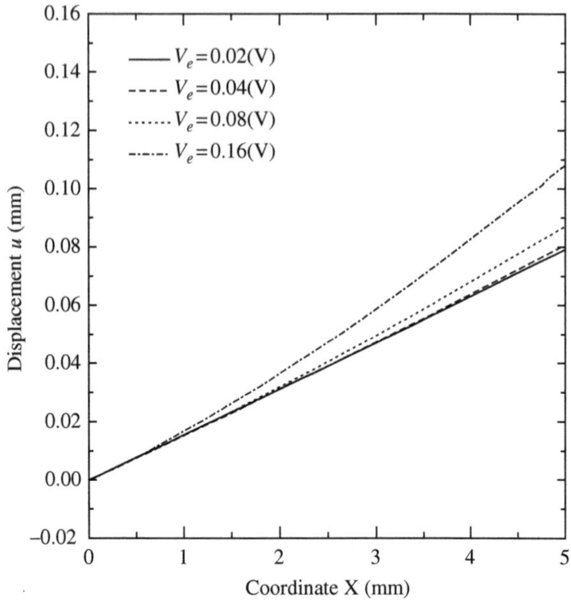

Fig. 3.9 Effect of the externally applied electric field on the variation of the displacement distribution of the hydrogel

is found from the figures that the diffusive Na^+ and Cl^- ionic concentrations are distributed almost symmetrically if the electric voltage V_e is very low, and this phenomenon will diminish when the voltage V_e becomes high. With increment of the applied voltage V_e, the variations of diffusive Na^+ and Cl^- concentrations on the hydrogel–solution interface near the anode are always smaller than those near the cathode, and the distributive gradients of the Na^+ and Cl^- concentrations increase within the hydrogels. It is shown from Fig. 3.8 that the gradient of distributive electric potential increases in both the hydrogels and surrounding solution as the applied voltage increases. Figure 3.9 demonstrates that, with the increase of applied voltage V_e, the difference of the ionic concentrations on the hydrogel–solution interface increases and the displacement of the hydrogel strip increases as well.

Figure 3.10 presents the influence of the externally applied electric field V_e on the variation of the average curvature Ka with the fixed charge density c_0^f, where $c^* =$ 1 (mol/m^3), $V_e = 0.02$, 0.1 and 0.2 (V), respectively. It is seen that the average curvature Ka increases with the fixed charge density c_0^f for a given applied voltage V_e. The presently simulated phenomena are validated by the experiment (Homma et al., 2000). Furthermore, in order to investigate the influence of the externally applied electric field V_e on the variation of average curvature Ka with the bath solution concentration c^*, Fig. 3.11 is depicted when $c_0^f = 10$(mol/m^3) and $V_e = 0.02$, 0.1 and 0.2 (V), respectively. It is observed that an optimal c^* value appears when the hydrogel strip reaches the largest bending deformation. This means that the bending deformation of the hydrogel strip decreases with increasing c^*, after the concentration of bath solution c^* is larger than the optimal value. The present simulations also

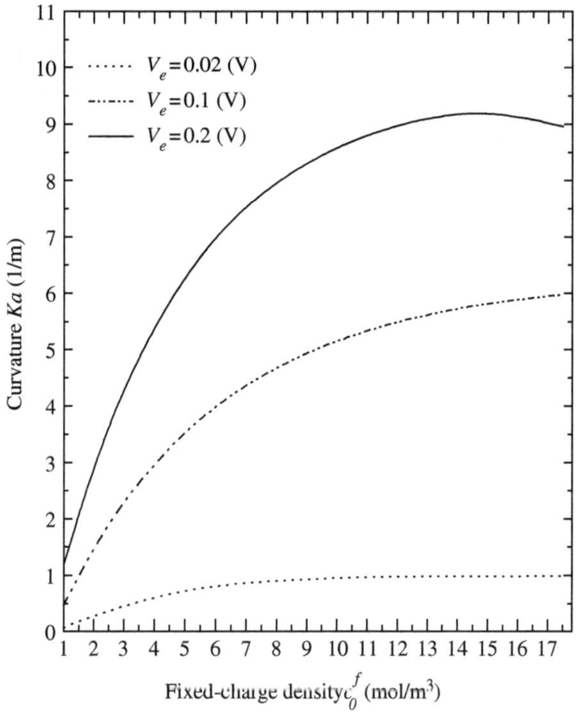

Fig. 3.10 Effect of the externally applied electric field on the variation of the average curvature *Ka* against the fixed charge density

agree well with the experimental phenomena (Homma et al., 2000, 2001; Sun et al., 2001; Fei et al., 2002).

Figure 3.12 illustrates the relation between the average curvature *Ka* and the thickness *h* of hydrogel strip subject to different levels of externally applied electric fields, $V_e = 0.02$, 0.1 and 0.2 (V), respectively, where $c_o^f = 10\,(\text{mol/m}^3)$ and $c^* = 1\,(\text{mol/m}^3)$. It is predicted that the average curvature *Ka* of the hydrogel strip decreases exponentially with increasing the thickness *h* of the hydrogel strip, which is in consistence with the experiment (Homma et al., 2000, 2001).

3.3.3.2 Influence of Initially Fixed Charge Density of Hydrogel

For the ionized hydrogels with capability of responding to electric stimulus, the fixed charge density has significant effect on the responsive performance of the hydrogels subject to externally applied electric field. Figures 3.13, 3.14, 3.15 and 3.16 demonstrate the influence of the fixed charge density c_0^f on the distributions of the diffusive ionic concentrations, the electric potential and the hydrogel displacement, where $c^* = 1$ (mol/m^3), $V_e = 0.2$ (V) and $c_0^f = 2,4,8$ (mol/m^3), respectively. It is known from Fig. 3.13 that the diffusive Na^+ ionic concentration within the

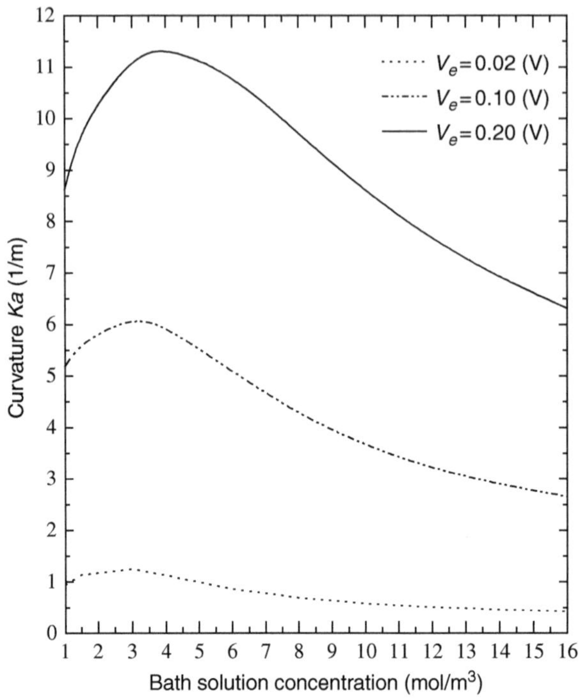

Fig. 3.11 Effect of the externally applied electric field on the variation of the average curvature *Ka* against the bath solution concentration

hydrogels increases obviously with the fixed charge density c_0^f. This phenomenon results from the fact that the present anion fixed charge groups ($z^f = -1$) attached on the polymeric network chains of the hydrogel strip attract the mobile cations Na^+ to compensate the electric potential. It is seen further from Fig. 3.14 that the influence of the fixed charge density c_0^fon the distribution of Cl^- ionic concentration is relatively small due to the anion fixed charge groups. Figure 3.15 is plotted for analysis of the effect of c_0^fon the distributive electric potential. With the increase of the fixed charge density c_0^f, the more mobile ions diffuse into the hydrogels, the higher conductivity the hydrogel achieves. This results in the smaller gradient of the electric potential distributed in the hydrogels, compared with that in the surrounding solution. Figure 3.16 indicates that, with the increase of the fixed charge density c_0^f, the difference of ionic concentrations increases on the hydrogel–solution interfaces, leading to the larger deformation of the hydrogels.

Figure 3.17 is achieved to discuss the deformation of the hydrogel strip with $h = 5\times10^{-3}$ (m) and $c^* = 1$ (mol/m^3), where the variation of the average curvature *Ka* against the externally applied electric field V_e is illustrated for different fixed charge densities c_0^f= 1, 5 and 10 (mol/m^3). It is observed that, with the increase of the externally applied electric voltage V_e, the differences of both the diffusive ionic

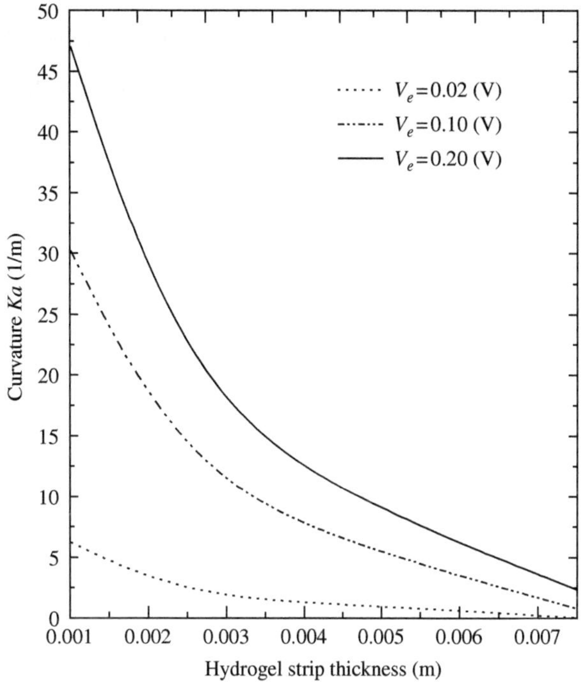

Fig. 3.12 Effect of the externally applied electric field on the variation of the average curvature *Ka* against the thickness of the hydrogel strip

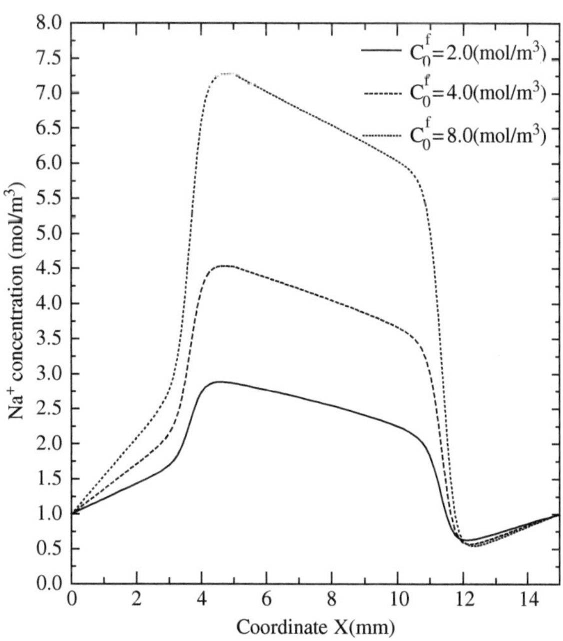

Fig. 3.13 Effect of the fixed charge density on the variation of the diffusive Na^+ concentration distribution

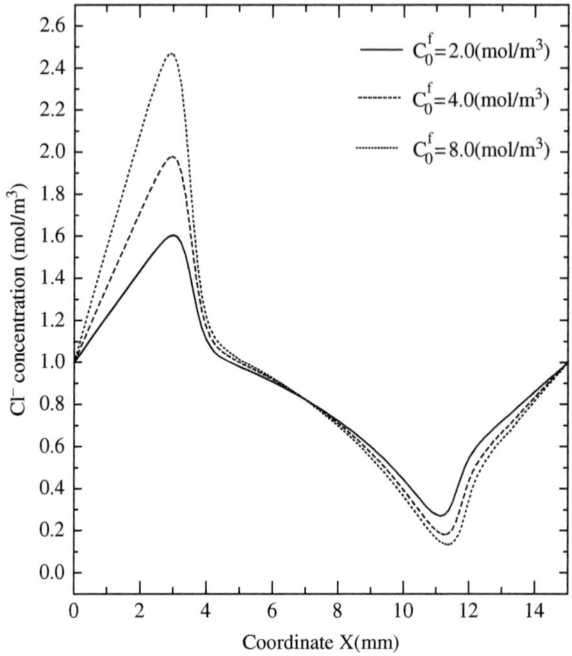

Fig. 3.14 Effect of the fixed charge density on the variation of the diffusive Cl^- concentration distribution

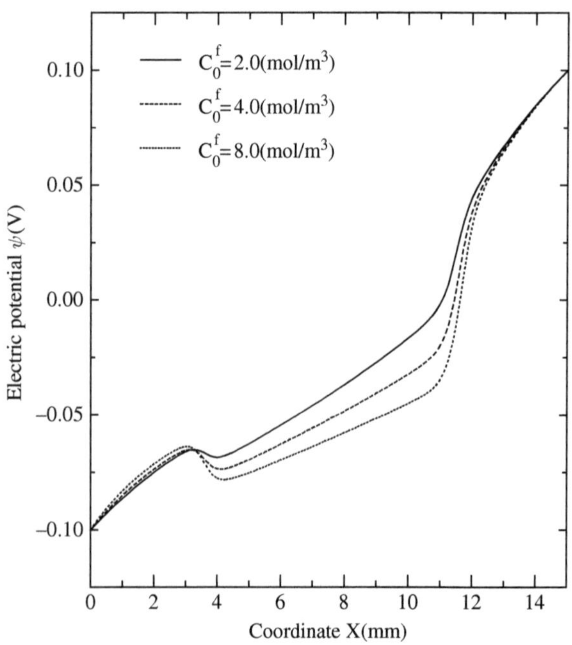

Fig. 3.15 Effect of the fixed charge density on the variation of the electric potential distribution

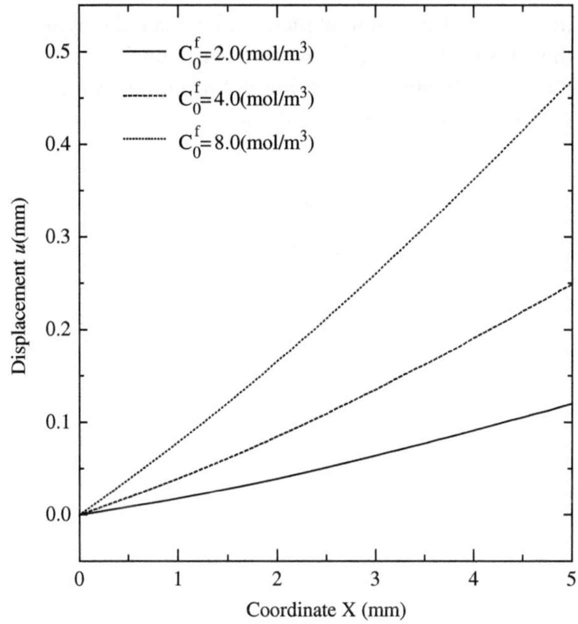

Fig. 3.16 Effect of the fixed charge density on the variation of displacement distribution of the hydrogel

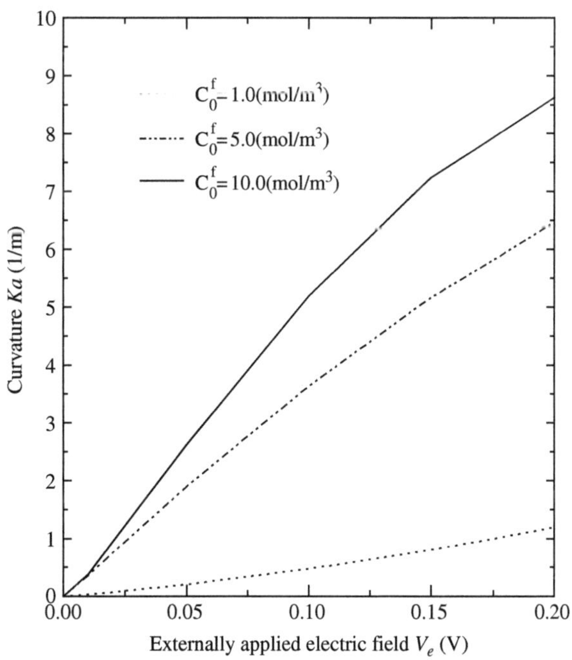

Fig. 3.17 Effect of the fixed charge density on the variation of the average curvature Ka against the externally applied electric field

concentrations and the electric potential increase between the hydrogel strip and the surrounding bath solution. As such, the average curvature Ka increases rapidly with the applied electric voltage and the bending deformation increases greatly, which are in a good match with the experimental phenomena (Homma et al., 2000; Sun et al., 2001; Fei et al., 2002).

3.3.3.3 Influence of Concentration of Bath Solution

The influence of the concentration of surrounding bath solution c^* on the distributive profiles of the diffusive ionic concentrations and the electric potential as well as the hydrogel displacement is shown in Figs. 3.18, 3.19, 3.20 and 3.21, where $c_0^f = 2\,(\text{mol/m}^3)$, $V_e = 0.2$ (V) and $c^* = 1, 2, 4, 8$ (mol/m^3), respectively. Figures 3.18 and 3.19 demonstrate that the increment of the bath solution concentration c^* makes the diffusive Na^+ and Cl^- concentrations increase in both the interior hydrogels and the exterior surrounding solution. It is also found in Fig. 3.20 that, when the bath solution concentration c^* reaches certain value much higher than the fixed charge density c_0^f, the change in the diffusive ion concentrations due to the effect of the fixed charge groups may be neglected, if compared with the distribution of bath solution concentration c^*. Then the conductivity of the

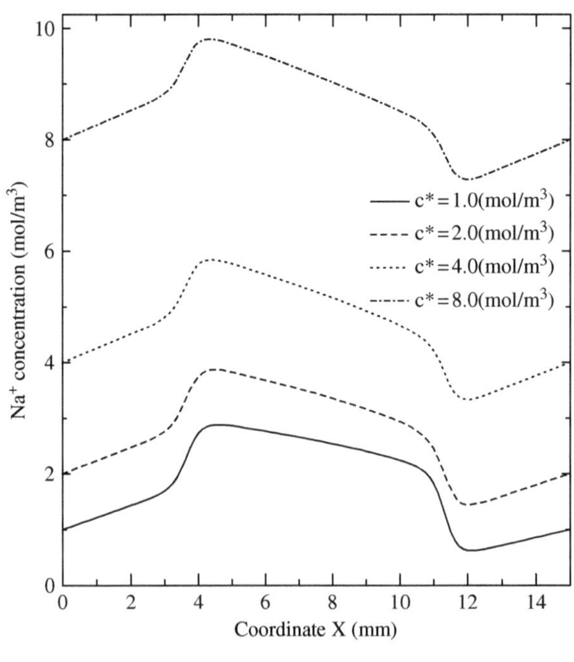

Fig. 3.18 Effect of the concentration of surrounding solution on the variation of the diffusive Na^+ concentration distribution

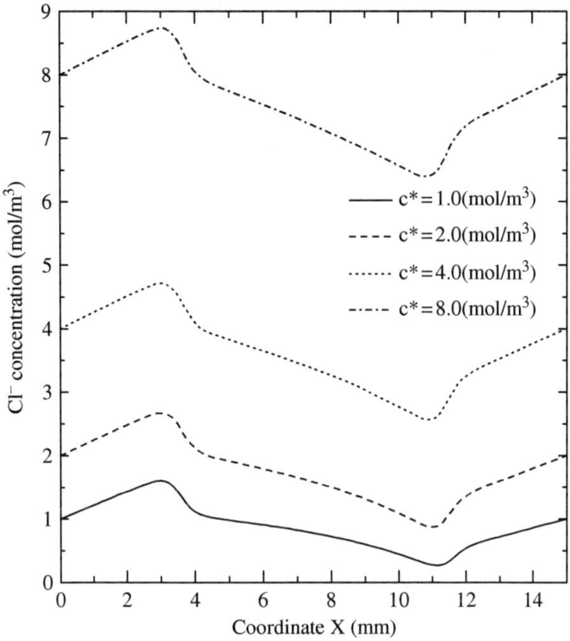

Fig. 3.19 Effect of the concentration of surrounding solution on the variation of the diffusive Cl^- concentration distribution

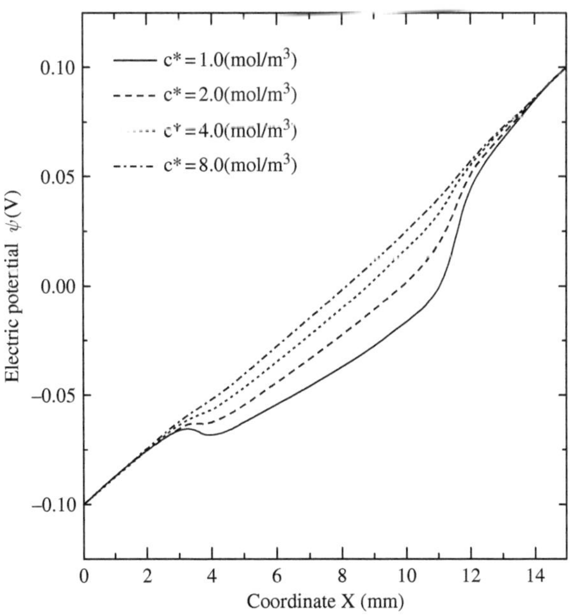

Fig. 3.20 Effect of the concentration of surrounding solution on the variation of the electric potential distribution

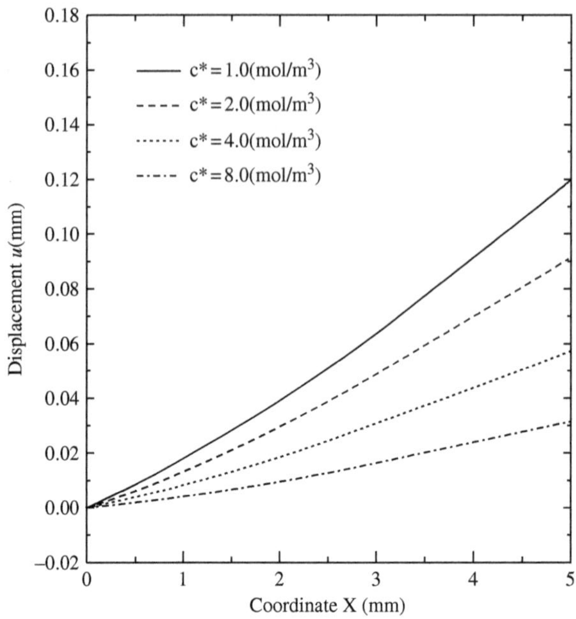

Fig. 3.21 Effect of the concentration of surrounding solution on the variation of the displacement distribution of the hydrogel

hydrogels is almost equal to that of the surrounding solutions, which makes the electric potential distributed quasilinearly over whole computational domain. It is also seen from Fig. 3.21 that the difference of the ionic concentrations on the hydrogel–solution interfaces decreases with the increase of c^*, and then the displacement of the hydrogel strip decreases as well.

3.3.3.4 Influence of Ionic Valence of Bath Solution

Figures 3.22, 3.23, 3.24 and 3.25 are plotted for discussion of the influences of the ionic valence z_k on the distributions of the diffusive ionic concentrations, the electric potential and the hydrogel displacement, where $c_0^f = 2$ (mol/m^3), $c^* = 1$ (mol/m^3), $V_e = 0.2$ (V) and $|z_k| = 1$, 2 and 3, respectively. It is found from Fig. 3.22 that the cation concentration decreases clearly within the hydrogels as the ionic valence $|z_k|$ increases. The corresponding distribution of the ionic concentration in the environmental solution has significant variation near the anode but insignificant variation near the cathode. Figure 3.23 shows that the corresponding concentration of the anion within the hydrogel grows with increasing $|z_k|$. In the surrounding solutions, the distribution of the anion concentration is similar to that of the cation concentration. It is also found from Figs. 3.24 and 3.25 that, when $|z_k|$varies from 1 to 2, the changes of the electric potential and the hydrogel displacement are much more distinct than those when $|z_k|$varies from 2 to 3.

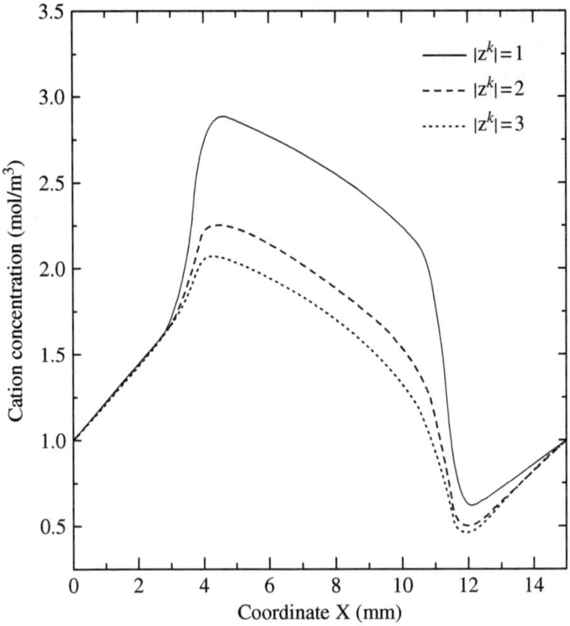

Fig. 3.22 Effect of the valence of surrounding solution on the variation of the diffusive cation concentration distribution

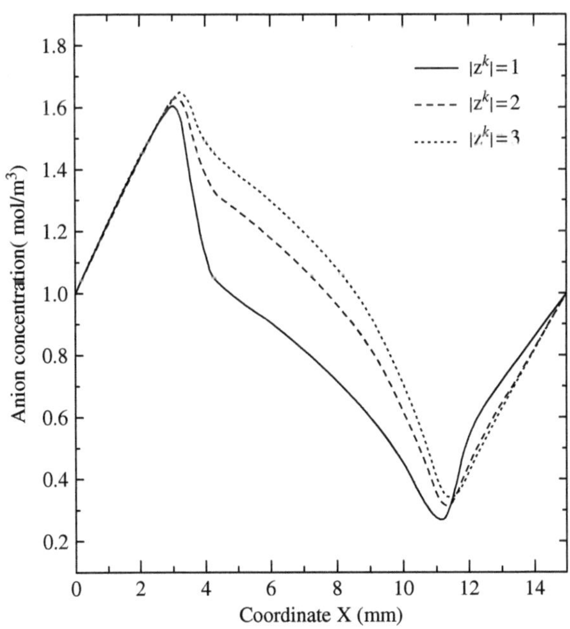

Fig. 3.23 Effect of the valence of surrounding solution on the variation of the diffusive anion concentration distribution

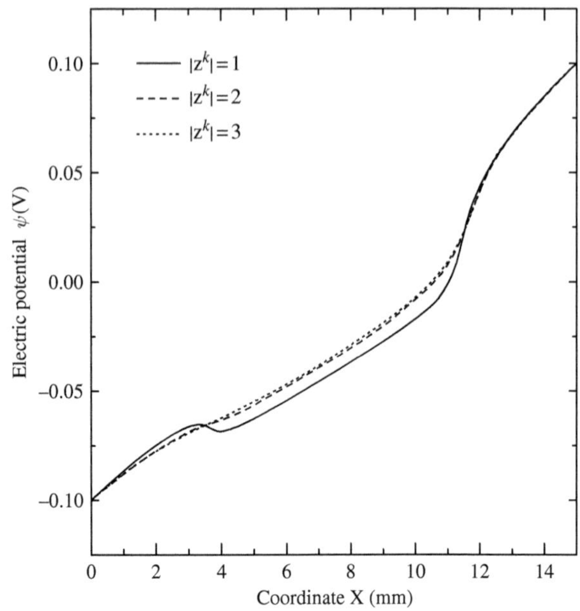

Fig. 3.24 Effect of the valence of surrounding solution on the variation of the electric potential distribution

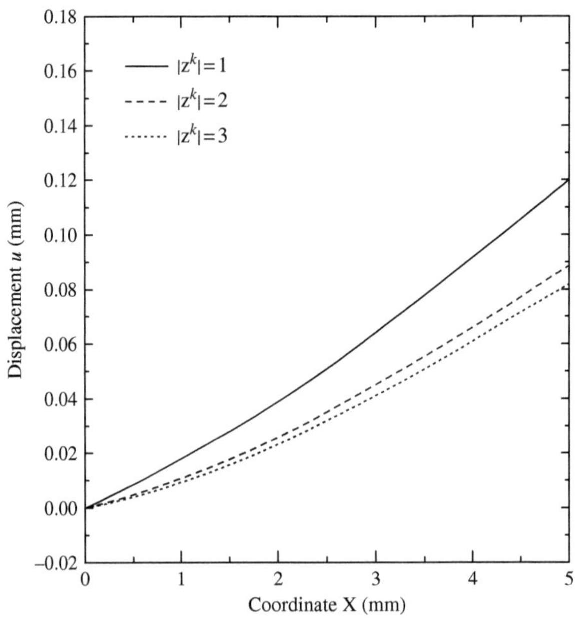

Fig. 3.25 Effect of the valence of surrounding solution on the variation of the displacement distribution of the hydrogel

3.4 Transient Simulation for Kinetics of Hydrogel

In this section, the transient simulation for the kinetics of the electric stimulus-responsive hydrogels is conducted by the developed MECe model. Similarly, after the discretization of the MECe governing equations, a numerical comparison is made between the presently simulated results and experimental data extracted from open literature. It is followed by the detailed discussions of the kinetic variations of several important characteristics of the system.

3.4.1 Numerical Implementation

For transient analysis of the electric-sensitive hydrogels, the time term is of critical importance in the governing equations of the MECe model. By the non-dimensional nonlinear partial differential governing equations of the MECe model expressed by Eqs. (3.68), (3.69), (3.70) and (3.71), non-dimensional MECe transient governing equations in one dimension can be written as

$$L_{\text{ref}}^2 \frac{\partial \bar{c}^k}{\partial t} = D_k \left(\frac{\partial^2 \bar{c}^k}{\partial \bar{x}^2} + \alpha z^k \frac{\partial \bar{c}^k}{\partial \bar{x}} \frac{\partial \bar{\psi}}{\partial \bar{x}} + \alpha z^k \bar{c}^k \frac{\partial^2 \bar{\psi}}{\partial \bar{x}^2} \right) \quad (k = +, -) \tag{3.94}$$

$$\frac{\partial^2 \bar{\psi}}{\partial \bar{x}^2} + \frac{F_c^2 L_{\text{ref}}^2 c_{\text{ref}}}{\varepsilon \varepsilon_0 RT\alpha} \left(\sum_{k=+,-} z^k \bar{c}^k + z^f \bar{c}^f \right) = 0 \tag{3.95}$$

$$(3\lambda_s + 2\mu_s) \frac{\partial^2 \bar{u}}{\partial \bar{x}^2} - \beta RT c_{\text{ref}} \frac{\partial \bar{p}}{\partial \bar{x}} = 0 \tag{3.96}$$

$$\frac{L_{\text{ref}}^2 f_{ws}}{c_{\text{ref}} RT} \frac{\partial \bar{u}_x}{\partial t} = \beta \left(\phi^w \frac{\partial^2 \bar{p}}{\partial \bar{x}^2} + 2 \frac{\partial \phi^w}{\partial \bar{x}} \frac{\partial \bar{p}}{\partial \bar{x}} \right) + \alpha \left[2 \frac{\partial \phi^w}{\partial \bar{x}} \frac{\partial \bar{\psi}}{\partial \bar{x}} \sum_{k=+,-} z^k \bar{c}^k + \phi^w \left(\frac{\partial \bar{\psi}}{\partial \bar{x}} \sum_{k=+,-} \left(z^k \frac{\partial \bar{c}^k}{\partial \bar{x}} \right) + \frac{\partial^2 \bar{\psi}}{\partial \bar{x}^2} \sum_{k=+,-} z^k \bar{c}^k \right) \right] \tag{3.97}$$

in which all the non-dimensional variables are defined in Eq. (3.67).

Using the well-known θ-weighted finite difference scheme ($0.5<\theta<1.0$), Eq. (3.94) can be discretized first in temporal domain as follows:

$$\bar{c}_{(n+1)}^k - \bar{c}_{(n)}^k = \frac{\Delta t D_k}{L_{\text{ref}}^2} \left[\theta \left(\frac{\partial^2 \bar{c}_{(n+1)}^k}{\partial \bar{x}^2} + \alpha z^k \frac{\partial \bar{c}_{(n+1)}^k}{\partial \bar{x}} \frac{\partial \bar{\psi}_{(n+1)}}{\partial \bar{x}} + \alpha z^k \bar{c}_{(n+1)}^k \frac{\partial^2 \bar{\psi}_{(n+1)}}{\partial \bar{x}^2} \right) + (1-\theta) \left(\frac{\partial^2 \bar{c}_{(n)}^k}{\partial \bar{x}^2} + \alpha z^k \frac{\partial \bar{c}_{(n)}^k}{\partial \bar{x}} \frac{\partial \bar{\psi}_{(n)}}{\partial \bar{x}} + \alpha z^k \bar{c}_{(n)}^k \frac{\partial^2 \bar{\psi}_{(n)}}{\partial \bar{x}^2} \right) \right] \tag{3.98}$$

where the subscripts n and (n+1) denote time variable at $t = t_n$ and at subsequent time $t = t_{n+1}$, respectively, and $\Delta t = t_{n+1} - t_n$ is the time step.

By the meshless Hermite-cloud method, the governing equation (3.98) is discretized further in spatial domain as follows:

$$
\begin{aligned}
&\sum_{j=1}^{N_{\text{total}}} N_j(\bar{x}_i)\bar{c}^k_{j(n+1)} - \sum_{m=1}^{N_{\text{total}}}\left(\bar{x}_i - \sum_{j=1}^{N_{\text{total}}} N_j(\bar{x}_i)\bar{x}_j\right) M_m(\bar{x}_i)\bar{c}_{xm(n+1)} \\
&\quad - \frac{\Delta t D_k \theta}{L^2_{ref}}\left\{\sum_{j=1}^{N_{\text{total}}} N_{xxj}(\bar{x}_i)\bar{c}^k_{j(n+1)}\right. \\
&\quad + \alpha z^k \sum_{m=1}^{N_{\text{total}}} M_m(\bar{x}_i)\bar{c}^k_{xm(n+1)} \sum_{m=1}^{N_{\text{total}}} M_m(\bar{x}_i)\bar{\psi}_{xm(n+1)} \\
&\quad + \alpha z^k \sum_{j=1}^{N_{\text{total}}} N_{xxj}(\bar{x}_i)\bar{\psi}_{j(n+1)}\left[\sum_{j=1}^{N_{\text{total}}} N_j(\bar{x}_i)\bar{c}^k_{j(n+1)}\right. \\
&\quad \left.\left. - \sum_{m=1}^{N_{\text{total}}}\left(\bar{x}_i - \sum_{j=1}^{N_{\text{total}}} N_j(\bar{x}_i)\bar{x}_j\right) M_m(\bar{x}_i)\bar{c}_{xm(n+1)}\right]\right\} \\
&= \sum_{j=1}^{N_{\text{total}}} N_j(\bar{x}_i)\bar{c}^k_{j(n)} - \sum_{m=1}^{N_{\text{total}}}\left(\bar{x}_i - \sum_{j=1}^{N_{\text{total}}} N_j(\bar{x}_i)\bar{x}_j\right) M_m(\bar{x}_i)\bar{c}_{xm(n)} \\
&\quad + \frac{\Delta t D_k (1-\theta)}{L^2_{ref}}\left\{\sum_{j=1}^{N_{\text{total}}} N_{xxj}(\bar{x}_i)\bar{c}^k_{j(n)} + \alpha z^k \sum_{m=1}^{N_{\text{total}}} M_m(\bar{x}_i)\bar{c}^k_{xm(n)} \sum_{m=1}^{N_{\text{total}}} M_m(\bar{x}_i)\bar{\psi}_{xm(n)}\right. \\
&\quad + \alpha z^k \left[\sum_{j=1}^{N_{\text{total}}} N_j(\bar{x}_i)\bar{c}^k_{j(n)} - \sum_{m=1}^{N_{\text{total}}}\left(\bar{x}_i - \sum_{j=1}^{N_{\text{total}}} N_j(\bar{x}_i)\bar{x}_j\right)\right. \\
&\quad \left.\left. M_m(\bar{x}_i)\bar{c}_{xm(n)}\right] \sum_{j=1}^{N_{\text{total}}} N_{xxj}(\bar{x}_i)\bar{\psi}_{j(n)}\right\}
\end{aligned}
\tag{3.99}
$$

Similarly, the other three governing equations (3.95), (3.96) and (3.97) can also be discretized in both temporal and spatial domains as follows:

$$
\begin{aligned}
&\sum_{j=1}^{N_{\text{total}}} N_{xxj}(\bar{x}_i)\bar{\psi}_{j(n+1)} + \frac{F_c^2 L^2_{\text{ref}} c_{\text{ref}}}{\varepsilon\varepsilon_0 RT\alpha}\left\{z^f \bar{c}^f + \sum_{k=+,-} z^k \left[\sum_{j=1}^{N_{\text{total}}} N_j(\bar{x}_i)\bar{c}^k_{j(n+1)}\right.\right. \\
&\quad \left.\left. - \sum_{m=1}^{N_{\text{total}}}\left(\bar{x}_i - \sum_{j=1}^{N_{\text{total}}} N_j(\bar{x}_i)\bar{x}_j\right) M_m(\bar{x}_i)\bar{c}_{xm(n+1)}\right]\right\} = 0
\end{aligned}
\tag{3.100}
$$

$$
(3\lambda_s + 2\mu_s)\sum_{j=1}^{N_{\text{gel}}} N_{xxj}(\bar{x}_i)\bar{u}_{j(n+1)} - \beta RT c_{\text{ref}} \sum_{m=1}^{N_{\text{gel}}} M_m(\bar{x}_i)\bar{p}_{xm(n+1)} = 0 \tag{3.101}
$$

$$
\begin{aligned}
&\frac{L_{\text{ref}}^2 f_{ws}}{c_{\text{ref}}RT}\sum_{m=1}^{N_{\text{gel}}} M_m(\bar{x}_i)\bar{u}_{xm(n+1)} - \Delta t\theta\beta\left[\phi^w \sum_{j=1}^{N_{\text{gel}}} N_{xxj}(\bar{x}_i)\bar{p}_{j(n+1)}\right.\\
&\quad\left.+2\frac{\partial\phi^w}{\partial\bar{x}}\sum_{m=1}^{N_{\text{gel}}} M_m(\bar{x}_i)\bar{p}_{xm(n+1)}\right] - \alpha\Delta t\theta\left\{2\frac{\partial\phi^w}{\partial\bar{x}}\left[\sum_{m=1}^{N_{\text{gel}}} M_m(\bar{x}_i)\bar{\psi}_{xm(n+1)}\right]\right.\\
&\quad\times\left[\sum_{k=+,-} z^k \sum_{j=1}^{N_{\text{gel}}}\left(N_j(\bar{x}_i)\bar{c}^k_{j(n+1)} - \sum_{m=1}^{N_{\text{gel}}}\left(\bar{x}_i - \sum_{j=1}^{N_{\text{gel}}} N_j(\bar{x}_i)\bar{x}_j\right) M_m(\bar{x}_i)\bar{c}_{xm(n+1)}\right)\right]\\
&\quad+\phi^w\left[\sum_{m=1}^{N_{\text{gel}}} M_m(\bar{x}_i)\bar{\psi}_{xm(n+1)}\right]\left(\sum_{k=+,-} z^k \sum_{m=1}^{N_{\text{gel}}} M_m(\bar{x}_i)\bar{c}^k_{xm(n+1)}\right)\\
&\quad+\sum_{j=1}^{N_{\text{gel}}} N_{xxj}(\bar{x}_i)\bar{\psi}_{j(n+1)}\sum_{k=+,-} z^k \sum_{j=1}^{N_{\text{gel}}}\left(N_j(\bar{x}_i)\bar{c}^k_{j(n+1)}\right.\\
&\quad\left.\left.\left.-\sum_{m=1}^{N_{\text{gel}}}\left(\bar{x}_i - \sum_{j=1}^{N_{\text{gel}}} N_j(\bar{x}_i)\bar{x}_j\right) M_m(\bar{x}_i)\bar{c}_{xm(n+1)}\right)\right]\right\}\\
&= \frac{L_{\text{ref}}^2 f_{ws}}{c_{\text{ref}}RT}\sum_{m=1}^{N_{\text{gel}}} M_m(\bar{x}_i)\bar{u}_{xm(n)} + \Delta t(1-\theta)\beta\left[\phi^w \sum_{j=1}^{N_{\text{gel}}} N_{xxj}(\bar{x}_i)\bar{p}_{j(n)}\right.\\
&\quad\left.+2\frac{\partial\phi^w}{\partial\bar{x}}\sum_{m=1}^{N_{\text{gel}}} M_m(\bar{x}_i)\bar{p}_{xm(n)}\right]\\
&\quad+\alpha\Delta t(1-\theta)\left\{2\frac{\partial\phi^w}{\partial\bar{x}}\left[\sum_{m=1}^{N_{\text{gel}}} M_m(\bar{x}_i)\bar{\psi}_{xm(n)}\right]\left[\sum_{k=+,-} z^k \sum_{j=1}^{N_{\text{gel}}}\left(N_j(\bar{x}_i)\bar{c}^k_{j(n)}\right.\right.\right.\\
&\quad\left.\left.-\sum_{m=1}^{N_{\text{gel}}}\left(\bar{x}_i - \sum_{j=1}^{N_{\text{gel}}} N_j(\bar{x}_i)\bar{x}_j\right) M_m(\bar{x}_i)\bar{c}_{xm(n)}\right)\right]\\
&\quad+\phi^w\left[\sum_{m=1}^{N_{\text{gel}}} M_m(\bar{x}_i)\bar{\psi}_{xm(n)}\right]\left(\sum_{k=+,-} z^k \sum_{m=1}^{N_{\text{gel}}} M_m(\bar{x}_i)\bar{c}^k_{xm(n)}\right)\\
&\quad+\sum_{j=1}^{N_{\text{gel}}} N_{xxj}(\bar{x}_i)\bar{\psi}_{j(n)}\sum_{k=+,-} z^k \sum_{j=1}^{N_{\text{gel}}}\left(N_j(\bar{x}_i)\bar{c}^k_{j(n)}\right.\\
&\quad\left.\left.\left.-\sum_{m=1}^{N_{\text{gel}}}\left(\bar{x}_i - \sum_{j=1}^{N_{\text{gel}}} N_j(\bar{x}_i)\bar{x}_j\right) M_m(\bar{x}_i)\bar{c}_{xm(n)}\right)\right]\right\}
\end{aligned}
\tag{3.102}
$$

In addition, the following auxiliary equations are required by the Hermite theorem:

$$\sum_{j=1}^{N_{\text{total}}} N_{xj}(\bar{x}_i)\bar{c}_j^k - \left[\sum_{j=1}^{N_{\text{total}}} N_{xj}(\bar{x}_i)\bar{x}_j\right] \sum_{m=1}^{N_{\text{total}}} M_m(\bar{x}_i)\bar{c}_{xm}^k = 0 \tag{3.103}$$

$$\sum_{j=1}^{N_{\text{total}}} N_{xj}(\bar{x}_i)\bar{\psi}_j - \left[\sum_{j=1}^{N_{\text{total}}} N_{xj}(\bar{x}_i)\bar{x}_j\right] \sum_{m=1}^{N_{\text{total}}} M_m(\bar{x}_i)\bar{\psi}_{xm} = 0 \tag{3.104}$$

$$\sum_{j=1}^{N_{\text{gel}}} N_{xj}(\bar{x}_i)\bar{u}_j - \left[\sum_{j=1}^{N_{\text{gel}}} N_{xj}(\bar{x}_i)\bar{x}_j\right] \sum_{m=1}^{N_{\text{gel}}} M_m(\bar{x}_i)\bar{u}_{xm} = 0 \tag{3.105}$$

$$\sum_{j=1}^{N_{\text{gel}}} N_{xj}(\bar{x}_i)\bar{p}_j - \left[\sum_{j=1}^{N_{\text{gel}}} N_{xj}(\bar{x}_i)\bar{x}_j\right] \sum_{m=1}^{N_{\text{gel}}} M_m(\bar{x}_i)\bar{p}_{xm} = 0 \tag{3.106}$$

3.4.2 Model Validation with Experiment

In terms of the experiments of the kinetics of the electric-sensitive hydrogels responsive to externally applied electric field, extensive search of literature has thus far found one work only published by Shiga et al. (1990), who measured the end point displacement D of the hydrogel strip in a conveniently observing manner in their experiments, as shown in Fig. 3.1, instead of the displacement u at the edge point a of the hydrogel domain between points a and b along x-axis for the present one-dimensional simulations. A relation is thus required between these two displacements D and u for comparison of the simulated results with the experimental data. The experimental data (Shiga et al., 1990) employed as the input data for the present computation include $T = 298$ (K), $R = 8.314$ (J/mol·K), $F = 9.648 \times 10^4$ (C/mol), $\varepsilon_0 = 8.854 \times 10^{-12}(\mathrm{C}^2/\mathrm{N}\cdot\mathrm{m}^2)$, ε=80, $\phi_0^w = 0.8$, $L = 5.0\times10^{-2}$ (m), $h = 5.0\times10^{-3}$ (m), $V_e = 3.0$ (V), $c_0^f = 35.3$ (mol/m^3), $c^* = 35.3$ (mol/m^3), $3\lambda + 2\mu = 1.8 \times 10^4$(Pa) that is adjustable for determination of reference configuration. The experimentally measured displacement D and numerically computed displacement u are tabulated in Table 3.1, at time $t = 1$, 2, 3, 4 and 5 (min), respectively. For construction of the relation between the displacements D and u, the least square method is applied for both the displacements D and u at time $t = 1$, 3 and

Table 3.1 Displacements D and u at the end point of the hydrogel strip and at the edge point a of the one-dimensional hydrogel domain, respectively

Time (min)	1	2	3	4	5
Displacement D (mm)	2.1	3.1	4.1	5.0	6.1
Displacement u (mm)	0.45	0.97	1.54	1.94	2.27

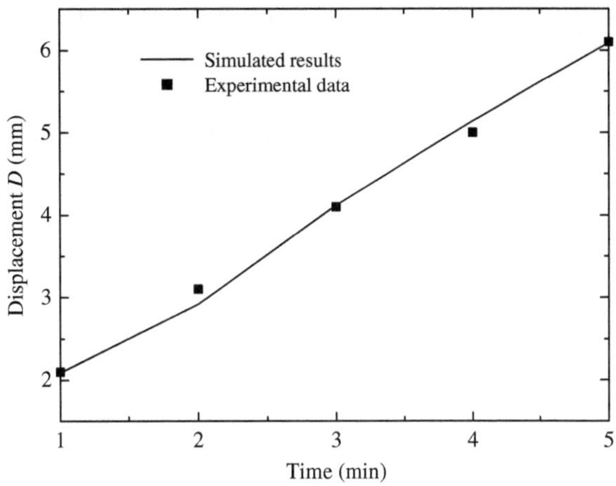

Fig. 3.26 Comparison between the transient simulations and the experimental data (Shiga et al., 1990)

5 (min) with best fitting. Based on the data in Table 3.1, the relation between the displacements D and u is constructed as

$$D = 1.58 + 0.93u + 0.47u^2 \tag{3.107}$$

By substituting the displacements D and u at time $t = 2$ and 4 (min) into Eq. (3.107), the relative discrepancies are computed and they are generally smaller than 6%, which validates the relation (Eq. (3.107)) between the displacements D and u acceptable.

Figure 3.26 demonstrates the comparison of the experimental end point displacement D with the corresponding simulated end point displacement D computed by Eq. (3.107), where very good agreement is achieved. This confirms that the presently developed MECe model is applicable for the kinetic analysis of the electric stimulus-responsive hydrogels.

3.4.3 Parameter Studies

In the previous steady-state simulations for equilibrium of the electric-sensitive hydrogels, the attentions are given to the influences of several important material properties and environmental conditions on the responsive equilibrium characteristics of the hydrogels, for example, the distributions of the diffusive ionic concentrations, the electric potential and the hydrogel displacement. In the present transient analysis for kinetics of the electric-sensitive hydrogels, the attentions are given to the variations of distributive profiles of these properties and conditions with time. In the following transient simulations, several input data taken are $T = 298$ (K), $R = 8.314$ (J/mol·K), $F = 9.648\times10^4$ (C/mol), $\phi_0^w = 0.8$, $3\lambda + 2\mu = 1.2 \times 10^5$ (Pa),ε_0=8.854×10^{-12} (C^2/N.m^2), ε=80 $z^f = -1$, $L =$

1.5×10^{-2} (m), $h = 5\times10^{-3}$ (m), $f_{ws} = 7.0\times10^{-16}$ (N·s/m^4) and $D_k = 1.0\times10^{-7}$ (m^2/s).

3.4.3.1 Variation of Ionic Concentration Distribution with Time

Figures 3.27, 3.28, 3.29, 3.30, 3.31, 3.32, 3.33, 3.34, 3.35, 3.36, 3.37, 3.38, 3.39 and 3.40 present the variations of the diffusive ionic concentration distributions with time under various combinations of the electric fields, the fixed charge densities and the bath solution concentrations. The figures depict the kinetics of the electric stimulus-responsive hydrogels and the kinetic characteristics of the ionic diffusion. It is observed that the diffusive ionic concentrations are distributed symmetrically over whole computational domain when the external electric field is not imposed initially at time $t = 0$, which results from the steady-state simulations without the external electric field ($V_e = 0$). However, once the electric voltage is applied on the system, the distributive profiles of the diffusive ionic concentrations are no longer symmetric. With the increase of time, the diffusive ions redistribute continuously in both the interior hydrogel and the exterior bath solution, and the differences of the diffusive ionic concentrations over the hydrogel–solution interfaces become larger and larger. It is expected that the ionic diffusion and convection will reach the equilibrium state after certain time, which is dependent on many effects including material properties and environmental conditions, for example, the fixed charge

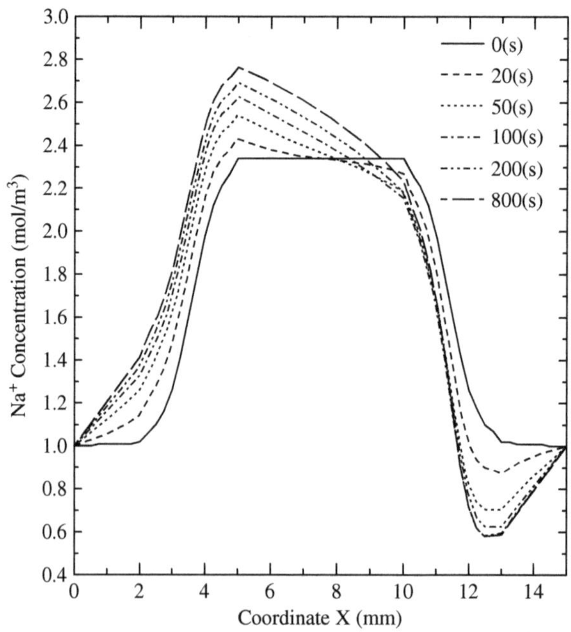

Fig. 3.27 Variation of the diffusive cation Na$^+$ concentration distribution with time for $V_e = 0.2$(V), $c_0{}^f = 2$ (mol/m^3) and $c^* = 1$ (mol/m^3)

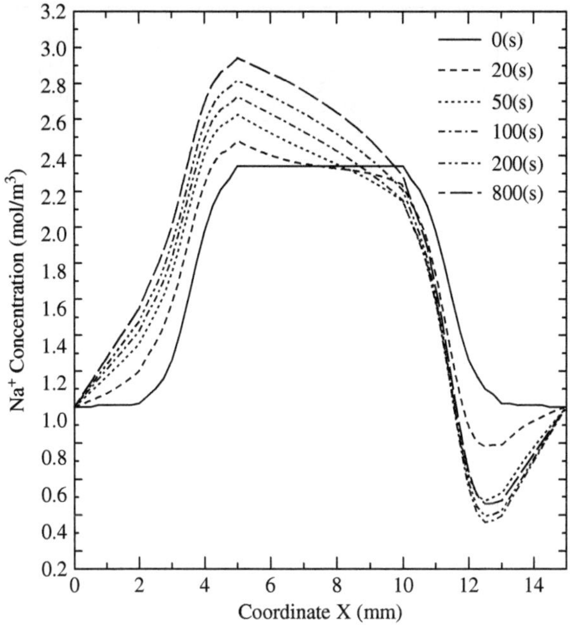

Fig. 3.28 Variation of the diffusive cation Na^+ concentration distribution with time for $V_e = 0.3$(V), $c_0{}^f = 2$ (mol/m^3) and $c^* = 1$ (mol/m^3)

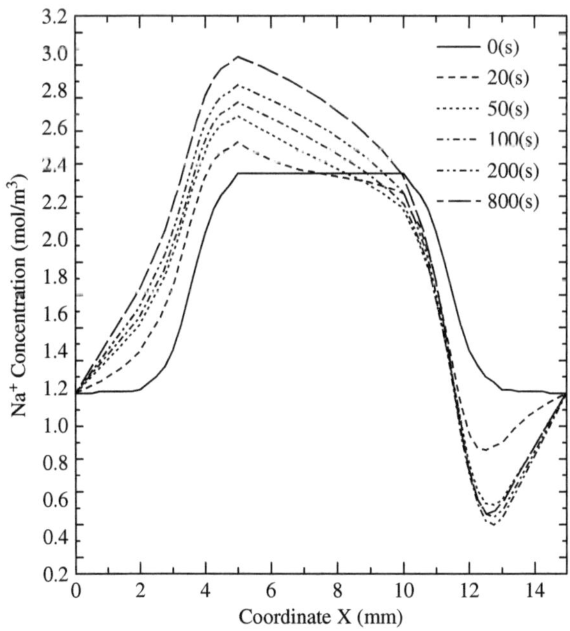

Fig. 3.29 Variation of the diffusive cation Na^+ concentration distribution with time for $V_e = 0.4$(V), $c_0{}^f = 2$ (mol/m^3) and $c^* = 1$ (mol/m^3)

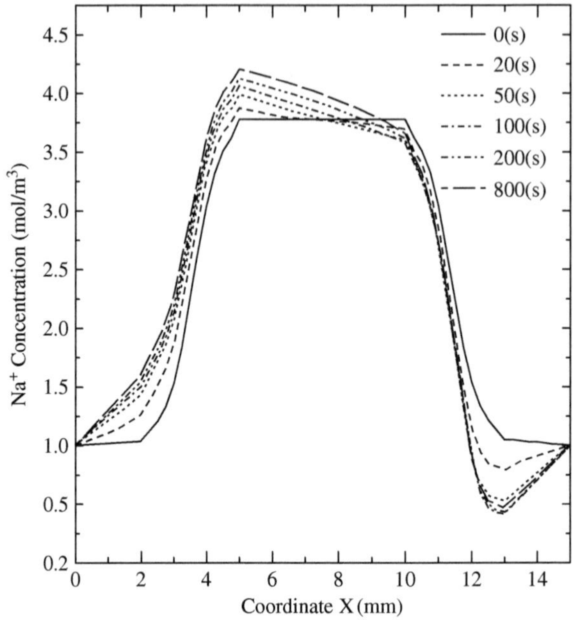

Fig. 3.30 Variation of the diffusive cation Na^+ concentration distribution with time for V_e = 0.2(V), $c_0{}^f = 4$ (mol/m^3) and $c^* = 1$(mol/m^3)

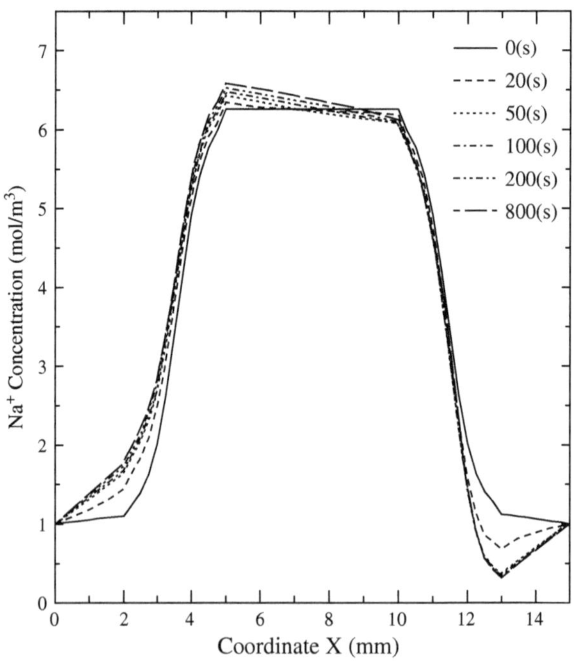

Fig. 3.31 Variation of the diffusive cation Na^+ concentration distribution with time for V_e = 0.2(V), $c_0{}^f = 8$ (mol/m^3) and $c^* = 1$ (mol/m^3)

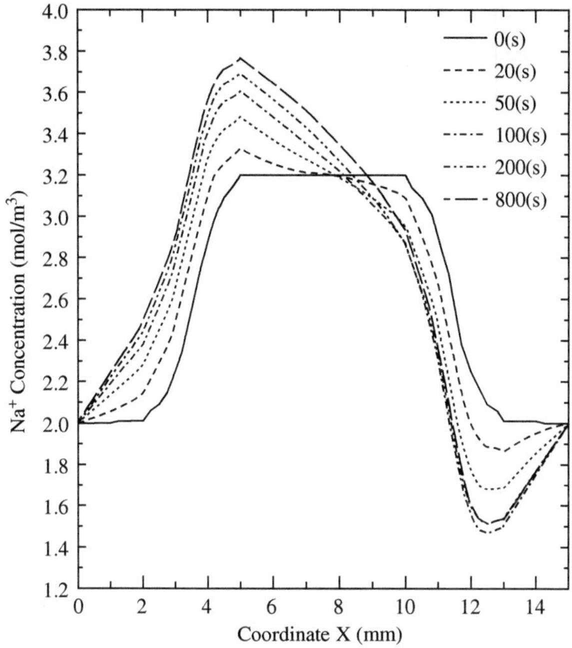

Fig. 3.32 Variation of the diffusive cation Na^+ concentration distribution with time for $V_e = 0.2$ (V), $c_0{}^f = 2$ (mol/m^3) and $c^* = 2$ (mol/m^3)

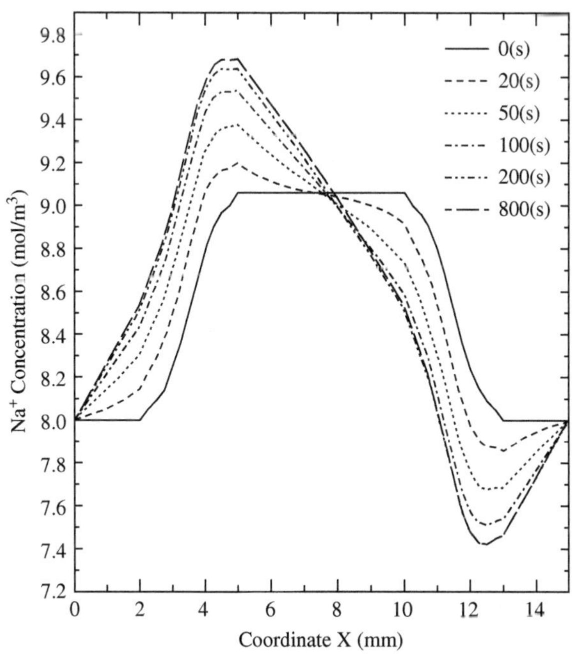

Fig. 3.33 Variation of the diffusive cation Na^+ concentration distribution with time for $V_e =$ 0.2(V), $c_0{}^f = 2$ (mol/m^3) and $c^* = 8$ (mol/m^3)

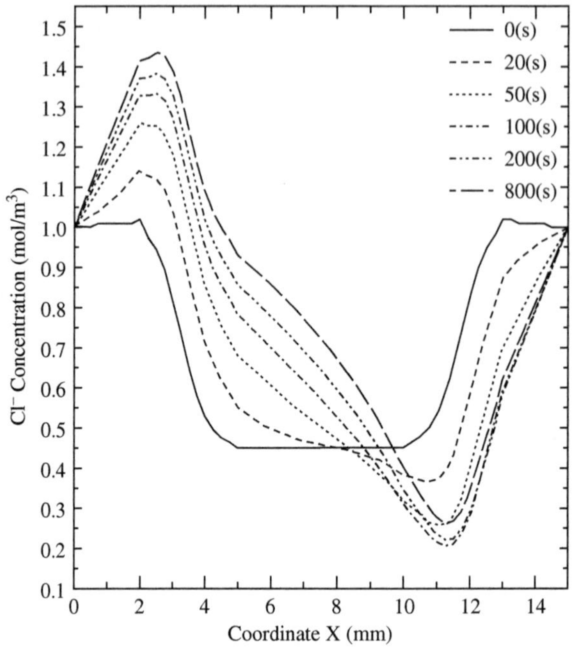

Fig. 3.34 Variation of the diffusive anion Cl^- concentration distribution with time for $V_e = 0.2$ (V), $c_0{}^f = 2$ (mol/m^3) and $c^* = 1$ (mol/m^3)

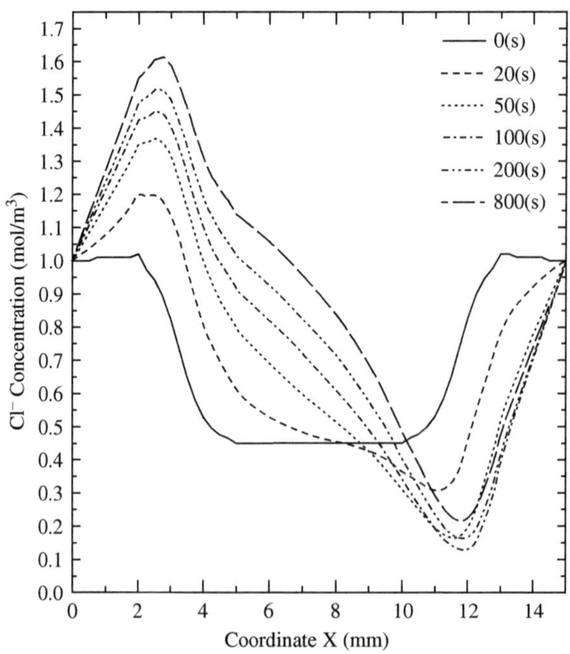

Fig. 3.35 Variation of the diffusive anion Cl^- concentration distribution with time for $V_e = 0.3$ (V), $c_0{}^f = 2$ (mol/m^3) and $c^* = 1$ (mol/m^3)

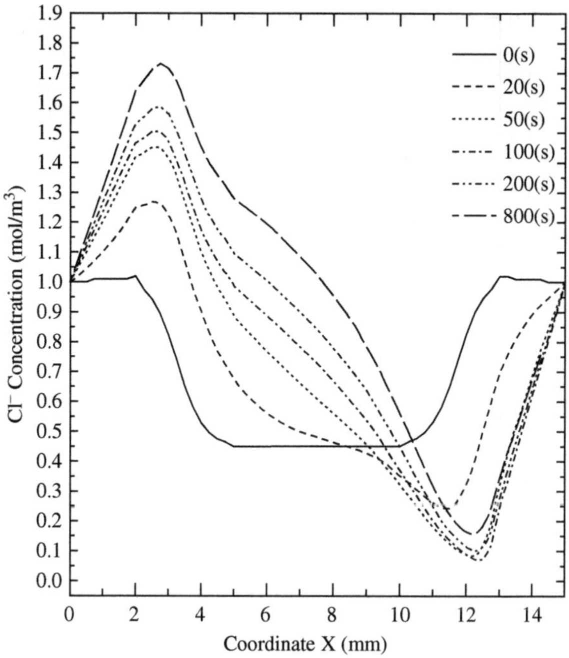

Fig. 3.36 Variation of the diffusive anion Cl^- concentration distribution with time for $V_e = 0.4$ (V), $c_0{}^f = 2$ (mol/m^3) and $c^* = 1$ (mol/m^3)

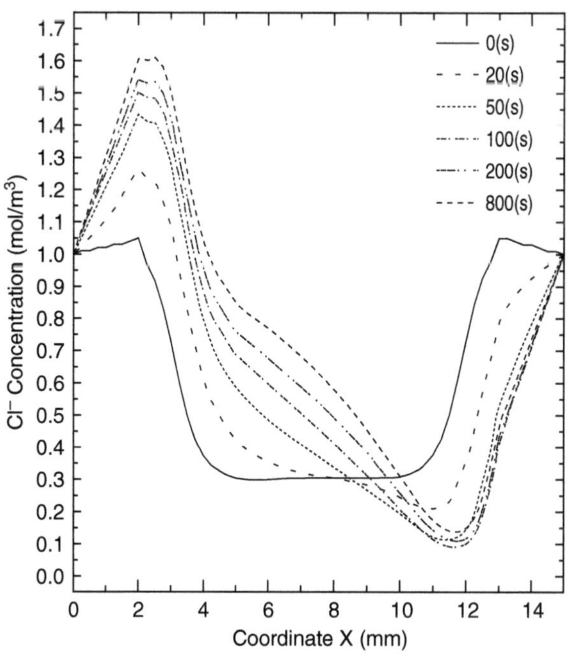

Fig. 3.37 Variation of the diffusive anion Cl^- concentration distribution with time for $V_e = 0.2$ (V), $c_0{}^f = 4$ (mol/m^3) and $c^* = 1$ (mol/m^3)

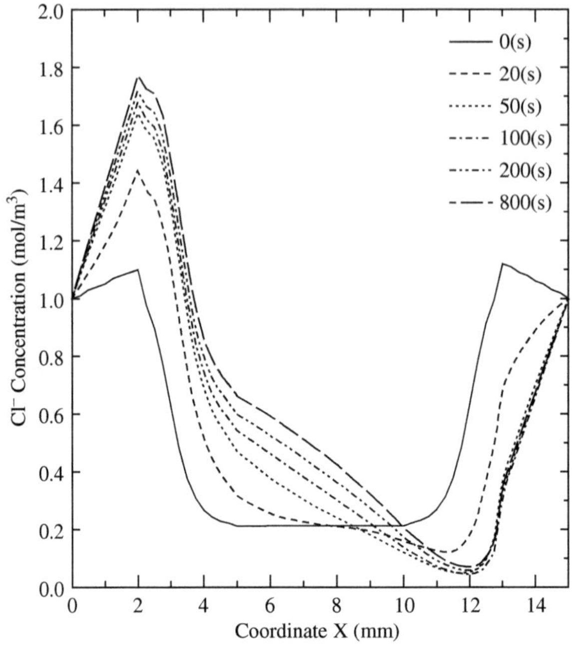

Fig. 3.38 Variation of the diffusive anion Cl^- concentration distribution with time for $V_e = 0.2$ (V), $c_0^f = 8$ (mol/m^3) and $c^* = 1$ (mol/m^3)

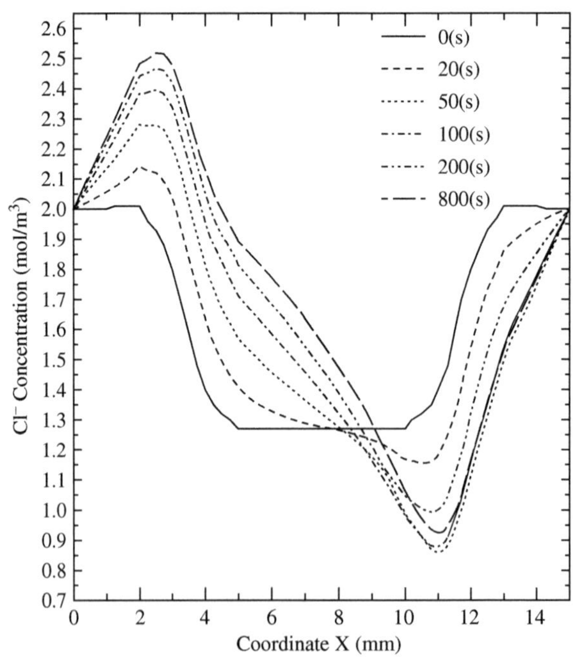

Fig. 3.39 Variation of the diffusive anion Cl^- concentration distribution with time for $V_e = 0.2$ (V), $c_0^f = 2$ (mol/m^3) and $c^* = 2$ (mol/m^3)

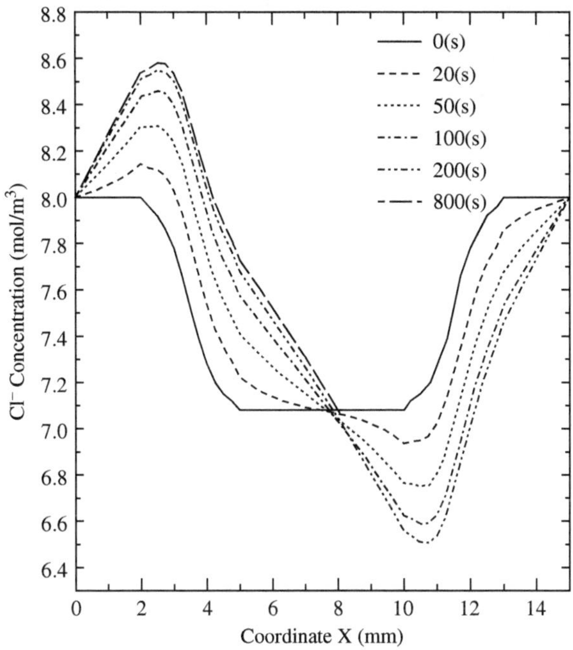

Fig. 3.40 Variation of the diffusive anion Cl^- concentration distribution with time for $V_e = 0.2$ (V), $c_0^f = 2$ (mol/m^3) and $c^* = 8$ (mol/m^3)

density, the externally applied electric field and the bath solution concentration. In brief, the trends of kinetics of the diffusive ionic concentration distributions shown in Figs. 3.27, 3.28, 3.29, 3.30, 3.31, 3.32, 3.33, 3.34, 3.35, 3.36, 3.37, 3.38, 3.39 and 3.40 are in good agreement with the FEM results conducted by Wallmersperger et al. (2004).

It is also found from Figs. 3.27, 3.28 and 3.29 and 3.34, 3.35 and 3.36 that, if other parameters are given at a given time, the peak concentrations of the diffusive ionic species on the hydrogel–solution interface near the cathode increase with the enlargement of the applied electric voltage, while those near the anode decrease. This means that the difference of the diffusive ionic concentrations between the two hydrogel–solution interfaces increase with the applied electric field at a given time. The present phenomena of the transient simulations are consistent with those in steady-state studies, where the difference of the diffusive ionic concentrations in equilibrium state increases as well with increment of the applied electric voltage.

Variations of the diffusive ionic concentration distributions with time under the various fixed charge densities are illustrated in Figs. 3.27, 3.30, 3.31, 3.34, 3.37 and 3.38, from which the similar characteristics are observed to the steady-state simulations. With the change of the fixed charge densities, the changes of distributive profiles of the diffusive cation Na^+ are significant at a given time, while those of the diffusive anion Cl^- are insignificant. Probably this results from the negative valence of the fixed charge groups. By comparison of Figs. 3.27, 3.32, 3.33, 3.34, 3.39 and 3.40, it is obviously known that the gradient of the kinetics of the diffusive

Fig. 3.41 Variation of the distributive electric potential with time for $V_e = 0.2$ (V), $c_0{}^f = 2$ (mol/m^3) and $c^* = 1$ (mol/m^3)

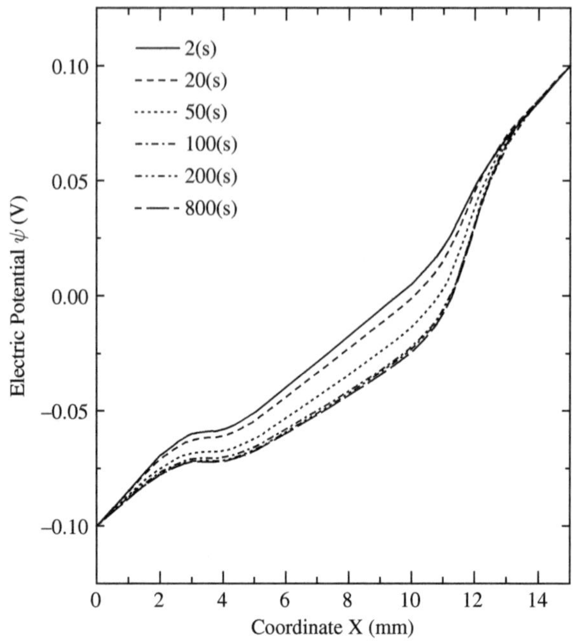

ion concentrations increases within the hydrogels as the bath solution concentration increases.

3.4.3.2 Variation of Electric Potential Distribution with Time

Figures 3.41, 3.42, 3.43, 3.44, 3.45, 3.46 and 3.47 show the kinetic variations of the distributive electric potential under different environmental conditions. It is presented that the downward step of the electric potential distributed within the hydrogels generally becomes larger with the increase of time. After time about 100 (s), the downward step changes gradually. It is found further from Figs. 3.41, 3.42 and 3.43 that, with the increase of externally applied electric voltage, the downward step of electric potential distributed over the hydrogel–solution interface near the cathode diminishes gradually, while that near the anode increases. It is also seen from Figs. 3.41, 3.44 and 3.45 that, with the increase of the fixed charge density, the variation of the electric potential distributions with time becomes small. Probably the reason is that the higher density the fixed charge groups have, the more mobile ions diffuse into the hydrogels. This makes the conductivity of the hydrogels much closer to that of the surrounding bath solution.

3.4.3.3 Variation of Hydrogel Displacement Distribution with Time

Figures 3.48, 3.49, 3.50, 3.51, 3.52, 3.53 and 3.54 present the variations of the distributive displacement of the hydrogel with time. The trend of the changing

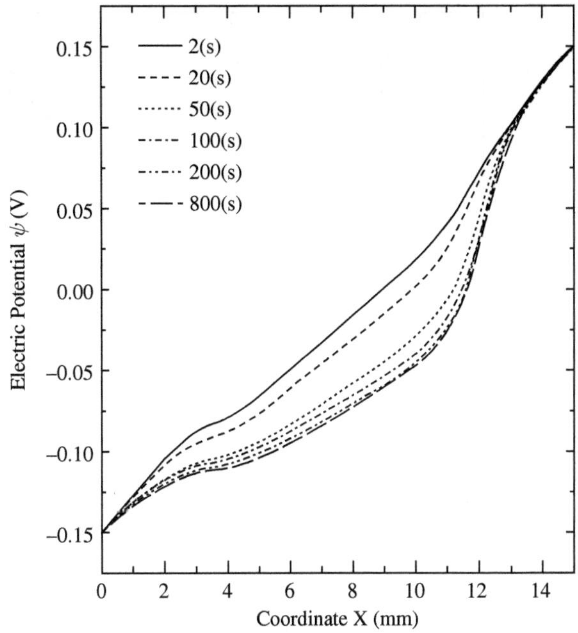

Fig. 3.42 Variation of the distributive electric potential with time for $V_e = 0.3$ (V), $c_0^f = 2$ (mol/m^3) and $c^* = 1$ (mol/m^3)

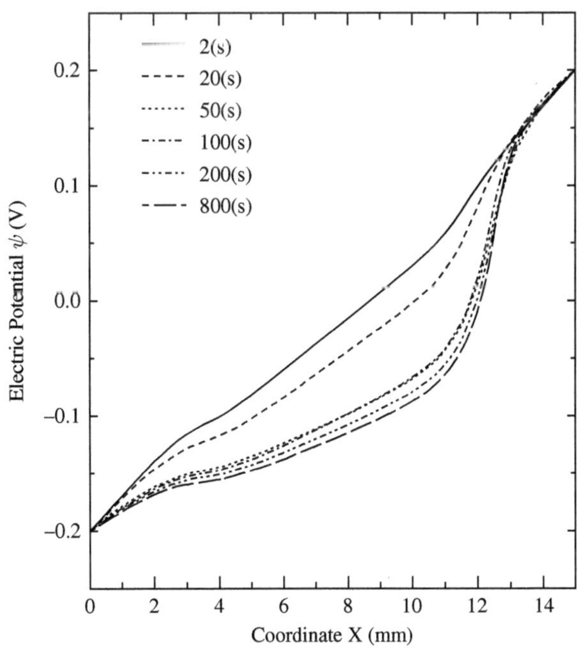

Fig. 3.43 Variation of the distributive electric potential with time for $V_e = 0.4$ (V), $c_0^f = 2$ (mol/m^3) and $c^* = 1$ (mol/m^3)

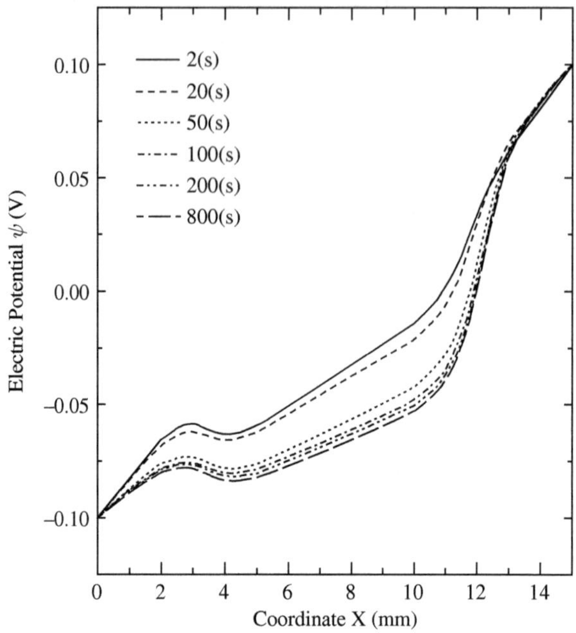

Fig. 3.44 Variation of the distributive electric potential with time for $V_e = 0.2$ (V), $c_0{}^f = 4$ (mol/m^3) and $c^* = 1$ (mol/m^3)

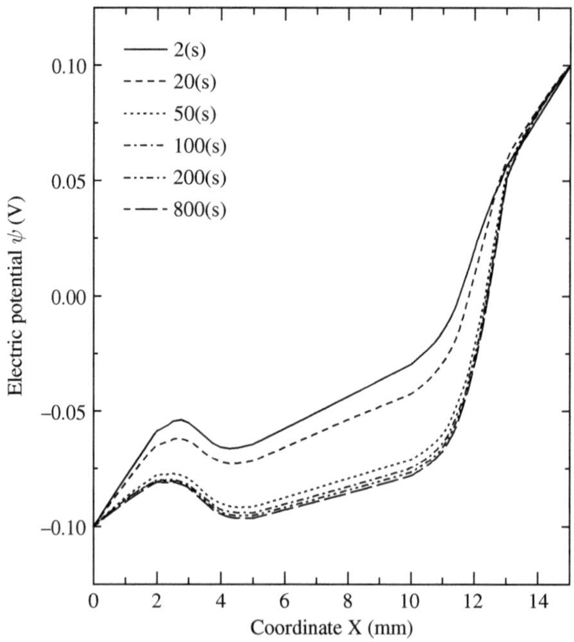

Fig. 3.45 Variation of the distributive electric potential with time for $V_e = 0.2$ (V), $c_0{}^f = 8$ (mol/m^3) and $c^* = 1$ (mol/m^3)

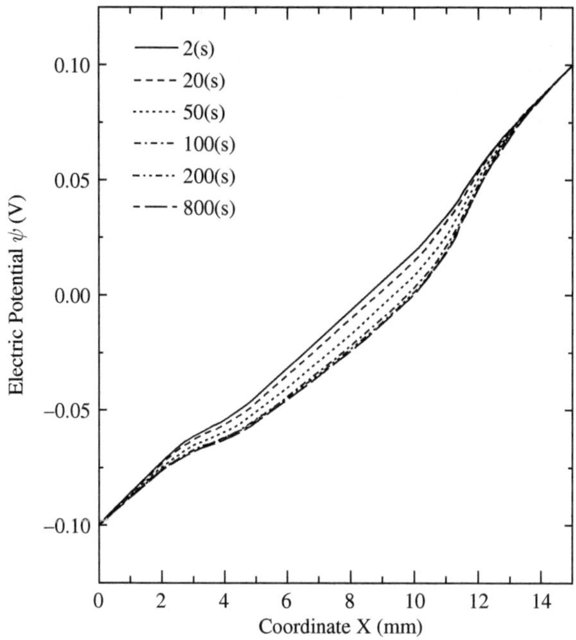

Fig. 3.46 Variation of the distributive electric potential with time for $V_e = 0.2$ (V), $c_0{}^f = 2$ (mol/m^3) and $c^* = 2$ (mol/m^3)

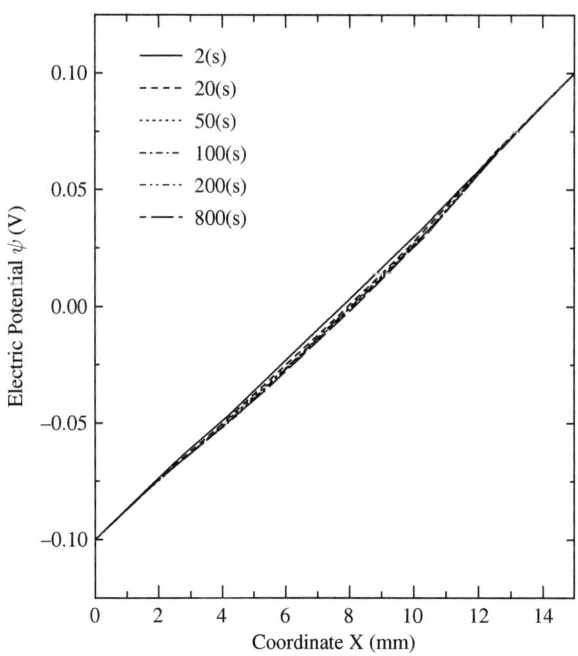

Fig. 3.47 Variation of the distributive electric potential with time for $V_e = 0.2$ (V), $c_0{}^f = 2$ (mol/m^3) and $c^* = 8$ (mol/m^3)

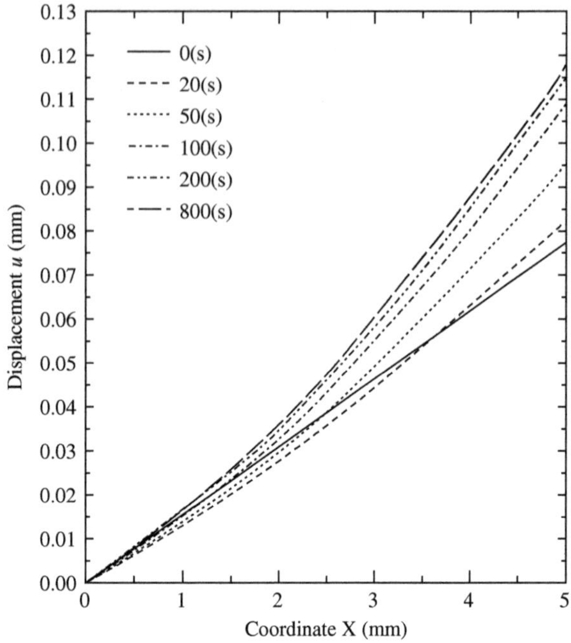

Fig. 3.48 Variation of the distributive displacement of the hydrogel with time for $V_e = 0.2$ (V), $c_0{}^f = 2$ (mol/m^3) and $c^* = 1$ (mol/m^3)

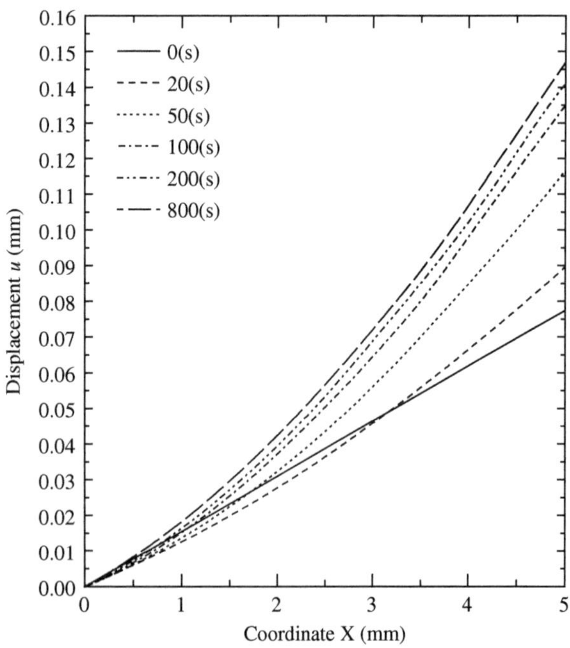

Fig. 3.49 Variation of the distributive displacement of the hydrogel with time for $V_e = 0.3$ (V), $c_0{}^f = 2$ (mol/m^3) and $c^* = 1$ (mol/m^3)

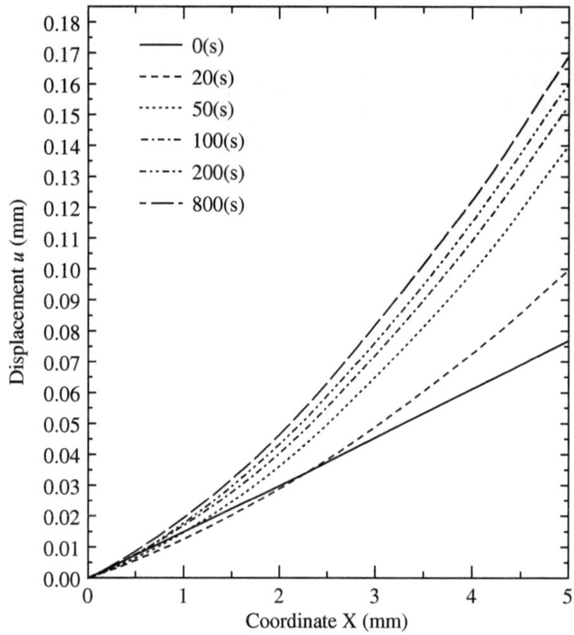

Fig. 3.50 Variation of the distributive displacement of the hydrogel with time for $V_e = 0.4$ (V), $c_0^f = 2$ (mol/m^3) and $c^* = 1$ (mol/m^3)

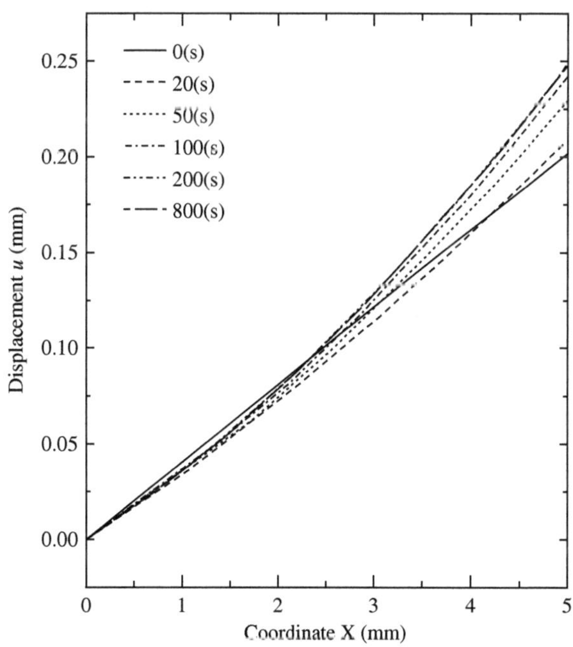

Fig. 3.51 Variation of the distributive displacement of the hydrogel with time for $V_e = 0.2$ (V), $c_0^f = 4$ (mol/m^3) and $c^* = 1$ (mol/m^3)

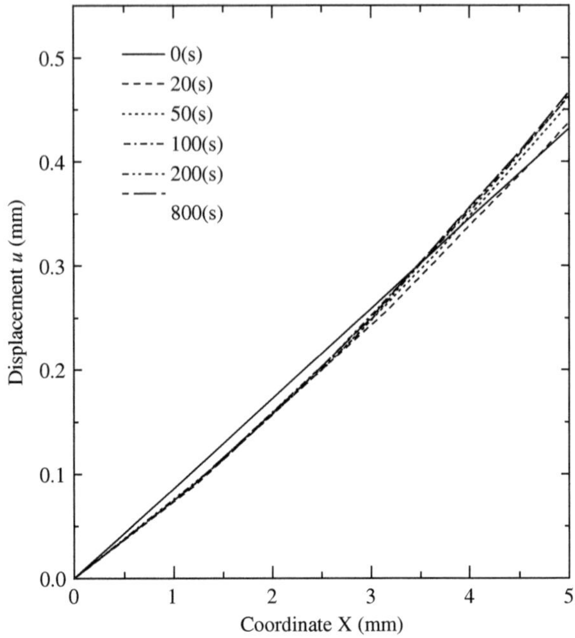

Fig. 3.52 Variation of the distributive displacement of the hydrogel with time for $V_e = 0.2$ (V), $c_0{}^f = 8$ (mol/m^3) and $c^* = 1$ (mol/m^3)

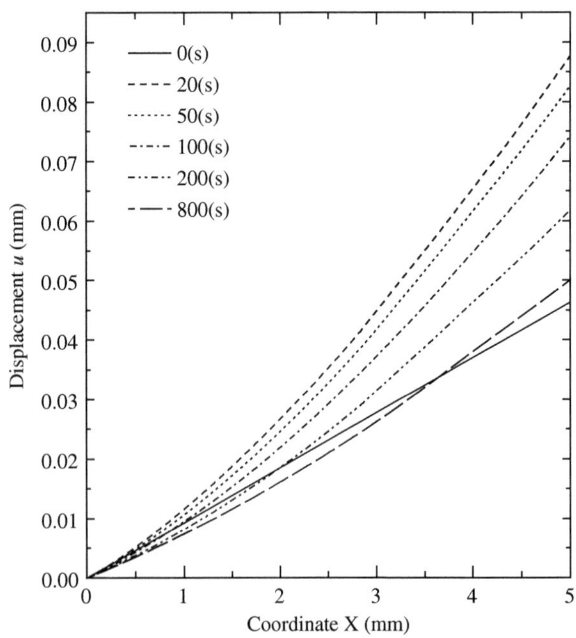

Fig. 3.53 Variation of the distributive displacement of the hydrogel with time for $V_e = 0.2$ (V), $c_0{}^f = 2$ (mol/m^3) and $c^* = 2$ (mol/m^3)

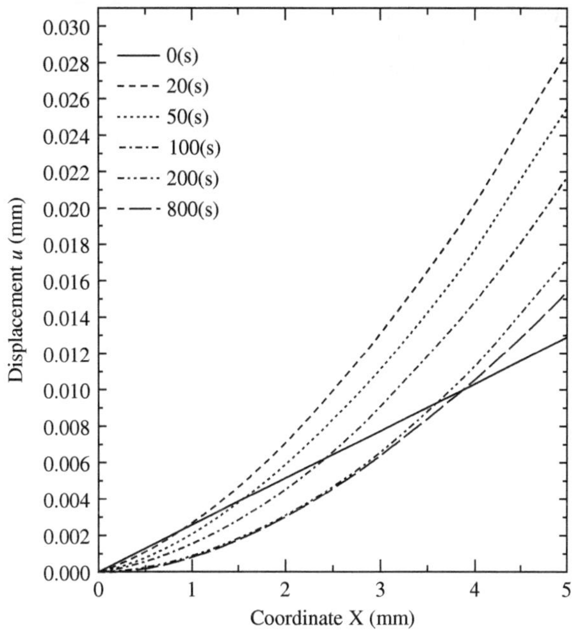

Fig. 3.54 Variation of the distributive displacement of the hydrogel with time for $V_e = 0.2$ (V), $c_0{}^f = 2$ (mol/m^3) and $c^* = 8$ (mol/m^3)

distributions of the hydrogel displacement is generally similar to that of the electric potential distributions, i.e., the displacement enlarges with time. The displacement initially increases rapidly, and then increases gradually. It is also known from these figures that the hydrogel displacement distributes linearly along the x-coordinate direction before the external electric field is applied. Once the electric voltage is imposed, however, the hydrogel displacement immediately demonstrates the non-linear distribution, i.e., the hydrogel displacement distributes nonlinearly over the hydrogel domain subject to externally applied electric field, due to the electric and chemical coupled effects in the considered system. This is observed obviously from Figs. 3.48, 3.49, 3.50, 3.51, 3.52, 3.53 and 3.54, where the distributive curves are linear at $t = 0$, subsequently no longer linear and become nonlinear.

3.4.3.4 Variation of Hydrogel Average Curvature with Time

The average curvature Ka is an important physical parameter to measure the overall extent of the hydrogel deformation. It is defined as $Ka = 2(e_1 - e_2)/(h(2 + e_1 + e_2))$ in Sect. 3.3.2 and is discussed here in detail. It is seen from Fig. 3.55 that the average curvature Ka at a given voltage initially increases rapidly with time, and then approaches the stable equilibrium state after a critical time, which varies with the change of applied voltages. The critical time decreases with the increase of the electric voltage. It may be explained that the drag force imposed on the diffusive ions increases as the electric voltage applied increases, which leads to faster diffusion of mobile species, reaching the equilibrium state shortly. Based on the discussion of

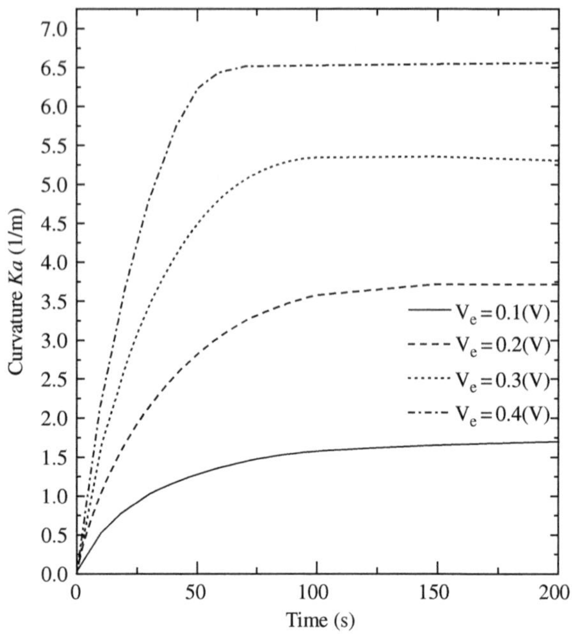

Fig. 3.55 Effect of the externally applied electric field V_e on the variation of the distributive average curvature *Ka* with time

Fig. 3.55, it is concluded that the higher electric voltage at a given time makes the larger average curvature. It may easily be understood because a strong electric field result in the larger difference of the ionic concentrations, which induces the larger deformation of the hydrogels.

Figure 3.56 shows the influence of the fixed charge density on the distributive profiles of the average curvature with time, where a critical time is observed from the variation of the average curvature with time. The critical time decreases with increasing the fixed charge density. A possible reason is that the fixed charge group is another driving source for ionic diffusion besides the electric field. It is found that the average curvature at a given time increases with the increase of the fixed charge density. This is consistent with the previous steady-state studies, where the increase of the fixed charge density causes larger deformation of the hydrogel.

Figure 3.57 demonstrates the influence of the bath solution concentrations on the variation of the average curvature with time. It is seen that the critical time changes slightly with the bath solution concentrations, which implies that the influence of the bath solution concentrations on the critical time is insignificant in comparison with the effects of the electric field and the fixed charge density. It is also illustrated in Fig. 3.57 that the average curvature at a given time decreases as the bath solution concentration increases. This is consistent with the previous studies in Sect. 3.3.3.3, i.e., the variation of the diffusive ionic concentrations because of the effect of the fixed charge groups could be negligibly small if the bath solution concentration is much higher than the fixed charge density.

Fig. 3.56 Effect of the fixed charged density $c_0{}^f$ on the variation of the distributive average curvature *Ka* with time

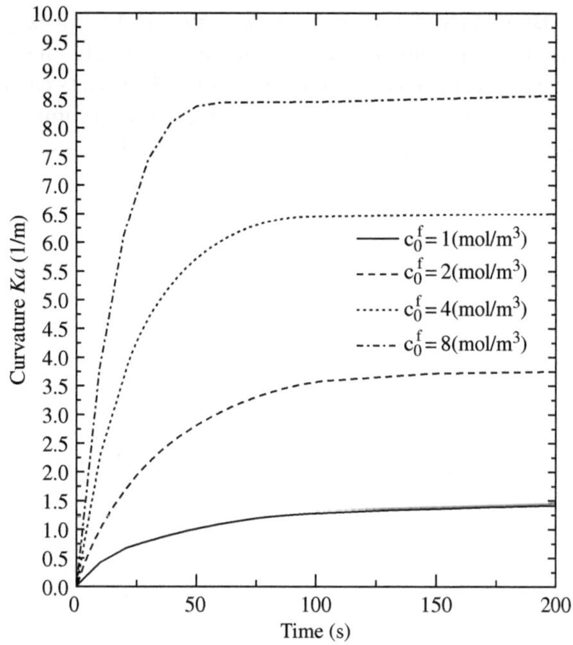

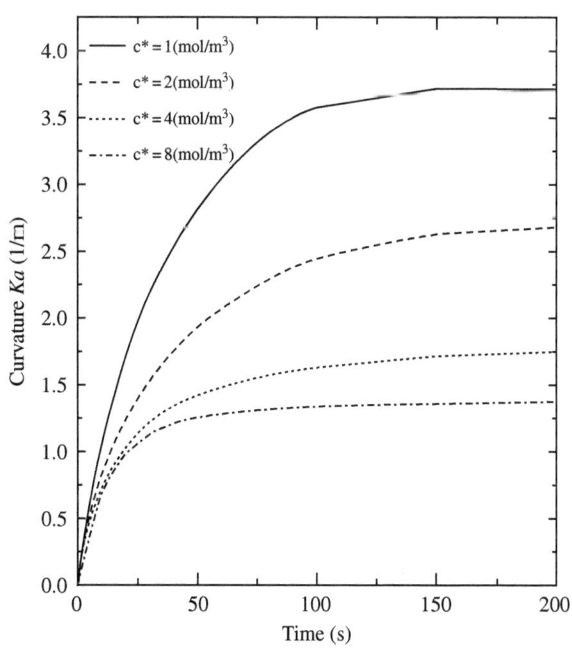

Fig. 3.57 Effect of the concentration of surrounding solution c^* on the variation of the distributive average curvature *Ka* with time

By comparison of Figs. 3.55, 3.56 and 3.57, it is generally concluded that the average curvature *Ka* initially increases rapidly, and then increases gradually. It reveals that the electric-sensitive hydrogels possess capability of responding quickly to the externally electric triggers, which is a very important feature of the hydrogels and makes the hydrogels very suitable as biosensors/bioactuators for BioMEMS applications.

3.5 Remarks

This chapter focuses on the analysis of the responsive characteristics of the electric-sensitive hydrogels via the simulation by the MECe model, when the hydrogels are immersed into a bath solution subject to an externally applied electric field. The novel multiphysics model termed the multi-effect-coupling electric-stimulus (MECe) model has been developed mathematically with consideration of chemo-electro-mechanical coupled effects. For numerical solution of the MECe model consisting of nonlinear coupled partial differential governing equations, the meshless technique so-called the Hermite-cloud method is employed in the present computational simulations. After examination of the MECe model through numerical comparison with experimental data extracted from open literature, the simulations are carried out for analysis of the equilibrium and kinetics swelling of the electric stimulus-responsive hydrogels. Influences of several important material properties and environmental conditions are discussed on the ionic diffusion, the electric potential and the deformation of the smart hydrogels.

As a key external stimulus to the electric-sensitive hydrogels, the externally applied electric voltage plays a critically important role in the response behaviours of the hydrogels. Due to the drag force of the electric field, the mobile ion species in the bath solution diffuse into the hydrogels and then produce a difference of the diffusive ionic concentrations over the hydrogel–solution interfaces, which makes the hydrogel deformed. For a given applied electric voltage, the difference of the diffusive ionic concentrations increases with time, and the hydrogel mixture finally reaches the equilibrium state after certain time here called the critical time. It is concluded that the critical time decreases and the hydrogels deform greatly as the electric voltage applied increases. This reveals a significant influence of the externally applied electric field on the responsive characteristics of the hydrogels.

Another key effect is the fixed charge groups attached onto the polymeric matrix network chains of the hydrogels. The fixed charge groups with negative valence will attract the mobile cations into the hydrogel mixture from the bath solution, resulting in a fluid pressure and making the hydrogel deformed. With the increase of the fixed charge density, the concentration of cations within the hydrogels has a dramatic variation while that of anions changes little. The critical time for the kinetics response of the hydrogels decreases with increasing the fixed charge density since the attraction of the fixed charge on the mobile ions strengthens.

We should also pay attention to the characteristics of surrounding bath solution, such as the concentration and composition of ionic species in the solution. The effect of bath solution concentrations is revealed by the counteractive function to the fixed charge density. With increment of bath solution concentrations, the effect of attraction of the fixed charge groups on the diffusive mobile ions becomes insignificant and the conductivity of bath solutions is almost identical in the whole computational domain. This results in a quasilinear distribution of the electric potential and decreases the critical time of the kinetics of the hydrogels. On the other hand, the ionic valence of bath solution can also affect the hydrogel deformation. A bath solution with higher ionic valence will cause a larger difference of the diffusive ion concentrations and then larger displacement of the hydrogels.

Finally, it should be noted that the responding time of the electric-sensitive hydrogels to the externally applied electric trigger is generally always very short, normally shorter than 2 min in the cases of present simulations. The simulated phenomena agree well with the experimental findings and validate the great promise of the electric-sensitive hydrogels in applications of biotechnology and bioengineering.

However, one-dimensional simulations are involved only in this chapter for the parameter studies. The smart hydrogel actually deforms always in all three directions. Therefore, for analysis of the responsive characteristics of the electric-sensitive hydrogels in more accuracy, it is necessary to make the two-dimensional or three-dimensional analyses in the future. In addition, the present discussions focus on the stimulus of externally applied electric field only. In fact, there are considerable variety and complexity of environmental stimuli. Other stimuli are also important triggers for the responsive behaviours of smart polymeric hydrogels, such as the solution pH, temperature and chemical reactions.

References

J.O'M. Bockris, B.E. Conway, E. Yeager (Eds.) (1983). *Comprehensive Treatise of Electrochemistry*, Vol. 6, *Electrodics: Transport*, New York: Plenum Press.

J. Fei, Z. Zhang, L. Gu. (2002). Bending behavior of electroresponsive poly(vinyl alcohol) and poly(acrylic acid) semi-interpenetrating network hydrogel fibres under an electric stimulus. *Polymer International*, 51, 502–509.

M. Homma, Y. Seida, Y. Nakano. (2000). Evaluation of optimum condition for designing high-performance electro-driven polymer hydrogel systems. *Journal of Applied Polymer Science*, 75, 111–118.

M. Homma, Y. Seida, Y. Nakano. (2001). Effect of ions on the dynamic behavior of an electrodriven ionic polymer hydrogel membrane. *Journal of Applied Polymer Science*, 82, 76–80.

Y.C. Hon, M.W. Lu, W.M. Xue, X. Zhou. (1999). A new formulation and computation of the triphasic model for mechano-electrochemical mixtures. *Computational Mechanics*, 24, 155–165.

W.M. Lai, J.S. Hou, V.C. Mow. (1991). A triphasic theory for the swelling and deformation behaviors of articular cartilage. *ASME Journal of Biomechanical Engineering*, 113, 245–258.

H. Li, T.Y. Ng, J.Q. Cheng, K.Y. Lam. (2003). Hermite-cloud: A novel true meshless method. *Computational Mechanics*, 33, 30–41.

T. Shiga, T. Karauchi. (1990). Deformation of polyelectrolyte gels under the influence of electric field. *Journal of Applied Polymer Science*, 39, 2305–2320.

T. Tanaka, I. Nishio, S.T. Sun, N.S. Ueno. (1982). Collapse of gels in an electric field. *Science*, 218, 467–469.

S. Sun, A.F.T. Mak. (2001). The dynamical response of a hydrogel fiber to electrochemical stimulation. *Journal of Polymer Science part B: Polymer Physics*, 39, 236–246.

T. Wallmersperger, B. Kroplin, R.W. Gulch. (2004). Coupled chemo-electro-mechanical formulation for ionic polymer gels – numerical and experimental investigations. *Mechanics of Materials*, 36, 411–420.

X. Zhou, Y.C. Hon, S. Sun, A.F.T. Mak. (2002). Numerical simulation of the steady-state deformation of a smart hydrogel under an external electric field. *Smart Materials and Structures*, 11, 459–467.

Chapter 4
Multi-Effect-Coupling pH-Electric-Stimuli (MECpHe) Model for Smart Hydrogel Responsive to pH-Electric Coupled Stimuli

4.1 Introduction

In general, the mechanism of the response of smart hydrogel to the externally applied electric field stimulus is similar to that to the pH stimulus of environmental solution (Tanaka, 1982; Qiu and Park, 2001). When an electric field is imposed on the bath solution where the hydrogel is immersed, the mobile ions diffuse between the hydrogel and the surrounding solution. The diffusion gives rise to the difference of ionic concentrations between the interior hydrogel and exterior solution, because of the fixed charge groups bound to the crosslinked macromolecular chains. As a result, an osmotic pressure is generated due to the concentration difference, which drives the swelling or shrinking of the hydrogel. Then the deformation of the hydrogel results in the redistribution of the diffusive ions and the fixed charge groups within the hydrogel, which causes new ionic diffusion and then the hydrogel deforms again. The recurrent kinetics continues and finally stops when the hydrogel reaches an equilibrium state, by converting the chemical energy to the mechanical one.

Literature search reveals that most of the studies on the smart hydrogels generally focus on a single environmental stimulus. However, it is expected that the hydrogel for a given application might experience the multiple coupled stimuli. One of examples is the poly(*N*-iso-propylacrylamide) (PNIPAAm) hydrogel, which is sensitive to both the surrounding temperature and the pH (Dong and Hoffman, 1990, 1991; Brazel and Peppas, 1996; Shin et al., 2003). Other examples include the typical pH-sensitive hydrogels that are often sensitive to electric field, such as poly(2-acrylamido-2-methylpropane sulfonic acid) (PAMPS), poly(acrylic acid) (PAA) and poly(methacrylic acid-*co*-vinyl alcohol) PMAA/PVA IPN hydrogel (Schreyer et al., 2000; Jin and Hsieh, 2005). If the hydrogel is subjected to a single stimulus, we often face the problems such as the long response time and limited deformation. In order to overcome these limitations, the increase of surface area of the hydrogel is one of possible solutions, e.g. by making the hydrogels thinner and smaller. Unfortunately, this approach generally compromises the mechanical strength and leads to more fragile systems (Schreyer et al., 2000; Qiu and Park, 2001). An alternative is to perturb the hydrogel with multiple coupled stimuli, which shortens the

H. Li, *Smart Hydrogel Modelling*, DOI 10.1007/978-3-642-02368-2_4,

response time and increases the swelling ratios. Kim et al. (2003, 2004a, b) demonstrated experimentally the improvement for the hydrogels subject to the changes in both environmental pH and externally applied electric field simultaneously, which reduces the response time of swelling and shrinking down to a few seconds or less. This makes fast response actuators and artificial muscles more feasible (Sawahata et al., 1990; Otake et al., 2002; Shihinpoor, 2003).

With the emergence of various applications that relies on the swelling behaviour of the hydrogels, a convincing need comes for physically accurate theories and numerical simulations capable of capturing, analysing and predicting the behaviour. In order to design micro-actuating and sensing devices based on the smart hydrogels, it is critical to understand the relation between the swelling time and the environmental stimuli. The theories and models should provide sufficient physical understanding of the various processes associated with the swelling of the hydrogels (Dolbow et al., 2005). The performance characteristics of the hydrogels can thus be optimized.

This chapter focuses on the hydrogels responding to the coupled stimuli, namely the surrounding solution pH coupled with externally applied electric voltage. The multi-effect-coupling pH-electric-stimuli (MECpHe) model is developed first for the hydrogels responsive to the pH-electric coupled stimuli. After discretization of the MECpHe governing equations and boundary conditions, the present MECpHe model is examined by comparison between the simulation results and the experiments published. It is followed by parameter studies on the influences of several important hydrogel material properties and environmental solution conditions on the response characteristics of the smart hydrogels (Fig. 4.1).

4.2 Development of the MECpHe Model

The hydrogels responsive to the pH-electric coupled stimuli are able to convert chemical energy to mechanical one. They are often employed for the controlled drug delivery, where the solution pH and the externally applied electric field affect the release pattern (Qiu and Park, 2001). For simulation of the responsive behaviours of the hydrogels subject to the pH-electric coupled stimuli, a novel model is presented in this chapter, called the multi-effect-coupling pH-electric-stimuli (MECpHe) model. The contributions of the presently developed MECpHe model include the reformulations of the fixed charge density and the large deformation of the hydrogels.

For simplicity of describing the flux of the kth ionic species in solution, the convective transport of the ionic species is neglected here. If the Nernst–Planck theory is considered for ionic transportation, the flux of the kth ionic species in the system consisting of the hydrogel and the surrounding solution can be characterized by

$$\boldsymbol{J}_k = -[D_k]\left(\mathbf{grad}(c_k) + \tfrac{z_k F}{RT} c_k \mathbf{grad}(\psi) + c_k \mathbf{grad}(\ln \gamma_k)\right)(k = 1,2,\ldots,N_{\text{ion}}) \tag{4.1}$$

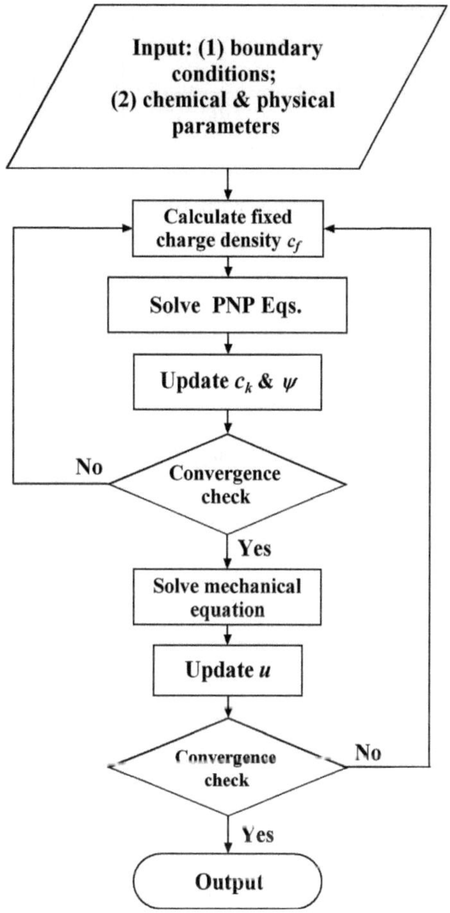

Fig. 4.1 Computational flowchart of the MECpHe model

where $\boldsymbol{J}_k$, $[D_k]$, c_k and z_k are the flux (mM/s), the diffusivity tensor, the concentration (mM) and the valence number of the kth diffusive ionic species. ψ is the electrostatic potential (V) and γ_k is the chemical activity coefficient. N_{ion} is the number of total diffusive species in the system. F, R and T are the Faraday's constant (9.6487×10^4 C/mol), the universal gas constant (8.314 J/mol·K) and the absolute temperature (K), respectively.

The three terms on the right-hand side of Eq. (4.1) represent the diffusive flux due to the concentration gradient, the migration flux due to the gradient of electrical potential and the chemical flux associated with chemical activity coefficient.

According to the law of mass conservation, the change of the species k contained in volume with respect to time t can be characterized by the difference between the fluxes entering and leaving the reference volume. If the chemical flux is neglected, the Nernst–Planck type of the mass conservation is derived as

$$\frac{\partial c_k}{\partial t} + \mathrm{div}(\boldsymbol{J}_k) = \frac{\partial c_k}{\partial t} + \mathrm{div}\left\{-[D_k]\left(\mathbf{grad}(c_k) + \frac{z_k F}{RT} c_k \mathbf{grad}(\psi)\right)\right\} = 0 \quad (k = 1,2,\ldots,N_{\text{ion}}) \tag{4.2}$$

which is coupled with the following Poisson equation to describe the spatial distribution of the electric potential in the domain:

$$\nabla^2 \psi = -\frac{F}{\varepsilon \varepsilon_0}\left(\sum_k z_k c_k + z_f c_f\right) \tag{4.3}$$

where ε is the relative dielectric constant of the surrounding medium and ε_0 is the vacuum permittivity or dielectric constant (8.85418×10^{-12} C^2/Nm^2). z_f is the valence of the fixed charge groups attached onto the polymeric network chains of the hydrogel. For example, $z_f = -1$ if the carboxylic acid groups are used as the fixed charges on the polymer chains. It is well known that the electroneutrality and constant field hypotheses are in fact the special cases of the Poisson equation.

Based on the Langmuir adsorption isotherm theory (Grimshaw et al., 1990), a relation between the fixed charge and the diffusive hydrogen ion concentration is presented to complete the Poisson–Nernst–Planck (PNP) system, whereby the concentration of the fixed charge group is determined by

$$c_f = c_f^0 - c_f^b = c_f^0 - \frac{c_f^0 \cdot c_{H^+}}{K + c_{H^+}} = \frac{c_f^0 \cdot K}{K + c_{H^+}} \tag{4.4}$$

where c_f and K are the concentration and the dissociation constant of the fixed charge groups attached onto the polymeric network chains within the hydrogel. c_f^0 is the total concentration of the ionizable groups in the hydrogel, c_{H^+} is the concentration of diffusive hydrogen ions H^+ within the hydrogel, and H is termed the local hydration of the hydrogel.

In the present MECpHe model, the total concentration of the fixed charge groups within the hydrogel at the relaxed state is defined as

$$c_{f,s}^0 = \frac{n}{V^s} \tag{4.5}$$

Then the total concentration of the ionizable groups in the hydrogel is obtained by

$$c_f^0 = \frac{n}{V} = \frac{n}{V^s + V^w} = \frac{n}{V^s}\frac{V^s}{V^s + V^w} = \frac{n}{V^s}\frac{V^s}{V^s + HV^s} = \frac{n}{V^s}\frac{V^s}{(H+1)V^s} = \frac{c_{f,s}^0}{H+1} \tag{4.6}$$

Substituting Eq. (4.6) into Eq. (4.4), one can have

$$c_f = \frac{c_f^0 \cdot K}{K + c_H} = \frac{1}{H+1} \cdot \frac{c_{f,s}^0 \cdot K}{K + c_H} \tag{4.7}$$

where the local hydration of the hydrogel is defined as $H = V^w/V^s$, namely

$$1 + H = 1 + \frac{V^w}{V^s} = \frac{V^s + V^w}{V^s} = \frac{V}{V^s} = \frac{1}{\phi^s} = \frac{1}{1 - \phi^w} \tag{4.8}$$

The volume fractions of the interstitial water and polymeric solid phases are thus written as

$$\phi^w = \frac{H}{1 + H} \tag{4.9}$$

$$\phi^s = \frac{1}{1 + H} \tag{4.10}$$

Since the volume fraction of the ion species ϕ^i is negligibly small when compared with ϕ^w and ϕ^s, the saturation equation is simplified to

$$\phi^w + \phi^s \approx 1 \tag{4.11}$$

The relation between the volume fractions of the interstitial water and polymeric network solid phases is then given by

$$\phi^w \approx 1 - \phi^s = 1 - \frac{V^s}{V} = 1 - \frac{V^s}{V_0}\frac{V_0}{V} = 1 - \phi_0^s \cdot J \tag{4.12}$$

where $J = dV_0/dV$ is the volume ratio of apparent polymeric network matrix solid phase and may be formulated by the Green strain tensor $\boldsymbol{E}$ of the apparent polymeric solid phase as follows (Hon et al., 1999):

$$\frac{1}{J} = \sqrt{1 + 2F_1(\boldsymbol{E}) + 4F_2(\boldsymbol{E}) + 8F_3(\boldsymbol{E})} \tag{4.13}$$

where $\mathrm{F}_1(\boldsymbol{E}) = tr(\boldsymbol{E})$, $F_2(\boldsymbol{E})$ and $F_3(\boldsymbol{E})$ are the first, second and third invariants of Green strain tensor $\boldsymbol{E}$, respectively.

The Green strain tensor can be expressed in terms of displacement gradients (Belytschko et al., 2001),

$$E_{ij} = \frac{1}{2}\left(\frac{\partial u_i}{\partial X_j} + \frac{\partial u_j}{\partial X_i} + \frac{\partial u_k}{\partial X_i}\frac{\partial u_k}{\partial X_j}\right) \tag{4.14}$$

where X_i and X_j are the components of the position vector in the initial configuration, u_i, u_j and u_k are the displacements. In one-dimensional case,

$$E_{11} = \frac{1}{2}\left(\frac{\partial u_1}{\partial X_1} + \frac{\partial u_1}{\partial X_1} + \frac{\partial u_1}{\partial X_1}\frac{\partial u_1}{\partial X_1}\right) = \frac{1}{2}\left[2\frac{du}{dX} + \left(\frac{du}{dX}\right)^2\right] \tag{4.15}$$

Three invariants of the deformation gradient tensor are defined as Lai et al. (1974),

$$F_1 = E_{11} + E_{22} + E_{33} \tag{4.16}$$

$$F_2 = \begin{vmatrix} E_{11} & E_{12} \\ E_{21} & E_{22} \end{vmatrix} + \begin{vmatrix} E_{11} & E_{13} \\ E_{31} & E_{33} \end{vmatrix} + \begin{vmatrix} E_{22} & E_{23} \\ E_{32} & E_{33} \end{vmatrix} \tag{4.17}$$

$$F_3 = \left|E_{ij}\right| = \begin{vmatrix} E_{11} & E_{12} & E_{13} \\ E_{21} & E_{22} & E_{23} \\ E_{31} & E_{32} & E_{33} \end{vmatrix} \tag{4.18}$$

For one-dimensional case,

$$F_1 = E_{11} = \frac{1}{2}\left[2\frac{du}{dX} + \left(\frac{du}{dX}\right)^2\right] = \frac{du}{dX} + \frac{1}{2}\left(\frac{du}{dX}\right)^2 \tag{4.19}$$

Using Eqs. (4.9) and (4.12), one can have

$$\phi^w = \frac{H}{1+H} = 1 - \phi_0^s J \tag{4.20}$$

The local hydration of the hydrogel H is rewritten as

$$H = \frac{1 - \phi_0^S J}{\phi_0^S J} \tag{4.21}$$

Substituting Eqs. (4.13) and (4.21) into Eq. (4.7), the density of fixed charge groups is finally derived as follows:

$$c_f = \frac{c_{m0}^s \cdot K \cdot \phi_0^s}{(K + c_H)\sqrt{1 + 2F_1(\boldsymbol{E}) + 4F_2(\boldsymbol{E}) + 8F_3(\boldsymbol{E})}} \tag{4.22}$$

By substituting Eq. (4.19) into the Eq. (4.22), the one-dimensional form of the fixed charge density is obtained as

$$c_f = \frac{c_{m0}^s \cdot K \cdot \phi_0^s}{(K + c_H)\sqrt{1 + 2\frac{du}{dX} + \left(\frac{du}{dX}\right)^2}} \tag{4.23}$$

As well known, the first Piola–Kirchhoff stress tensor **P** is a kind of expatriate, living partially in the deformed (current) configuration $\boldsymbol{x}$ and partially in the reference (initial) configuration $\boldsymbol{X}$ where $\boldsymbol{x} = \boldsymbol{X}+\mathbf{u}$, and it is unable to measure. Because of the absence of symmetry in the first Piola–Kirchhoff stress tensor **P**, it is seldom used in constitutive equations. However, the second Piola–Kirchhoff stress tensor **S** is symmetric and is often used as the stress measure for large deformation. The relation between the first Piola–Kirchhoff stress tensor **P** and the second

Piola–Kirchhoff stress tensor **S** is written as

$$\mathbf{P} = \mathbf{S}\mathbf{F}^{\mathrm{T}} \tag{4.24}$$

where **F** is the deformation gradient tensor and defined as

$$\mathbf{F} = \mathbf{I} + \nabla\mathbf{u} \tag{4.25}$$

The second Piola–Kirchhoff stress tensor **S** is given by

$$\mathbf{S} = \mathbf{C}\mathbf{E} - p_{\text{osmotic}}\,\mathbf{I} \tag{4.26}$$

where **C** is the material tensor and **E** is the Green–Lagrangian strain tensor used as the strain measure

$$\mathbf{E} = \frac{1}{2}(\mathbf{F}^{\mathrm{T}}\mathbf{F} - \mathbf{I}) \tag{4.27}$$

The nonlinear mechanical governing equation for large deformation of the smart hydrogel is finally written as follows:

$$\nabla \cdot [(\mathbf{C}\mathbf{E} - p_{\text{osmotic}}\,\mathbf{I})\mathbf{F}^{\mathrm{T}}] = 0 \tag{4.28}$$

For one-dimensional analysis,

$$(\lambda_s + 2\mu_s)\left[\frac{d^2u}{dX^2} + 3\frac{du}{dX}\frac{d^2u}{dX^2} + \frac{3}{2}\left(\frac{du}{dX}\right)^2\frac{d^2u}{dX^2}\right] - \frac{dp_{\text{osmotic}}}{dX} = 0 \tag{4.29}$$

So far the development of MECpHe model has been completed. It is composed of the Nernst–Planck diffusion equation (4.2) for the diffusive ion concentrations, the Poisson equation (4.3) with the fixed charge density (4.22) or (4.23) for the electric potential and nonlinear mechanical equation (4.28) or (4.29) for the large displacement of the smart hydrogel.

The MECpHe governing equations are associated with the boundary conditions of the diffusive ion concentrations and the electric potential, which are imposed at the edges of the surrounding solution

$$c|_{\text{Anode}} = c|_{\text{Cathode}} = c^* \tag{4.30}$$

$$\psi|_{\text{Anode}} = 0.5V_e \text{ and } \psi|_{\text{Cathode}} = -0.5V_e \tag{4.31}$$

where c^* is the initial ionic concentration of the bath solution and V_e the externally applied electric voltage.

Boundary condition of the mechanical deformation is imposed at the hydrogel–solution interfaces

$$(\lambda_s + 2\mu_s)\left[\frac{du}{dX} + \frac{1}{2}\left(\frac{du}{dX}\right)^2\right] = RT\sum_{k=1}^{N_{\text{ion}}}(c_k - c_k^0) \quad \text{at} \quad X = (L \pm h)/2 \quad (4.32)$$

In order to prevent the hydrogel from undergoing rigid-body motion, a point constraint is requested in the middle of the hydrogel

$$u = 0 \quad \text{at} \quad X = L/2 \quad (4.33)$$

4.3 Numerical Implementation

For computational simplification, non-dimensional variables are defined as

$$\bar{x} = \frac{X}{L_{\text{ref}}}, \bar{u} = \frac{u}{L_{\text{ref}}}, \bar{c}_k = \frac{c_k}{c_{\text{ref}}}, \bar{c}_f = \frac{c_f}{c_{\text{ref}}}, \bar{\psi} = \frac{\psi}{\psi_{\text{ref}}} = \frac{F\psi}{\eta RT}, \bar{p}_{\text{osmotic}} = \frac{p_{\text{osmotic}}}{\xi c_{\text{ref}} RT} \quad (4.34)$$

The non-dimensional form of the partial differential governing equations of the MECpHe model are then written as

$$\frac{d^2\bar{c}_k}{d_{\bar{x}}^2} + \eta z_k \frac{d\bar{c}_k}{d\bar{x}}\frac{d\bar{\psi}}{d\bar{x}} + \eta z_k \bar{c}_k \frac{d^2\bar{\psi}}{d\bar{x}^2} = 0 \quad (k = 1,2,\ldots,N_{\text{ion}}) \quad (4.35)$$

$$\frac{d^2\bar{\psi}}{d_{\bar{x}}^2} = -\frac{F^2}{\varepsilon\varepsilon_0 RT}\frac{L_{\text{ref}}^2 c_{\text{ref}}}{\eta}\left(z_f\bar{c}_f + \sum_{k=1}^{N_{\text{ion}}} z_k\bar{c}_k\right) \quad (4.36)$$

$$\left[\frac{d^2\bar{u}}{d\bar{x}^2} + 3\frac{d\bar{u}}{d\bar{x}}\frac{d^2\bar{u}}{d\bar{x}^2} + \frac{3}{2}\left(\frac{d\bar{u}}{d\bar{x}}\right)^2\frac{d^2\bar{u}}{d\bar{x}^2}\right] - \frac{\xi c_{\text{ref}} RT}{f_1}\frac{d\bar{p}_{\text{osmotic}}}{d\bar{x}} = 0 \quad (4.37)$$

The non-dimensional osmotic pressure is given as

$$\bar{p}_{\text{osmotic}} = \frac{1}{\xi}\sum_k(\bar{c}_k - \bar{c}_k^0) \quad (4.38)$$

By the Hermite-cloud method (Li et al., 2003), the unknown variables are discretized as

$$\bar{c}_k(\bar{x}_i) = \sum_{j=1}^{N_{\text{total}}} N_j(\bar{x}_i)\bar{c}_{kj} + \sum_{m=1}^{N_{\text{total}}}\left(\bar{x}_i - \sum_{j=1}^{N_{\text{total}}} N_j(\bar{x}_i)\bar{x}_j\right) M_m(\bar{x}_i)\bar{c}_{km,x} \quad (4.39)$$

$$\overline{\psi}(\bar{x}_i) = \sum_{j=1}^{N_{\text{total}}} N_j(\bar{x}_i)\overline{\psi_j} + \sum_{m=1}^{N_{\text{total}}}\left(\bar{x}_i - \sum_{j=1}^{N_{\text{total}}} N_j(\bar{x}_i)\bar{x}_j\right) M_m(\bar{x}_i)\bar{\psi}_{m,x} \quad (4.40)$$

$$\overline{u}(\overline{x}_i) = \sum_{j=1}^{N_{\text{gel}}} N_j(\overline{x}_i)\overline{u}_j + \sum_{m=1}^{N_{\text{gel}}} \left(\overline{x}_i - \sum_{j=1}^{N_{\text{gel}}} N_j(\overline{x}_i)\overline{x}_j \right) M_m(\overline{x}_i)\bar{u}_{m,x} \tag{4.41}$$

The non-dimensional form of the nonlinear coupled partial differential governing equations of the MECpHe model and the auxiliary conditions for one-dimensional steady-state simulation are finally discretized as

$$\begin{aligned} &\sum_{j=1}^{N_{\text{total}}} N_{j,xx}(\overline{x}_i)\overline{c}_{kj} + \eta z_k \left[\sum_{m=1}^{N_{\text{total}}} M_m(\overline{x}_i)\overline{c}_{km,x} \right] \left[\sum_{m=1}^{N_{\text{total}}} M_m(\overline{x}_i)\overline{\psi}_{m,x} \right] + \\ &+\eta z_k \left[\sum_{j=1}^{N_{\text{total}}} N_j(\overline{x}_i)\overline{c}_{kj} - \sum_{m=1}^{N_{\text{total}}} \left(\overline{x}_i - \sum_{j=1}^{N_{\text{total}}} N_j(\overline{x}_i)\overline{x}_j \right) M_m(\overline{x}_i)\overline{c}_{km,x} \right] \\ &\left[\sum_{j=1}^{N_{\text{total}}} N_{j,xx}(\overline{x}_i)\overline{\psi}_j \right] = 0 \end{aligned} \tag{4.42}$$

$$\begin{aligned} &\sum_{j=1}^{N_{\text{total}}} N_{j,xx}(\overline{x}_i)\overline{\psi}_j + \frac{F^2}{\varepsilon\varepsilon_0 RT} \frac{L_{\text{ref}}^2 c_{\text{ref}}}{\eta} \left\{ z_f c_f + \sum_{k=1}^{N_{\text{ion}}} z_k \left[\sum_{j=1}^{N_{\text{total}}} N_j(\overline{x}_i)\overline{c}_{kj} \right.\right. \\ &\left.\left. - \sum_{m=1}^{N_{\text{total}}} \left(\overline{x}_i - \sum_{j=1}^{N_{\text{total}}} N_j(\overline{x}_i)\overline{x}_j \right) M_m(\overline{x}_i)\overline{c}_{km,x} \right] \right\} = 0 \end{aligned} \tag{4.43}$$

$$\begin{aligned} &f_1 \left\{ \sum_{j=1}^{N_{\text{gel}}} N_{j,xx}(\overline{x}_i)\overline{u}_j + 3 \left[\sum_{m=1}^{N_{\text{gel}}} M_m(\overline{x}_i)\overline{u}_{m,x} \right] \left[\sum_{j=1}^{N_{\text{gel}}} N_{j,xx}(\overline{x}_i)\overline{u}_j \right] \right. \\ &\left. +\frac{3}{2} \left[\sum_{m=1}^{N_{\text{gel}}} M_m(\overline{x}_i)\overline{u}_{m,x} \right] \left[\sum_{m=1}^{N_{\text{gel}}} M_m(\overline{x}_i)\overline{u}_{m,x} \right] \left[\sum_{j=1}^{N_{\text{gel}}} N_{j,xx}(\overline{x}_i)\overline{u}_j \right] \right\} \\ &-\frac{1}{\xi} \sum_k \left[\sum_{m=1}^{N_{\text{gel}}} M_m(\overline{x}_i)\overline{c}_{km,x} \right] = 0 \ \ (k = 1,2,\ldots,N_{\text{ion}}) \end{aligned} \tag{4.44}$$

$$\sum_{j=1}^{N_{\text{total}}} N_{xj}(\overline{x}_i)\overline{c}_{kj} - \left[\sum_{j=1}^{N_{\text{total}}} N_{xj}(\overline{x}_i)\overline{x}_j \right] \sum_{m=1}^{N_{\text{total}}} M_m(\overline{x}_i)\overline{c}_{km,x} = 0 \tag{4.45}$$

$$\sum_{j=1}^{N_{\text{total}}} N_{xj}(\overline{x}_i)\overline{\psi}_j - \left[\sum_{j=1}^{N_{\text{total}}} N_{xj}(\overline{x}_i)\overline{x}_j \right] \sum_{m=1}^{N_{\text{total}}} M_m(\overline{x}_i)\overline{\psi}_{m,x} = 0 \tag{4.46}$$

$$\sum_{j=1}^{N_{\text{gel}}} N_{xj}(\overline{x}_i)\overline{u}_j - \left[\sum_{j=1}^{N_{\text{gel}}} N_{xj}(\overline{x}_i)\overline{x}_j \right] \sum_{m=1}^{N_{\text{gel}}} M_m(\overline{x}_i)\overline{u}_{m,x} = 0 \tag{4.47}$$

where N_{total} is the number of the total scattered points covering the whole computational domain including both the hydrogel and the surrounding solution, and N_{gel} is the number of the total points for discretization of the smart hydrogel domain only (Fig. 4.1).

4.4 Model Validation with Experiment

For examination of the MECpHe model, a comparison is conducted between the numerically simulated results and experimentally measured bending data extracted from the published work (Kim et al., 2004b), where Kim et al. (2004b) measured experimentally the deformation of the PMAA/PVA IPN hydrogel strip which was immersed in the pH buffer solution, as shown in Fig. 4.2. The carboxylic acid group in PMAA is in the form of R–COOH when the pH is low, and then R–COOH is dissociated to $R{-}COO^{-}$ as the pH increases. The hydrogel strip swells uniformly before the electric voltage is imposed (Kim et al., 2004b). When the electric field is applied, the distributions of ionic concentrations are no longer uniform, which lead to the unequal concentration differences at the two interfaces between the hydrogel and surrounding solution near the anode and cathode, respectively. The unequal concentration differences result in the unequal osmotic pressure at the two interfaces near the anode and cathode, which makes the hydrogels bent (Wallmersperger and Kroeplin, 2001; Homma et al., 2001).

The studied hydrogels contain the carboxylic groups and undergo drastic change in swelling capacity with pH of the external media. The minimum swelling in the media of low pH may be attributed to the –COOH groups bound along the macromolecular chains in the network matrix remain almost unionized because pK_α of the acrylic acid is 5.4. This results in almost nil osmotic swelling pressure

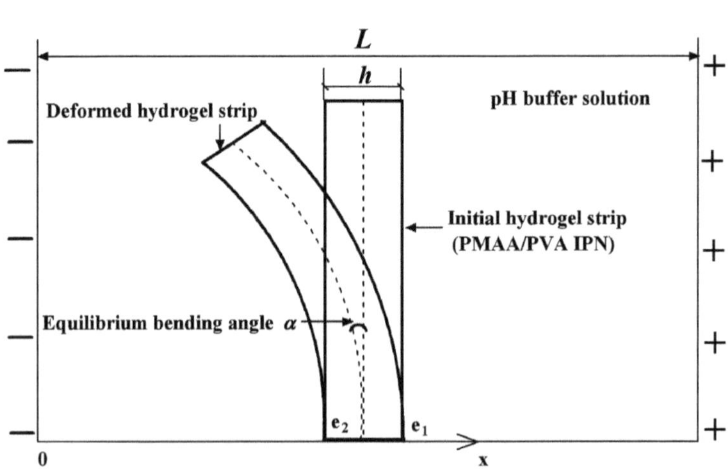

Fig. 4.2 Schematic diagram of a hydrogel strip immersed in pH buffer solution subject to an externally applied electric field

since no difference of mobile counterion concentrations occurs over the hydrogel–solution interfaces. Moreover, there occur H-bonding interactions occur among the carboxylic groups within the hydrogel, which provides a compact H-bonded structure to the hydrogel. This ultimately restricts the movements of polymeric segments and highly discourages the solvent entrance. However, when the hydrogel is placed in the medium with pH higher than pK_α, the ionization of $-COOH$ groups not only enlarges the osmotic swelling pressure but also causes relaxation of polymeric chains due to repulsion among similarly charged $R-COO^-$ groups bound along the macromolecular chains of the hydrogel network. This generates the extensive swelling of the hydrogel as indicated by higher swelling ratio of the hydrogel (Bajpai and Dubey, 2005).

The input parameters used for examination of the MECpHe model include $R = 8.314\,\mathrm{J/mol \cdot K}$, $F = 9.648\times10^4$ C/mol, $T = 298$ K, $\varepsilon_0 = 8.854\times10^{-12}\,\mathrm{C^2/N \cdot m^2}$, ε=80, $c^* = 137.1$ mM, $z_f = -1$, $c_f^0 = 200$ mM, the initial volume fraction of the interstitial water phase $\phi_0^w = 0.8$. In general, the elastic modulus of the polymer PMAA varies with the pH of buffer solution (Yin et al., 1997; Bashir et al., 2002; Rong et al., 2004). The elastic modulus taken here is 3 MPa according to the experiment (Yin et al., 1997). The distance between the two carbon electrodes $L = 30$ mm, and the hydrogel strip is tailored in $20\times5\times0.2\,\mathrm{mm^3}$. The simulation results are illustrated in Fig. 4.3 for the bending behaviour of the hydrogel subject to the electric voltage $V_e = 15$ V coupled with solution pH stimulus. In order to measure the bending deformation of the hydrogel, an equilibrium bending angle (EBA) is defined as $\alpha = 15L_0(e_1 \quad e_2)/\pi h$ in the unit of degree (Kim et al., 2004b), where e_1 and e_2 are the strains of the hydrogel strip at the two ends in thickness direction, L_0 and h are the length and thickness of the hydrogel strip, respectively. As the electric voltage increases to higher level such as $V_e = 15$ V, the profile of α increasing gradually with

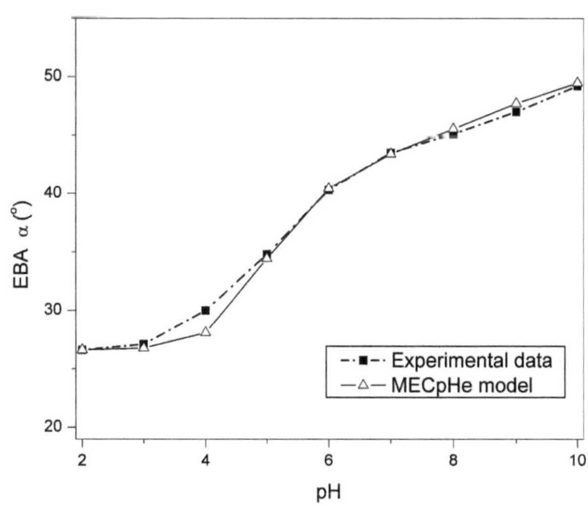

Fig. 4.3 Comparison of numerically simulated results with experimental data (Kim et al., 2004b)

buffer solution pH value may be divided into three stages. When the pH is lower than pH=4.0 or higher than pH=6.0, the α increases at gradual rate. However, it will increase rapidly in the range from pH=4.0 to 6.0. Basically the mechanism of deformation of the hydrogel may be explained by Flory's osmotic pressure theory (Shiga and Kurauchi, 1990; Yang and Engberts, 2000). If external electric field is applied, mobile ions move towards their counter electrode. As a result, a gradient of diffusive ion concentration is developed, which generates the osmotic pressure due to the difference of diffusive ion concentrations over the interfaces between the hydrogel and the surrounding solution. The cations Na^+ diffuse into the hydrogel more than the anions Cl^- and move towards the cathode. Since the increase of the osmotic pressure at the interface near the anode is larger than that at the interface near the cathode, the hydrogel near the anode swells greater than that near the cathode, which results in the bending towards the cathode, as shown in Fig. 4.2. Figure 4.3 demonstrates that the simulation results agree well with the experimental data (Kim et al., 2004b), which validates the MECpHe model with capability of efficiently simulating the hydrogels responsive to the pH-electric coupled stimuli.

4.5 Parameter Studies by Steady-State Simulation for Equilibrium of Hydrogel

For further understanding of the influences of hydrogel material properties and environmental conditions on the responsive characteristics of the hydrogels subject to the surrounding solution pH and electric field coupled stimuli, several simulations are carried out with the input parameters, $R = 8.314$ J/mol·K, $F = 9.648 \times 10^4$ C/mol, $T = 298$ K, $\varepsilon_0 = 8.854 \times 10^{-12}\,\mathrm{C^2/N \cdot m^2}$, ε=80, $c^* = 4.0$ mM, $c_f^s = 10.0$ mM, $z_f = -1$, $\phi_0^w = 0.8$, $L = 2400\,\mu$m, $h = 800\,\mu$m, and the Young's modulus taken is 3.0 MPa. The simulation results are discussed for analysis of the effects of several important material and environmental parameters on the distributive variations of responsive characteristics of the diffusive ionic concentrations c_k, the electric potential ψ, the fixed charge density c_f and the mechanical deformation of the hydrogel strip, including the displacement u, the swelling ratio R_s and the average curvature K_a that is defined as $K_a = 2(e_1 - e_2)/[h(2 + e_1 + e_2)]$ (e_1 and e_2 are the strains of the hydrogel strip at the two ends in the thickness direction) at the middle point of the hydrogel thickness for measurement of the bending deformation of the hydrogel (Li et al., 2007).

4.5.1 Influence of Solution pH Coupled with External Electric Voltage

Figures 4.4 and 4.5 show the influence of surrounding solution pH on the distribution of diffusive ionic species concentrations of the system in response to the stimulus of the electric voltage $V_e = 0$, 0.08, 0.16 and 0.32 V, respectively. If no external electric field is applied, i.e. $V_e = 0$, the distributions of the diffusive ionic

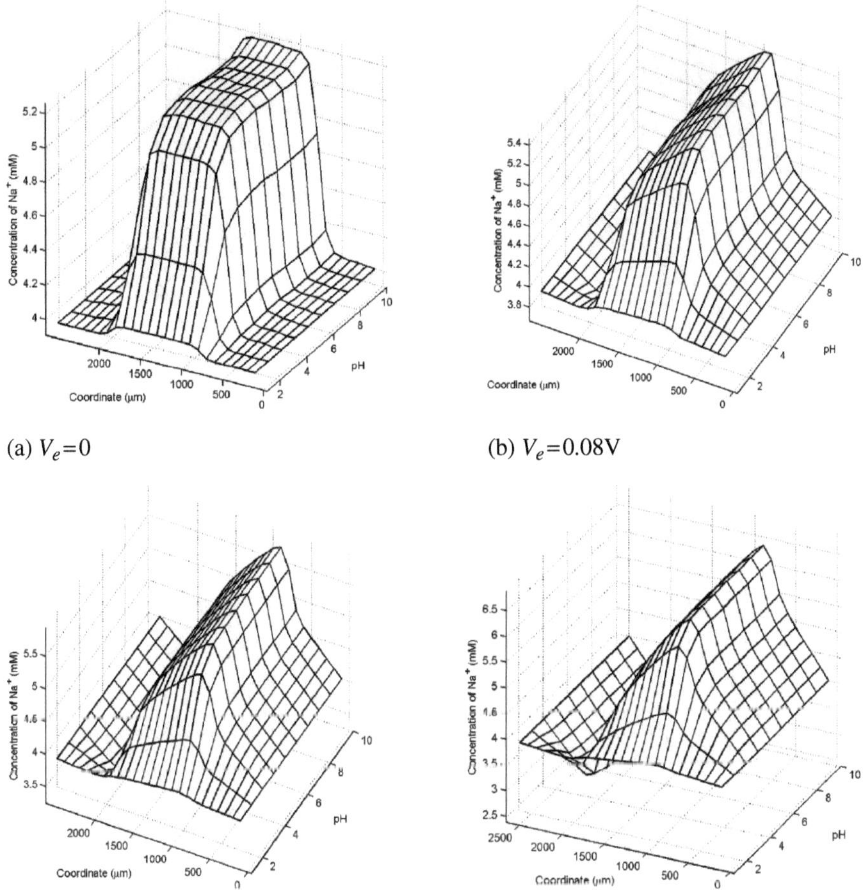

(a) V_e=0

(b) V_e=0.08V

(c) V_e=0.16V

(d) V_e=0.32V

Fig. 4.4 Coupled effect of the solution pH and external electric voltage V_e on the distributive profiles of the diffusive Na^+

concentrations are simulated and shown in Figs. 4.4a and 4.5a. It is observed from the figures that the electroneutrality phenomenon exists in the bath solution and hydrogel strip. The concentrations of the diffusive ionic species Na^+ and Cl^- are distributed uniformly within the hydrogels and symmetrically in the whole system domain. The same profiles of the concentration differences are observed over the two hydrogel–solution interfaces near the anode and cathode. This makes the hydrogel strip swell uniformly without bending deformation. Once the electric field is applied however, such as $V_e = 0.16$ V, the distributions of the diffusive ionic species Na^+ and Cl^- concentrations are no longer uniform in the hydrogel and bath solution, and also no longer symmetric in the whole domain. It is noted that the simulated hydrogel is assumed to be charged negatively. As the electric field is applied, the

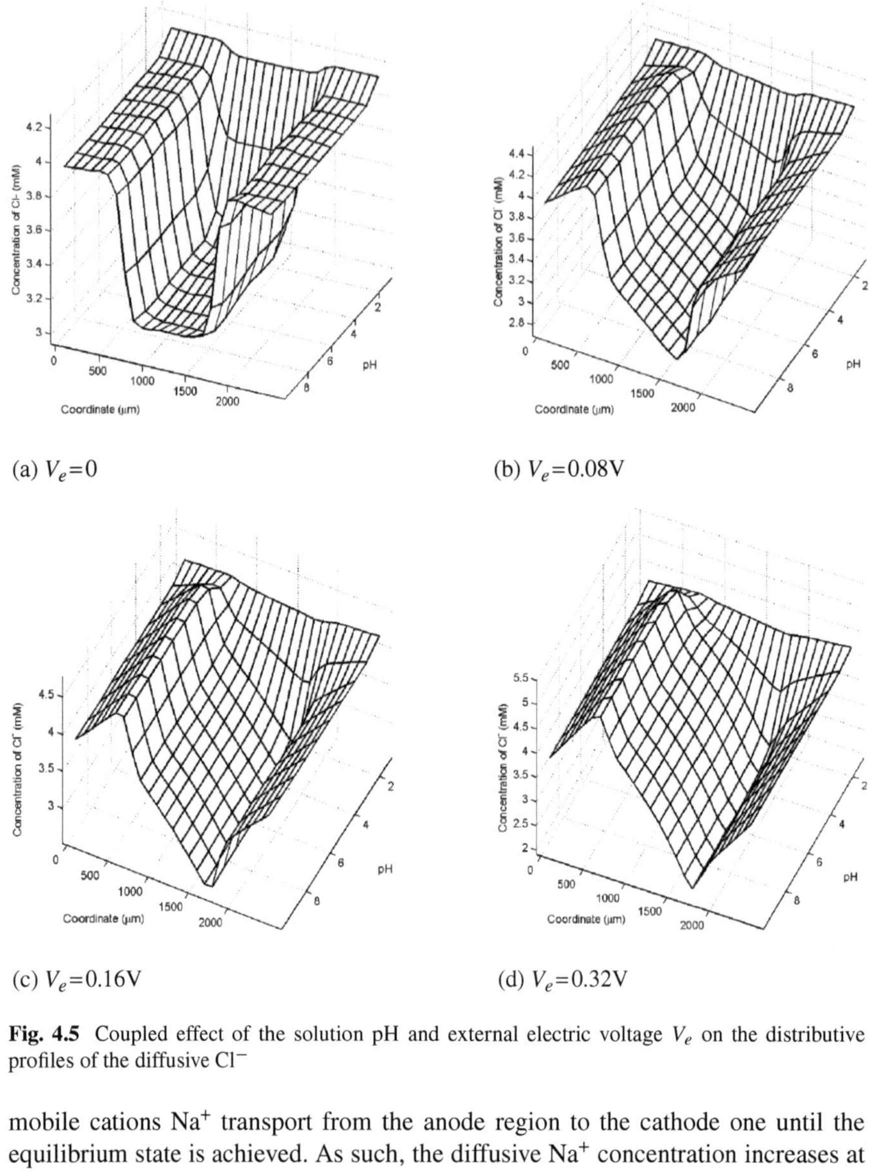

(a) V_e=0 (b) V_e=0.08V

(c) V_e=0.16V (d) V_e=0.32V

Fig. 4.5 Coupled effect of the solution pH and external electric voltage V_e on the distributive profiles of the diffusive Cl^-

mobile cations Na^+ transport from the anode region to the cathode one until the equilibrium state is achieved. As such, the diffusive Na^+ concentration increases at the hydrogel edge near the cathode and decreases near the anode. When the electric current is constant in equilibrium state, it is understood that the electroneutrality conserves everywhere in the solution, and the global flow of all the ions across the boundary yields a null current. As a result, the Cl^- concentration also increases at the hydrogel edge near the cathode and decreases near the anode. With increasing the distance from the cathode, the concentrations of the diffusive ionic species Na^+ and Cl^- decrease within the hydrogels. It can also be observed from Figs. 4.4b–d and 4.5b–d that the differences of diffusive ionic concentrations between the interior hydrogel and the exterior bath solution near the anode are larger than those near the cathode. Along the pH-axis direction, the distributive profile of the ionic concentration forms a slant for the concentration of cation Na^+ over the hydrogel–solution interface near the anode and a trench near the cathode, while the distributive profile of the anion Cl^- concentration forms a trench near the anode and a slant near the cathode. Moreover, the distributive profiles of the diffusive ionic concentrations change significantly from pH=1.0 to 6.0, while the distributions vary gradually from pH=6.0 to 10.0. It is also known that the simulations shown in Figs. 4.4 and 4.5 are consistent qualitatively with the experimental phenomena (Doi et al., 1992).

Figure 4.6a–d demonstrates the coupled effect of the solution pH and electric voltage V_e on the distribution of the electric potential ψ in the system. If $V_e = 0$, the distribution of electric potential ψ is symmetric in the whole domain and also uniform in the hydrogel strip, as shown in Fig. 4.6a. When an electric voltage is applied to the system, as shown in Fig. 4.6b–d, the distribution of the electric potential ψ becomes unsymmetrical and non-uniform, where the gradients of electric potential ψ are observed over the hydrogel–solution interfaces. The large difference of the electric potential ψ occurs near the anode, which enlarges concentration difference over the interfaces near the anode. Furthermore, relatively compared with the concentration difference near the cathode, the larger concentration difference near the anode results in the higher osmotic pressure near the anode. In addition, the collapse within the hydrogel domain diminishes with the increase of electric voltage. As such, the unequal osmotic pressures between the anode and the cathode cause the hydrogel bending (Homma et al., 2001).

Figure 4.7a–d illustrates the influence of surrounding solution pH on the distribution of the fixed charge density c_f of the hydrogel responding to different levels of the electric voltages. It is seen from these figures that the distribution of the fixed charge density c_f within the hydrogel decreases when the electric voltage is applied to the system. The increase of the electric voltage makes the hydrogel strip swell greatly, the fixed charge groups redistribute within the hydrogel and the density of the fixed charge groups decreases. This causes the hydrogel strip to swell and bend. Along the pH-axis direction, it is also seen that the distributive profiles of the fixed charge groups increase rapidly from pH=1.0 to 6.0, and then gradually from pH 6.0–10.0.

Figure 4.8a–d shows the influence of the buffer solution pH on the distribution of displacement u of the hydrogel under different electric voltages. It is observed from the figures that the displacement of the hydrogel strip changes dramatically in response to the pH-electric coupled stimuli. In the range of pH 2–6, if an electric voltage is applied, such as $V_e = 0.32$ V, the hydrogel strip swells obviously and the displacement is about 10 times larger than that of the hydrogel strip without electric stimulus. When pH is higher than 6, the effect of solution pH on the displacement of the hydrogel is insignificant and the displacement is distributed almost uniformly. This is a good example to demonstrate that the smart hydrogels responsive to the solution pH and electric voltage coupled stimuli can achieve larger deformation and perform better mechanical strength. Therefore, they have a great potential as linear actuators for applications in BioMEMS (Schreyer et al., 2000).

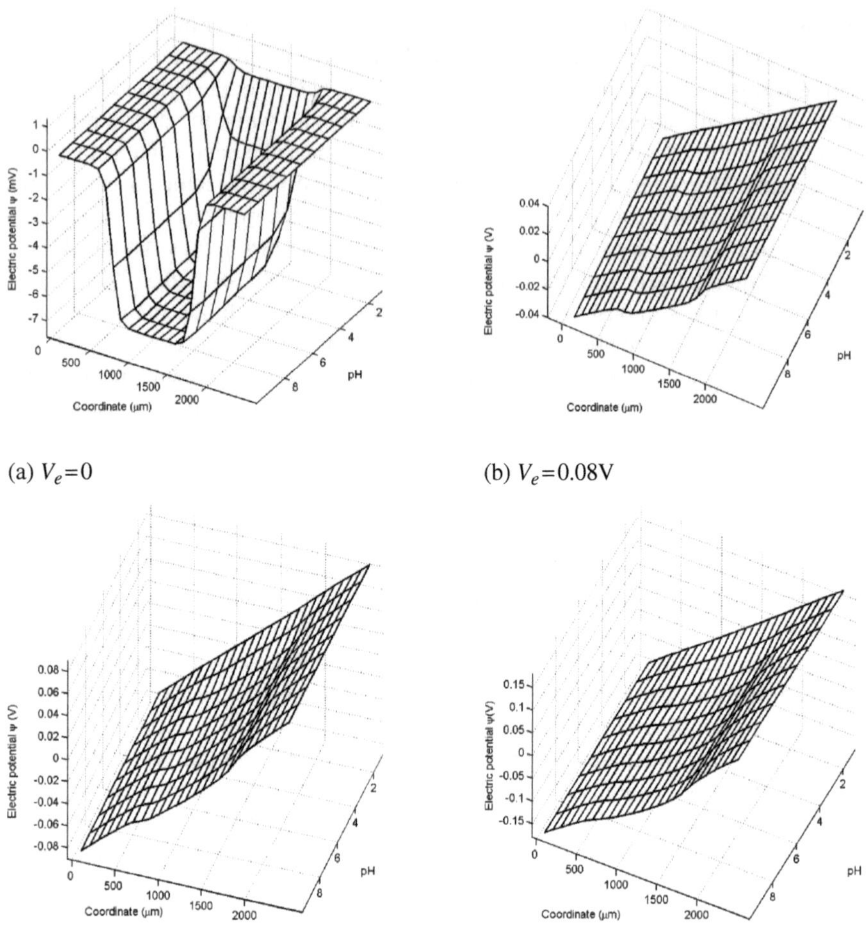

(a) V_e=0

(b) V_e=0.08V

(c) V_e=0.16V

(d) V_e=0.32V

Fig. 4.6 Coupled effect of the solution pH and external electric voltage V_e on the distributive profiles of the electric potential ψ

Figures 4.9, 4.10, 4.11, 4.12 and 4.13 are plotted for analysis of the coupled effect of the solution pH and electric voltage V_e on the variation of swelling ratio R_s and average curvature K_a of the hydrogel strip. Figures 4.9 and 4.10 illustrate the coupled effect of the solution pH and external electric voltage V_e stimuli on the swelling ratio R_s, where the swelling ratio R_s increases with the applied voltage V_e. Theoretically it is known that the increment of the applied voltage V_e amplifies the osmotic pressure and makes the hydrogels swell greatly. It is also seen that the swelling ratio R_s of the hydrogel strip changes dramatically with the pH-electric coupled stimuli, especially in the range of pH 1.0–6.0. In the range of pH 6.0–9.0, however, the effect of the solution pH on the swelling ratio R_s is insignificant.

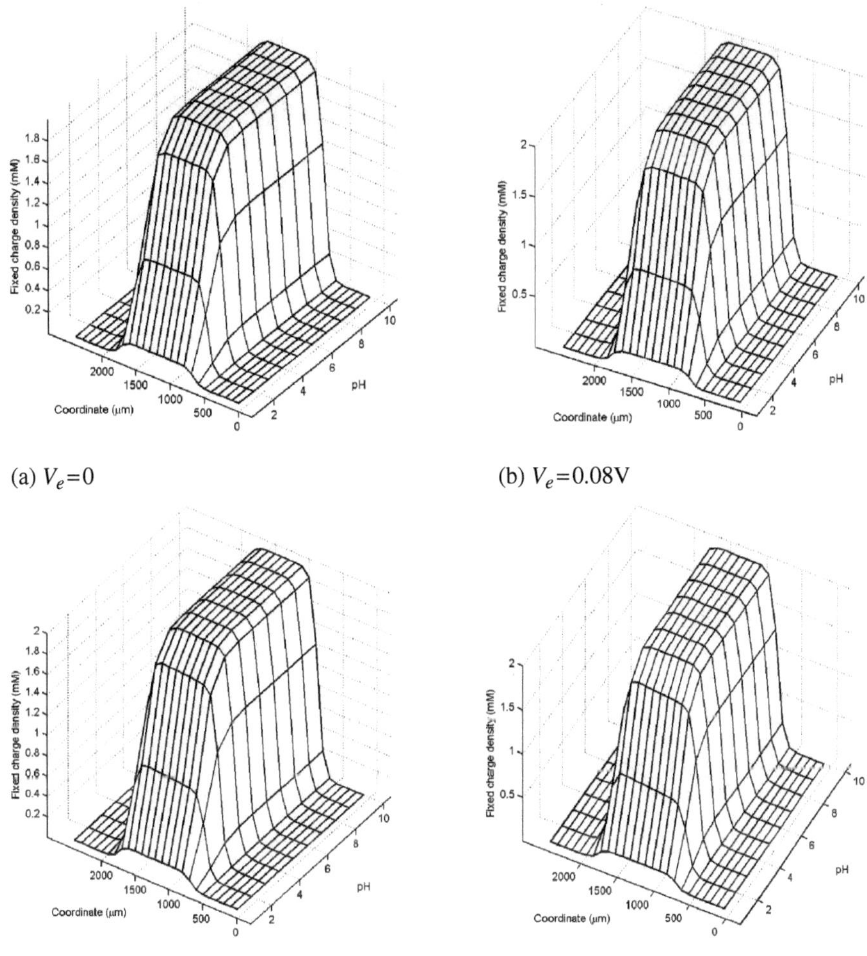

(a) V_e=0

(b) V_e=0.08V

(c) V_e=0.16V

(d) V_e=0.32V

Fig. 4.7 Coupled effect of the solution pH and external electric voltage V_e on the distributive profiles of the fixed charge density c_f

Figures 4.11 and 4.12 illustrate the coupled influence of the solution pH and electric voltage on the average curvature K_a of the hydrogel. It is observed that the average curvature K_a of the hydrogel increases with the externally applied voltage V_e, especially under higher electric voltage V_e. Then the hydrogel in equilibrium state gains larger bending deformation, which is different from the performance behaviour of the hydrogel responsive to the single stimulus, the solution pH only. At the same level of the electric voltage, the average curvature K_a increases dramatically from pH 1.0 to 4.0. However, it gains tiny increase in the range of pH 5.0–9.0. The deformation of the hydrogel may be produced by changes of two physical

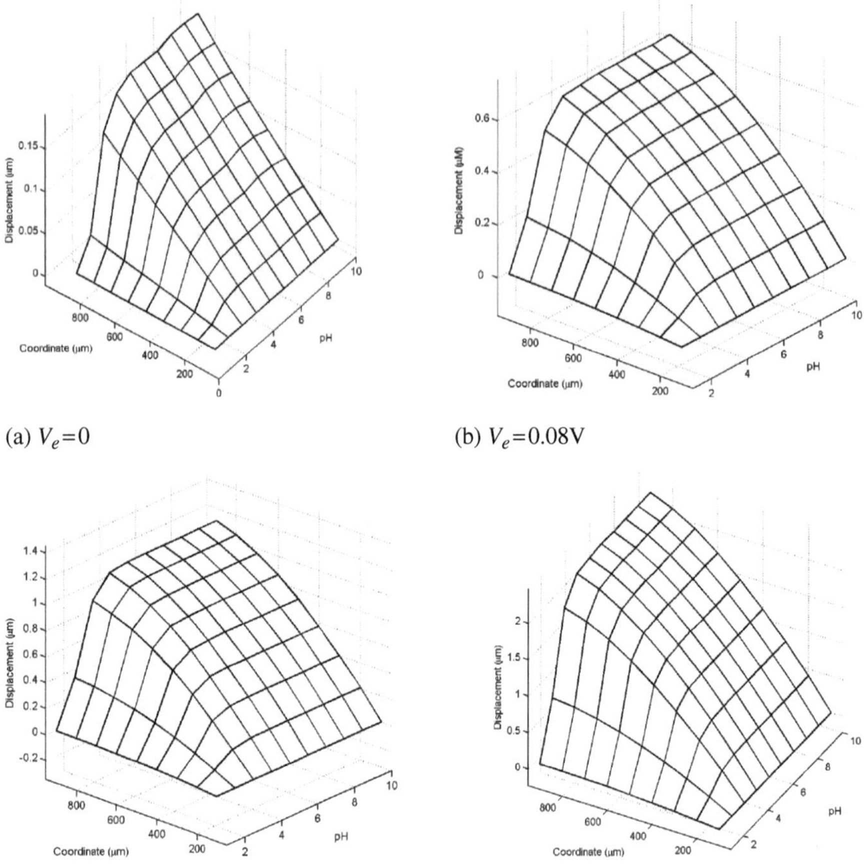

(a) V_e=0

(b) V_e=0.08V

(c) V_e=0.16V

(d) V_e=0.32V

Fig. 4.8 Coupled effect of the solution pH and external electric voltage V_e on the distributive profiles of displacement u of the hydrogel strip

parameters (Kim et al., 2003; Lam et al., 2006). One is the change of the osmotic pressure due to the difference of the diffusive ion concentrations between the interior hydrogel and the exterior solution. The pressure is the main driving force to make the hydrogel have swelling and bending deformations. The other is the change of the applied electric voltage. The increase of the electric voltage results in the larger difference of the diffusive ionic concentrations over the hydrogel–solution interfaces. The concentration differences lead to the differences of the osmotic pressures at the two interfaces of the hydrogel, which makes the hydrogel bent. As a result, the increase of the electric voltage V_e makes the swelling ratio R_s enlarge, as shown in Fig. 4.13. The predictions based on the present simulations are consistent with the experimental phenomena (Bajpai, 2001).

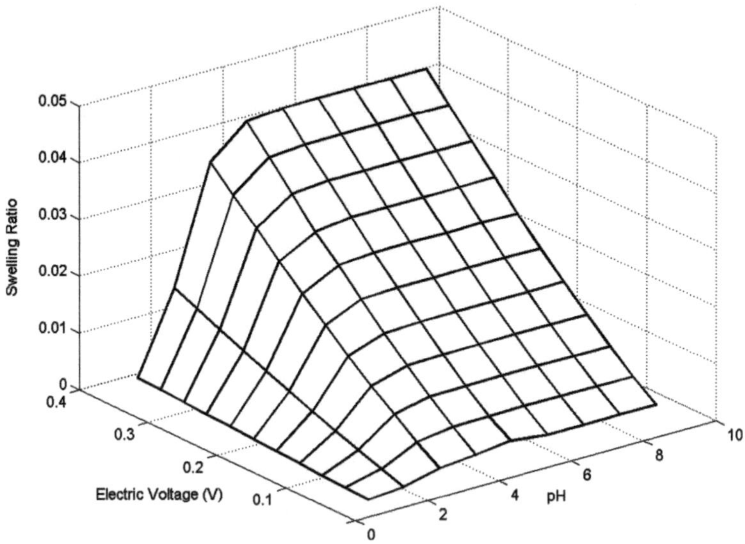

Fig. 4.9 Coupled effect of the solution pH and external electric voltage V_e on the variation of swelling ratio R_s

Fig. 4.10 Variation of the swelling ratio R_s of the hydrogel strip responding to the coupled stimuli of the solution pH and external electric field V_e

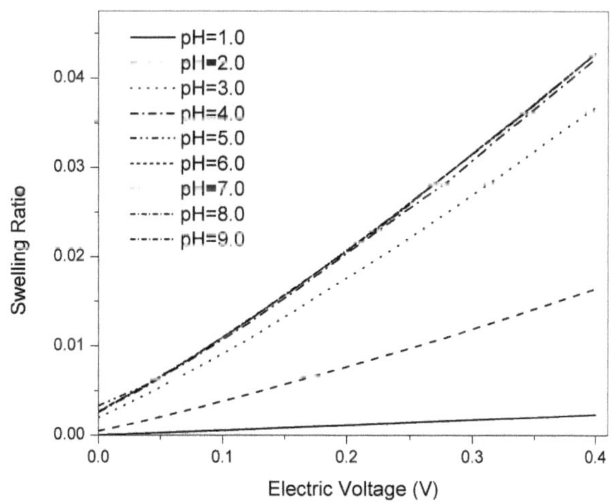

4.5.2 *Influence of Initially Fixed Charge Density of Hydrogel*

For discussion of the effect of the initially fixed charge density on the responsive behaviours of the hydrogels subject to the solution pH and electric field coupled stimuli, several numerical simulations are carried out with the input parameters required by the MECpHe model, $R = 8.314$ J/mol·K, $F = 9.648 \times 10^4$ C/mol, $\varepsilon_0 = 8.854 \times 10^{-12}\,\mathrm{C^2/N \cdot m^2}$, ε=80, $T = 298$ K, $c^* = 4.0$ mM, $|z_k| = 1$, $c_f^s = 10.0$

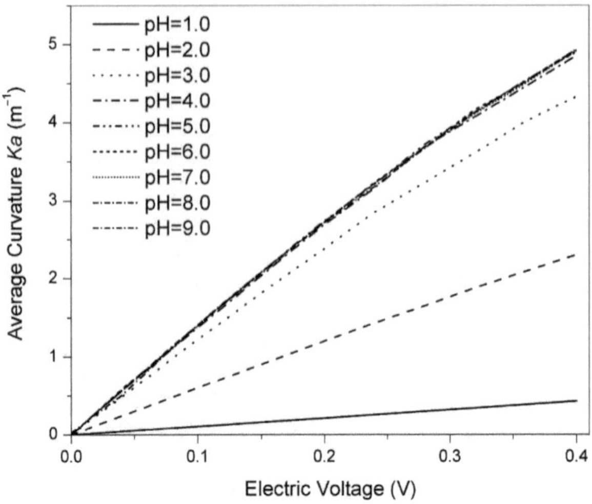

Fig. 4.11 Variation of the average curvature K_a of the hydrogel strip responding to the coupled stimuli of the solution pH and external electric field V_e

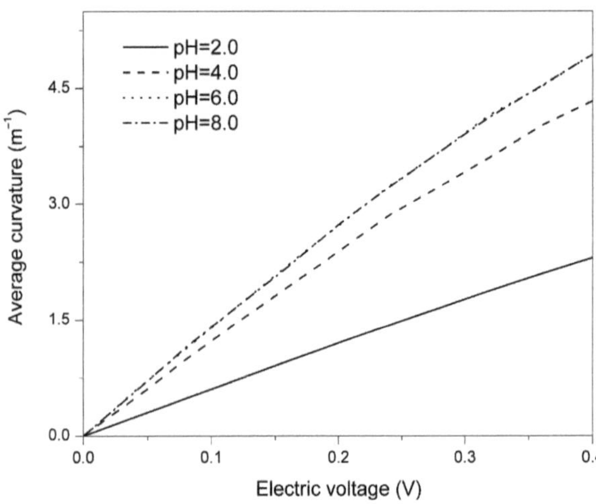

Fig. 4.12 Variation of the average curvature K_a of the hydrogel strip responding to the coupled stimuli of the solution pH and external electric field V_e

mM, $z_f = -1$, $K = 10^{-2.1}$ mM, $\phi_0^w = 0.8$, $L = 2400\,\mu\text{m}$, $h = 800\,\mu\text{m}$ and the Young's modulus taken is 3.0 MPa.

Figures 4.14 and 4.15 are plotted for the distributions of the mobile ions Na^+ and Cl^- concentrations with the coupled effect of surrounding solution pH at the external electric voltage $V_e = 0.16$ V for different levels of the initially fixed charge densities: $c_f^0 = 2.0, 4.0, 8.0 \text{and} 16.0\,\text{mM}$, respectively. As shown in Fig. 4.14a–d, as

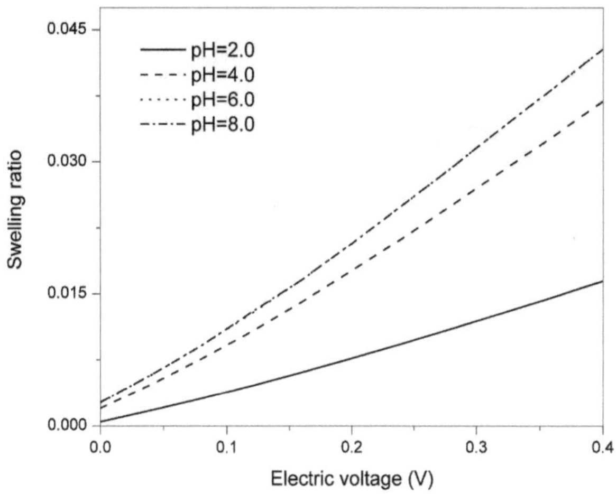

Fig. 4.13 Variation of the swelling ratio R_s of the hydrogel strip responding to the coupled stimuli of the solution pH and external electric field V_e

the initially fixed charge density c_f^0 increases, the distribution of the cation Na^+ concentration increases within the hydrogel, and the gradient of the ionic concentration within the hydrogel increases as well, which enhances the osmotic pressure to swell the hydrogel strip. The increase of the initially fixed charge density c_f^0 enlarges the concentration of the cation within the hydrogels, which makes main contribution into the increment of the ionic concentration differences over the hydrogel–solution interfaces. Subsequently, this leads to the increase of the osmotic pressure and then the swelling of the hydrogel. The simulation results are consistent with the experimental phenomena (Fei et al., 2002).

Figures 4.16 and 4.17 show the influence of the environmental solution pH at the external electric voltage $V_e = 0.16\,\text{V}$ for different initially fixed charge densities c_f^0 on the distributive profiles of the electric potential ψ and fixed charge density c_f, where $c_f^0 = 2.0, 4.0, 8.0 \text{and} 16.0\,\text{mM}$, respectively. It is found that the collapse of the distributive electric potential ψ within the hydrogel increases as the initially fixed charge density c_f^0 increases, while the gradient of the electric potential ψ within the hydrogel diminishes, as shown in Fig. 4.16a–d. It is also found in Fig. 4.17a–d that the fixed charge density increases obviously.

Figures 4.18, 4.19, 4.20, 4.21 and 4.22 focus on the coupled effect of the solution pH and electric voltage V_e as well as initially fixed charge density c_f^0 on the mechanical deformation of the pH-electric-sensitive hydrogel.

Figure 4.18a–d illustrates the distributive profiles of displacement u of the hydrogel strip with the effect of surrounding solution pH at the electric voltage $V_e = 0.16\,\text{V}$ for different levels of the initially fixed charge densities c_f^0, respectively. It is seen that the displacement u increases dramatically with increment of the initially fixed charge density c_f^0. For example, the displacement increases about 10 times as c_f^0 changes from 2 to 16.0 mM.

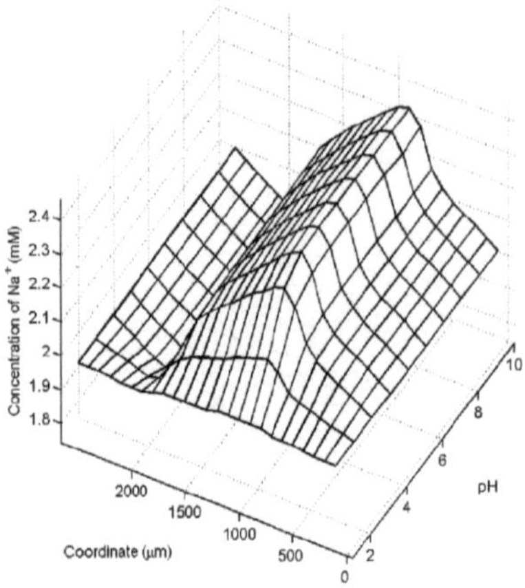

(a) $c_f^0 = 2.0\,\mathrm{mM}$

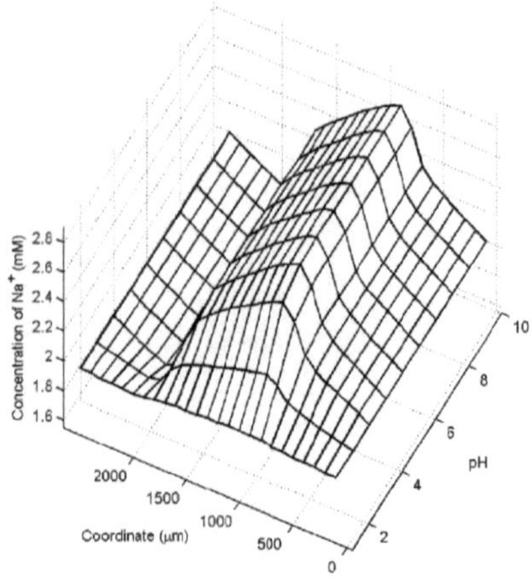

(b) $c_f^0 = 4.0\,\mathrm{mM}$

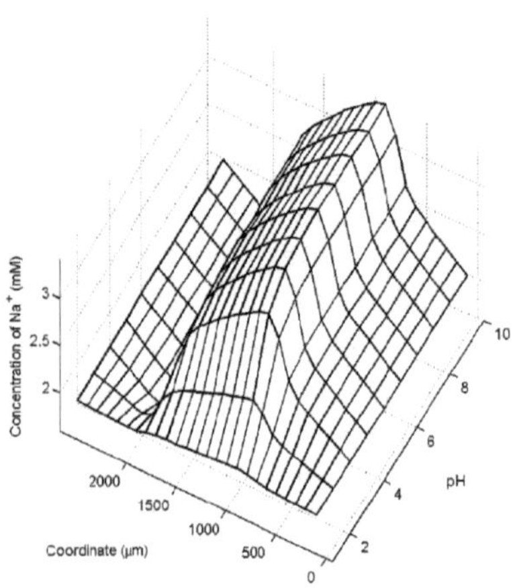

(c) $c_f^0 = 8.0\,\mathrm{mM}$

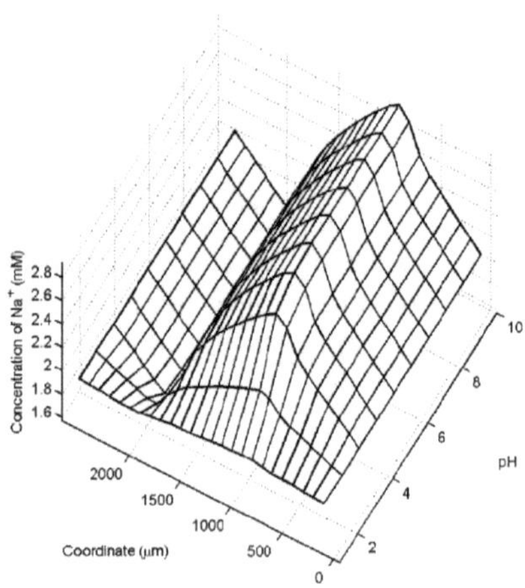

(d) $c_f^0 = 16.0\,\mathrm{mM}$

Fig. 4.14 Distributive profiles of the diffusive Na^+ subject to the effect of the solution pH coupled with the initially fixed charge density c_f^0 ($V_e = 0.16\,\mathrm{V}$)

Figures 4.19 and 4.22 illustrate the coupled effects of the solution pH and the electric voltage V_e as well as the initially fixed charge density c_f^s on the variations of the average curvature K_a, where the solution pH varies from 1.0 to 10.0 and the electric voltages V_e from 0.08 to 0.4 V, and the initially fixed charge densities $c_f^0 = 2.0, 4.0, 8.0 \text{and} 16.0\,\mathrm{mM}$, respectively. It is observed from the figures that the average curvature K_a of the hydrogel strip increases with the externally applied voltage V_e. The greater electric voltage V_e increases, the larger equilibrium bending deformation the hydrogel strip gains, which is different from the responsive

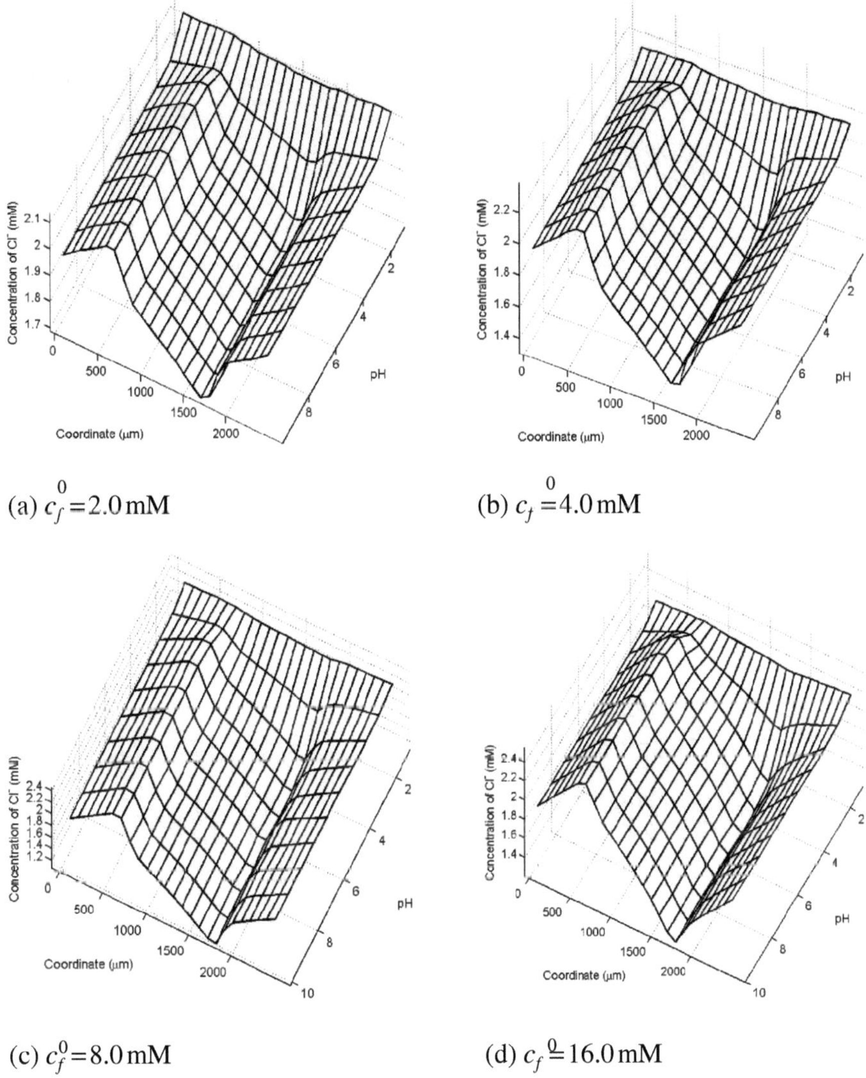

Fig. 4.15 Distributive profiles of the diffusive Cl^- subject to the effect of the solution pH coupled with the initially fixed charge density c_f^0 ($V_e = 0.16$ V)

characteristics of the hydrogel strip to a single stimulus of the solution pH. For the given electric voltage V_e, the average curvature K_a increases dramatically from pH 1.0 to 4.0. However, it reaches almost constant state in the range of pH 5.0–9.0. The deformation of the hydrogel may be produced by changes of the osmotic pressure and the conformation of the crosslinked polymer chains (Kim et al., 2003). The osmotic pressure results from the difference of diffusive ionic concentrations between the hydrogel and the bathing solution, which dominates the driving source

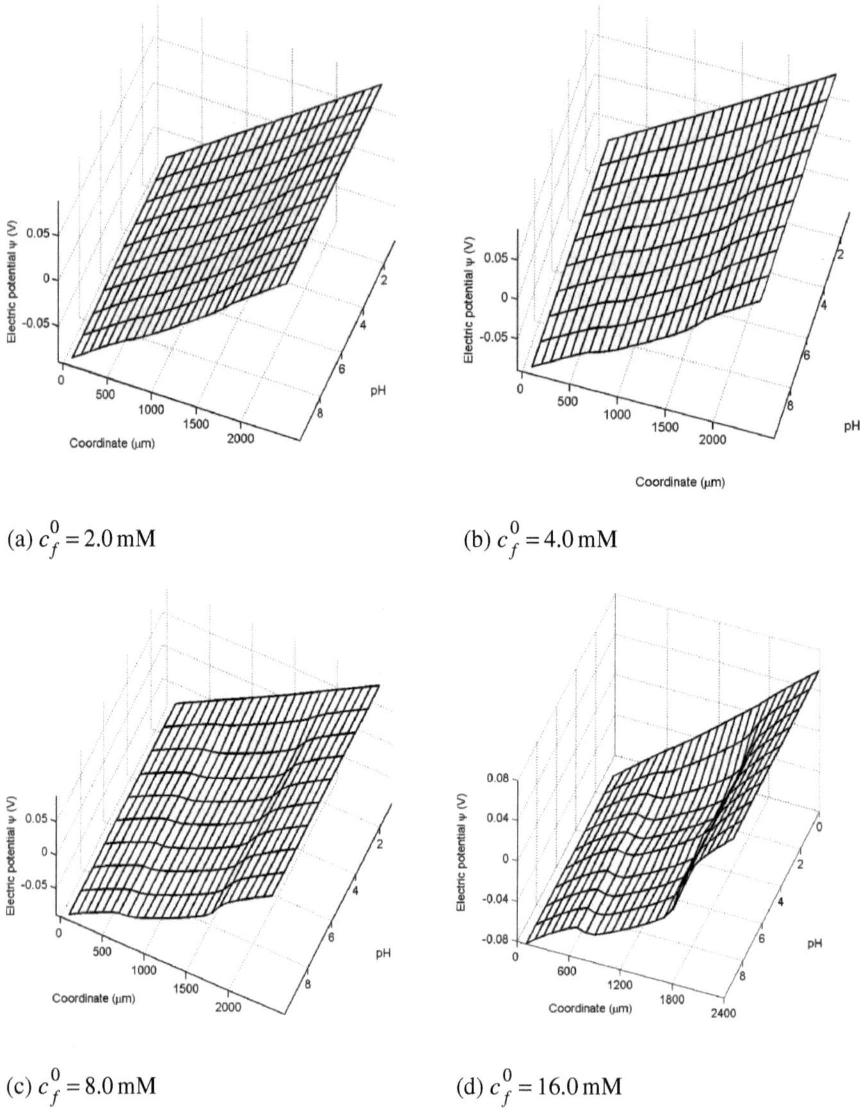

Fig. 4.16 Distributive profiles of the electric potential ψ subject to the effect of the solution pH coupled with the initially fixed charge density c_f^0 ($V_e = 0.16\,\mathrm{V}$)

for swelling and bending deformations of the hydrogel strip. The increase of electric voltage enlarges the difference of the diffusive ionic concentrations over the interfaces of the hydrogel. The differences of the ionic concentrations cause the differences of the osmotic pressures along the interfaces of the hydrogel, which makes the hydrogel strip bent. In other words, the increase of electric voltage V_e enlarges the average curvature K_a. The change of the conformation of crosslinked polymer chains results from the pH change in the hydrogel. The increase of pH

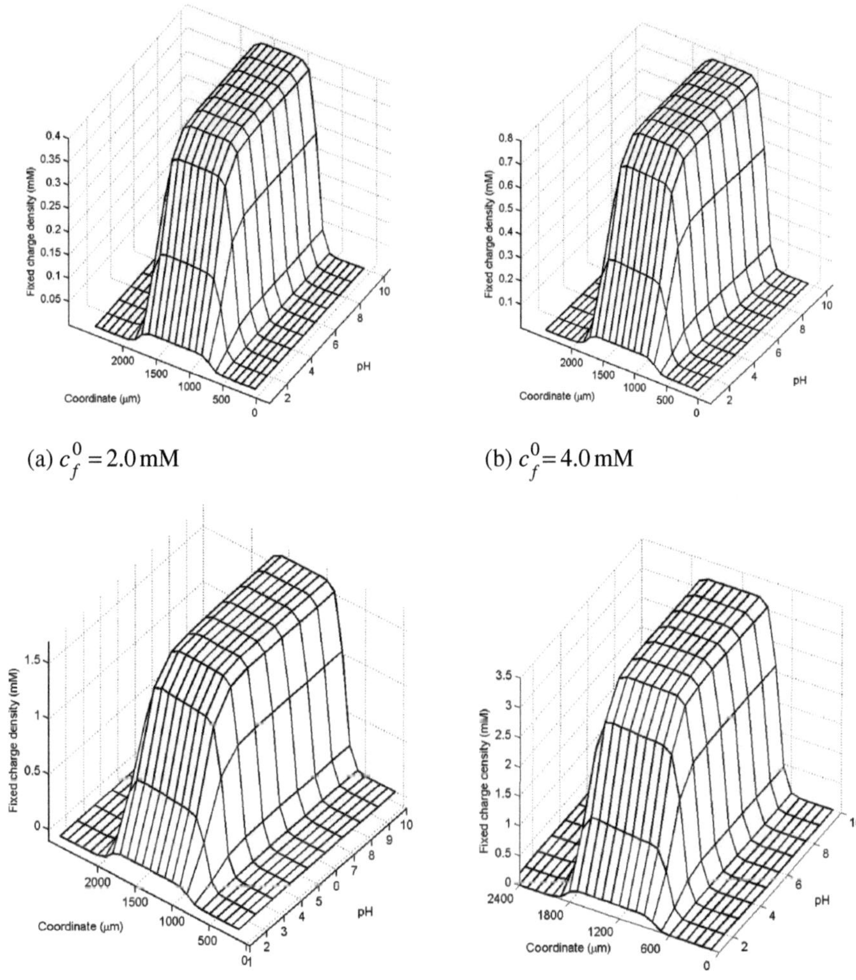

(a) $c_f^0 = 2.0\,\text{mM}$ (b) $c_f^0 = 4.0\,\text{mM}$

(c) $c_f^0 = 8.0\,\text{mM}$ (d) $c_f^0 = 16.0\,\text{mM}$

Fig. 4.17 Distributive profiles of the fixed charge density c_f subject to the effect of the solution pH coupled with the initially fixed charge density c_f^0 ($V_e = 0.16\,\text{V}$)

within the hydrogel leads to the change of the carboxylic acid groups fixed on the polymer chains from $R - COOH$ to $R - COO^-$. The change from weak polyelectrolyte to strong one means the increase of the electrorepulsive interaction between the carboxylic acid groups as the fixed charges. Therefore, the conformation of the polymeric chains of the hydrogel is changed from the compact state to the expanded one, and then the hydrogel swells. A close look into the figures reveals that the significant increase of the equilibrium swelling and bending of the hydrogel in

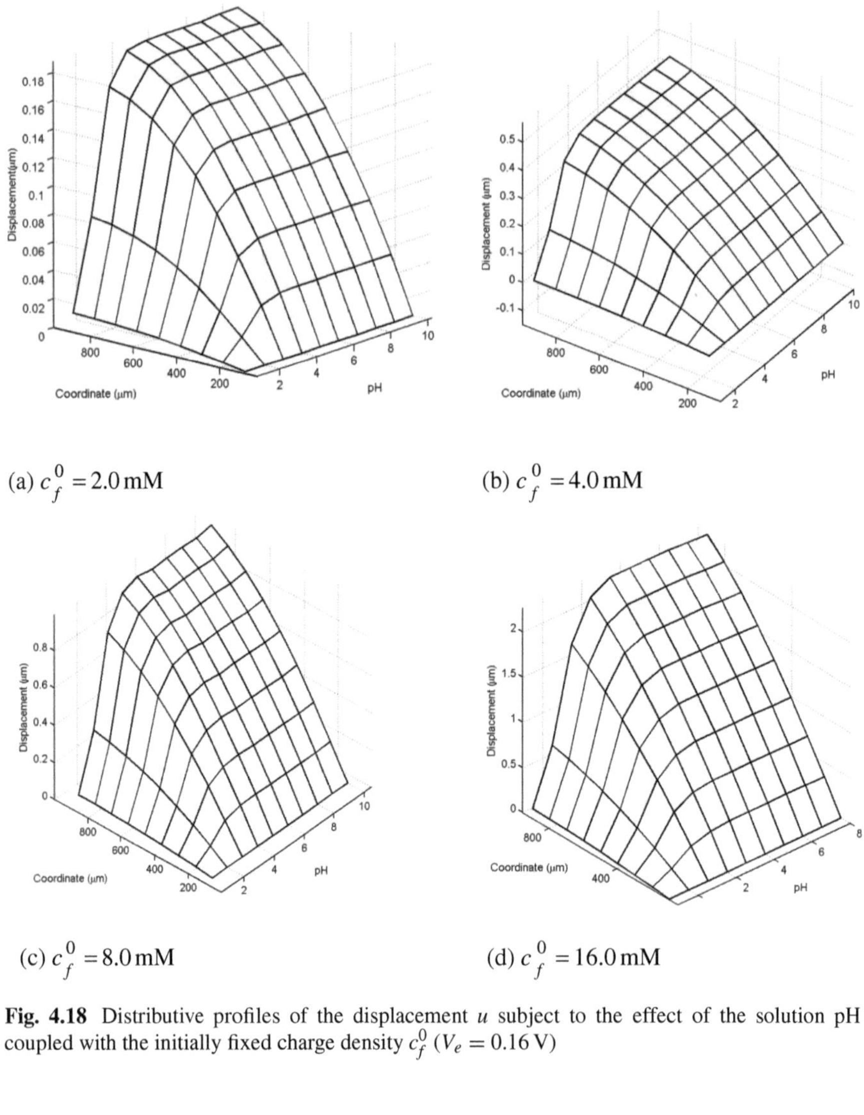

(a) $c_f^0 = 2.0\,\text{mM}$ (b) $c_f^0 = 4.0\,\text{mM}$

(c) $c_f^0 = 8.0\,\text{mM}$ (d) $c_f^0 = 16.0\,\text{mM}$

Fig. 4.18 Distributive profiles of the displacement u subject to the effect of the solution pH coupled with the initially fixed charge density c_f^0 ($V_e = 0.16\,\text{V}$)

micro-scale occurs in the range of solution pH 2.0–4.0, which is close to the dissociation constant of the fixed charge groups, $pK_a = 2.1$. The carboxylic groups at pH $\geq$4.0 are totally ionized, which may be the reason for the tiny increase of the average curvature K_a in the range of pH 5.0–9.0. The predictions by the present simulations are consistent well with the experimental phenomena (Bajpai, 2001), where it was reported that the local pH gradient attributed to water electrolysis may be additional factor to influence the bending deformation (Lam et al., 2006).

Figures 4.20 and 4.21 are plotted for analysis of the coupled effect of the initially fixed charge density c_f^0 and electric voltage V_e as well as solution pH on variation of the swelling ratio R_s. It is shown from the figures that the swelling ratio R_s increases

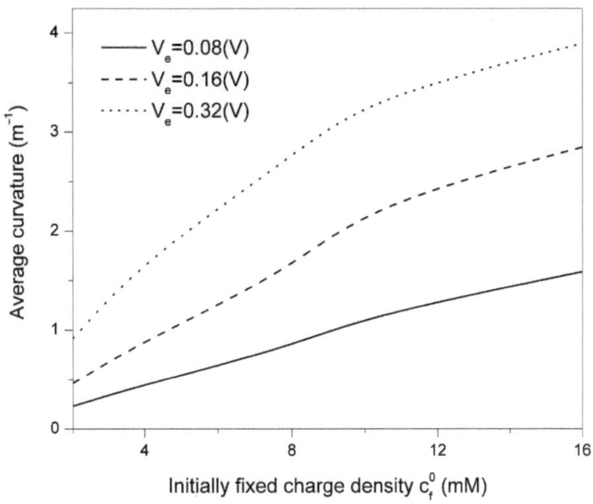

Fig. 4.19 Coupled effect of the initially fixed charge density c_f^0 and electric voltage V_e on variation of the average curvature K_a (pH 4.0)

Fig. 4.20 Coupled effect of the initially fixed charge density c_f^0 and electric voltage V_e on variation of the swelling ratio R_s (pH 4.0)

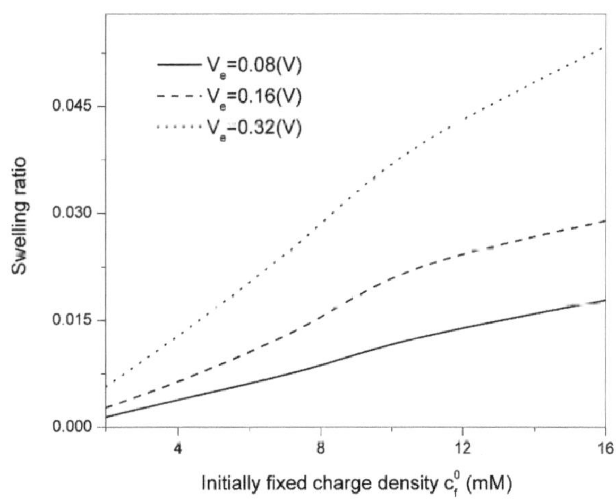

with the initially fixed charge density c_f^0. The present simulation results qualitatively agree well with the experimental observations (Fei et al., 2002).

4.5.3 Influence of Ionic Strength

For the pH-electric-sensitive hydrogel as promising materials used as sensors, actuators and artificial muscles for biomedical engineering applications, the mechanical properties of the smart hydrogels are particularly concerned. It is known that the mechanical force is generated by the osmotic pressure, which is governed by the

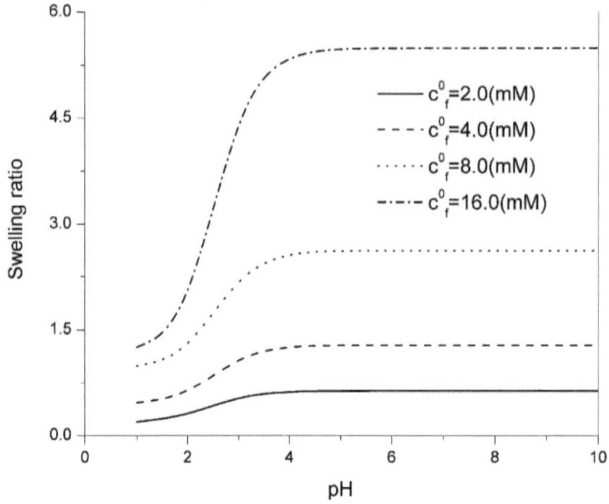

Fig. 4.21 Coupled effect of the initially fixed charge density c_f^0 and solution pH on variation of the swelling ratio R_s ($V_e = 0.4$ V)

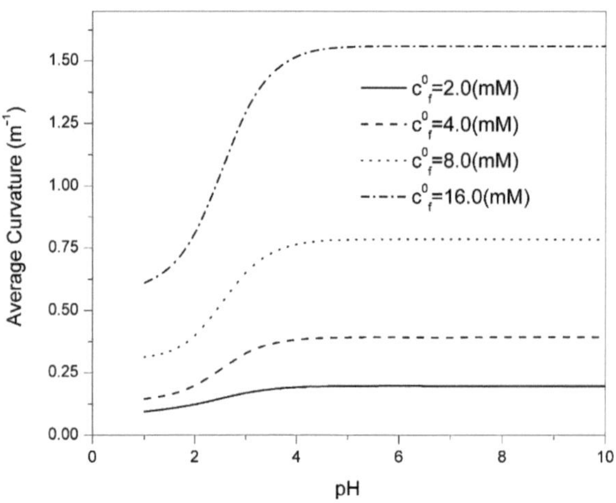

Fig. 4.22 Coupled effect of the initially fixed charge density c_f^0 and solution pH on variation of the average curvature K_a ($V_e = 0.4$ V)

ionic transport through the system, and is affected by the hydrogel architecture (Chiarellim and Rossi, 1996; Carlson et al., 2003). In order to investigate the effect of the ionic strength I ($I = 0.5 \times \sum_k c_k z_k^2$) of environmental solution on the bending deformation of the hydrogels, several simulations are conducted numerically with the input parameters for the MECpHe model, $R = 8.314$ J/mol·K, $F = 9.648 \times 10^4$ C/mol, $T = 298$ K, $\varepsilon = 80$, $\varepsilon_0 = 8.854 \times 10^{-12}$ C^2/N·m^2, $z_f = -1$, $K = 10^{-2.1}$ mM, $\phi_0^w = 0.8$, $L = 2400$ μm, $h = 800$ μm and the Young's modulus is equal to 3.0 MPa.

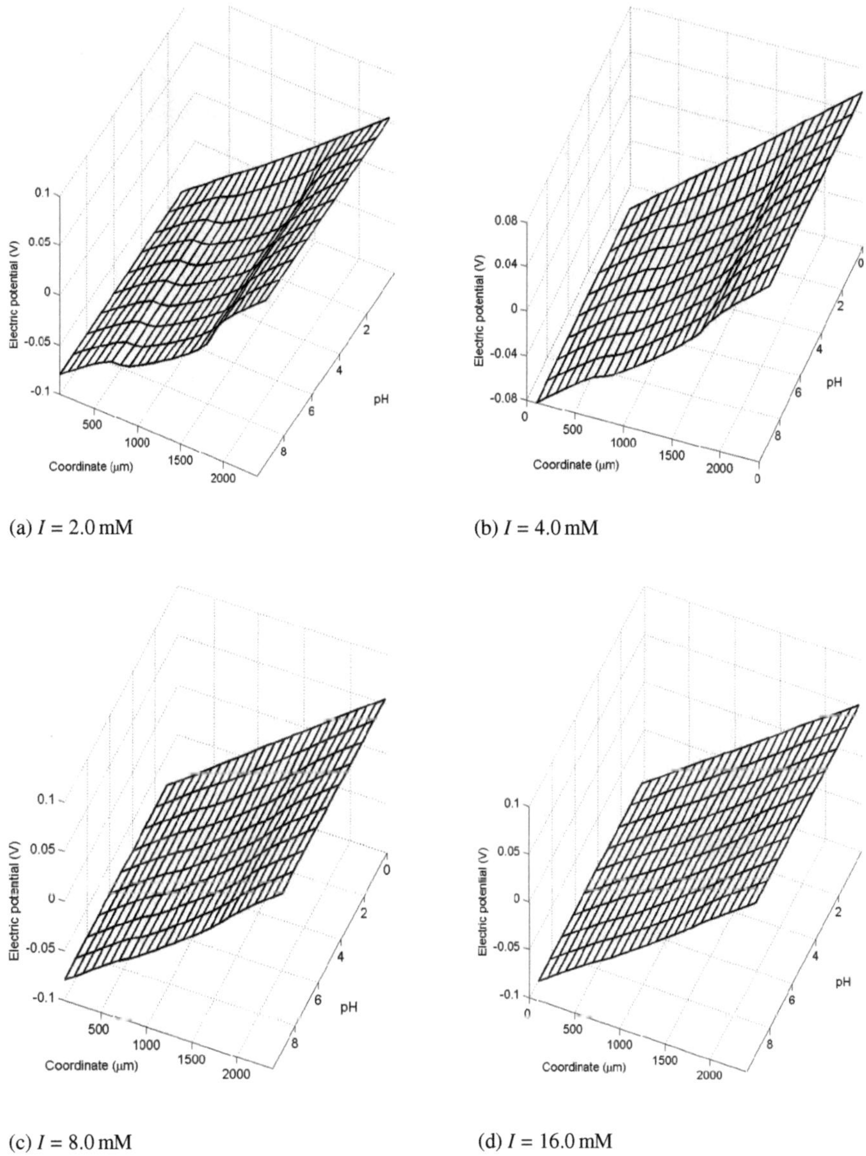

Fig. 4.23 Distributive profiles of the electric potential ψ subject to the effect of the solution pH coupled with the ionic strength I ($V_e = 0.16\,\text{V}$)

Figures 4.23 and 4.24 illustrate the distributive profiles of the electric potential ψ and fixed charge density c_f with coupled influences of solution pH at the external electric voltage $V_e = 0.16\,\text{V}$ for different levels of the ionic strengths $I = 2.0, 4.0, 8.0$ and $16.0\,\text{mM}$. It is found that, as the ionic strength I of bathing solution increases, the gradient of electric potential ψ increases within the hydrogel, while the collapse

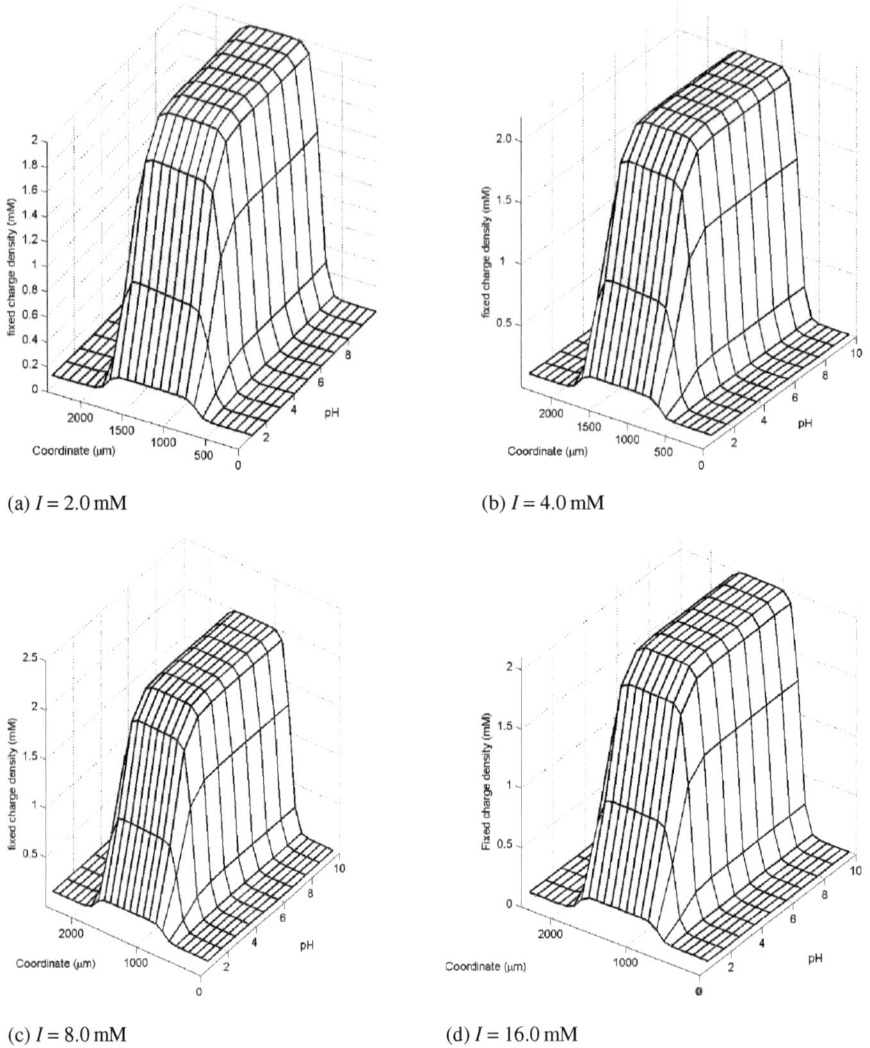

(a) $I = 2.0\,\text{mM}$ (b) $I = 4.0\,\text{mM}$

(c) $I = 8.0\,\text{mM}$ (d) $I = 16.0\,\text{mM}$

Fig. 4.24 Distributive profiles of the fixed charge density c_f subject to the effect of the solution pH coupled with the ionic strength I ($V_e = 0.16\,\text{V}$)

of the electric potential ψ diminishes. The potential ψ distributes linearly in the whole domain covering the hydrogel and bathing solution, especially at high ionic strength I. Theoretically the increase of the ionic strength of environmental solution makes the more mobile ions diffuse into the hydrogel, then the conductivity of hydrogels becomes almost equal to that of surrounding solution, which results in the electric potential ψ distributing quasilinearly over the whole domain. It is known from Fig. 4.24 that the fixed charge density c_f and the corresponding gradient in the hydrogel increase with the ionic strength I. It is noted that the dissociation of carboxylic groups fixed on the polymeric chains of the hydrogel imparts the ionic

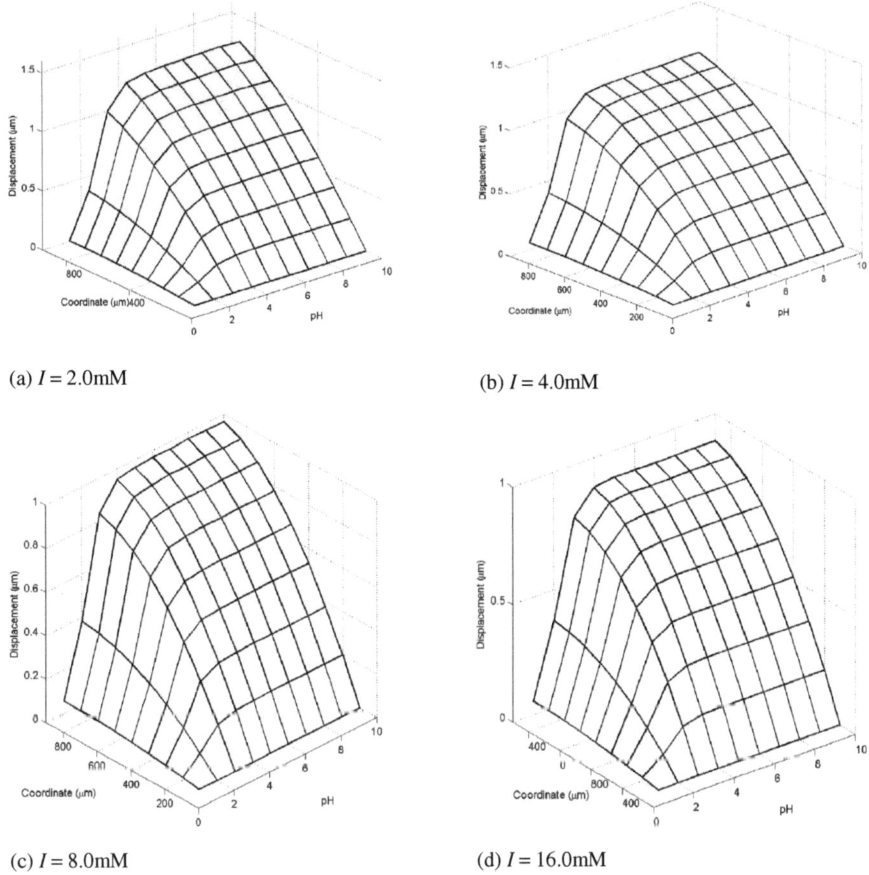

Fig. 4.25 Distributive profiles of the displacement u subject to the effect of the solution pH coupled with the ionic strength I ($V_e = 0.16\,\mathrm{V}$)

character to the hydrogels and affects the ion osmotic swelling pressure (Bajpai, 2001). In fact, the enhancement of the ionic strength I makes more contribution to the increase of the concentrations in the surrounding solution, relatively compared with the contribution to that in the hydrogels. This reduces the difference of diffusive ion concentrations over the hydrogel–solution interfaces. The reduction of concentration differences decreases the osmotic pressure. From the experimental swelling phenomena of polyelectrolyte gels, it is also observed that the osmotic pressure ultimately reduces the equilibrium swelling capacity of the hydrogels (Khare and Peppas, 1995), and the hydrogels thus shrink.

Figures 4.25, 4.26, 4.27, 4.28, 4.29, 4.30, 4.31, 4.32, 4.33 and 4.34 show the coupled influences of the solution pH and electric voltage V_e as well as initially fixed charge density c_f^0 on the mechanical deformation of the pH-electric-sensitive hydrogel.

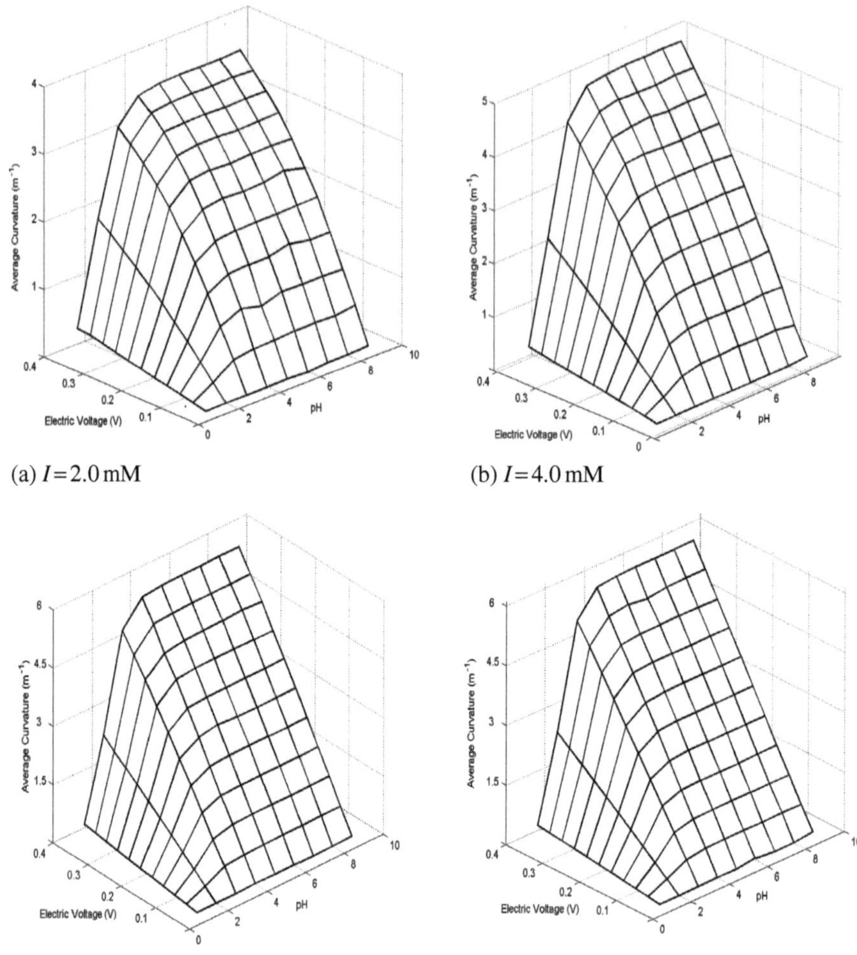

(a) $I = 2.0\,\text{mM}$ (b) $I = 4.0\,\text{mM}$

(c) $I = 8.0\,\text{mM}$ (d) $I = 16.0\,\text{mM}$

Fig. 4.26 Variations of the average curvature K_a subject to the effect of the solution pH coupled with the electric voltage V_e at various ionic strengths I

Figure 4.25a–d illustrate the distributive profiles of displacement u with the coupled influences of solution pH at the electric voltage $V_e = 0.16\,\text{V}$ for various ionic strengths I. The displacement u of the hydrogel strip decreases with the increase of ionic strength I of bathing solution. As mentioned above, the increase of I will reduce the osmotic pressure and thus cause the hydrogel deswelling. However, the average curvature K_a of the hydrogel increases with the enhancement of ionic strength I. Figure 4.26a–d shows the coupled influences of the solution pH, the electric voltage V_e and the solution ionic strength I on the variation of the average curvature K_a, where $c_f^s = 10.0$ mM, the solution pH ranges from 1.0 to 9.0,

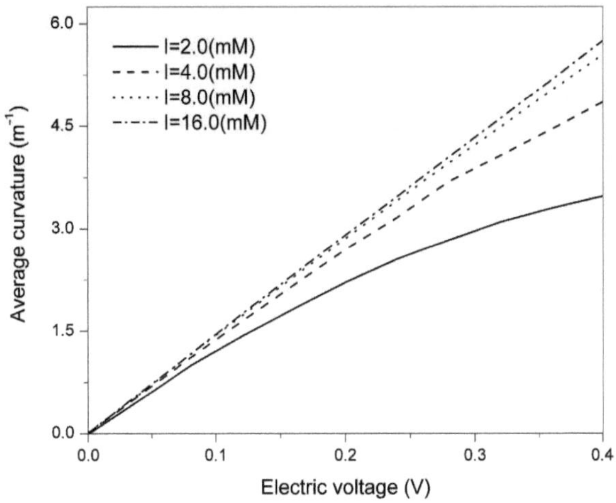

Fig. 4.27 Coupled effect of the ionic strength I and electric voltage V_e on variation of the average curvature K_a (pH 4.0)

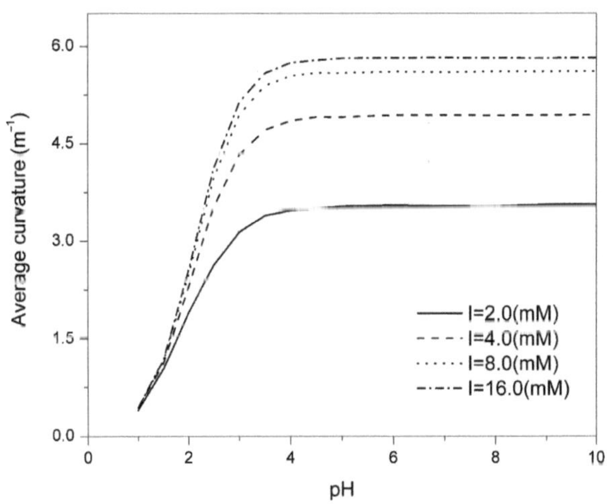

Fig. 4.28 Coupled effect of the ionic strength I and solution pH on variation of the average curvature K_a ($V_e = 0.4$ V)

the electric voltages V_e from 0 to 0.4 V and the solution ionic strengths $I = 2.0$, 4.0, 8.0 and 16.0 mM, respectively. Figures 4.27, 4.28, 4.29 and 4.30 illustrate the influence of the ionic strength I on the average curvature K_a for various levels of the solution pH and electric voltage V_e. It is observed that the average curvature K_a of the hydrogel increases with the externally applied voltage V_e, i.e., the hydrogel gains larger equilibrium bending deformation at higher electric voltage V_e. But it is different from the characteristics of the hydrogels in response to the single stimulus of solution pH. The average curvature K_a under the same electric voltage

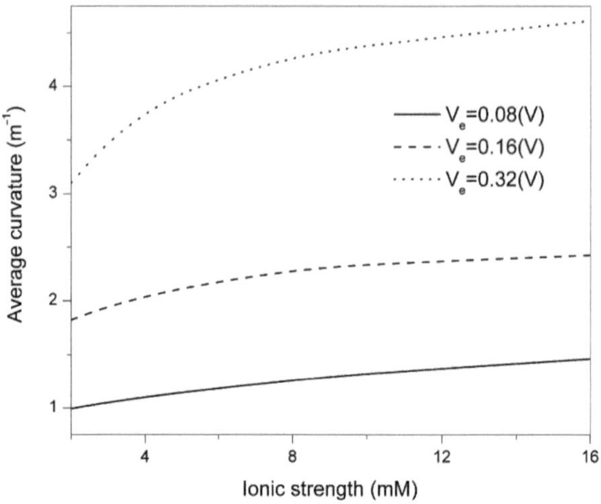

Fig. 4.29 Coupled effect of the ionic strength I and electric voltage V_e on variation of the average curvature K_a (pH 4.0)

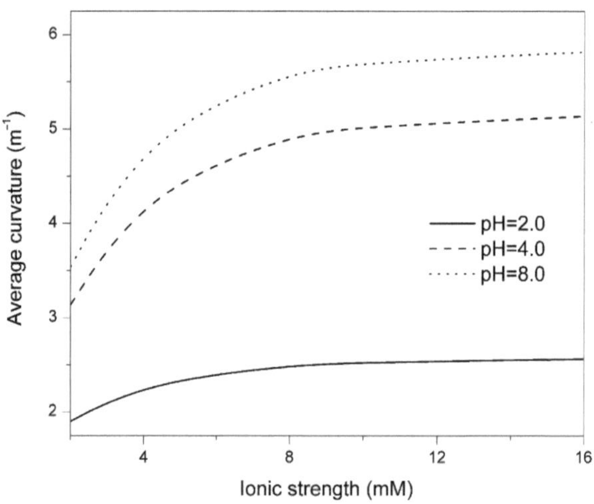

Fig. 4.30 Coupled effect of the ionic strength I and solution pH on variation of the average curvature K_a ($V_e = 0.4$ V)

increases significantly from pH=1.0 to 4.0. However, it gains the slight increment in the range of pH 5.0–9.0. The deformation of the hydrogel may be mainly driven by the change in the osmotic pressure (Shiga et al., 1990), which is generated by the difference of the diffusive ion concentrations between the interior hydrogels and the exterior solution. The increase of the electric voltage results in the larger ionic concentration differences over the interfaces between the hydrogel and the bathing solution. The differences of the diffusive ion concentrations lead to the difference

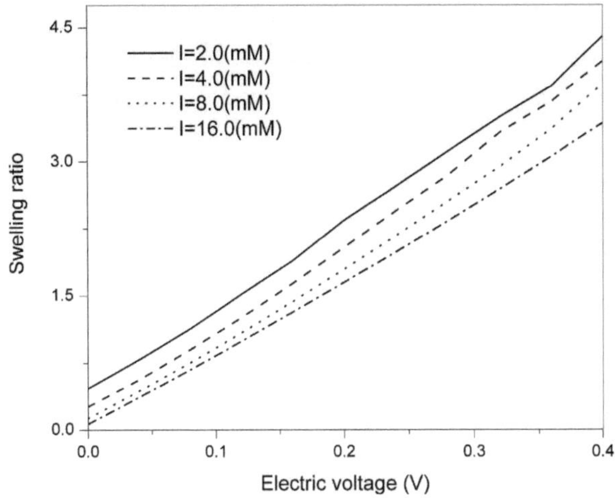

Fig. 4.31 Coupled effect of the ionic strength I and electric voltage V_e on variation of the swelling ratio R_s (pH 4.0)

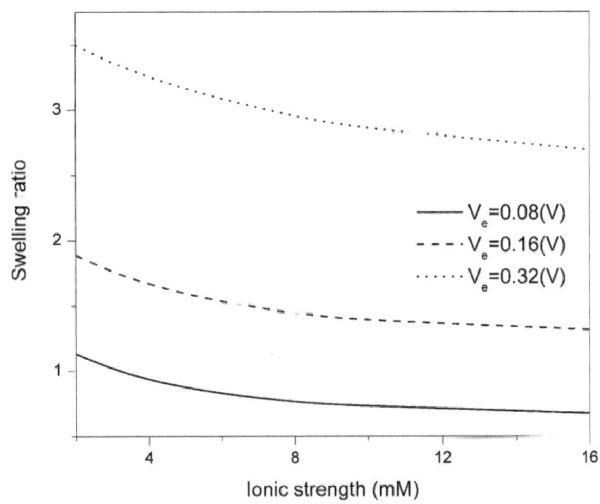

Fig. 4.32 Coupled effect of the ionic strength I and electric voltage V_e on variation of the swelling ratio R_s (pH 4.0)

of the osmotic pressures at the two sides of the hydrogel, which makes the hydrogel bend. Therefore, the increase of electric voltage V_e enlarges the average curvature K_a, as shown in the Fig. 4.26a–d. The present simulations reveal that the significant increase of the equilibrium swelling and bending of the hydrogel occurs in the range of solution pH 2.0–4.0, which is associated with the dissociation constant $pK_a = 2.1$ of the hydrogels. If the surrounding solution pH is larger than 4.0, the carboxylic groups are totally ionized, which could be the reason for the tiny increase of K_a in

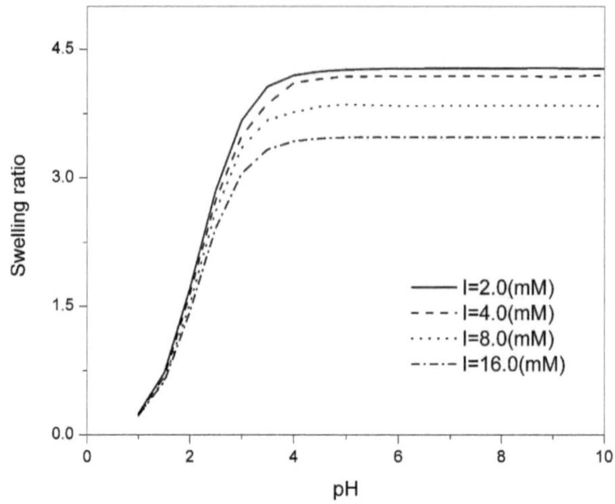

Fig. 4.33 Coupled effect of the ionic strength I and solution pH on variation of the swelling ratio R_s ($V_e = 0.4$ V)

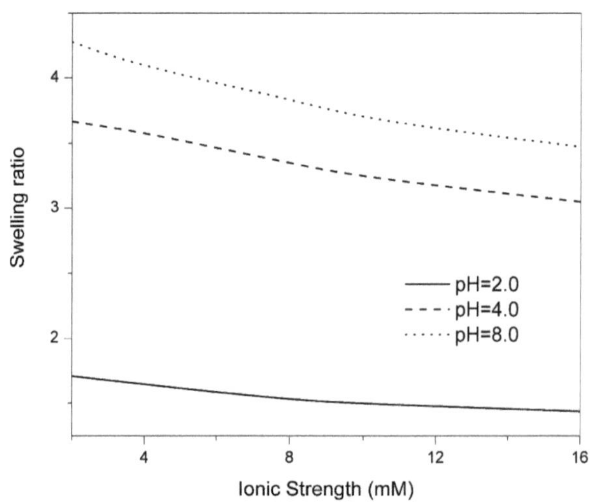

Fig. 4.34 Coupled effect of the ionic strength I and solution pH on variation of the swelling ratio R_s ($V_e = 0.4$ V)

the range of pH 5.0–9.0. The simulation results are consistent with the experimental phenomena (Bajpai, 2001). It is found from the figures that the average curvature K_a of the hydrogel increases with the ionic strength I subject to the coupled stimuli of the solution pH and externally applied electric voltage V_e, in which the increment of K_a is not linear and will slow down with the ionic strength I. It is known that the osmotic pressure decreases with the increase of the ionic strength I of the system medium, and ultimately reduces the equilibrium swelling capacity of the hydrogels

(Lam et al., 2006). However, the increase of the ionic strength or electrolyte concentration induces the increase of mobile counterions diffusing into the hydrogel. Due to the fixed charge groups existing in the hydrogel subject to electric field, the redistribution of the counterions enlarges the difference of the diffusive ionic concentrations at both the ends of the hydrogel, which amplifies the difference of the osmotic pressures over the two interfaces between the hydrogel and the surrounding solution. Therefore, the bending deformation of the hydrogel increases further and the average curvature K_a increases as well. The influence of ionic strength I may be weakened by the shielding effect of the fixed charge groups (Kim et al., 2004a, b), and the average curvature K_a increases gradually at higher levels of the ionic strength I, as shown in Fig. 4.29. The simulations agree well with the published experimental studies (Kim et al., 2004a, b).

Figures 4.31, 4.32, 4.33 and 4.34 are plotted for the coupled effect of the ionic strength I and electric voltage V_e as well as solution pH on the variation of swelling ratio R_s. It is found that the swelling ratio R_s decreases with increment of the ionic strength I of environmental solution. In fact, the increase of the ionic strength I makes more contribution to the increase of the concentration in the surrounding solution, relatively compared with the contribution to that in the hydrogels. This reduces the difference of diffusive ion concentrations over the hydrogel–solution interfaces, and thus it decreases the osmotic pressure, which makes the hydrogels shrink. The details of how the ionic strength I influences the pH sensitivity of the hydrogels require more comprehensive investigations, particularly in the case of multivalent ions (Mao and McShane, 2006).

4.5.4 Influence of Ionic Valence

In order to gain further insight into the effects of various important parameters of the hydrogel material properties and environmental conditions on the responsive behaviours of the hydrogels to the coupled stimuli of the solution pH and electric field, several numerical simulations are carried out with the input parameters required by the MECpHe model, $R = 8.314$ J/mol· K, $F = 9.648\times10^4$ C/mol, $\varepsilon_0 = 8.854\times10^{-12}\mathrm{C}^2/\mathrm{N}\cdot\mathrm{m}^2$, $T = 298$ K, ε=80, $c^* = 4.0$ mM, $c_f^0 = 10.0$ mM, $z_f = -1$, $K = 10^{-2.1}$ mM, $\phi_0^w = 0.8$, $L = 2400\,\mu$m, $h = 800\,\mu$m and the Young's modulus of 3.0 MPa is taken. The simulation results are discussed below for the influence of the ionic valence $|z_k|$ of the surrounding solution on the responsive distributions of the diffusive ionic concentrations c_k, the electric potential ψ, the fixed charge density c_f and the mechanical deformation of hydrogel strip.

Figures 4.35 and 4.36 show the coupled effects of solution pH at the external electric voltage $V_e = 0.16$ V for various levels of the ionic valences $|z_k| = 1, 2\text{and}3$ on the distributive profiles of mobile cation and anion concentrations. It is noted that the distributive concentrations of diffusive cation species within the hydrogel reduce as $|z_k|$ increases. In order to maintain the electroneutrality within the hydrogel, the cations diffuse into the hydrogel to neutralize the fixed charges ($c_f^0 = 10.0$ mM, z_f

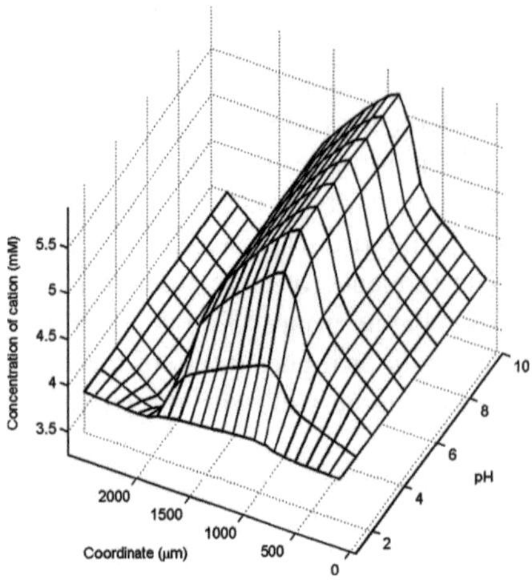

(a) $|z_k|=1$

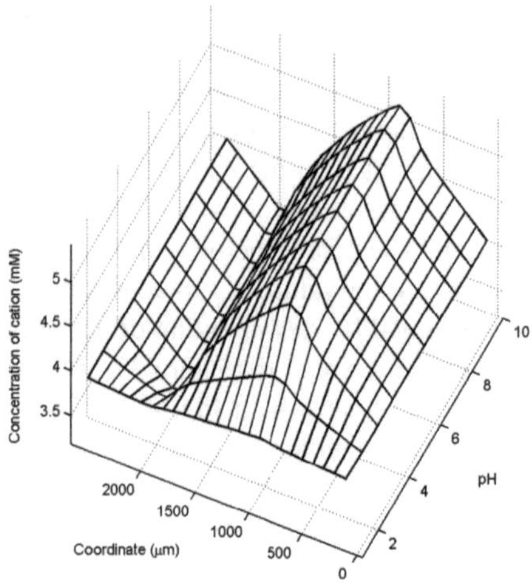

(b) $|z_k|=2$

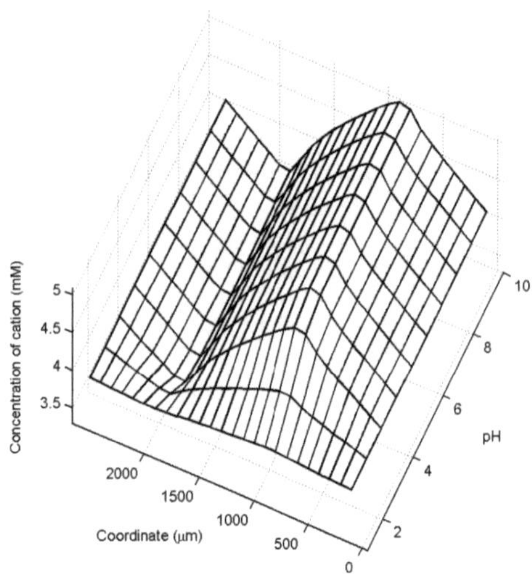

(c) $|z_k|=3$

Fig. 4.35 Distributive profiles of cation subject to the effect of the solution pH coupled with the ionic valence ($V_e = 0.16\,\text{V}$)

$= -1$) bound on the polymeric network chains. For example, if the ionic exchange process is considered, it needs 10.0 mM monovalent cations to neutralize the fixed charge groups bound on the polymer chains. However, it needs 5.0 mM divalent cations only. Therefore, when the surrounding solution is changed from monovalent electrolyte to bivalent or trivalent one, the difference of diffusive cation concentrations decreases significantly between the hydrogel and the solution, the variation of the concentration distributions is relatively small for the diffusive anions. This gives rise to the smaller osmotic pressure and then makes the hydrogels shrink.

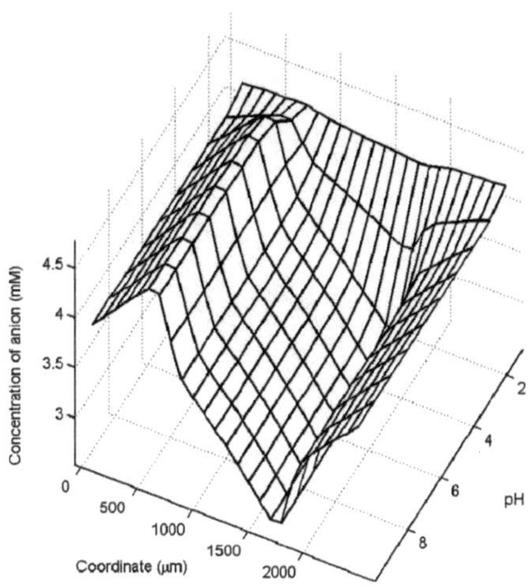

(a) $|z_k|=1$

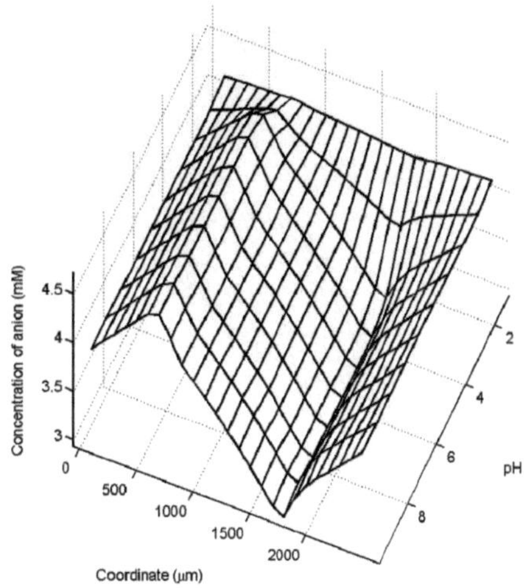

(b) $|z_k|=2$

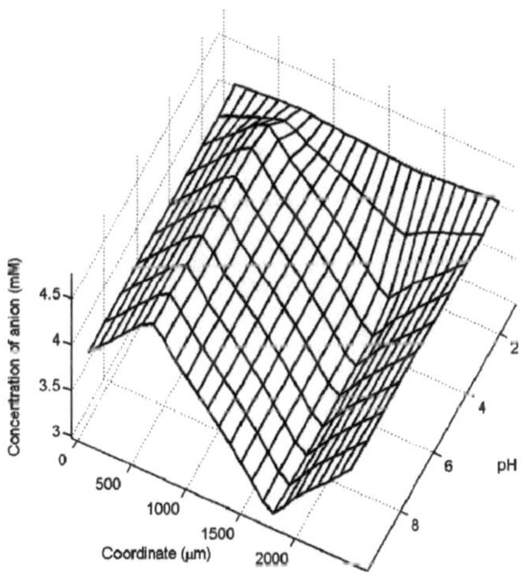

(c) $|z_k|=3$

Fig. 4.36 Distributive profiles of anion subject to the effect of the solution pH coupled with the ionic valence ($V_e = 0.16\,\mathrm{V}$)

Figures 4.37 and 4.38 show the distributive profiles of the electric potential ψ and fixed charge density c_f subject to the influences of bathing solution pH at the external electric voltage $V_e = 0.16\,\mathrm{V}$ for different ionic valences $|z_k| = 1, 2$ and 3. It is found that the gradient of electric potential ψ within the hydrogel increases as the ionic valence $|z_k|$ increases, while the collapse of ψ diminishes. The electric potential ψ distributes more and more linearly over the whole system domain covering the hydrogel and bathing solution, especially at larger ionic valence.

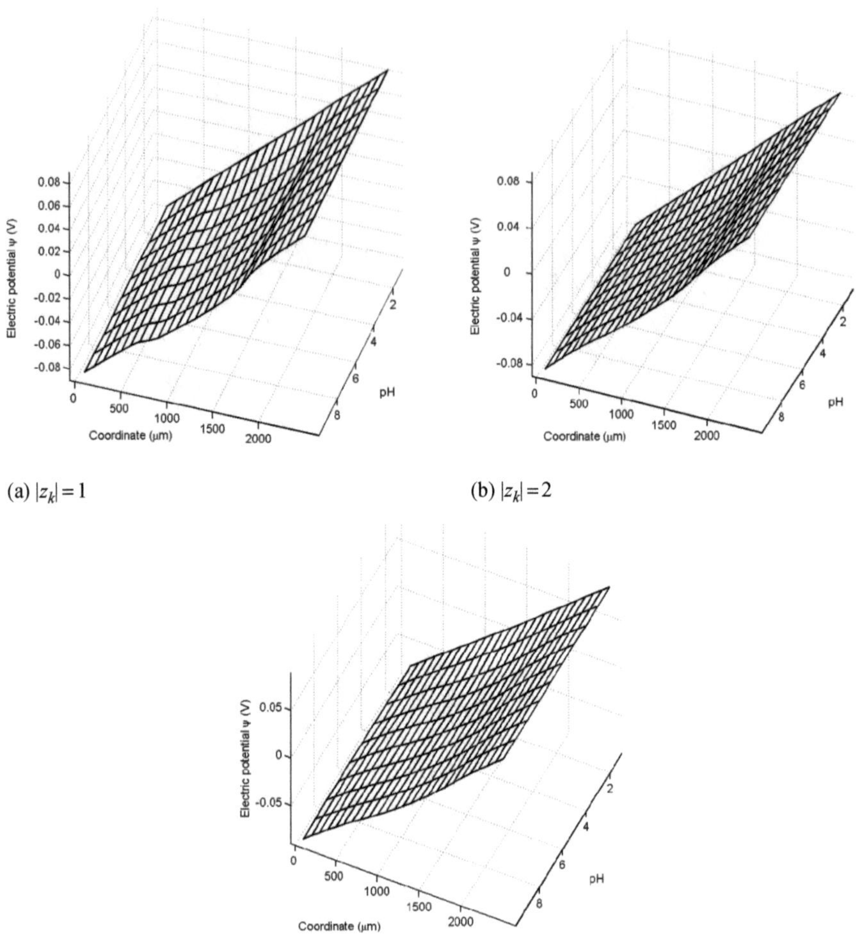

(a) $|z_k| = 1$

(b) $|z_k| = 2$

(c) $|z_k| = 3$

Fig. 4.37 Distributive profiles of the electric potential ψ subject to the effect of the solution pH coupled with the ionic valence ($V_e = 0.16\,\mathrm{V}$)

Figures 4.39, 4.40, 4.41, 4.42 and 4.43 demonstrate the coupled influences of the solution pH and electric voltage V_e as well as ionic valence $|z_k|$ on the mechanical deformation of the pH-electric-sensitive hydrogel.

Figure 4.39a–c illustrates the variation of displacement u of the hydrogel strip. It is noted that when the surrounding solution is changed from monovalent electrolyte to bivalent or trivalent one, the osmotic pressure decreases dramatically over the interface between the hydrogel and the bathing solution. As a result, the displacement u of the hydrogel strip decreases, especially when the monovalent electrolyte is changed to the divalent one, as shown in the Fig. 4.39.

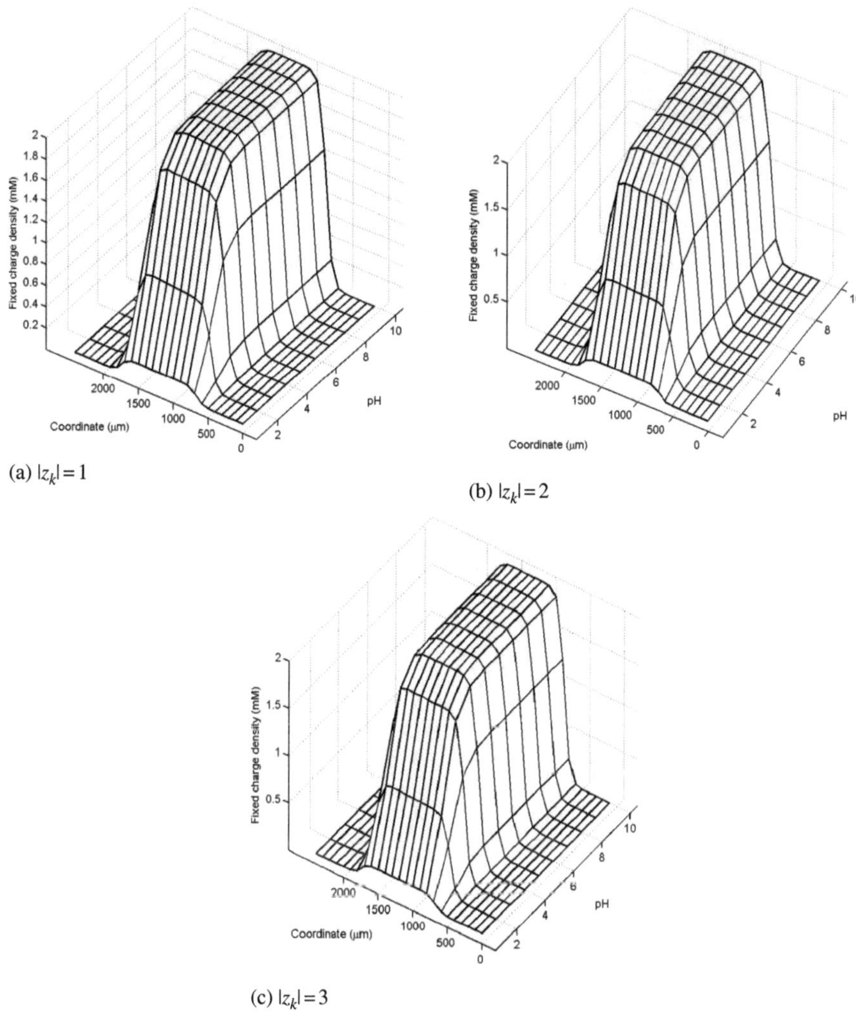

Fig. 4.38 Distributive profiles of the fixed charge density c_f subject to the effect of the solution pH coupled with the ionic valence ($V_e = 0.16\,\text{V}$)

Figures 4.40 and 4.41 are plotted for analysis of the coupled effect of the ionic valence $|z_k|$ and electric voltage V_e as well as solution pH on the variation of the average curvature K_a. It is found that the average curvature K_a increases slightly with increment of the ionic valence $|z_k|$. However, it is also seen that the swelling ratio R_s decreases with increment of the ionic valence $|z_k|$, as shown in Figs. 4.42 and 4.43. Similar results were also reported elsewhere (Horkay et al., 2001; Bajpai and Dubey, 2005). Based on the above discussions, it is concluded that the presence of divalent ions in the intestinal fluid within the hydrogel may cause a great volume

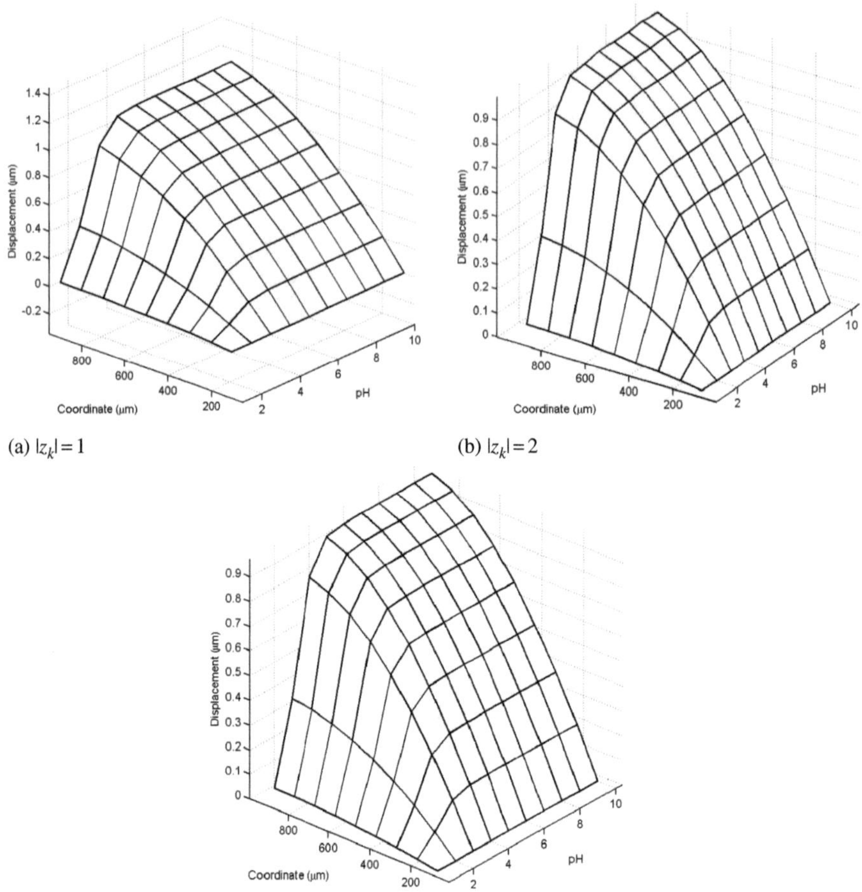

(a) $|z_k|=1$

(b) $|z_k|=2$

(c) $|z_k|=3$

Fig. 4.39 Distributive profiles of the displacement u subject to the effect of the solution pH coupled with the ionic valence ($V_e = 0.16\,\text{V}$)

transition in the swollen hydrogel. This in turn may affect the drug releasing capacity of the hydrogel if it is used as drug delivery system.

4.6 Remarks

In this chapter, the responses of the equilibrium characteristics of the smart hydrogels have been modelled and simulated when they are stimulated by the environmental solution pH coupled with the externally applied electric field. The simulation results obtained by the present MECpHe model agree well with published experimental data qualitatively and quantitatively. The influences of important hydrogel

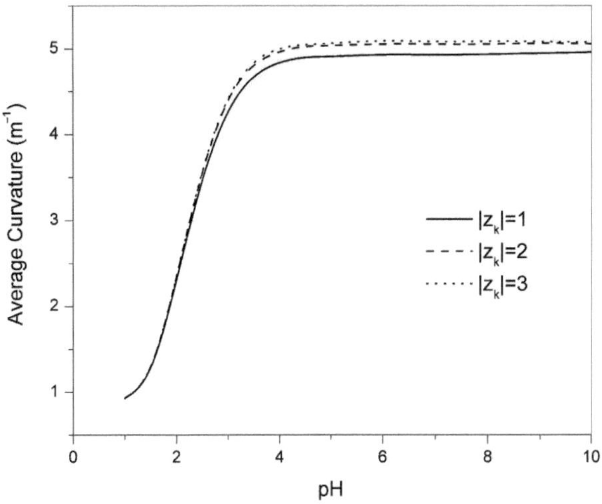

Fig. 4.40 Coupled effect of the ionic valence and solution pH on variation of the average curvature K_a ($V_e = 0.4$ V)

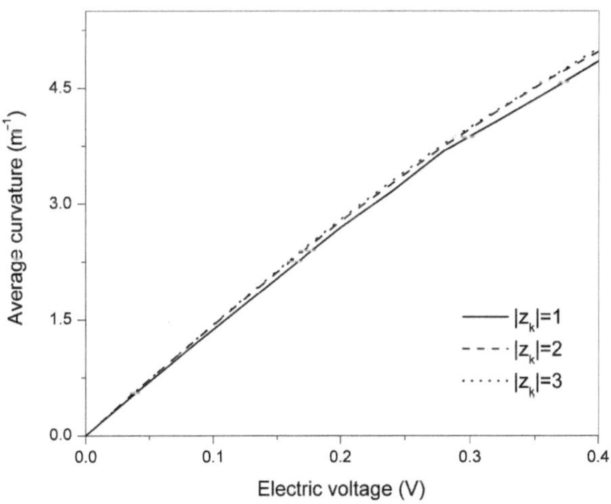

Fig. 4.41 Coupled effect of the ionic valence and electric voltage V_e on variation of the average curvature K_a (pH 4.0)

material properties and environmental conditions on the degree of swelling are investigated. It is concluded that the swelling deformation of the smart hydrogels can be improved further by increasing the externally applied electric voltage, or by increasing the fixed charge density or by lowering the ionic strength of the surrounding solution.

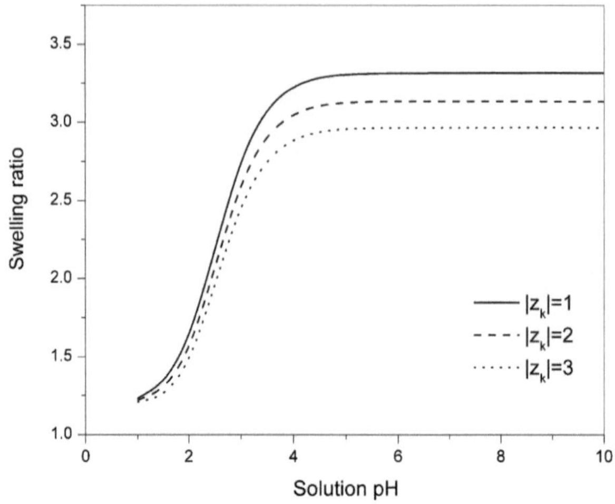

Fig. 4.42 Coupled effect of the ionic valence and solution pH on variation of the swelling ratio R_s ($V_e = 0.4\,\mathrm{V}$)

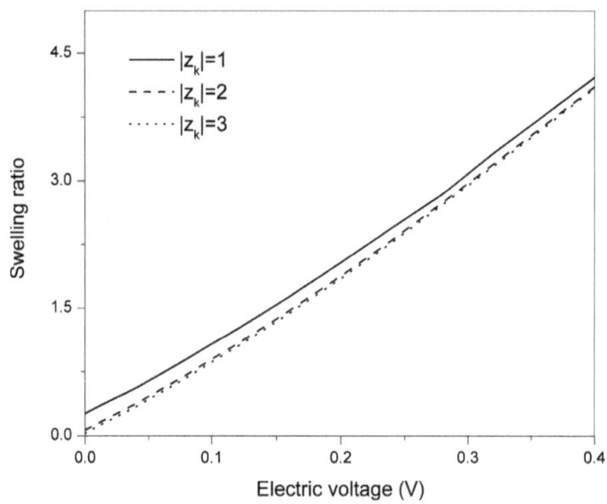

Fig. 4.43 Coupled effect of the ionic valence and electric voltage V_e on variation of the swelling ratio R_s (pH=4.0)

The MECpHe model can be employed theoretically not only to explain the experimental phenomena, such as the distributions of diffusive ionic concentrations, electric potential and the polymeric network deformation, but also to support the design and optimization of BioMEMS devices based on the smart hydrogels responsive to pH-electric coupled stimuli. The developed model is able to predict helpful information, such as the distribution of diffusive ionic concentration in both the hydrogel and bathing solution, the distribution of electric potential in both the hydrogel and surrounding solution, the degree of equilibrium swelling, the displacement and bending curvature of the pH-electric-sensitive hydrogels. The modelling and simulation work would be useful for the design and optimization of the sensor, actuator, micro-fluidic valve and drug delivery system, which is based on the pH-electric-sensitive hydrogels.

References

S.K. Bajpai. (2001). Swelling–deswelling behavior of poly(acrylamide-*co*-maleic acid) hydrogels. *Journal of Applied Polymer Science*, 80, 2782–2789.

S.K. Bajpai, S. Dubey. (2005). In vitro dissolution studies for release of vitamin B12 from poly(*N*-vinyl-2-pyrrolidone-*co*-acrylic acid) hydrogels. *Reactive and Functional Polymers*, 62, 93–104.

R. Bashir, J.Z. Hilt, O. Elibol, A. Gupta, N.A. Peppas. (2002). Micromechanical cantilever as an ultrasensitive pH microsensor. *Applied Physics Letters*, 81, 3091–3093.

T. Belytschko, W.K. Liu, B. Moran. (2001). Nonlinear Finite Elements for Continua and Structures, New York: John Wiley & Sons.

C.S. Brazel, N.A. Peppas. (1996). Pulsatile local delivery of thrombolytic and antithrombotic agents using poly (*N*-isopropylacrylamide-*co*-methacrylic acid) hydrogels. *Journal of Controlled Release*, 39, 57–64.

K.Y. Carlson, L.A. Setton, A. Chilkoti. (2003). Swelling and mechanical behaviors of chemically cross-linked hydrogels of elastin-like polypeptides. *Biomacromolecules*, 4, 572–580.

P. Chiarellim, D. De Rossi. (1996). Bi-phasic elastodynamics of electron conducting polymers. *Polymer Gels and Network*, 4, 499–508.

M. Doi, M. Matsumoto, Y. Hirose. (1992). Deformation of ionic polymer gels by electric fields. *Macromolecules*, 25, 5504–5511.

J. Dolbow, E. Fried, H. Ji. (2005). A numerical strategy for investigating the kinetic response of stimulus-responsive hydrogels. *Computer Methods in Applied Mechanics and Engineering*, 194, 4447–4480.

L.C. Dong, A.S. Hoffman. (1990). Synthesis and application of thermally reversible heterogels for drug delivery. *Journal of Controlled Release*, 13, 21–31.

L.C. Dong, A.S. Hoffman. (1991). A novel approach for preparation of pH-sensitive hydrogels for enteric drug delivery. *Journal of Controlled Release*, 15, 141–152.

J. Fei, Z. Zhang, L. Gu. (2002). Bending behavior of electroresponsive poly(vinyl alcohol) and poly(acrylic acid) semi-interpenetrating network hydrogel fibres under an electric stimulus. *Polymer International*, 51, 502–509.

P.E. Grimshaw. (1989). Electrical control of solute transport across polyelectrolyte membranes. Ph.D Thesis, Massachusetts Institute of Technology.

M. Homma, Y. Seida, Y. Nakano. (2001). Effect of ions on the dynamic behavior of an electrodriven ionic polymer hydrogel membrane. *Journal of Applied Polymer Science*, 82, 76–80.

Y.C. Hon, M.W. Lu, W.M. Xue, X. Zhou. (1999). A new formulation and computation of the triphasic model for mechano-electrochemical mixtures. *Computational Mechanics*, 24, 155–165.

F. Horkay, I. Tasaki, P.J. Basser. (2001). Effect of monovalent-divalent cation exchange on the swelling of polyacrylate hydrogels in physiological salt solutions. *Biomacromolecules*, 2, 195–199.

X. Jin, Y.L. Hsieh. (2005). pH-responsive swelling behavior of poly(vinyl alcohol)/poly(acrylic acid) bi-component fibrous hydrogel membranes. *Polymer*, 46, 5149–5160.

A.R. Khare, N.A. Peppas. (1995). Swelling/deswelling of anionic copolymer gels. *Biomaterials*, 16, 559–567.

S.J. Kim, S.J. Park, S.I. Kim. (2003). Synthesis and characteristics of interpenetrating polymer network hydrogels composed of poly(vinyl alcohol) and poly(*N*-isopropylacrylamide). *Reactive and Functional Polymers*, 55, 61–67.

S.J. Kim, C.K. Lee, S.I. Kim. (2004a). Electrical/pH responsive properties of poly(2-acrylamido-2-methylpropane sulfonic acid)/hyaluronic acid hdyrogels. *Journal of Applied Polymer Science*, 92, 1731–1736.

S.J. Kim, S.G. Yoon, S.M. Lee, S.H. Lee, S.I. Kim. (2004b). Electrical sensitivity behavior of a hydrogel composed of polymethacrylic acid/poly(Vinyl alcohol). *Journal of Applied Polymer Science*, 91, 3613–3617.

W.M. Lai, D. Rubin, E. Krempl. (1974). Introduction to Continuum Mechanics, Oxford: Pergamon Press.

K.Y. Lam, H. Li, T.Y. Ng, R.M. Luo. (2006). Modelling and simulation of the deformation of multi-state hydrogels subjected to electrical stimuli. *Engineering Analysis with Boundary Elements*, 30, 1011–1017.

H. Li, R.M. Luo, K.Y. Lam. (2007). Modelling and simulation of deformation of hydrogels responding to electric stimulus. *Journal of Biomechanics*, 40, 1091–1098.

H. Li, T.Y. Ng, J.Q. Cheng, K.Y. Lam. (2003). Hermite-cloud: A novel true meshless method. *Computational Mechanics*, 33, 30–41.

J.S. Mao, M.J. McShane. (2006). Transduction of volume change in pH-sensitive hydrogels with resonance energy transfer. *Advanced Materials*, 18, 2289–2293.

M. Otake, Y. Kayami, M. Inaba, H. Inoue. (2002). Motion design of a starfish-shaped gel robot made of electro-active polymer gel. *Robotics Autonomous Systems*, 40, 185–191.

Y. Qiu, K.N. Park. (2001). Environment-sensitive hydrogels for drug delivery. *Advanced Drug Delivery Reviews*, 53, 321–339.

W.Z. Rong, A.E. Pelling, A. Ryan, J.K. Gizewski, S.K. Friedlander. (2004). Complementary TEM and AFM force spectroscopy to characterize the nanomechanical properties of nanoparticle chain aggregates. *Nano Letters*, 4, 2287–2292.

K. Sawahata, M. Hara, H. Yasunaga, Y. Osada. (1990). Electrically controlled drug delivery system using polyelectrolyte gels. *Journal of Controlled Release*, 14, 253–262.

H.B. Schreyer, N. Gebhart, K.J. Kim, M. Shahinpoor. (2000). Electrical activation of artificial muscles containing polyacrylonitrile gel fibers. *Biomacromolecules*, 1, 642–647.

M. Shihinpoor. (2003). Ionic polymer-conductor composites as biomimetic sensors, robotic actuators and artificial muscles-a review. *Electrochimica Acta*, 48, 2343–2353.

T. Shiga, T. Kurauchi. (1990). Deformation of polyelectrolyte gels under the influence of electric field. *Journal of Applied Polymer Science*, 39, 2305–2320.

B.C. Shin, S.S. Kim, J.K. Ko, J. Jegal, B.M. Lee. (2003). Gradual phase transition of poly(*N*-isopropylacrylamide-*co*-acrylic acid) gel induced by electric current. *European Polymer Journal*, 39, 579–584.

T. Tanaka, I. Nishio, S.T. Sun, N.S. Ueno. (1982). Collapse of gels in an electric field. *Science*, 218, 467–469.

T. Wallmersperger, B. Kroeplin. (2001). Modelling and Analysis of the Chemistry and Electromechanics. In: *Electroactive Polymer Actuators as Artificial Muscles*, Y. Bar-Cohen (Ed.) SPIE Press, pp. 285–307.

Y.J. Yang, J. Engberts. (2000). Stimuli response of polysoap hydrogels in aqueous solution and DC electric fields. *Colloids and Surfaces A: Physicochmeical and Engineering Aspects*, 169, 85–94.

W.S. Yin, J. Li, T.R. Gu, J.P. Wu. (1997). Design and synthesis of conducting secondary crosslinked interpenetrating polymer network. *Journal of Applied Polymer Science*, 63, 13–16.

Chapter 5
Multi-Effect-Coupling Thermal-Stimulus (MECtherm) Model for Temperature-Sensitive Hydrogel

5.1 Introduction

In this chapter, two models are developed mathematically for steady-state simulation of the temperature-sensitive ionized hydrogel and for transient simulation of the temperature-sensitive neutral hydrogel respectively, based on the analysis of the fundamental interactions during the swelling or shrinking of the smart hydrogel. One is a novel multiphysics model, termed the multi-effect-coupling thermal-stimulus (MECtherm) model, which consists of a transcendental equation and the nonlinear coupled Poisson–Nernst–Planck partial differential equations, and it is developed for simulation of the volume phase transition of ionized temperature-sensitive hydrogel at swelling equilibrium state. The other is a transient model for kinetics of temperature-sensitive neutral hydrogel, which is contributed mainly by Erik Birgersson (2008) who is one of the important collaborators of the author.

A literature search reveals that many studies in past decades were carried out for the thermal stimulus-responsive hydrogels. However, most of them were experiment based. Few studies involved the modelling and simulation of the responsive behavior of the hydrogels, especially for the ionized hydrogels. For example, Lele et al. (1995) proposed a statistical thermodynamic model with consideration of hydrogen bond interaction for prediction of swelling equilibrium of the PNIPA hydrogel–water system. Otake et al. (1989) presented a model with effects of hydrophobic hydration and interaction for the thermally induced discontinuous shrinkage of ionized hydrogel. For analysis of discontinuous volume phase transition, Erman and Flory (1986) made the assumption that the polymer–solvent interaction parameter depends on the volume fraction of solid-phase polymer network. Later works include the molecular thermodynamic model which Hino and Prausnitz (1998) proposed with combination of the impressible lattice-gas model (Birshtein and Pryamitsyn, 1991) and the interpolated affine model (Wolf, 1984) for simulation of the volume phase transition of PNIPA hydrogel. However, it is still difficult for these models to capture experimental data well, although they may provide qualitatively prediction of volume phase transition of temperature-sensitive hydrogels.

H. Li, *Smart Hydrogel Modelling*, DOI 10.1007/978-3-642-02368-2_5,

5.2 Development of the MECtherm Model

In this section, a multiphysics model is developed mathematically with chemo-electro-thermo-mechanical coupled energy domains, which is called the multi-effect-coupling thermal-stimulus (MECtherm) model, in order to simulate the variations of volume phase transition with environmental temperature, mobile ion concentration and electric potential for the swelling equilibrium of thermal stimulus-responsive hydrogel when immersed in solution. The present model incorporates the steady-state Nernst–Planck equations simulating the distribution of diffusive ionic species and Poisson equation simulating the electric potential.

For analysis of the volume phase transition of the ionized temperature-sensitive hydrogels, usually it is necessary to investigate four fundamental interactions, namely hydrogen bond, hydrophobic, electrostatic and the van der Waals interactions (Shirota et al., 1998). The competitive balance between the repulsive and the attractive interactions results in the volume phase transition (Li and Tanaka, 1992). According to Flory's mean field theory (Flory, 1953) for swelling equilibrium of the hydrogels, the four interactions mentioned for the volume phase transition of temperature-sensitive hydrogels may be presented mathematically in the form of three contributions to the change of free energy, namely the polymer–solvent mixing, the elastic deformation of the polymer network matrix and the osmotic pressure due to the gradients of ionic concentrations. Polymer–solvent mixing contributes to either attractive or repulsive force, depending upon the relation between the entropy change and the heat associated with the mixing. The elastic deformation of hydrogel is balanced by the mechanical elastic restoring force of solid-phase network due to the polymer elasticity. As one of driving expansion forces, the osmotic pressure is generated by the concentration difference of mobile ions between the interior hydrogel and the exterior solution. It is noted that the charged groups attached onto the polymer chains play an essential role in the volume phase transition of the ionized hydrogel (Tanaka et al., 1980). When the hydrogel is immersed in the electrolyte solution, the negatively charged groups attached onto the polymer chains are compensated by the diffusive cations from the solution into the hydrogel. Consequently the cation concentration increases within the hydrogel prior to the volume change. This unequal distribution of the solute induces the osmotic pressure to drive the swelling of the ionic hydrogel. As a result, the volume phase transition of thermal stimulus-responsive hydrogel can be predicted generally by the thermodynamic equilibrium theorem. In the presently developed MECtherm model, these three fundamental contribution forces are considered in the swelling equilibrium, and two forms of the polymer–solvent interaction parameters are employed.

5.2.1 Free Energy

According to the thermodynamics theory, the swelling equilibrium of the ionized temperature stimulus-responsive PNIPA hydrogel may be determined by the finial temperature field and the initial conditions including initial temperature, fixed charge density, effective crosslink density and the polymeric network volume fraction. Based on Flory's mean field theory (Flory, 1953), the total change of free energy, ΔG_{gel}, within the ionized thermal-sensitive hydrogel can be expressed as

$$\Delta G_{\text{gel}} = \Delta G_{\text{Mixing}} + \Delta G_{\text{Elastic}} + \Delta G_{\text{Ion}}, \tag{5.1}$$

where ΔG_{Mixing}, $\Delta G_{\text{Elastic}}$ and ΔG_{Ion} denote the mixing, elastic deformation and ionic contributions to the change of free energy, respectively. By differentiating Eq. (5.1) with respect to the number of solvent molecules, the chemical potential of the solvent within the swollen hydrogel is obtained as

$$\Delta \mu_{\text{gel}} = \Delta \mu_{\text{Mixing}} + \Delta \mu_{\text{Elastic}} + \Delta \mu_{\text{Ion}}. \tag{5.2}$$

When the swelling hydrogel reaches equilibrium state, the chemical potential of the solvent within the hydrogel will be equal to that of the solvent in surrounding solution, i.e.

$$\Delta \mu_{\text{Mixing}} + \Delta \mu_{\text{Elastic}} + \Delta \mu_{\text{Ion}} - \Delta \mu_{\text{Ion}}^{*} = 0. \tag{5.3}$$

where $\Delta \mu_{\text{Ion}}^{*}$ represents the chemical potential of solvent in the external solution.

By Flory–Huggins lattice theory (Flory, 1953), the change of mixing chemical potential induced by changing the solvent–solvent contact into the solvent–polymer contact may be written as

$$\Delta \mu_{\text{Mixing}} = k_B T \upsilon^{-1} (\phi + \ln (1 - \phi) + \chi \phi^2), \tag{5.4}$$

where k_B is Boltzmann constant, T is absolute temperature, υ is the molar volume of the solvent, ϕ is the volume fraction of polymeric network at swelling equilibrium state and χ is the polymer–solvent interaction parameter.

It is well known that the interaction parameter χ not only depends on the absolute temperature T but also depends on the volume fraction of polymeric network ϕ (Moerkerke et al., 1995; Shirota et al., 1998; Hino and Prausnitz, 1998). In the case of swollen hydrogels with low volume fraction of polymeric network below the lower critical solution temperature (LCST), the polymer–solvent interaction parameter is defined as the following form:

$$\chi = \chi_1(T) + \chi_2 \phi = (\delta h - \delta s T)/(k_B T) + \chi_2 \phi, \tag{5.5}$$

in which χ_2 is an experimental parameter. δs and δh are the changes of entropy and enthalpy per monomeric unit of the network, respectively. The following numerical studies in this chapter will validate that the parameter χ expressed by Eq. (5.5) is suitable for simulation of the PNIPA hydrogels at swelling state. Furthermore, in the case of shrunken hydrogels with high volume fraction of polymer network above LCST, the interaction parameter χ is defined as

$$\chi = F(T) P(\phi), \tag{5.6}$$

where $F(T)$ and $P(\phi)$ are the functions of absolute temperature and polymeric network volume fraction, respectively. $P(\phi)$ is given by Bae et al. (1993) as

$$P(\phi) = (1 - b\phi)^{-1} \tag{5.7}$$

in which b is an empirical parameter, and $b = -0.65$ to $b = 0.65$ is taken in the present studies. For $F(T)$, one can have the expression given by Hino and Prausnitz (1998) as

$$F(T) = \frac{z}{2}\left(\frac{\zeta + 2\zeta_{12}}{RT} + 2\ln\left(\frac{1 + s_{12}}{1 + s_{12}\exp\left(\zeta_{12}/(RT)\right)}\right)\right), \tag{5.8}$$

where z is the lattice coordination number (here z=6), ζ is the interchange energy, ζ_{12} is the difference between the segmental interaction energy for specific interactions and that for non-specific interactions. R is the gas constant, and s_{12} is the degeneracy ratio of non-specific interactions to that of specific interactions. In addition, it is also noted that in numerical implementation the transformation between Eqs. (5.5) and (5.6) is determined by detecting the volume phase transition, when the difference of polymer volume fractions between the previous and current iterating steps is much larger than the specified convergence region.

In order to characterize the contribution of elastic deformation to the change of chemical potential, the affine model is given by Flory (1953) as

$$\Delta\mu_{\text{Elastic}} = k_B T v_e((\phi/\phi_0)^{1/3} - (\phi/2\phi_0)) \tag{5.9}$$

where v_e is the effective crosslink density, ϕ_0 is the initial volume fraction of polymeric network in the pre-gel solution and $\phi_0\ /\phi$ is the volume swelling ratio.

For the ionic contribution to the change of chemical potential, usually it is determined by the concentration difference of the mobile ions between the inside and the outside hydrogel (Flory, 1953), namely

$$\Delta\mu_{\text{Ion}} - \Delta\mu^*_{\text{Ion}} = -k_B T \sum_{j=1}^{N}(c_j - c_j^*), \tag{5.10}$$

where N denotes the number of different mobile ion species, c_j and c_j^* are the concentrations of the jth mobile ion species in the interior hydrogel and exterior bathing solution, respectively.

By substituting Eqs. (5.4), (5.9) and (5.10) into Eq. (5.3), the swelling equilibrium governing equation is obtained in the following form of transcendental equation:

$$\upsilon^{-1}(\phi + \ln(1-\phi) + \chi\phi^2) + v_e((\phi/\phi_0)^{1/3} - (\phi/2\phi_0)) - \sum_{j=1}^{N}(c_j - c_j^*) = 0 \tag{5.11}$$

When the hydrogel is immersed into pure water, where the mobile ion concentrations of the external solution are equal to zero, Eq. (5.11) is then simplified into the transcendental equation as

$$\upsilon^{-1}(\phi + \ln(1-\phi) + \chi\phi^2) + v_e((\phi/\phi_0)^{1/3} - (\phi/2\phi_0)) - c_f^0\phi/2\phi_0 = 0 \quad (5.12)$$

where c_f^0 is the fixed charge density at the reference state (ϕ=ϕ_0).

5.2.2 Poisson–Nernst–Planck Formulation

In order to incorporate the effects of mobile ion concentration and electric potential into simulation of the temperature-sensitive hydrogel, Poisson–Nernst–Planck formulation is required. If the contributions of migration and diffusion into the transport of mobile ions are considered only during the thermal swelling of the hydrogels, the steady-state Nernst–Planck equation for the *j*th mobile ion species can be expressed by Samson et al. (1999):

$$D_j\nabla^2 c_j + \frac{FD_j z_j}{RT}\left(\nabla c_j \nabla\psi + c_j\nabla^2\psi\right) = 0 \; (j = 1,2,\ldots,N) \quad (5.13)$$

where F is Faraday constant, D_j is the diffusive coefficient, z_j is the valence of the *j*th mobile ion species, c_j is the concentration of the *j*th mobile ion species and ψ is the electric potential.

In order to establish the relation between the mobile ion concentration and the electric potential, the following Poisson equation is required (Samson et al., 1999):

$$\nabla^2\psi = -\frac{F}{\varepsilon\varepsilon_0}\left(z_f c_f + \sum_{j=1}^{N}(z_j c_j)\right) \quad (5.14)$$

where ε_0 is the permittivity for vacuum, ε is the dielectric constant of medium relative to vacuum (e.g. $\varepsilon = 80$ for water), z_f is the valence of fixed charge and c_f is the density of fixed charge.

For simulation of the ionized thermal stimulus-responsive hydrogels, it is generally assumed that the fixed charges attached onto the polymeric network chains distribute uniformly within the hydrogel during the thermal swelling, and the total amount of fixed charges is invariable. In other words, the fixed charge density $c_f = c_f^0\phi/\phi_0$ at the state of swelling equilibrium.

5.3 Numerical Implementation

For the isotropic swelling of the hydrogels, the elongation ratios along three principal axes are equal to each other, in which the displacement vector $\boldsymbol{u}$ may be expressed by the difference between the original position $\boldsymbol{x_0}(x_0, y_0, z_0)$ and the deformed position $\boldsymbol{x}(\alpha x_0, \alpha y_0, \alpha z_0)$, in which α is the linear volume swelling ratio, and $\alpha = (V/V_0)^{1/3} = (\phi_0/\phi)^{1/3}$, namely

$$u = x - x_0 = ((\phi_0/\phi)^{1/3} - 1)x_0 \tag{5.15}$$

In the present parameter studies, only a cylindrical hydrogel is simulated numerically. Therefore, due to the axisymmetry, it is reasonably approximated to one-dimensional computational domain along the radial direction covering both the hydrogel radius and the bathing solution. The steady-state Nernst–Planck equation is thus rewritten in the polar coordinates as

$$\frac{\partial^2 c_j}{\partial r^2} + \frac{1}{r}\frac{\partial c_j}{\partial r} + \frac{Fz_j}{RT}\left(c_j\frac{\partial^2 \psi}{\partial r^2} + \frac{c_j}{r}\frac{\partial \psi}{\partial r} + \frac{\partial c_j}{\partial r}\frac{\partial \psi}{\partial r}\right) = 0 \ (j = 1,2,\ldots,N) \tag{5.16}$$

and the Poisson equation is rewritten as

$$\frac{\partial^2 \psi}{\partial r^2} + \frac{1}{r}\frac{\partial \psi}{\partial r} = -\frac{F}{\varepsilon\varepsilon_0}\left(z_f c_f + \sum_{j=1}^{N}(z_j c_j)\right) \tag{5.17}$$

The radial displacement of the swelling hydrogel is given as

$$u_r = ((\phi_0/\phi)^{1/3} - 1)R_0 \tag{5.18}$$

where R_0 is the radius of the cylindrical hydrogel at the reference state.

Corresponding boundary conditions required are imposed at both the ends of one-dimensional computational domain. Due to axisymmetry of the present problem, the boundary conditions at one end point located at the centre of circle cross-sectional area are given as

$$\frac{\partial \psi}{\partial r} = 0 \text{ and } \frac{\partial c_j}{\partial r} = 0 \ (j = 1,2,\ldots,N) \quad \text{at} \quad r = 0 \tag{5.19}$$

The boundary conditions at the other end point located at the edge of surrounding solution region are given by

$$\psi = 0 \text{ and } c_j = c^* \ (j = 1,2,\ldots,N) \quad \text{at} \quad r = L \tag{5.20}$$

In order to facilitate the numerical implementation of the models developed above, several non-dimensional variables are introduced, including the non-dimensional radius $\xi = r/L_{\mathrm{ref}}$, the non-dimensional concentration of the jth ionic species $\overline{c_j} = c_j/c_{\mathrm{ref}}$, the non-dimensional fixed charge density $\overline{c_f} = c_f/c_{\mathrm{ref}}$ and the non-dimensional electrical potential $\overline{\psi} = (F\psi)/(\eta RT)$, where L_{ref}, c_{ref} and ψ_{ref} are the characteristic parameters.

After substituting the non-dimensional variables into Eqs. (5.16) and (5.17), the non-dimensional steady-state Poisson–Nernst–Planck system is finally derived as

$$\frac{\partial^2 \overline{c_j}}{\partial \xi^2} + \frac{1}{\xi}\frac{\partial \overline{c_j}}{\partial \xi} + \eta z_j \left(\overline{c_j}\frac{\partial^2 \overline{\psi}}{\partial \xi^2} + \frac{\overline{c_j}}{\xi}\frac{\partial \overline{\psi}}{\partial \xi} + \frac{\partial \overline{c_j}}{\partial \xi}\frac{\partial \overline{\psi}}{\partial \xi} \right) = 0 \ (j = 1,2,\ldots,N) \quad (5.21)$$

$$\frac{\partial^2 \overline{\psi}}{\partial \xi^2} + \frac{1}{\xi}\frac{\partial \overline{\psi}}{\partial \xi} = -\frac{F^2 L_{\text{ref}}^2 c_{\text{ref}}}{\varepsilon\varepsilon_0 \eta RT} \left(z_f \overline{c_f} + \sum_{j=1}^{N} (z_j \overline{c_j}) \right) \quad (5.22)$$

For simulation of response behavior of the ionized thermo-sensitive hydrogels when immersed in univalent electrolyte solution, it is required to solve iteratively a set of coupled nonlinear partial differential governing equations, consisting of the swelling equilibrium equation (5.11), Nernst–Plank equations (5.21) and Poisson equation (5.22). By following the computational flowchart as illustrated in Fig. 5.1, a guessed value of polymeric network volume fraction ϕ^* at a given temperature T is provided first. Using Newton's iterative technique for solution of the coupled nonlinear partial differential equations (5.21) and (5.22), the distributions of the electric potential and mobile ion concentrations are obtained at the temperature T. Subsequently, substituting the computed ionic concentrations into the swelling equilibrium equation (5.11), the corresponding volume fraction ϕ of the polymer network is computed, and then used as the guessed value ϕ^* in next iterative step. Following this way, the iterative recurrence is conducted until the polymer network volume fraction ϕ converges.

In summary, the multiphysics model has been developed in the chemo-electro-thermo mechanical coupled energy domains and termed the multi-effect-coupling thermal-stimulus (MECtherm) model. It is composed of the nonlinear Poisson–Nernst–Planck partial differential equations (5.13) and (5.14) coupled with the transcendental equation (5.11) of the swelling equilibrium that is based on the Flory's mean field theory. The model has the capability of simulating the volume phase transition of the ionized temperature stimulus-responsive hydrogels when they are immersed in the bathing solution with temperature change. This model is validated in the following section, and then the influences of several important environmental conditions and material properties will be simulated and discussed in detail.

Before the numerical simulation is carried out, it is necessary to discretize the presently developed MECtherm governing equations, which are solved by the Hermite-cloud method (Li et al., 2003) for simulation of swelling equilibrium of the temperature-sensitive hydrogels. In order to simulate the steady-state swelling equilibrium of temperature-sensitive hydrogels, the non-dimensional Nernst–Plank equations (5.21) and Poisson equation (5.22) are discretized in spatial domain, and then solved by the Hermite-cloud method (Li et al., 2003).

By the Hermite-cloud method (Li et al., 2003), the approximations of non-dimensional unknown variables, $\overline{c_j}$ and $\overline{\psi}$, and their derivatives are given as

$$\overline{c_j} = \sum_{n=1}^{N_T} N_n(\xi_i)\overline{c_n^j} + \sum_{m=1}^{N_S} \left(\xi_i - \sum_{n=1}^{N_T} N_n(\xi_i)\xi_n \right) M_m(\xi_i)\overline{c_{\xi m}^j} \quad (5.23)$$

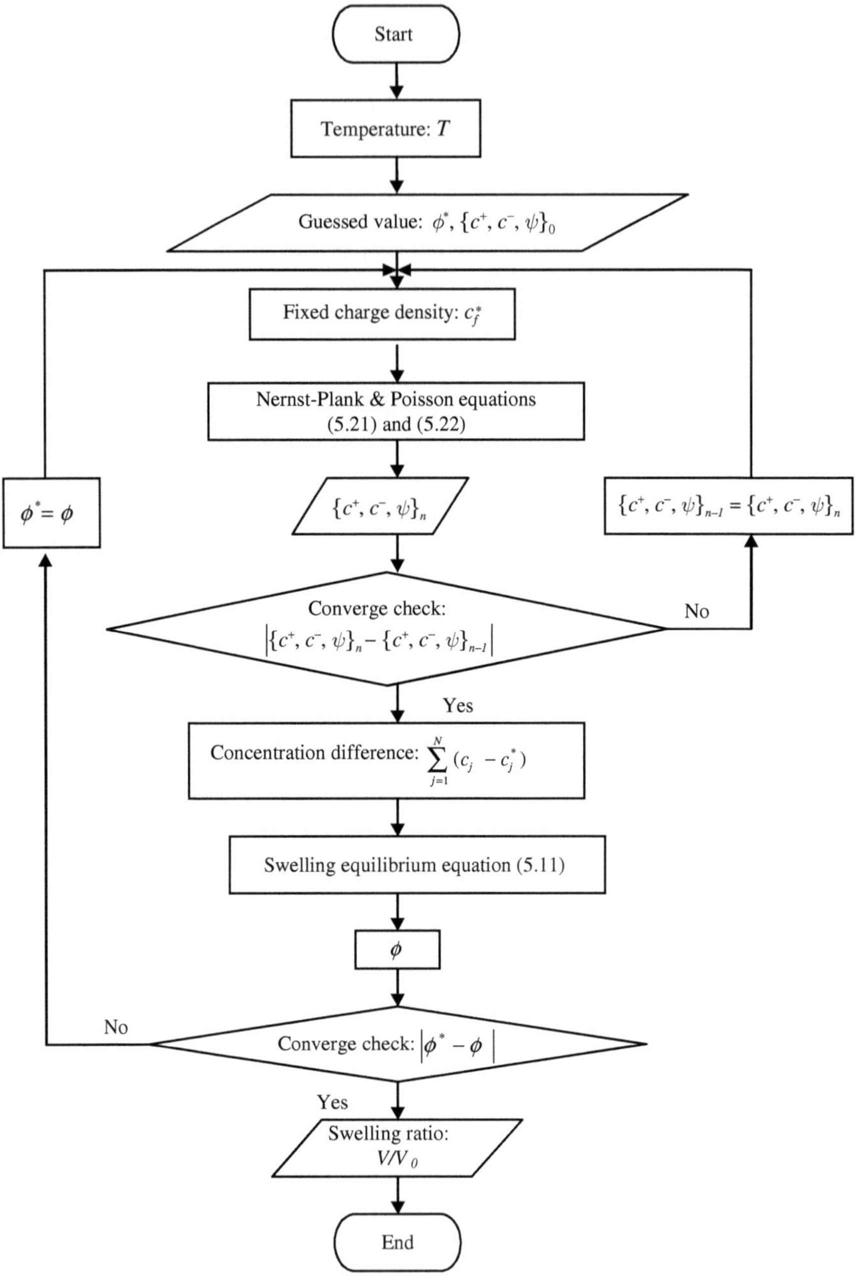

Fig. 5.1 Computational flowchart of the MECtherm model

$$\frac{\partial \overline{c_j}}{\partial \xi} = \sum_{m=1}^{N_S} M_m(\xi_i)\overline{c^j_{\xi m}} \tag{5.24}$$

$$\frac{\partial^2 \overline{c_j}}{\partial \xi^2} = \sum_{n=1}^{N_T} N_{n,\xi\xi}(\xi_i)\overline{c^j_n} \tag{5.25}$$

$$\overline{\psi} = \sum_{n=1}^{N_T} N_n(\xi_i)\overline{\psi}_n + \sum_{m=1}^{N_S}\left(\xi_i - \sum_{n=1}^{N_T} N_n(\xi_i)\xi_n\right)M_m(\xi_i)\overline{\psi}_{\xi m} \tag{5.26}$$

$$\frac{\partial \overline{\psi}}{\partial \xi} = \sum_{m=1}^{N_S} M_m(\xi_i)\overline{\psi}_{\xi m} \tag{5.27}$$

$$\frac{\partial^2 \overline{\psi}}{\partial \xi^2} = \sum_{n=1}^{N_T} N_{n,\xi\xi}(\xi_i)\overline{\psi}_n \tag{5.28}$$

Substituting Eqs. (5.23), (5.24), (5.25), (5.26), (5.27) and (5.28) into the governing equations (5.21) and (5.22), the approximations of Poisson–Nernst–Planck system are written as

$$\begin{aligned}
&\sum_{n=1}^{N_T} N_{n,\xi\xi}(\xi_i)\overline{c^j_n} + \frac{1}{\xi_i}\sum_{m=1}^{N_S} M_m(\xi_i)\overline{c^j_{\xi m}} + \eta z_j \sum_{m=1}^{N_S} M_m(\xi_i)\overline{c^j_{\xi m}} \sum_{m=1}^{N_S} M_m(\xi_i)\overline{\psi}_{\xi m} \\
&\quad + \eta z_j \left[\sum_{n=1}^{N_T} N_n(\xi_i)\overline{c^j_n} + \sum_{m=1}^{N_S} (\xi_i - \sum_{n=1}^{N_T} N_n(\xi_i)\xi_n)M_m(\xi_i)\overline{c^j_{\xi m}}\right] \\
&\quad \times \left[\sum_{n=1}^{N_T} N_{n,\xi\xi}(\xi_i)\overline{\psi}_n + \frac{1}{\xi_i}\sum_{m-1}^{N_S} M_m(\xi_i)\overline{\psi}_{\xi m}\right] = 0
\end{aligned} \tag{5.29}$$

$$\begin{aligned}
&\sum_{n=1}^{N_T} N_{n,\xi\xi}(\xi_i)\overline{\psi}_n + \frac{1}{\xi_i}\sum_{m=1}^{N_S} M_m(\xi_i)\overline{\psi}_{\xi m} = -\frac{F^2 L_{\text{ref}}^2 c_{\text{ref}}}{\varepsilon\varepsilon_0 \eta RT}\left\{z_f \overline{c_f}\right. \\
&\quad \left. + \sum_j^n \left\{z_j \left[\sum_{n=1}^{N_T} N_n(\xi_i)\overline{c^j_n} + \sum_{m=1}^{N_S}\left(\xi_i - \sum_{n=1}^{N_T} N_n(\xi_i)\xi_n\right)M_m(\xi_i)\overline{c^j_{\xi m}}\right]\right\}\right\}
\end{aligned} \tag{5.30}$$

Corresponding auxiliary equations are obtained as

$$\sum_{n=1}^{N_T} N_{n,\xi}(\xi_i)\overline{c^j_n} - \sum_{m=1}^{N_S}\left(\sum_{n=1}^{N_T} N_{n,\xi}(\xi_i)\xi_n\right)M_m(\xi_i)\overline{c^j_{\xi m}} = 0 \tag{5.31}$$

$$\sum_{n=1}^{N_T} N_{n,\xi}(\xi_i)\overline{\psi}_n - \sum_{m=1}^{N_S}\left(\sum_{n=1}^{N_T} N_{n,\xi}(\xi_i)\xi_n\right)M_m(\xi_i)\overline{\psi}_{\xi m} = 0 \tag{5.32}$$

As described in the computational flowchart shown in Fig. 5.1, after discretization of the Poisson–Nernst–Planck governing equations, the MECtherm model, consisting of the transcendental equation (5.11) and the nonlinear coupled Poisson–Nernst–Planck partial differential equations (5.21) and (5.22), will be solved by a hierarchical Newton iteration technique. In the inner iteration, the ionic concentrations c_j and the electric potential ψ are computed first simultaneously by the coupled equations (5.21) and (5.22). Then, substituting the computed results into the outer iteration, the polymeric network volume fraction ϕ is obtained by the transcendental equation (5.11). Subsequently, the computed ϕ is substituted into the inner iteration again for the next iterative step until all the computational variables converge to the required accuracy, including the ionic concentrations c_j, electric potential ψ and polymeric network volume fraction ϕ.

5.4 Model Validation with Experiment

For examination of the present MECtherm model, it is acceptable to compare the numerically computed results with the experimental data published in open literature. In this section, a numerical comparison is conducted with the experimentally measured swelling data found in a published work, where Hirotsu et al. (1987) carried out an experiment for the temperature-sensitive ionized poly(*N*-isopropylacrylamide) (PNIPA) hydrogels immersed in pure water subject to the change in temperature. The PNIPA hydrogel is well known as a typical thermo-shrinking polymeric mixture due to its distinctive property of unique alteration between hydrophilicity and hydrophobicity upon external stimulation of temperature. When surrounding temperature is lower than the corresponding lower critical solution temperature (LCST), the PNIPA hydrogel behaves like hydrophilic characteristics alluring more water since the hydrogen bonds form a stable shell around the hydrophobic groups. With the increase of external temperature, the hydrophobic characteristics unveil to free the entrapped water molecules from the network as the hydrogen bond interactions become weakened or destroyed. When the temperature reaches to or is higher than the LCST, the hydrophobic interactions become fully dominant and the hydrogel is thoroughly dehydrated. As the water releases, the polymer chains in the hydrogel collapse abruptly and the phase separation occurs within the PNIPA hydrogel, which is often called the volume phase transition.

For simulation of the ionized PNIPA hydrogels immersed in pure water, only the governing equation (5.12) is required and solved independently. For examination of the MECtherm model by comparison with the experimental data (Hirotsu et al., 1987), several parameters extracted from the experiment of the ionized PNIPA hydrogels are taken as the input data of the MECtherm model. They are the initial fixed charge density $c_f^0 = 8\,\text{mM}$, the corresponding valence $z_f = -1$ and the initial polymer volume fraction at the reference configuration $\phi_0 = 0.07$. In the presently simulated temperature region, it is reasonably assumed that the change of the water density is negligibly small, and then the effect of the change on the volume phase transition of PNIPA hydrogels is negligible. As such, one can take the water molar volume as the popularly specified constant $v = 18.0\,\text{cm}^3/\text{mol}$. In addition, the effective crosslink densities $v_e = 1.4 \times 10^{-5}\text{mol/cm}^3$

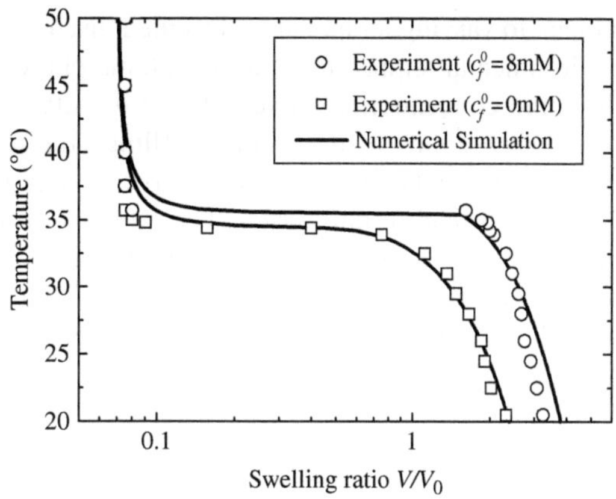

Fig. 5.2 Comparison of numerical simulations with the experimental swelling data (Hirotsu et al., 1987) for the temperature-sensitive PNIPA hydrogels in pure water

and $1.0 \times 10^{-5}\mathrm{mol/cm^3}$ are taken for the hydrogels with and without the fixed charges, respectively. The polymer solvent interaction parameter χ is calculated by Eqs. (5.5), (5.6), (5.7) and (5.8) with the given data (Hirotsu, 1987; Hino, 1998), where $\delta s = -4.717 \times 10^{-23}\mathrm{J/K}$, $\delta h = -1.246 \times 10^{-20}\mathrm{J}$, $\zeta = 0.698$ kcal/mol and $\zeta_{12}/\zeta = -7$.

Comparison of the simulation results with the experimental data is shown in Fig. 5.2 for the swelling equilibrium of the PNIPA hydrogels with and without the fixed charges, when immersed in the pure water within the temperature range of 20–50°C. As illustrated in the figure, the simulation results agree well with the experimental data. It is also known that the fixed charges attached onto the polymeric network chains enhance the swelling capability of the ionized hydrogels, and also increase the lower critical solution temperature (LCST), as compared with the hydrogels without the fixed charges. At temperature $T \approx 34.3$°C, the unionized PNIPA hydrogels undergo a continuous volume change, while the ionized PNIPA hydrogels exhibit a discontinuous volume phase transition at temperature $T \approx 35.6$°C. The temperature of volume phase transition of the ionized PNIPA hydrogel is about 1°C higher than that of the unionized hydrogel. As the temperature increases, the volume swelling ratios of both the hydrogels decrease due to the shrinking properties. Furthermore, the swelling equilibrium curves of the hydrogels tend to merge together at the temperature above 40.0°C, where the hydrogels are almost fully dehydrated.

5.5 Parameter Studies by Steady-State Simulation for Thermo-Sensitive Ionized Hydrogel

In this section, the simulations are carried out for prediction of the influences of several environmental conditions and material properties on the response behaviour of the hydrogels at the swelling equilibrium state, when immersed in the solution subject to varying temperature. Detailed discussions are conducted for the effect of environmental temperature on the volume phase transition and the distributions of the ionic concentrations, fixed charge density and electric potential. The present parameter studies focus on the swelling equilibrium behavior of the responsive PNIPA hydrogels with the fixed charges, when immersed in univalent electrolyte solution, instead of pure water. The effects of important environmental and material parameters on the volume phase transition of the PNIPA hydrogels at swelling equilibrium state are discussed in details. They include the initial fixed charge density c_f^0, bathing solution concentration c^*, effective crosslink density v_e and initial polymer volume fraction ϕ_0.

5.5.1 *Influence of Initially Fixed Charge Density*

For discussion of the influence of the initial fixed charge density c_f^0 on the swell equilibrium of the PNIPA hydrogels, Figs. 5.3, 5.4, 5.5, 5.6, and 5.7 are plotted for variation of the swelling ratio V/V_0 with the temperature and the distributions of the mobile ionic concentrations c_j and fixed charge density c_f as well as electric potential ψ at various c_f^0 and T. Theoretically, it is known that the hydrogels experience a dynamic volume change before the swelling equilibrium is finally achieved, when the thermo-sensitive ionized PNIPA hydrogels are immersed in the electrolyte bathing solution. After the hydrogels are placed into the electrolyte solution, the negatively charged groups attached onto the polymeric chains are compensated by the cations diffusing into the hydrogels from the external solution, which results in the increase of the cation concentration within the hydrogels prior to the volume

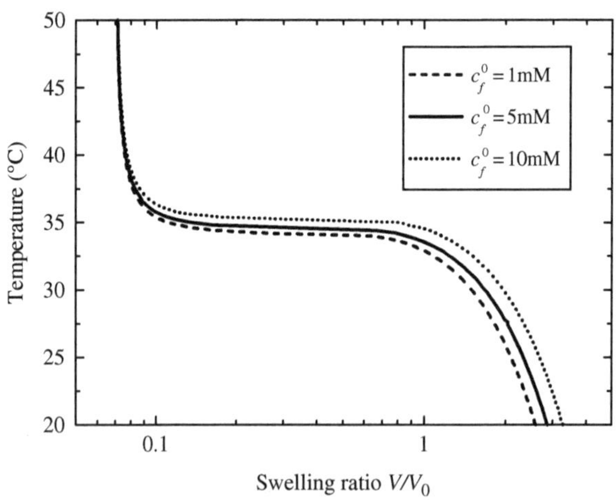

Fig. 5.3 Relation between the temperature and the swelling ratio V/V_0 of equilibrium volume for the ionized hydrogels with different initial fixed charge densities c_f^0 immersed in the univalent electrolyte solution c^*=20 mM

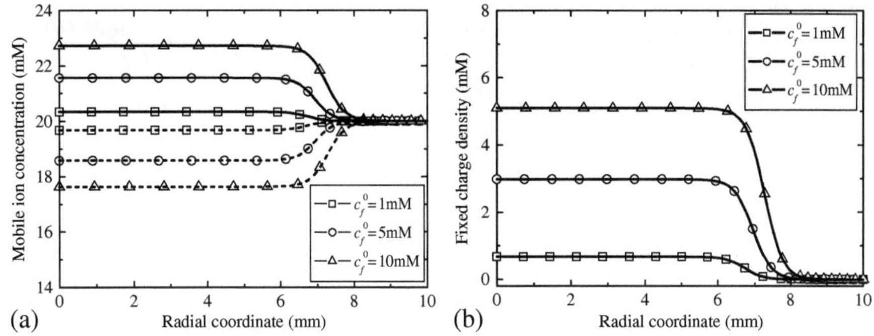

Fig. 5.4 Distributions of the mobile cation (*solid line*) and anion (*dash line*) concentrations (**a**) and the fixed charge densities (**b**) versus radial coordinate for the ionized hydrogels with different initial fixed charge densities c_f^0 at temperature T=30°C prior to volume phase transition

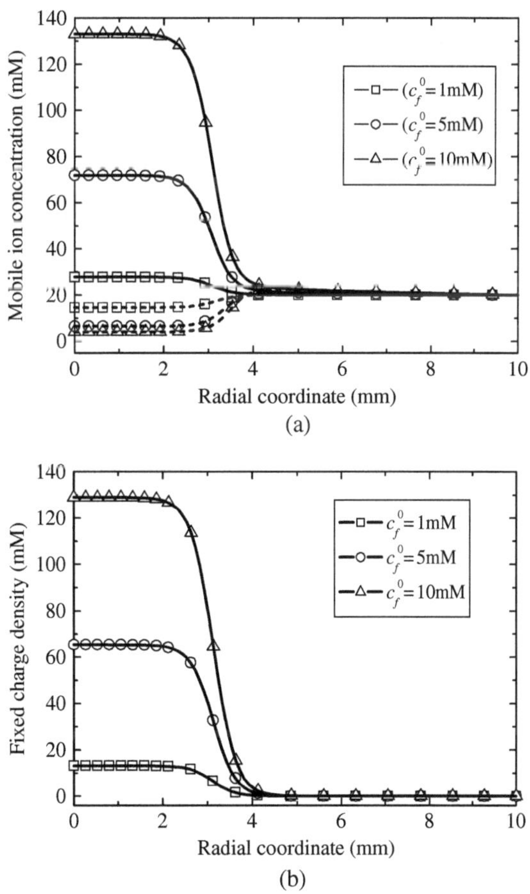

Fig. 5.5 Distributions of the mobile cation (*solid line*) and anion (*dash line*) concentrations (**a**) and the fixed charge densities (**b**) versus radial coordinate for the ionized hydrogels with different initial fixed charge densities c_f^0 at temperature T=40°C posterior to volume phase transition

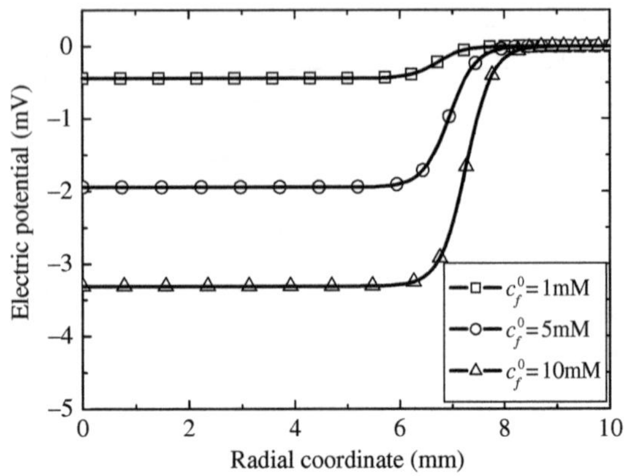

Fig. 5.6 Distributions of electric potential versus radial coordinate for the ionized hydrogels with different initial fixed charge densities c_f^0 at temperature T=30°C prior to volume phase transition

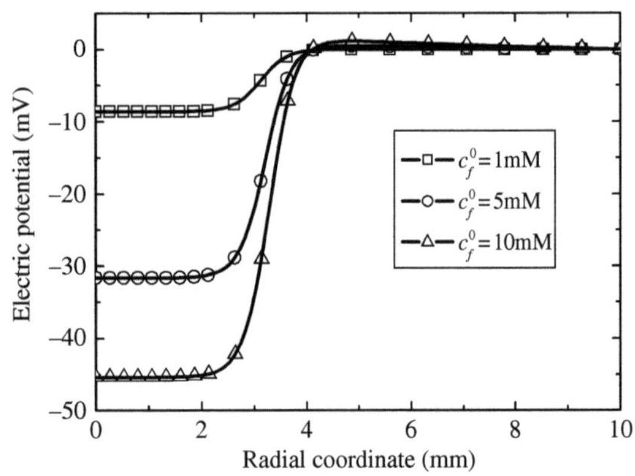

Fig. 5.7 Distributions of electric potential versus radial coordinate for the ionized hydrogels with different initial fixed charge densities c_f^0 at temperature T=40°C posterior to volume phase transition

change. Here it is reasonably assumed that the ionic concentration in the external solution remains constant because of sufficient solution provided. The diffusion of the cations develops a concentration gradient between the interior hydrogel and the exterior bathing solution, which is the source to generate the osmotic pressure driving the dynamic swelling of the hydrogels. The initial fixed charge density c_f^0 has significant effect on the dynamic swelling, since a higher concentration increases the number of cations that diffuse into the hydrogels to compensate the ionized polymer chains and enhance the magnitude of the concentration gradient. As such, as the initial fixed charge density c_f^0 increases, the concentration gradient enlarges. This will generate higher osmotic pressure, resulting in larger swelling of the hydrogels.

In order to investigate the influence of the initial fixed charge density c_f^0 on the swelling equilibrium behavior, three thermo-sensitive ionized PNIPA hydrogels with different initial fixed charge densities are simulated when they are immersed in univalent electrolyte solution, as shown in Fig. 5.3, where the case study corresponding to the initial fixed charge density c_f^0=1 mM is represented by the dashed line, c_f^0= 5 mM by the solid line and c_f^0=10 mM by the dotted line, and the univalent electrolyte bathing solution $c^* = 20$ mM. It is seen from the figure that the simulated LCST at the initial fixed charge density $c_f^0 = 5$ mM is higher than that at $c_f^0 = 1$ mM and lower than that at $c_f^0 = 10$ mM. The temperature range of volume phase transition for the ionized hydrogels with higher fixed charge density becomes broader than that for the hydrogels with lower density.

In order to investigate the influences of the environmental temperature T and the initial fixed charge density c_f^0 on the distributions of mobile ion concentrations c_j and fixed charge density c_f as well as electric potentials ψ in swelling equilibrium state, Figs. 5.4, 5.5, 5.6 and 5.7 are presented for the cylindrical ionized PNIPA hydrogels with various initial fixed charge densities $c_f^0 = 1$, 5 and 10 mM, and the concentration of univalent electrolyte bathing solution $c^* = 20$ mM.

Figure 5.4a demonstrates the distributive concentrations of the mobile cation and anion in both the interior hydrogel and the exterior bathing solution at temperature T=30°C and Fig. 5.5a at temperature T=40°C. Distributions of the fixed charge density c_f in the interior hydrogel and the exterior bathing solution are plotted in Fig. 5.4b at temperature T=30°C and in Fig. 5.5b at temperature T=40°C. Figs. 5.4a and 5.5a show that, for a given initial fixed charge density c_f^0, the mobile cation concentration within the PNIPA hydrogel at temperature T=40°C is larger than that at T=30°C, but the opposite conclusion is drawn for the mobile anion concentration. At a given temperature, the mobile cation concentration within the PNIPA hydrogel increases with the initial fixed charge density, while the opposite results are obtained for the mobile anion concentration. Increasing the fixed charge density enlarges the difference in mobile ion concentrations between the interior hydrogel and the exterior solution. From Fig. 5.4a and b, it is seen that the total concentrations of all the mobile ions (cation and anion) and fixed charge are just compensated despite the individual concentration differences of the mobile ions and fixed charge density. Similar phenomena at T=40°C are also observed in Fig. 5.5a and b.

The electric potential distributed in both the hydrogel and the bathing solution is plotted in Fig. 5.6 at temperature 30°C and in Fig. 5.7 at temperature 40°C. For a given initial fixed charge density c_f^0, the electric potential within the PNIPA hydrogel at temperature T=30°C is much higher than that at T=40°C. In the same manner, at a given temperature, the electric potential within the PNIPA hydrogel decreases with the increase of the initial fixed charge density c_f^0.

In addition, by comparison of the simulations at T=30°C with those at T=40°C as shown in Figs. 5.4, 5.5, 5.6 and 5.7, it is noted that the hydrogel–solution interface moves with the temperature. They also demonstrate that the volume fraction of polymeric network at temperature T=40°C is higher than that at T=30°C, due to the thermal shrinking characteristics of the ionized PNIPA hydrogels.

5.5.2 *Influence of Bath Solution Concentration*

For analysis of the influence of bathing solution concentration c^* on the response behavior of the PNIPA hydrogel at swelling equilibrium state, Figs. 5.8, 5.9, 5.10, 5.11 and 5.12 are presented for the relation between the swelling ratio V/V_0 and the

Fig. 5.8 Relation between the temperature and the swelling ratio V/V_0 of equilibrium volume for the ionized PNIPA hydrogels with initial fixed charge density c_f^0=5 mM immersed in pure water and different bathing solution concentrations c^*=5, 20 and 100 mM, respectively

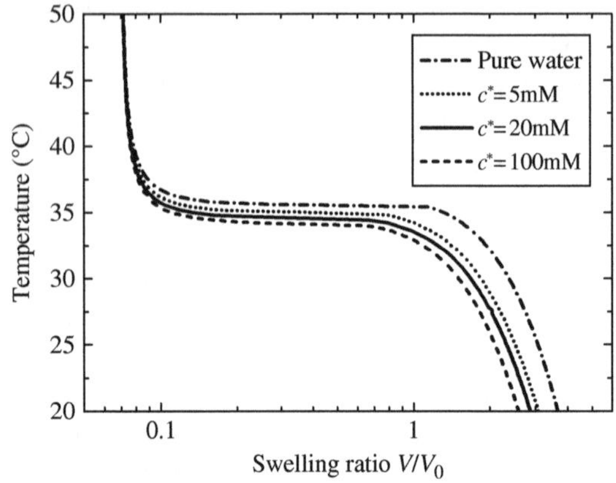

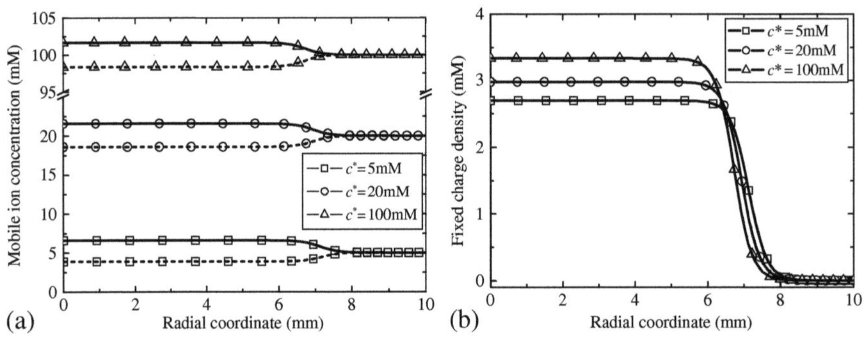

Fig. 5.9 Distributions of the mobile cation (*solid line*) and anion (*dash line*) concentrations (**a**) and the fixed charge densities (**b**) versus radial coordinate for the ionized PNIPA hydrogels with initial fixed charge density $c_f^0 = 5$ mM immersed in different bathing solution concentrations c^* at temperature T=30°C

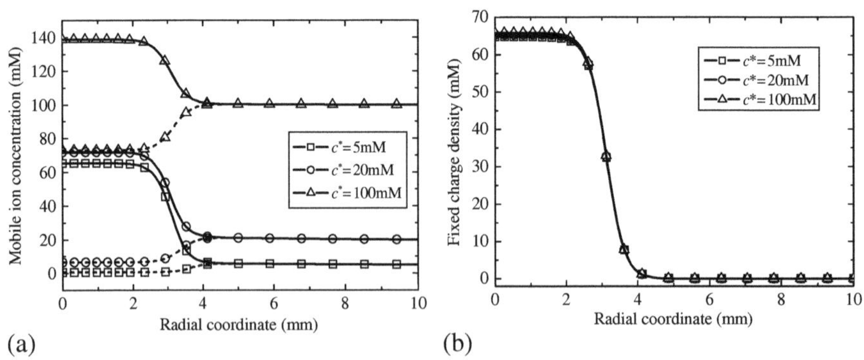

Fig. 5.10 Distributions of the mobile cation (*solid line*) and anion (*dash line*) concentrations (**a**) and the fixed charge densities (**b**) versus radial coordinate for the ionized PNIPA hydrogels with initial fixed charge density $c_f^0 = 5$ mM immersed in different bathing solution concentrations c^* at temperature T=40°C

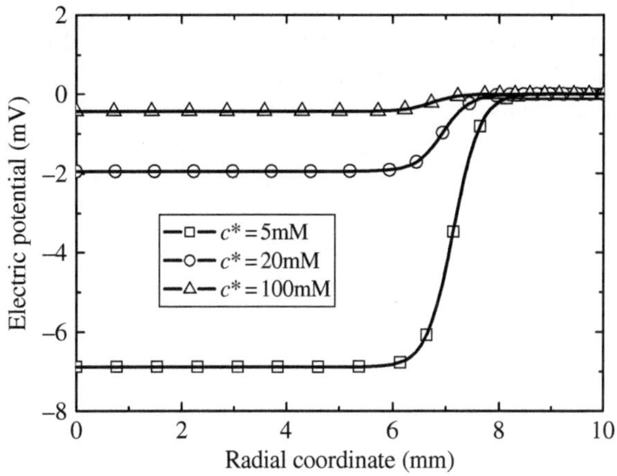

Fig. 5.11 Distributions of electric potential versus radial coordinate for the ionized PNIPA hydrogels with initial fixed charge density $c_f^0 = 5$ mM immersed in different bathing solution concentrations c^* at temperature T=30°C

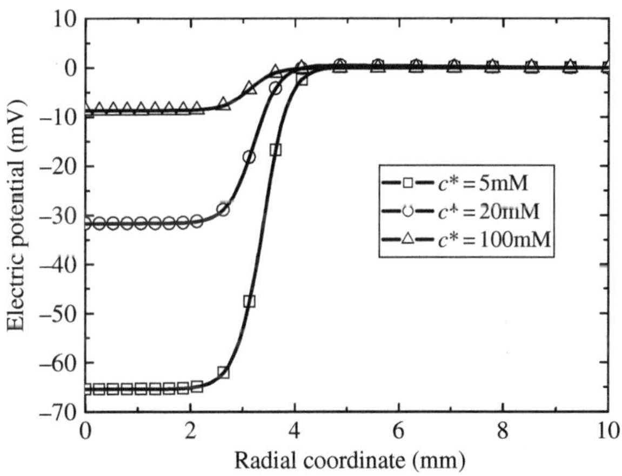

Fig. 5.12 Distributions of electric potential versus radial coordinate for the ionized PNIPA hydrogels with initial fixed charge density $c_f^0 = 5$ mM immersed in different bathing solution concentrations c^* at temperature T=40°C

temperature T, and the distributive variation of the mobile ion concentrations and fixed charge density as well as electric potential along the radial coordinate with various c^* and T. Theoretically, for a thermo-sensitive ionized PNIPA hydrogel immersed in pure water, the swelling equilibrium is achieved when the summation of total osmotic pressure is equal to zero. However, this equilibrium state may be destroyed simply by introducing the electrolyte into the pure water, and then a dynamic volume change will undergo until a new equilibrium is reached. At lower electrolyte concentration or in the special case of pure water, the negative charges of the PNIPA hydrogel are neutralized by counter hydrogen ions. As the electrolyte concentration increases, the hydrogen ions within the PNIPA hydrogel are replaced by the diffusive mobile cations from the bathing solution. If the electrolyte concentration increases further in the system, more and more cations and anions diffuse into the hydrogel from the external solution, and then the overall concentration of mobile ions within the hydrogel increases higher. This reduces the concentration difference between the interior hydrogel and exterior bathing solution. Consequently the driving force of swelling decreases gradually, thus the swelling ratio becomes small.

The thermo-sensitive ionized PNIPA hydrogel with the initial fixed charge density (c_f^0=5 mM) is simulated for discussion of the effect of ionic concentration of bathing solution on swelling behavior. Figure 5.8 shows the variation of the swelling ratio V/V_0 of the hydrogel with temperature T subject to different bathing solution concentrations c^*, where the pure water is indicated by a doted dash line, c^*=5 mM by a dotted line, c^*=20 mM by a solid line and c^*=100 mM by a dashed line, respectively. It is observed from the figure that the simulation results verify the above theoretical analysis, in which the PNIPA hydrogel placed in pure water has the larger swelling ratio than those in electrolyte solution. With the increase of the ionic concentration of bathing solution, the swelling ratio decreases, and the temperature of volume phase transition also decreases. After the temperature increases above 40°C, the hydrogels become completely dehydrated regardless of the ionic concentration of bathing solution.

In order to analyse the influences of the bathing solution concentration and environmental temperature on the distributive profiles of mobile ion concentrations, fixed charge density and electric potential at swelling equilibrium state, Figs. 5.9, 5.10, 5.11 and 5.12 are presented for the ionized cylindrical PNIPA hydrogel immersed in different univalent electrolyte bathing solutions.

Figures 5.9a and 5.10a demonstrate the distributive concentrations of the mobile cation and anion in the hydrogel and surrounding solution at temperature T=30°C and T=40°C, respectively. The distributions of fixed charge density at swelling equilibrium are plotted in Fig. 5.9b at T=30°C and Fig. 5.10b at T=40°C, respectively. It is known from Figs. 5.9a and 5.10a that, for a given electrolyte bathing solution, the mobile cation concentration in the PNIPA hydrogel at temperature T=40°C is higher than that at T=30°C, but the opposite trend is for the mobile anion concentration. At a given temperature, with increasing the bathing solution concentration c^*, the mobile cation concentration in the PNIPA hydrogel increases while the mobile anion concentration decreases. As the ionic concentration of bathing solution increases, the concentration difference of mobile ion species decreases between the hydrogel and the surrounding solution. Figures 5.9 and 5.10 also demonstrate that all the concentrations of mobile ions and fixed charge are compensated.

The electric potential distributed in the interior hydrogel and the exterior bathing solution is shown in Fig. 5.11 at T=30°C and Fig. 5.12 at T=40°C, respectively. For a given electrolyte bathing solution, the electric potential within the PNIPA hydrogel at temperature T=30°C is much higher than that at T=40°C. Similarly, for a given temperature T, the electric potential within the PNIPA hydrogel increases with the increase of the ionic concentration of electrolyte bathing solution, which is different from the phenomena studied above for the effect of the initial fixed charge c_f^0.

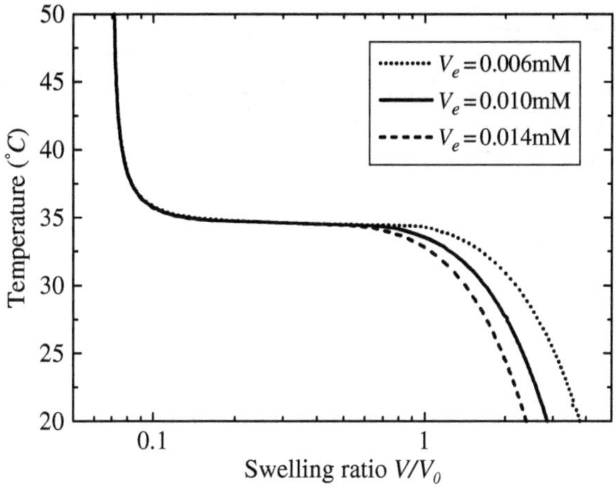

Fig. 5.13 Relation between the temperature and the swelling ratio V/V_0 of equilibrium volume for the ionized hydrogels with initial fixed charge density $c_f^0 = 5$ mM and different crosslink densities v_e immersed in the univalent electrolyte solution c^*20 mM

5.5.3 *Influence of Effective Crosslink Density*

In order to study the influence of the crosslink density, the relation between the temperature T and the swelling ratio V/V_0 is simulated and illustrated in Fig. 5.13, subjected to different crosslink densities v_e. Distributions of the diffusive ionic concentrations, fixed charge density and electric potential with different v_e and T are also illustrated in Figs. 5.14, 5.15, 5.16 and 5.17. As well known, usually the crosslink induces the formation of chemical bonds between linear polymeric

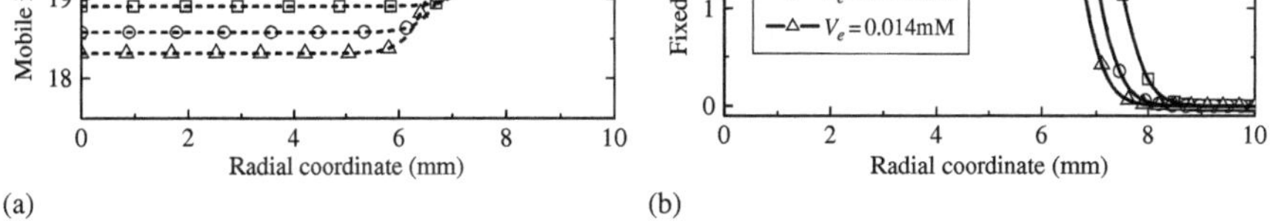

Fig. 5.14 Distributions of the mobile cation (*solid line*) and anion (*dash line*) concentrations (**a**) and the fixed charge densities (**b**) versus radial coordinate for the ionized PNIPA hydrogels with initial fixed charge density $c_f^0 = 5$ mM and different crosslink densities v_e immersed in the univalent electrolyte solution $c^*=20$ mM at temperature $T=30°C$

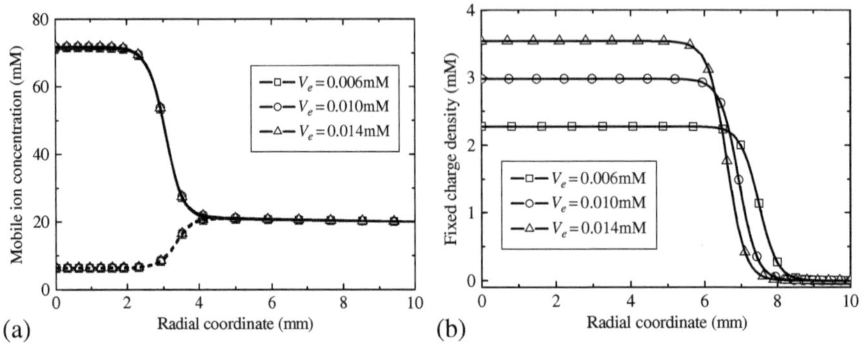

Fig. 5.15 Distributions of the mobile cation (*solid line*) and anion (*dash line*) concentrations (**a**) and the fixed charge densities (**b**) versus radial coordinate for the ionized PNIPA hydrogels with initial fixed charge density $c_f^0 = 5$ mM and different crosslink densities v_e immersed in the univalent electrolyte solution c^*=20 mM at temperature T=40°C

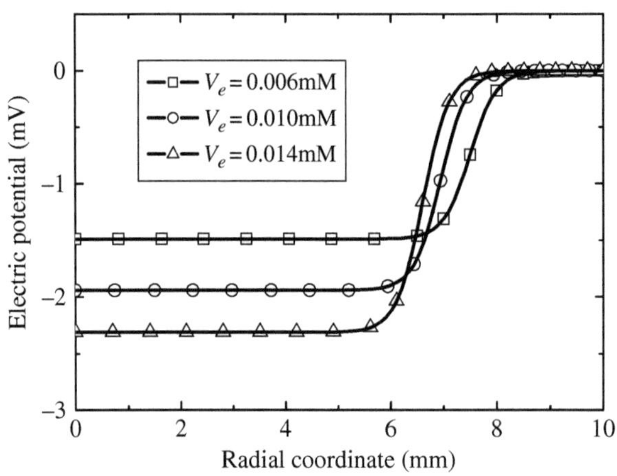

Fig. 5.16 Distributions of electric potential versus radial coordinate for the ionized PNIPA hydrogels with initial fixed charge density $c_f^0 = 5$ mM and different crosslink densities v_e immersed in the univalent electrolyte solutions c^*=20 mM at temperature T=30°C

molecules, and it also leads to forming infinite networks. The formed crosslinks divide long chain into the connected short chains and thus form numerous pores inside the hydrogel. Actually the hydrogel with large expansion contains a large amount of water in the pores. At swelling equilibrium state, the water uptake mostly depends on the crosslink density. As the crosslink density increases, the polymeric network becomes stiffer and the pore size becomes smaller, and then the water uptake decreases. Therefore, the increase of the crosslink density reduces the porous volume within the network hydrogel and provides smaller space for water uptake, which leads to a lower swelling ratio in equilibrium state.

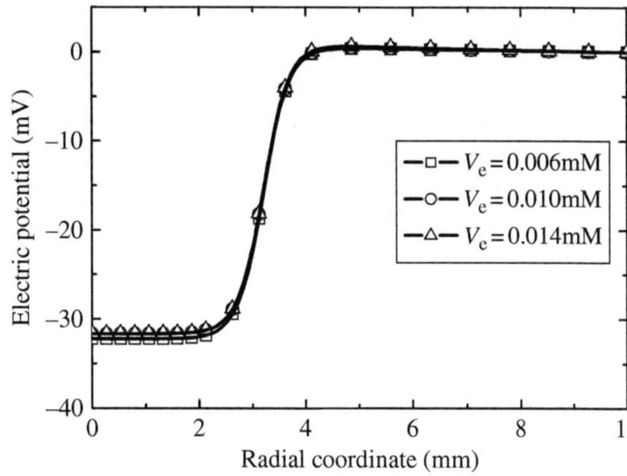

Fig. 5.17 Distributions of electric potential versus radial coordinate for the ionized PNIPA hydrogels with initial fixed charge density $c_f^0 = 5$ mM and different crosslink densities v_e immersed in the univalent electrolyte solution c^*=20 mM at temperature T=40°C

In the present simulation for study of the effect of effective crosslink density v_e on the swelling ratio V/V_0 at equilibrium state, three thermo-sensitive ionized PNIPA hydrogels are considered with different crosslink densities, when they are immersed in univalent electrolyte bathing solution, where the initial fixed charge density c_f^0=5 mM and the concentration of univalent electrolyte bathing solution c^* =20 mM. The hydrogel swelling ratio V/V_0 in equilibrium state versus the environmental temperature is shown in Fig. 5.13, where the hydrogel with the effective crosslink density v_e=0.006 mM is denoted by dotted line, v_e=0.01 mM by solid line and v_e=0.014 mM by dashed line. It is seen that the swelling ratio V/V_0 of the equilibrium thermo-sensitive PNIPA hydrogel has similar profiles as function of temperature, regardless of crosslink densities. The LCSTs or the volume phase transition temperatures have no shift for the hydrogel, in which all lie in the vicinity of 34.6°C. Figure 5.13 also shows that, if the temperature is below LCST, the volume swelling ratio V/V_0 of the equilibrium PNIPA hydrogel with the effective crosslink density v_e=0.010 mM is larger than that with larger v_e=0.014 mM and smaller than that with lower v_e=0.006 mM. The equilibrium volume changes for the hydrogel with low crosslink density are greater than those with high crosslink density at environmental temperature below LCST.

In order to demonstrate the influence of the temperature T and effective crosslink density v_e on the distributive characteristics of the mobile ion concentrations, fixed charge density and electric potential along the radial coordinate in swelling equilibrium state, Figs. 5.14, 5.15, 5.16 and 5.17 are plotted for the ionized cylindrical PNIPA hydrogel immersed in the univalent electrolyte bathing solution.

Distributive concentrations of the diffusive cation and anion species in both the hydrogel and bathing solution are illustrated in Fig. 5.14a at temperature T=30°C and Fig. 5.15a at temperature T=40°C. Correspondingly, the distributions of fixed charge density c_f at equilibrium state are plotted in Fig. 5.14b at T=30°C and Fig. 5.15b at T=40°C. For a given effective crosslink density v_e, Figs. 5.14a and 5.15a predict that the mobile cation concentration in the PNIPA hydrogel at temperature T=40°C is higher than that at T=30°C, which results from depleting water from the hydrogel due to the collapse. At temperature T=30°C, the mobile cation concentration in the PNIPA hydrogel enlarges with increasing the effective crosslink density ν_e. The opposite trends are observed for the mobile anion concentration. As the effective crosslink density ν_e increases, the diffusive anion concentration decreases. In addition, the increase of the effective crosslink density ν_e enlarges the difference of ionic concentrations between the hydrogel and exterior solution. At temperature T=40°C, all mobile ion concentrations in both the hydrogel and the solution are equal to each other for different crosslink densities ν_e because of the same swelling ratio, as portrayed in Fig. 5.13. Figures 5.14 and 5.15 demonstrate similar phenomena to those in Figs. 5.4 and 5.5, where all the concentrations of diffusive cation and anion species and fixed charge are compensated.

Distributions of the electric potential in both the hydrogel and exterior bathing solution are plotted in Fig. 5.16 at temperature T=30°C and in Fig. 5.17 at temperature T=40°C. For a given effective crosslink density ν_e, the electric potential in the PNIPA hydrogel at temperature T=30°C is much higher than that at T=40°C. At the environmental temperature T below the volume phase transition temperature, such as T=30°C, the electric potential in the PNIPA hydrogel decreases with increasing the effective crosslink density ν_e. At the environmental temperature above the volume phase transition temperature, such as T=40°C, the electrical potential under different effective crosslink densities ν_e remains the same in both the hydrogel and solution because of the same swelling ratio V/V_0.

5.5.4 Influence of Initial Volume Fraction of Polymeric Network

Initial polymer volume fraction ϕ_0 is prescribed at the reference state by the product of the molar volume and concentration of the monomeric unit during preparation of the hydrogel. Crosslinking agent induces the connection of these monomeric units to form polymeric network. Initially the hydrogel is in the dry state, solvent has to diffuse into the network to fulfil the free volume, which is not occupied by the polymer chains in the hydrogel. Influence of the initial polymer volume fraction ϕ_0 for a constant crosslink density is different from that of the effective crosslink density ν_e in the hydrogel.

For examination of influence of the initial polymer volume fraction ϕ_0 on the volume swelling ratio V/V_0 in equilibrium state, three thermo-sensitive ionized PNIPA hydrogels with different initial polymer volume fractions ϕ_0 are simulated when they are immersed in univalent electrolyte solution, where the initial fixed charge density c_f^0=5 mM, and the univalent electrolyte bathing solution c^* =20 mM. Figure 5.18 presents the response behavior of the hydrogel with different initial polymer volume fractions, ϕ_0=0.05 represented by dotted line, ϕ_0=0.07 by solid line and ϕ_0=0.09 by dashed line. It is found that the profiles of volume swelling ratio V/V_0 are quite different for the equilibrium thermo-sensitive PNIPA hydrogels with different initial polymer volume fractions. At the temperature below LCST, the volume swelling ratio V/V_0 of the equilibrium PNIPA hydrogel with ϕ_0=0.07 is larger than that with smaller ϕ_0=0.05 and is smaller than that with larger ϕ_0=0.09. In the contrast, the

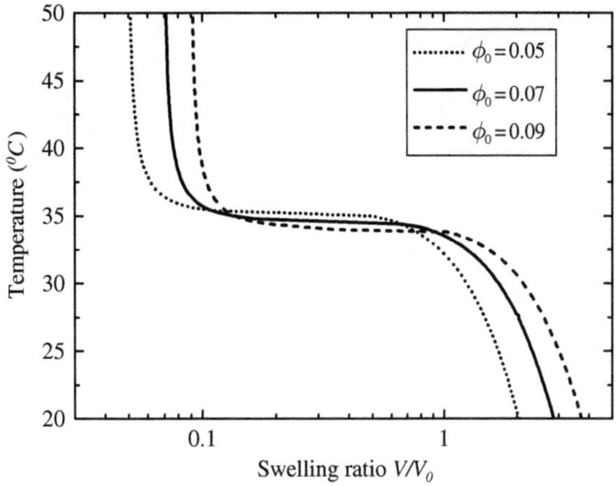

Fig. 5.18 Relation between the temperature and the swelling ratio V/V_0 of equilibrium volume for the ionized hydrogels with initial fixed charge density c_f^0 =5 mM and different initial polymer volume fractions ϕ_0 immersed in the univalent electrolyte solution c^*=20 mM

computed LCST for the initial polymer volume fraction ϕ_0=0.07 is greater than that for ϕ_0=0.09 and is lower than that for ϕ_0=0.05. This means that the computed LCST decreases with increasing the initial polymer volume fraction ϕ_0. Obviously, the volume change of the equilibrium hydrogel with large initial polymer volume fraction ϕ_0 is greater than that with small ϕ_0 at the temperature below LCST. At temperature above LCST, the volume swelling ratio for the equilibrium hydrogel with large initial polymer volume fraction ϕ_0 is smaller than that with low ϕ_0.

For discussion of the influence of the temperature T and initial polymer volume fraction ϕ_0 on the distributions of mobile ion species concentrations, fixed charge density and electric potential in swelling equilibrium state, Figs. 5.19, 5.20, 5.21 and 5.22 are plotted for the ionized PNIPA hydrogel immersed in the univalent electrolyte bathing solution.

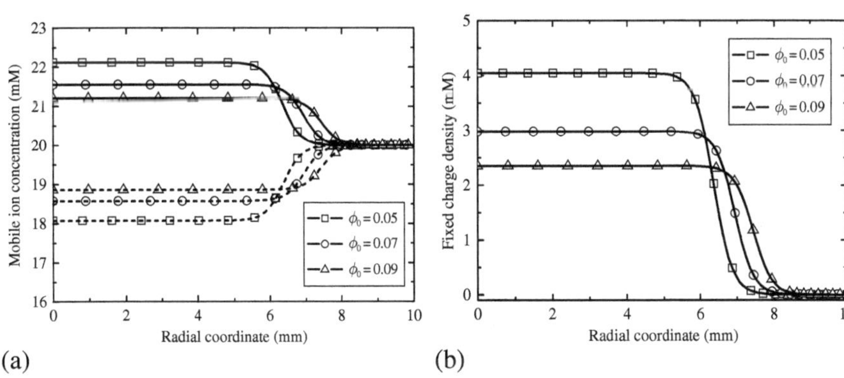

Fig. 5.19 Distributions of the mobile cation (*solid line*) and anion (*dash line*) concentrations (**a**) and the fixed charge densities (**b**) versus radial coordinate for the ionized PNIPA hydrogels with initial fixed charge density c_f^0 = 5 mM and different initial polymer volume fractions ϕ_0 immersed in the univalent electrolyte solution c^*=20 mM at temperature T=30°C

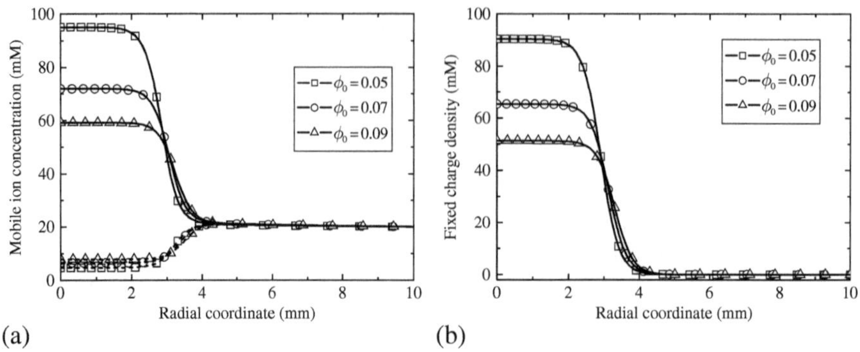

Fig. 5.20 Distributions of the mobile cation (*solid line*) and anion (*dash line*) concentrations (**a**) and the fixed charge densities (**b**) versus radial coordinate for the ionized PNIPA hydrogels with initial fixed charge density $c_f^0 = 5$ mM and different initial polymer volume fractions ϕ_0 immersed in the univalent electrolyte solution $c^* = 20$ mM at temperature $T = 40$°C

Fig. 5.21 Distributions of electric potential versus radial coordinate for the ionized PNIPA hydrogels with initial fixed charge density $c_f^0 = 5$ mM and different initial polymer volume fractions ϕ_0 immersed in the univalent electrolyte solution $c^* = 20$ mM at temperature $T = 30$°C

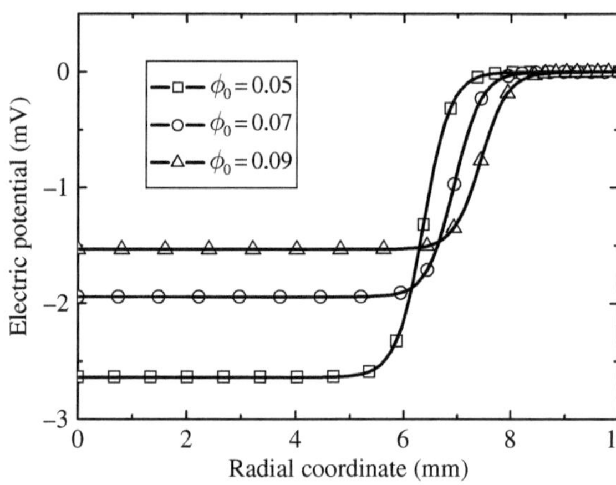

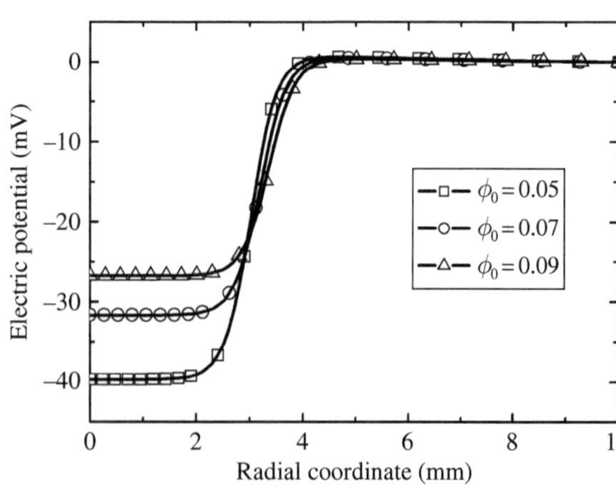

Fig. 5.22 Distributions of electric potential versus radial coordinate for the ionized PNIPA hydrogels with initial fixed charge density $c_f^0 = 5$ mM and different initial polymer volume fractions ϕ_0 immersed in the univalent electrolyte solution $c^* = 20$ mM at temperature $T = 40$°C

The distributive concentrations of the mobile cation and anion species in both the hydrogel and bathing solution are demonstrated in Fig. 5.19a at temperature T=30°C and Fig. 5.20a at temperature T=40°C. Distribution of fixed charge density in swelling state is plotted in Fig. 5.19b at T=30°C and Fig. 5.20b at T=40°C. For a given initial polymer volume fraction ϕ_0, Figs. 5.19a and 5.20a predict that the diffusive cation concentration in the PNIPA hydrogel at temperature T=40°C is higher than that at T=30°C. At a given temperature T, the mobile cation concentration in the PNIPA hydrogel becomes high for small initial polymer volume fraction ϕ_0. The opposite trends are observed for the diffusive anion concentration. As the initial polymer volume fraction ϕ_0 increases, the mobile anion concentration increases, and the difference of mobile ion concentrations decreases between the hydrogel and bathing solution.

Electric potential distributed in both the hydrogel and environmental solution is plotted in Fig. 5.21 at temperature T=30°C and in Fig. 5.22 at temperature T=40°C. For a given initial polymer volume fraction ϕ_0, the electric potential within the PNIPA hydrogel at temperature T=30°C is much higher than that at T=40°C. At a given temperature, the electric potential in the PNIPA hydrogel increases accordingly with the initial polymer volume fraction ϕ_0.

5.6 Transient Modelling of Temperature-Sensitive Neutral Hydrogel

This section will focus on the modelling of transient deformation of neutral hydrogel. A model that takes into account the conservations of momentum, heat and mass for the solid polymer and fluid phase is derived, non-dimensionalized and analysed in both Eulerian and Lagrangian frames of reference. Slow- and fast-response hydrogels subject to a step change in temperature or temperature gradient are studied. Model predictions shown agree well with experiments for uniform spherical swelling. Heat transfer found has a negligible impact on the kinetics, even for the fast-response hydrogel. For a constrained hydrogel slab, the evolution is non-uniform, with solvent entering at the sides and flowing into the interior and towards the corners. The fast-response hydrogel exhibits a solvent penetration in the form of a sharp front during deswelling, whereas the solvent penetration for the slow-response hydrogel and for the fast-response swelling is more akin to a diffusion process. Immersed in a temperature gradient, the studied hydrogel bends to a curved equilibrium shape, with solvent flowing from the cooler to the warmer sides due to the difference in osmotic pressure. The benefit of the scale analysis is conducted here to predict correctly, prior to numerical computation. Important characteristics such as stress, osmotic pressure and deformation time are also highlighted.

Literature search reveals that the stimuli-responsive hydrogels have attracted much attention over the last decade for their potential in a wide array of applications, ranging from drug delivery (Galaev and Mattiasson, 1999; Roy and Gupta, 2003), biosensors (Luo et al., 2008), tissue engineering (Hoffman, 2002) and artificial muscles (Mao et al., 2005) to micro-electro-mechanical system (MEMS) devices such as microvalves (Eddington and Beebe, 2004) and micro-fluidic controllers (Eddington and Beebe, 2004; Roy and Gupta, 2003). Other applications of the smart hydrogels include disposable diapers (Nishizawa et al., 1981), soil irrigation (Pohl, 2004), seed coating (Pamuk, 2004) and smart windows (Zrínyi et al., 2001), to name a few.

A hydrogel essentially comprises an elastic three-dimensional network of crosslinked polymers and a penetrating solvent, such that the hydrogel exhibits both solid-like and liquid-like properties: in the dry state, the hydrogel behaves like a solid; immersed in a liquid solvent; however, the hydrogel is able to absorb the solvent and swells until it reaches a swelling equilibrium, lending the hydrogel liquid properties. A stimuli-responsive hydrogel is capable of undergoing large reversible volume changes of up to a thousand-fold or more, when subjected to perturbations in its environment: alterations in pH, ionic strength and temperature as well as light irradiation and external electric field number among the most common stimuli (Roy and Gupta, 2003; Qiu and Park, 2001). The volume change and simultaneous alteration in the macroscopic properties of the hydrogel are mainly brought about by four fundamental molecular interactions (Shibayama and Tanaka, 1993): hydrogen bonding, hydrophobic interaction, electrostatic interaction and van der Waals interaction. These interactions give rise to an osmotic pressure and an elastic stress of the hydrogel, which determine the swelling properties.

Perhaps Kuhn was the first who demonstrated the volume change of the hydrogel in 1948 and showed that the hydrogel can convert chemical energy directly into mechanical energy (Osada and Gong, 1993). Since then, several stimuli have been identified and a vast number of tailor-made hydrogels have been synthesized, a short summary of which is found in the work of Roy and Gupta (2003). In tandem with experimental work, many mathematical models have been derived in an effort to understand and predict the deformation kinetics and swelling equilibrium of the hydrogel. Usually the latter is considered to follow Flory's mean field theory (Flory, 1953) or alterations thereof, see, e.g. Oliveira et al. (2004), Hino and Prausnitz (1998), Oh and Bae (1998), Otake et al. (1989) and Maurer and Prausnitz (1996). In contrast to the theories of swelling equilibria, there is no unified approach to characterize the kinetics of the deformation, since a wide range of physical phenomena have to be considered, requiring a blend of continuum mechanics, electrochemistry, transport phenomena, polymer science and polyelectrolyte hydrogel. Several models of varying degree of complexity were derived to predict the kinetics of the volume phase transition of the hydrogels. The first was one dimensional and based on diffusion equation, where Tanaka et al. (1973) derived a diffusion equation for the polymer displacement, now known as the collective diffusion equation. It was later modified by Tanaka and Fillmore (1979), who introduced a gel network diffusion coefficient comprising the ratio of the bulk modulus and friction coefficient, and then generalized by Li and Tanaka (1990). However, the collective diffusion equation was shown to fail for the poly(*N*-isopropylacrylamide) gel, when the shrinking occurred from below to above the critical temperature by Andersson et al. (1998). Wang et al. (1997) revised the collective diffusion model by adding a convective term. Another diffusion model was developed by Bisschops et al. (1998), based on the generalized Maxwell–Stefan equation (Taylor and Krishna, 1993) to predict the swelling behavior. Furthermore, Fick's second law was used, e.g. by Singh and Weber (1996). Colombo et al. (1996) and Grassi et al. (2000) applied a modified viscoelastic diffusion equation derived by Camera-Roda and Sarti (1990) to account for non-Fickian behaviour for the kinetics of the hydrogels. In addition to these diffusion-based models, simple exponential functions (Okajima et al., 2002) or empirical equations (Chern et al., 2004) were fitted to experimental deformation curves. Various two-fluid models were also derived and applied to the study of gels by Doi (1990), Sekimoto and Doi (1991), Sekimoto (1991), Achilleos et al. (2000a, b, 2001), Barrière and Leibler (2003) and Wolgemuth et al. (2004). Similar to the two-fluid models is the biphasic model derived by Netti and Travascio (2003). The advantage of the two-fluid or mixture theory over the aforementioned diffusion models is the more rigorous description of the solvent phase in addition to the deformation of the solid polymeric network phase. Recently, Cushman et al. (2004a, b) derived the models at both the mesoscale and macroscale of a swelling hydrogel. Other approaches (Dolbow et al., 2004; Ji et al., 2006; Dolbow et al., 2005) include the addition of a level-set function to track the interface between the swollen and the collapsed phases in the hydrogel. In summary, most of these models are limited to one-dimensional uniform studies, with the exception of Achilleos et al. (2000a, b, 2001), Dolbow et al. (2005) and Ji et al. (2005). While one-dimensional models are able to address the uniform swelling behavior, they cannot account for non-uniform deformations, which might arise due to, e.g. the shape of the hydrogel, such as slabs (Achilleos et al., 2000a, b, 2001). Non-uniform deformations can be expected to be important in applications, where the hydrogel is subjected to external flow field or gradient in temperature or other properties, e.g. in autonomous flow control systems (Beebe et al., 2000a).

This section primarily addresses the kinetics of thermo-sensitive neutral hydrogels, with the aim to develop a generic model for the (i) uniform and (ii) non-uniform responses of the neutral hydrogels subject to the change in temperature and (iii) the non-uniform deformation in a temperature gradient. The first, which can be captured with a one-dimensional model in spherical coordinates, allows for parameter adaptation and validation of the model with published experimental data. The second, which is studied in the form of a constrained three-dimensional slab, gives an insight into the non-uniform deformation response of the hydrogel as it adapts to the new temperature field. The third illustrates the ability of a temperature-sensitive hydrogel to sense the variation in temperature and respond upon these with a bending motion. The study is motivated by the fact that most of the existing models for thermo-sensitive hydrogels from the onset assume that the heat transfer is much faster than the mechanical deformation, while this is in line with experimental findings by Gehrke (1993) for poly(*N*-isopropylacrylamide) gels (PNIPA gels). There is no a priori guarantee that this is always the case. In fact, recent studies of fast-response hydrogels (Cheng et al., 2003; Zhang and Zhuo, 2000a, b; Chen et al., 1999) demonstrated that the mechanical deformation from a shrunken to swollen state can be reduced to minutes or less, which is of the same order of magnitude as the transient response to the increase of temperature that Gehrke (1993) measured. The present model considers both slow-response, i.e. with deformation time around

1 hour or more, and fast-response hydrogels. Furthermore, the present approach differs from most of the previous work, in which non-dimensionalization and scaling arguments are applied to ascertain the main features and parameters prior to numerical computation.

In the following subsections, the mathematic formulation is conducted first in an Eulerian (spatial) framework. The model consists of the biphasic conservations of mass, momentum and heat, allowing the finite uniform and non-uniform deformations of the hydrogel. The model has two important antecedents: first, in the limit of no deformation, Darcy's law is recovered and second, for a uniform spherical swelling, a diffusion equation is obtained, similar to the aforementioned diffusion-based models, such as the collective diffusion model. Special attention is given to the derivation of the stress tensor, both at the reference state, usually taken as the state of polymerization, and at an arbitrary equilibrium initial state. Apart from these, a non-dimensional analysis of the governing equations provides an indication of the qualitative features one would expect of the kinetics of a hydrogel; foremost of which are the time scales for heat transfer and the deformation of the hydrogel. Furthermore, the complexity of the governing equations can be reduced via scaling arguments. It is followed by a subsequent reformulation of the reduced model in Lagrangian (material) coordinates, which is solved numerically. The simulation results are then presented and discussed for the slow- and fast-response hydrogels during uniform and non-uniform deformations, where good agreements between the model predictions and experiments are demonstrated, and the conclusions are finally drawn. In addition, as mentioned at the beginning of this chapter, it should be pointed out that Erik Birgersson (Birgersson et al., 2008), who is one of the important collaborators of the author, has also made significant contributions to this research effort for transient modelling of the temperature-sensitive neutral hydrogels.

5.6.1 Model Formulation in Eulerian Frame

In this subsection, a transient phenomenological model is derived in an Eulerian frame for a neutral hydrogel subject to heat stimulus, based on the biphasic mixture theory. In short, the biphasic mixture theory, also commonly referred to as the poroelastic theory, characterizes the flow of a fluid through a porous medium, which itself undergoes a deformation. The three-dimensional polymer network and the penetrating fluid in the hydrogel are taken as the solid and fluid phases, respectively. The former is denoted by the superscript (p) and the latter by (f). The model accounts for the coupled transport phenomena as follows:

- *Conservation of mass*. Conservation of the mass for the binary mixture comprising the polymer network and penetrating solvent is solved for.
- *Momentum transfer*. Conservation of momentum is considered in terms of inertia, a reaction couple between the phases and the driving forces due to stress and osmotic pressure for both phases.

- *Conservation of phases*. Two phases are considered: a polymer comprising large macromolecular chains and a solvent phase of low molecular weight. The latter is highly mobile compared with the former.
- *Energy transfer*. The principal modes of energy transfer are convection, deformation and conduction through the polymer network and fluid. The present convection is associated with the fluid phase and the deformation with the solid phase. In addition, the transient accumulation of energy, energy dissipation and energy exchange between the phases due to friction are considered.

The main model assumptions and approximations include the following:

- *Local thermal equilibrium*. Local thermal equilibrium is assumed between the polymer and the penetrating fluid.
- *Incompressibility*. Both the polymer and liquid phases are treated as incompressible. However, the resulting mixture is highly compressible.
- *Free energy*. The free energy of the hydrogel provides the means by which the stress tensor and osmotic pressure for the hydrogel can be identified. Several expressions for the free energy of mixing and elasticity for polymer hydrogel systems have been derived from molecular thermodynamics, e.g. Oliveira et al. (2004), Hino and Prausnitz (1998), Oh and Bae (1998), Maurer and Prausnitz (1996) and Otake et al. (1989). Most of these are modifications of the original thermodynamic theory by Flory (1953); a short summary of the various thermodynamic models can be found in Wu et al. (2004). For the present purposes, the original Flory–Rehner theory (Flory, 1953) as applied by Shirota et al. (1998) is sufficient.
- *Hyperelastic solid*. The hydrogel is treated as an isotropic hyperelastic material, which together with the elastic free energy provides an expression for the stress tensor.
- *Newtonian fluid*. The penetrating fluid is assumed to behave as a Newtonian fluid within the polymer network.

The hydrogels considered include the following:

- *Crosslinked poly(N-acryloyl pyrrolidine)*, prepared and analysed by Oh and Bae (1998), with 15% mol styrene (AS15) and 20% St (AS20). These can be classified as the slow-response hydrogels, displaying moderate swelling ratios of around 1–5. The swelling ratio is here defined as the ratio of the weight of water in the swollen hydrogel at a certain temperature to the weight of the hydrogel after drying, usually in vacuum.
- *Conventional poly(N-isopropylacrylamide)*, prepared and analysed by Cheng et al. (2003), termed NC000, which is one of the most extensively studied temperature-sensitive hydrogels. This is a slow-response hydrogel with large swelling ratio of 4–40.
- *Macroporous poly(N-isopropylacrylamide)*, polymerized with a 0.3 M aqueous sodium chloride solution (NC030) by Cheng et al. (2003) and polymerized with

(i)

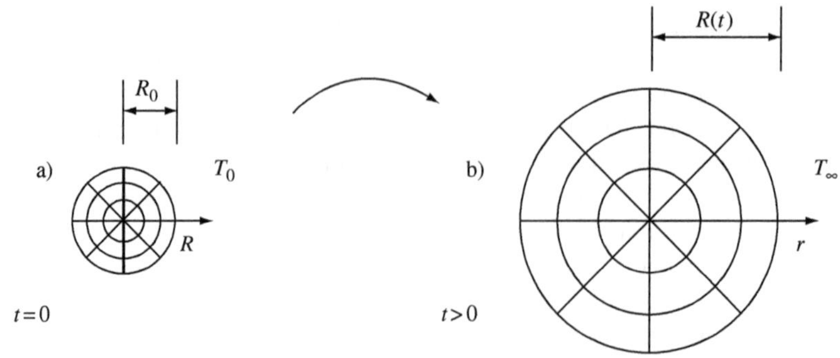

Fig. 5.23 Schematic of a spherical uniform response to a step change in temperature

polyethylene glycol (G8) by Zhang and Zhuo (2000a), exhibiting large swelling ratios of 4–65 and 4–90, respectively. These are considered as the fast-response hydrogels due to their large pore sizes.

Within this framework, and as already motivated at the beginning of this section, three deformation cases are considered due to the changes in temperature:

(i) *Transient uniform response to a temperature change.* Most of the experimental studies on the swelling equilibrium and kinetics of hydrogels are for cylindrical or spherical hydrogels immersed in a solvent of uniform temperature. Upon the changes in temperature, the subsequent deformation of the hydrogel as it approaches the new equilibrium can be studied. The uniform swelling is considered, as illustrated in Fig. 5.23, in a uniform temperature field for both slow-response and fast-response hydrogels, for which a one-dimensional spherical model in the radial direction is sufficient. A symmetry boundary condition at the centre of the sphere and a free surface at $r=R(t)$ are prescribed. As observed later, the model predictions agree well with experimental findings.

(ii) *Transient non-uniform response to a temperature change.* In addition to the uniform swelling, the non-uniform deformation behaviour of a hydrogel slab to a step change in temperature is investigated, as depicted in Fig. 5.24, which is constrained in the normal direction (z-direction), such that it only needs to consider deformations in the streamwise (x-direction) and spanwise (y-direction) directions. The computational domain is further reduced to one quarter by invoking symmetry of the hydrogel slab along the centre lines, with symmetry boundary conditions along the x- and y-axis and free surface conditions for the remaining two boundaries.

(iii) *Transient deformation in a temperature gradient.* A slender hydrogel is immersed in a solvent with uniform temperature T_0 at $t=0$. For $t>0$, a temperature gradient is applied over the solution and hydrogel as shown in Fig. 5.25. Analogous to the behaviour of polyelectrolyte hydrogels exposed to electric field (Li et al., 2004a, b; Wallmersperger et al., 2004; Zhou et al., 2002; Snita

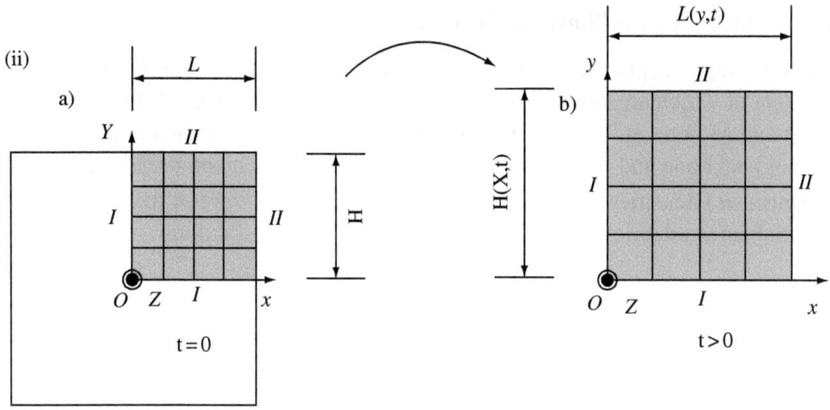

Fig. 5.24 Schematic of a non-uniform response to a step change in temperature

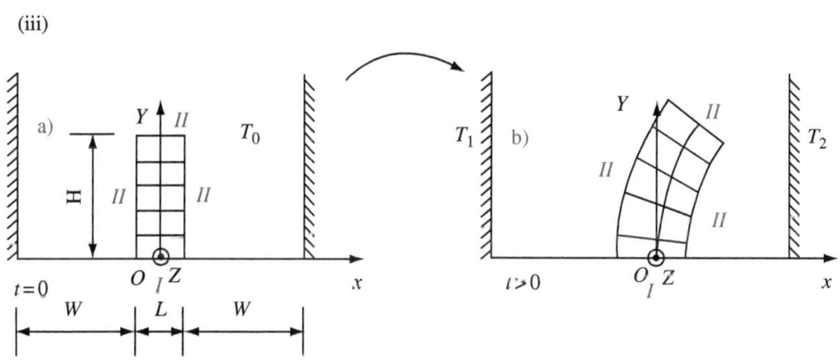

Fig. 5.25 Schematic of the deformation in a temperature gradient

et al., 2001), the neutral hydrogel is expected to bend in the temperature field. The non-uniform deformation in a temperature field is studied for the macroporous hydrogel NC030, which is constrained in the normal direction, similar to the above case (ii). In general, the deformed hydrogel can be expected to affect the surrounding temperature field, especially so if the thermal properties of the hydrogel differ from the solvent counterparts. For the NC030, however, the polymer volume fraction is around $O(10^{-2})$ in the swollen state due to the macroporous nature, i.e. with a high solvent content, whence the thermal properties can be considered equal to the surrounding solvent as a first approximation, since the heat conductivity contribution of the polymer phase will be negligible. The present work will focus on NC030 and assume a one-way coupling to study the deformation in a prescribed temperature gradient. Further, it will assume a priori that heat transfer is much faster than the deformation, whence it is not required to solve for the heat transfer. It will be shown later that this condition is indeed satisfied.

5.6.1.1 Basics of Two-Phase Mixture Theory

Several basic concepts are essential for modelling of two-phase mixture, the most important of which are summarized here, see, e.g. Bowen (1980), Atkin and Craine (1976) and Mow et al. (1980) for more details. In general, the two phases, i.e. the polymer backbone and the penetrating fluid, are treated as incompressible, with the true (intrinsic) densities $\rho_0^{(p)}$ and $\rho_0^{(f)}$, respectively. The mixture of these two phases requires the introduction of the polymer and fluid volume fractions:

$$\phi^{(p)} \equiv \phi = \frac{\nu^{(p)}}{\nu^{(p)} + \nu^{(f)}}, \quad \phi^{(f)} = 1 - \phi \tag{5.33}$$

where $\nu^{(i)}$and $\phi^{(i)}$ are the volume and volume fractions of the phase i (i=f, p), respectively. Further, the mixture density is given by

$$\rho = \rho^{(p)} + \rho^{(f)} \tag{5.34}$$

where $\rho^{(i)}$ is the apparent (superficial) density of the phase i, defined as

$$\rho^{(p)} = \phi\rho_0^{(p)}, \quad \rho^{(f)} = (1 - \phi)\rho_0^{(f)} \tag{5.35}$$

Dividing the mass conservation of each phase gives

$$\frac{\partial\rho^{(i)}}{\partial t} + \nabla \cdot (\rho^{(i)}\mathbf{v}^{(i)}) = 0 \tag{5.36}$$

where $v^{(i)}$ is the velocity of the phase i. By $\rho_0^{(i)}$, adding the equations for each phase results in

$$\nabla \cdot \mathbf{v} = 0 \tag{5.37}$$

where $\mathbf{v}$ is the volume-averaged velocity and defined as

$$\mathbf{v} = \phi\mathbf{v}^{(p)} + (1 - \phi)\mathbf{v}^{(f)} \tag{5.38}$$

For later use, it is noted that the substantial derivative with respect to the phase i is defined as

$$\frac{D^{(i)}}{Dt} = \frac{\partial^{(i)}}{\partial t} + (\mathbf{v}^{(i)} \cdot \nabla) \tag{5.39}$$

5.6.1.2 Governing Equations

In the hydrogel, the conservation of mass for the polymer phase is considered as

$$\frac{\partial\phi}{\partial t} + \nabla \cdot (\phi\mathbf{v}) = -\nabla \cdot [\phi(1 - \phi)(\mathbf{v}^{(p)} - \mathbf{v}^{(f)})] \tag{5.40}$$

where the volume-averaged velocity, Eq. (5.38), is introduced into the mass conservation for the polymer, Eq. (5.36). The left-hand side (LHS) of Eq. (5.40) comprises the accumulation and bulk convection of the polymer phase, and the right-hand side (RHS) considers diffusion.

The conservation of the mixture is

$$\nabla \cdot \mathbf{v} = 0 \tag{5.41}$$

Conservation of momentum for the polymer and fluid phases are given, respectively, by

$$\phi \rho_0^{(p)} \frac{D^{(p)} \mathbf{v}^{(p)}}{Dt} = \nabla \cdot \sigma^{(p)} + \pi^{(p)} \tag{5.42}$$

$$(1-\phi) \rho_0^{(f)} \frac{D^{(f)} \mathbf{v}^{(f)}}{Dt} = \nabla \cdot \sigma^{(f)} + \pi^{(f)} \tag{5.43}$$

where $\sigma^{(i)}$ is the stress tensor of the phase i, and $\pi^{(i)}$ is the reaction couple of the drag force between the two phases. The LHS of the momentum equations, Eqs. (5.42) and (5.43), describes the inertia, and the RHS comprises the driving forces in terms of gradients in stress and osmotic pressure as well as the drag force. The contribution of osmotic pressure at this stage is not readily visible, but will appear when the constitutive relations are introduced in the following subsection.

Conservation of energy for each phase is given by Lustig et al. (1992) and Huyghe and Janssen (1999):

$$\rho^{(i)} \frac{D^{(i)} \Xi^{(i)}}{Dt} = \sigma^{(i)} : \nabla \mathbf{v}^{(i)} - \nabla \cdot \mathbf{q}^{(i)} + \rho^{(i)} \hat{\Xi}^{(i)} \tag{5.44}$$

where $\Xi^{(i)}$, $\hat{\Xi}^{(i)}$ and $\mathbf{q}^{(i)}$ are the internal energy, the energy interaction with other phases and conductive heat flux of the phase i, respectively. The accumulation of internal energy and convective energy transfer are captured by the LHS. Heat dissipation is characterized by the first term on the RHS, followed by the conductive heat flux and the energy exchange with other phases. The total energy balance for both phases (Lustig et al., 1992; Huyghe and Janssen, 1999) is defined as

$$\sum_i \left(\rho^{(i)} \hat{\Xi}^{(i)} + \mathbf{v}^{(i)} \cdot \pi^{(i)} \right) = 0 \tag{5.45}$$

In order to remove the energy interaction parameters, Eqs. (5.44) and (5.45) are combined, which yields a more convenient form for the energy balance

$$\begin{aligned} \phi \rho_0^{(p)} \frac{D^{(p)} \Xi^{(p)}}{Dt} + (1-\phi) \rho_0^{(f)} \frac{D^{(f)} \Xi^{(f)}}{Dt} &= \sigma^{(p)} : \nabla \mathbf{v}^{(p)} + \sigma^{(f)} : \nabla \mathbf{v}^{(f)} \\ &- \nabla \cdot (\mathbf{q}^{(p)} + \mathbf{q}^{(f)}) - \mathbf{v}^{(p)} \cdot \pi^{(p)} - \mathbf{v}^{(f)} \cdot \pi^{(f)} \end{aligned} \tag{5.46}$$

5.6.1.3 Constitutive Relations

Liquid Phase

Rates of strain and stress tensors. For the fluid phase, the rates of strain, $\mathbf{E}^{(f)}$, and stress tensors, $\sigma^{(f)}$, are defined as

$$\mathbf{E}^{(f)} = \frac{1}{2}[\nabla \mathbf{v}^{(f)} + (\nabla \mathbf{v}^{(f)})^T] \tag{5.47}$$

$$\sigma^{(f)} = -(1-\phi)p\mathbf{I} + 2\mu^{(f)}\mathbf{E}^{(f)} \tag{5.48}$$

where p is the total intrinsic fluid pressure, $\mathbf{I}$ is the unit second-order tensor and $\mu^{(f)}$ is the dynamic viscosity of the fluid. The total intrinsic fluid pressure can be expanded as

$$p = p\prime + p^{(\mathrm{osm})} \tag{5.49}$$

where $p^{(\mathrm{osm})}$ is the osmotic pressure, and p' is the remaining part of the intrinsic fluid pressure. In the absence of an osmotic pressure and a stationary solvent, the intrinsic pressure corresponds to the hydrostatic pressure in the fluid. Contrary to Barrière and Leibler (2003), who combined the elastic stress and osmotic effect into one stress tensor, we instead include the osmotic pressure in the overall intrinsic fluid pressure, similar to Atheshian et al. (2004).

Dynamic viscosity. The dynamic viscosity of the liquid, $\mu^{(f)}$, is temperature dependent and can be approximated by Gawin et al. (1999)

$$\mu^{(f)} = 0.6612(T-229)^{-1.562} \tag{5.50}$$

Thermal properties. The thermal conductivity, $\mathbf{k}^{(f)}$, in the fluid phase is given by

$$\mathbf{k}^{(f)} = k^{(f)}\mathbf{I} \tag{5.51}$$

where

$$k^{(f)}(T) = c_1T^2 + c_2T + c_3 \tag{5.52}$$

which is a parameter adapted function to tabulated thermal conductivities of water (IAPS, 1998). The specific heat, $C_p^{(f)}$, can be considered as constant for the temperature interval considered here (283 K$\leq T \leq$353 K) (Bird et al., 2002). The enthalpy, $\Theta^{(f)}$, and internal energy, $\Xi^{(f)}$, for the incompressible fluid (Bird et al., 2002) are related through

$$\Xi^{(f)} = \Theta^{(f)} - \frac{p}{\rho_0^{(f)}} \tag{5.53}$$

and the enthalpy (Bird et al., 2002) is given by

$$d\Theta^{(f)} = C_p^{(f)}dT + \frac{dp}{\rho_0^{(f)}} \tag{5.54}$$

Polymer Phase

Finite deformation and stress tensor. For the polymer phase, the stress tensor is defined as

$$\sigma^{(p)} = -\phi p\mathbf{I} + \sigma_{eff}^{(p)} \tag{5.55}$$

where $\sigma_{\text{eff}}^{(p)}$ is the Cauchy stress tensor for the polymer phase. In order to determine the Cauchy stress, it is assumed that the polymer behaves as an isotropic hyper-elastic material to account for finite deformation, since the hydrogel can undergo significant volume change. Further, the elasticity for the hydrogel is given by the deformation from a stress-free state to its stressed counterpart. In line with common practice, see, e.g. Oliveira et al. (2004), the stress-free state is taken to be the polymerization state, whence it is required to account for this reference state when defining the stress tensor. For this purpose, as illustrated in Fig. 5.26, it is convenient to consider the deformation in the reference frame, $\mathbf{X}_{\text{ref}} = (X_{\text{ref}}, Y_{\text{ref}}, Z_{\text{ref}}) \in R^3$, in an initial Lagrangian (material) frame of reference, $\mathbf{X} = (X,Y,Z) \in R^3$, and the Eulerian (spatial) counterpart, $\mathbf{x} = (x,y,z) \in R^3$; these in turn are related via the mappings $\mathbf{x} = \chi_{\text{ref}}(\mathbf{X}_{\text{ref}},t) = \chi(\mathbf{X},t)$ and $\mathbf{X} = \chi_0(\mathbf{X}_{\text{ref}},t)$. The deformation gradient tensors between the various frames can now be introduced as $\mathbf{F}_{\text{ref}}(\mathbf{X}_{\text{ref}},t) = \partial\chi_{\text{ref}}(\mathbf{X}_{\text{ref}},t)/\partial\mathbf{X}_{\text{ref}}$, $\mathbf{F}_0(\mathbf{X}_{\text{ref}},t) = \partial\chi_0(\mathbf{X}_{\text{ref}},t)/\partial\mathbf{X}_{\text{ref}}$, $\mathbf{F}(\mathbf{X},t) = \partial\chi(\mathbf{X},t)/\partial\mathbf{X}$. The displacement field in the Eulerian frame of reference is $\mathbf{u}(\mathbf{x},t) = (u_1,u_2,u_3) = \mathbf{x}-\mathbf{X}(\mathbf{x},t)$ and in the initial frame $\mathbf{U}(\mathbf{X},t) = (U_1,U_2,U_3) = \mathbf{x}(\mathbf{X},t) - \mathbf{X}$.

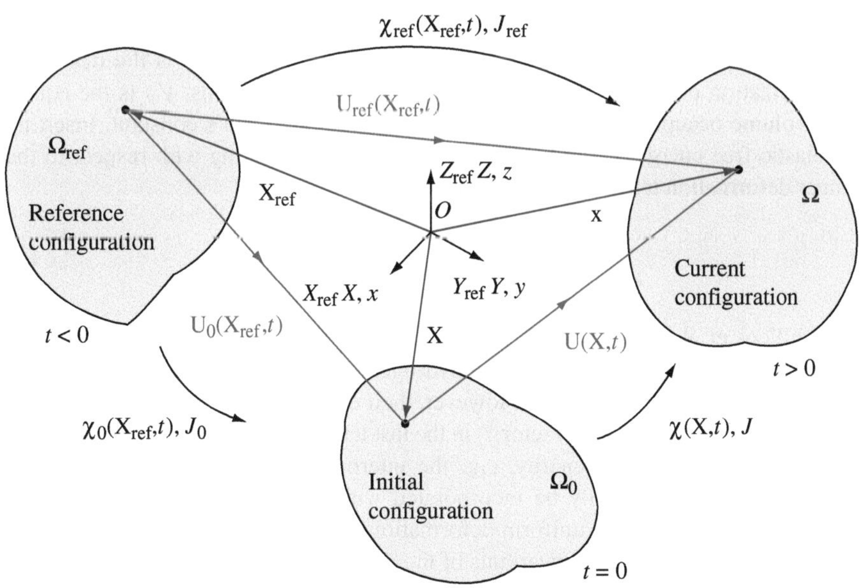

Fig. 5.26 Schematic of the reference (Ω_{ref}), initial (Ω_0) and current (Ω) configurations, the coordinates and the displacement fields

For an isotropic hyperelastic material, the Cauchy stress tensor (Holzapfel, 2000) relative to the reference frame $\mathbf{X}_{\text{ref}}$ is given by

$$\sigma_{\text{eff}}^{(p)} = 2J_{\text{ref}}^{-1}\mathbf{b}_{\text{ref}}\frac{\partial\Psi(\mathbf{b}_{\text{ref}})}{\partial\mathbf{b}_{\text{ref}}} \tag{5.56}$$

where $J_{\text{ref}} = \det\mathbf{F}_{\text{ref}} = \phi_{\text{ref}}/\phi$ is the volume ratio, $\mathbf{b}_{ref} = \mathbf{F}_{ref}\mathbf{F}_{ref}^{T}$ is the Finger deformation tensor, also commonly referred to as the left Cauchy–Green tensor and $\Psi(\mathbf{b}_{\text{ref}})$ is the free energy of elasticity as a function of the Finger deformation tensor. The elastic free energy for a hydrogel is usually expressed according to either the phantom network or the affine network theory (Baker et al., 1994). In short, these two theories consider the idealized cases, with the main difference that the former allows for free fluctuations of crosslinks in the network, whereas the latter suppresses these. We shall make use of the affine network theory, which for isotropic swelling of a tetrafunctional network (Lele et al., 1997) yields

$$\Psi = \frac{k_B T}{2V_m N_x}[3J_{\text{ref}}^{\frac{2}{3}} - 3 - \ln J_{\text{ref}}] \tag{5.57}$$

which can be rewritten in terms of the invariants of the Finger deformation tensor, namely $I_1(\mathbf{b}_{\text{ref}}) = \text{tr}\mathbf{b}_{\text{ref}}$, $I_2(\mathbf{b}_{\text{ref}}) = 0.5[(\text{tr}\mathbf{b}_{\text{ref}})^2 - \text{tr}(\mathbf{b}_{\text{ref}}^2)]$, $I_3(\mathbf{b}_{\text{ref}}) = \det\mathbf{b}_{\text{ref}}$, where tr denotes the trace of the indicated tensor, thus

$$\Psi = \frac{k_B T}{2V_m N_x}\left[I_1 - 3 - \frac{1}{2}\ln I_3\right] \tag{5.58}$$

for an isotropic deformation, where $I_1 = 3J_{\text{ref}}^{(2/3)}$, $I_3 = J_{\text{ref}}^2$, N_x is the degree of polymerization of the subchains between the crosslinking points, V_m is the equivalent volume occupied by one monomer and k_B is Boltzmann's constant. Inserting the elastic free energy, given by Eq. (5.58), and differentiating with respect to the Finger deformation tensor, Eq. (5.56), we arrive at

$$\sigma_{\text{eff}}^{(p)} = \frac{k_B T}{V_m N_x J_{\text{ref}}}\left[\mathbf{b}_{\text{ref}} - \frac{1}{2}\mathbf{I}\right] \tag{5.59}$$

Achilleos et al. (2000a, b, 2001) used a similar approach to find the stress tensor for the swelling of polyelectrolyte hydrogels and showed good agreement with experimental swelling behaviour. However, their expression differs from Eq. (5.59), in which they do not have the factor ½ in the last term inside the bracket on the RHS.

Extensions to the free energy, e.g. the interpolation expression by Hino and Prausnitz (1998), can easily be incorporated within the framework of the model for uniform as well as non-uniform deformations, provided that the free energy can be expressed in terms of the invariants of the deformation.

The effective stress tensor in Eq. (5.59) is defined in terms of the reference state (Ω_{ref}), as shown in Fig. 5.26, whence the deformation of the hydrogel requires the deformation from this state. For the present purposes, however, we are not primarily interested in the reference state, but rather in the deformation of a hydrogel from an arbitrary equilibrium state to another, without having to return to the reference frame, i.e. to the state of polymerization of the hydrogel. In other words, we would like to rephrase the stress tensor $\sigma_{\text{eff}}^{(p)}$ in terms of the initial configuration. This can be accomplished by invoking a uniform deformation from the reference (Ω_{ref}) to the initial frame (Ω_0), with $\mathbf{F}_0 = J_0^{1/3}\mathbf{I}$, where $J_0 = \det \mathbf{F}_0 = \phi_{ref}/\phi_0$ is the volume ratio. The deformation gradient tensor $\mathbf{F}_{\text{ref}}$ can then be expanded in terms of the initial configuration as $\mathbf{F}_{ref} = J_0^{1/3}\mathbf{F}$, whence

$$\sigma_{\text{eff}}^{(p)} = \frac{k_B T}{V_m N_x J}\left[J_0^{-\frac{1}{3}}\mathbf{b} - \frac{1}{2}J_0^{-1}\mathbf{I}\right] \tag{5.60}$$

where $J_{\text{ref}}=J_0 J$ is employed. Henceforth, it is required only to consider the current (Ω) and the initial (Ω_0) configurations, provided that the initial conditions originate from a uniform deformation from the reference state. One more advantage of this approach is that the reference state, commonly approximated as that of the polymerization, might require $\phi_{\text{ref}} \to 1$ for polymerizations in gas (Hirotsu, 1993), for which the osmotic pressure $p^{(\text{osm})} \to \infty$ (defined later in Eq. (5. 75)).

For later use, it is noted that the Lagrangian counterpart in the initial configuration of the Cauchy stress is given by the first Piola–Kirchhoff stress tensor and can be found from the Piola transformation (Holzapfel, 2000) as the following:

$$\mathbf{P}_{\text{eff}}^{(p)} = J\sigma_{\text{eff}}^{(p)}\mathbf{F}^{-T} \tag{5.61}$$

Deformation gradient tensors. For a spherical hydrogel with a uniform swelling, the deformation can be expressed as $r=r(R_{\text{ref}})$, $\theta=\Theta_{\text{ref}}$, $\phi=\Phi_{\text{ref}}$ due to the spherical symmetry, whence the deformation gradient tensor (Chung et al., 1986; Ogden, 1997) reduces to

$$\mathbf{F}_{\text{ref}} = r'\mathbf{e}_r \otimes \mathbf{e}_{R_{\text{ref}}} + \frac{r}{R_{\text{ref}}}\mathbf{e}_\theta \otimes \mathbf{e}_{\Theta_{\text{ref}}} + \frac{r}{R_{\text{ref}}}\mathbf{e}_\varphi \otimes \mathbf{e}_{\Phi_{\text{ref}}} \tag{5.62}$$

and volume ratio

$$J_{\text{ref}} = \det \mathbf{F}_{\text{ref}} = \frac{\phi_{\text{ref}}}{\phi} = r'\frac{r^2}{R_{\text{ref}}^2}. \tag{5.63}$$

Here $[\cdot]'$ denotes differentiation with respect to R_{ref}. Equation (5.63) can be integrated once, which yields the relation between the deformed and reference radii at equilibrium as

$$r(R_{\text{ref}}) = J_{\text{ref}}^{\frac{1}{3}}R_{\text{ref}} \tag{5.64}$$

For a hydrogel slab, we will consider the case where the hydrogel is constrained in the out-of-plane direction ($\mathbf{e}_{Z_{\text{ref}}}$,$\mathbf{e}_Z$,$\mathbf{e}_z$), while being able to slide freely, which allows us to reduce the geometry from three dimensions to two. This assumption is in line with the experiments and modelling conducted by Achilleos et al. (2000a). The deformation gradient tensor then reduces to

$$\mathbf{F}_{\text{ref}}=\frac{\partial x}{\partial X_{\text{ref}}}\mathbf{e}_x\otimes\mathbf{e}_{X_{\text{ref}}}+\frac{\partial x}{\partial Y_{\text{ref}}}\mathbf{e}_x\otimes\mathbf{e}_{Y_{\text{ref}}}+\frac{\partial y}{\partial X_{\text{ref}}}\mathbf{e}_y\otimes\mathbf{e}_{X_{\text{ref}}}+\frac{\partial y}{\partial Y_{\text{ref}}}\mathbf{e}_y\otimes\mathbf{e}_{Y_{\text{ref}}}+\frac{\partial z}{\partial Z_{\text{ref}}}\mathbf{e}_z\otimes\mathbf{e}_{Z_{\text{ref}}} \tag{5.65}$$

For the case (iii), we further impose the constraint $\mathbf{u}=\mathbf{0}$ at $\mathbf{x}=\mathbf{0}$, in order to prevent a translational movement of the hydrogel in the temperature gradient.

Heat conductivity. For the polymer, an isotropic heat conductivity is assumed as a first approximation, i.e.

$$\mathbf{k}_p=k_p\mathbf{I} \tag{5.66}$$

where k_p is the heat conductivity of the polymer phase.

Internal energy and enthalpy. For the incompressible polymer, it is assumed (Godovsky, 1992) that

$$\Xi^{(p)} \approx \Theta^{(p)} \tag{5.67}$$

and further that

$$d\Theta^{(p)} \approx C_p^{(p)}dT \tag{5.68}$$

Total mass of polymer. For later verification purposes, it is noted that the total mass of the polymer in the hydrogel will remain constant during the deformation, i.e. at any given time, the following Eq. (5.69) has to be satisfied

$$\int\limits_{v(t)} \phi dV = \int\limits_{v_{\text{ref}}} \phi_{\text{ref}} dV_{\text{ref}} \tag{5.69}$$

Mixture

Free energy of mixing. The free energy for the hydrogel is usually taken to follow the framework of Flory–Huggins theory (Flory, 1953), accounting for the entropy and energy changes of the mixture. The free energy functional (Onuki, 1989) of the system can be written as

$$F[\phi,T] = \int \left[f(\phi,T) + \frac{\gamma}{2}(\nabla\phi)^2\right]d\mathbf{x} \tag{5.70}$$

where $f(\phi,T)$ is the free energy per volume of the mixture, $F[\phi,T]$ is the Gibbs free energy and γ is a material parameter. For a neutral polymer, the following form for the free energy (Oh and Bae, 1998) per volume is assumed:

$$f(\phi,T) = \frac{k_BT}{V_m}\left[(1-\phi)\ln(1-\phi) + \phi\int_{\phi\prime}^{1} \chi(T,\phi\prime)d\phi\prime\right] \tag{5.71}$$

with $\chi(T,\phi)$ is the polymer–solvent interaction parameter. Several expressions for the interaction $\chi(T,\phi)$ were derived, e.g. by Oh and Bae (1998), Shirota et al. (1998) and Hirotsu (1991), who employed the expressions for the interaction $\chi(T,\phi)$, which requires three adjustable model parameters. We choose to define the interaction parameter (Shirota et al., 1998; Hirotsu, 1991) as

$$\chi(T,\phi) = -\frac{\Delta\Gamma}{k_B} + \frac{\Delta\Theta}{k_B T} + \chi_2\phi \tag{5.72}$$

where $\Delta\Gamma$ and $\Delta\Theta$ are the entropy and enthalpy contributions to the polymer–solvent interaction parameter, respectively, and χ_2 is a less well-defined parameter to express the volume fraction dependence. In addition to the temperature and volume fraction dependence, Kato (1997) determined experimentally that the interaction parameter also hinges on the hydrostatic pressure. The hydrostatic pressures that Kato measured, however, were of the order of 10^6 N/m^2 and beyond, i.e. larger than the atmospheric pressure considered here, whence we only consider temperature and volume fraction dependence.

Osmotic pressure. Araki and Tanaka (2001) related the divergence of the osmotic pressure to the free energy functional, which for the gradient of the osmotic pressure can be rewritten as

$$\nabla p^{(\mathrm{osm})} = \phi\nabla\frac{\delta F[\phi,T]}{\delta\phi} \tag{5.73}$$

from which the osmotic pressure can be found as

$$p^{(\mathrm{osm})} = \left[\phi\frac{\partial f}{\partial\phi} - f - \phi\gamma\nabla^2\phi + \frac{\gamma}{2}(\nabla\phi)^2\right] \tag{5.74}$$

By combining the osmotic pressure with the free energy per volume, Eq. (5.71), we finally arrive at

$$p^{(\mathrm{osm})} = \left[-\frac{k_B T}{V_m}(\phi + \chi\phi^2 + \ln(1-\phi)) - \gamma(\phi\nabla^2\phi - \frac{1}{2}(\nabla\phi)^2)\right] \tag{5.75}$$

Friction between polymer and liquid phases. The friction force between the liquid and polymer phases is given by Holmes (1986) as

$$\pi^{(p)} = -\pi^{(f)} = p\nabla\phi - \zeta(\mathbf{v}^{(p)} - \mathbf{v}^{(f)}) \tag{5.76}$$

where ζ is the friction coefficient, also referred to as the diffusive drag coefficient, which is defined (Holmes et al., 1985) as

$$\zeta = \frac{\mu^{(f)}(1-\phi)^2}{\kappa} \tag{5.77}$$

where κ is the hydrodynamic permeability of the hydrogel. It is noted that we retain the dynamic viscosity in the friction coefficient explicitly in Eq. (5.77). As we shall find later, the definition of ζ in Eq. (5.77), and of the friction force as a linear function of the relative velocity between the two phases, Eq. (5.76) will allow us to recover Darcy's law for a stationary solid from the momentum equations. Higher order terms, such as the quadratic dependence on velocity in the Darcy–Forchheimer equation (Nield and Bejan, 1998), are negligible due to the low velocities inside the hydrogels, as seen later in the analysis.

Permeability. The permeability of the hydrogel is important for the deformation kinetics as it determines the resistance that the penetrating solvent experiences inside the polymer network. A low permeability gives rise to slow deformation, whereas a highly permeable network can adapt to perturbations in the environment more easily. The former is typically the case for most hydrogels, e.g. for PNIPA gel. The latter is of interest in any fast-response-dependent applications, such as artificial organs (Zhang and Zhuo, 2000a, b). A typical response rate for an ordinary PNIPA gel is around 1 hour or more, in stark contrast to a few minutes for fast-responsive hydrogels (Zhang and Zhuo, 2000a, b; Chen et al., 1999). Two important parameters affecting the permeability are the degree of crosslinking and pore size in the polymer network. Experiments on the permeability of polyacrylamide hydrogels by Tokita and Tanaka (1991) and Grattoni et al. (2001) have shown that the permeability to flow through the gel can be described by a power law relation $\kappa \propto \phi^{-3/2}$ and that the friction factor is related to the dynamic viscosity; the latter is included in the definition of the friction factor, as described in Eq. (5.77). In general, however, the constitutive relation for the permeability can be expected to be more complex and highly dependent on degree of crosslinking, as the studies on soft permeable tissues and gels have shown (Holmes et al., 1985; Holmes and Mow, 1990). Grattoni et al. (2001) also found that the permeability is velocity dependent. For the present purposes, it is noted that the permeability is a function of the volume fraction of polymer ϕ, which is assumed to follow an exponential function for the slow-response hydrogels AS15 and AS20, i.e.

$$\kappa = \iota_1 \phi^{-3/2} \tag{5.78}$$

and neglect the contribution from the velocity dependence. Here, ι_1 is a parameter which requires adaptation to experimental kinetic data. For the fast-response hydrogels, G0 and NC030, the volume fraction spans several magnitudes of volume fraction, down to $O(10^{-2})$. In light of the low polymer volume fraction in the macroporous hydrogels, a constant permeability is prescribed, i.e.

$$\kappa = \iota_1 \tag{5.79}$$

As shown later, the assumption of constant permeability, due to the low volume fraction of polymer, gives good agreement with the experiments, which is expected since a lower volume fraction leads to less fluid–solid interaction. For the conventional poly(*N*-isopropylacrylamide) gel NC000, a constant permeability is considered as a first approximation similar to the macroporous counterparts, as the dependence of the polymer volume fraction in Eq. (5.78) cannot account for low polymer volume fractions, i.e. as $\phi \to 0$. It is possible to define composite expressions for the permeability of NC000 as a function of the polymer volume fraction similar to Barrière and Leibler (2003), but we refrain from doing so as we are not primarily interested in NC000. On a final note, we could also have defined the friction coefficient in analogy with the multicomponent mass transfer theory (Taylor and Krishna, 1993; Wesselingh and Krishna, 2000), which prescribes an exponential function for the diffusion coefficient in polymer solutions, based on the free volume theory (Wesselingh and Bollen, 1997; Wesselingh and Krishna, 2000). However, in view of the explicit appearance of the dynamic viscosity in experiments (Tokita and Tanaka, 1991; Grattoni et al., 2001), the hydrodynamic analogue, Eq. (5.77), seems more suitable.

Effective thermal properties. The effective heat conductivity of the hydrogel comprises two parts. One originates from the heat conduction in the penetrating liquid and the other from conduction through the polymer. The heat conduction tensor **k** may depend on several of the state variables, such as the temperature and the deformation, e.g. the Finger tensor **b**. For example, a non-uniform deformation could lead to an orientation of the polymer backbone, accompanied by a change of heat conductivity in the direction of and perpendicular to the elongation. While the most general form can be expected to be a function of these, an additive function for the total effective heat conductivity is defined as

$$\mathbf{k} = \phi \mathbf{k}^{(p)} + (1 - \phi)\mathbf{k}^{(f)} \tag{5.80}$$

which is consistent with the earlier definitions of the apparent density and mixture velocity. Similarly, for the effective heat conductivity, we prescribe

$$C_p = \phi C_p^{(p)} + (1 - \phi) C_p^{(f)} \tag{5.81}$$

Total mixture stress tensor. As found later, it is convenient to introduce the total mixture stress tensor, defined as

$$\sigma = -p\mathbf{I} + \sigma_{\text{eff}}^{(p)} \tag{5.82}$$

Swelling ratio. The swelling ratio, ξ, of the hydrogel is given by

$$\xi = \frac{m^{(f)}}{m^{(p)}} = \frac{\int_v (1 - \phi)\rho_0^{(f)} dV}{\int_v \phi \rho_0^{(p)} dV} \tag{5.83}$$

which for a uniform deformation at equilibrium reduces to $\xi_\infty = (\phi_\infty^{-1} - 1)\rho_0^{(f)} / \rho_0^{(p)}$. Here, $m^{(f,p)}$ is the total weight for the solvent and polymer phases, respectively. The total mass of the dry hydrogel is usually measured after vacuum drying, whence there is most likely some residual water left in the hydrogel. However, the mass of the dry gel is assumed to be equal to the mass of the polymer.

Water retention. The water retention, ψ, for the hydrogel is defined as

$$\psi = \frac{m^{(f)}}{m_0^{(f)}} = \frac{\int_v (1-\phi)dV}{\int_{v_0} (1-\phi_0)dV_0} \tag{5.84}$$

5.6.1.4 Boundary and Initial Conditions

As illustrated in Figs. 5.23, 5.24 and 5.25, the boundary and initial conditions for the three cases (i–iii) are the following:

(i) *Uniform spherical deformation*:

Symmetry. For (r=0, t>0):

$$\frac{\partial T}{\partial r} = 0, \;\; u_r = v_r = 0 \tag{5.85}$$

Free surface. For (r=R(t), t>0):

$$\sigma_{rr} = -p_\infty, \;\; p^f = p_\infty, \;\; T = T_\infty \tag{5.86}$$

Initial conditions. For ($0 \le r \le R_0$, t=0):

$$p^f = p_\infty, \;\; u_r = 0, \;\; T = T_0 \tag{5.87}$$

where u_r is the radial displacement, v_r is the radial mixture velocity component and σ_{rr} is the component of the mixture Cauchy stress tensor in the radial direction.

(ii) *Non-uniform deformation*:

Symmetry (I):

$$\mathbf{u} \cdot \mathbf{n} = \mathbf{v} \cdot \mathbf{n} = \nabla T \cdot \mathbf{n} = (\mathbf{n} \cdot \nabla)(\mathbf{t} \cdot \mathbf{u}) = 0 \tag{5.88}$$

Free surface (II):

$$\sigma \cdot \mathbf{n} = (-p_\infty \mathbf{I}) \cdot \mathbf{n}, \;\; p^f = p_\infty, \;\; T = T_\infty \tag{5.89}$$

Initial conditions. For ($0 \le x \le L(0,0)$, $0 \le y \le H(0,0)$, t=0):

$$p^f = p_\infty, \;\; \mathbf{u} = \mathbf{0}, \;\; T = T_0 \tag{5.90}$$

(iii) *Deformation in a temperature gradient*:

Symmetry (I):

$$\mathbf{u} \cdot \mathbf{n} = \mathbf{v} \cdot \mathbf{n} = (\mathbf{n} \cdot \nabla)(\mathbf{t} \cdot \mathbf{u}) = 0 \tag{5.91}$$

Free surface (II):

$$\sigma \cdot \mathbf{n} = (- p_\infty \mathbf{I}) \cdot \mathbf{n}, \;\; p\prime = p_\infty \tag{5.92}$$

Point constraint. For (x=0,y=0):

$$\mathbf{u} = \mathbf{0} \tag{5.93}$$

Initial conditions. For ($-0.5L(0,0) \leq x \leq 0.5L(0,0)$, $0 \leq y \leq H(0,0)$, t=0):

$$p\prime = p_\infty, \;\; \mathbf{u} = \mathbf{0}, \;\; T = T_0 \tag{5.94}$$

The prescribed temperature gradient is given by $T_\infty(x) = -(T_1 - T_2)/Lx + (T_1 + T_2)/2$.

It is noted that the time-dependent boundary conditions in the Eulerian frame require to keep track of the motion of the interface as the hydrogel deforms. Later, the model in Lagrangian coordinates will be formulated at the initial state, whence it no longer explicitly has to account for the motion of the surface.

5.6.2 Analysis

In an effort to elucidate the relative importance of the mechanisms for the momentum transfer due to deformation of the hydrogel and convection of the solvent as well as mass and heat transfer a priori to computational simulations, we shall proceed by studying the relevant scales and non-dimensionless numbers for the hydrogel model. In doing so, we will not only provide some of the means by which we can interpret the computational results but also allow for simplifications of the governing equations.

5.6.2.1 Non-dimensionalization

One can introduce

$$\tilde{\mathbf{x}} = \frac{\mathbf{x}}{l}, \tilde{\mathbf{X}} = \frac{\mathbf{X}}{l}, \tilde{\mathbf{v}} = \frac{\mathbf{v}}{U}, \tilde{\mathbf{v}}^{(f)} = \frac{\mathbf{v}^{(f)}}{U_f}, \tilde{\mathbf{v}}^{(p)} = \frac{\mathbf{v}^{(p)}}{U_p}, \tilde{t} = \frac{t}{[t]}, \tilde{p} = \frac{p - p_\infty}{\Delta p}$$

$$\tilde{p}^{(\mathrm{osm})} = \frac{p^{(\mathrm{osm})}}{[p^{(\mathrm{osm})}]}, \tilde{p}\prime = \frac{p\prime - p_\infty}{\Delta p}, \tilde{T} = \frac{T - T_\infty}{\Delta T}, \tilde{\zeta} = \frac{\zeta}{[\zeta]}, \tilde{\sigma}^{(p)}_{\mathrm{eff}} = \frac{\sigma^{(p)}_{\mathrm{eff}}}{[\sigma^{(p)}_{\mathrm{eff}}]}, \tilde{\mathbf{k}} = \frac{\mathbf{k}}{[k]}$$

$$\tilde{k}^{(f)} = \frac{k^{(f)}}{[k]}, \tilde{\mu}^{(f)} = \frac{\mu^{(f)}}{[\mu^{(f)}]}, \tilde{\kappa} = \frac{\kappa}{[\kappa]}, \tilde{\mathbf{E}}^{(f)} = \frac{\mathbf{E}^{(f)} l}{U}, \tilde{\pi} = \frac{\pi}{[\pi]},$$

where [$\cdot$] represents a typical scale, l is a representative length scale of the system, ΔT=$T_0 - T_\infty$ and Δp are scales for the temperature and pressure drop, respectively. At this stage, U, U_f, U_p, [t], [ζ], [π], [k], [$\sigma^{(p)}_{\mathrm{eff}}$], [$p^{(\mathrm{osm})}$] and Δp are still unknown.

Furthermore, it is necessary to consider two different time scales: one for the deformation of the hydrogel, $[t_{\mathrm{def}}]$, and the other for the heat transfer, $[t_{\mathrm{heat}}]$. Before proceeding to write down the non-dimensionalized equations, we will attempt to identify some of the scales for the deformation of a hydrogel not subject to an external flow field, i.e., all scales originate from the deformation of the hydrogel itself.

Returning to the definition of the volume-averaged velocity, Eq. (5.38), one can obtain $U{=}U_f{=}U_p$ in order to balance the volume-averaged and phase velocities, since ϕ and $(1{-}\phi)$ can be considered $O(1)$ quantities. The time scale for the deformation can then conveniently be chosen as $[t_{\mathrm{def}}]{=}l/U$, since the deformation of the hydrogel is the present primary concern, thus set $[t]{=}[t_{\mathrm{def}}]$. The scale for the polymer stress tensor can be secured from Eq. (5.60) as $[\sigma_{\mathrm{eff}}^{(p)}] = k_B T_\infty J_\infty^{-1}/(V_m N_x)$, where the volume ratio is included at the new equilibrium $J_\infty{=}\phi_{\mathrm{ref}}/\phi_\infty$, originating from the second term in the bracket on the RHS of Eq. (5.60), in order to account for changes in the volume fraction during deformation. It is noted that the scale for the stress is linked to the uniform deformation and as such cannot give any information on any shear stresses that might arise. In addition, the constitutive relations for the friction coefficient and permeability, Eqs. (5.77), (5.78) and (5.79) give $[\zeta] = [\mu^{(f)}]/[\kappa]$, $[\kappa] = \phi_0^{-3/2}\iota_1$ for AS15 and AS20, $[\kappa]{=}\iota_1$ for NC000, NC030 and G8, respectively. The initial polymer volume fraction, ϕ_0, is included for the two slow-response hydrogels, AS15 and AS20, to account for the dependence of the permeability on the polymer volume fraction. For the polymer–solvent interaction parameter, χ, it is assumed and verified later that it is $O(1)$. Further, the polymer phase heat conductivity is chosen as the typical scale, i.e. $[k]{=}k^{(p)}$ and Eq. (5.76) provides $[\pi]{=}[\zeta]U$.

Finally, the following dimensionless numbers are introduced:

$$\mathrm{Re} \equiv \frac{Ul\rho_0^{(p)}}{[\mu^{(f)}]}, \mathrm{Pr} \equiv \frac{C_p^{(p)}[\mu^{(f)}]}{[k]}, \mathrm{Br}^{(f)} \equiv \frac{[\mu^{(f)}]U^2}{[k]\Delta T}, \mathrm{Br}^{(p)} \equiv \frac{[\sigma_{\mathrm{eff}}^{(p)}]Ul}{[k]\Delta T}, \Pi \equiv \frac{[\zeta]U^2 l^2}{[k]\Delta T},$$

$$\lambda \equiv \frac{p_\infty Ul}{[k]\Delta T}, \quad \Lambda \equiv \frac{\gamma V_m}{k_B T_\infty l^2}, \quad T^* \equiv \frac{\Delta T}{T_\infty}, \quad P^* \equiv \frac{\Delta p}{p_\infty}, \quad C^* \equiv \frac{\rho_0^{(f)} C_p^{(f)}}{\rho_0^{(p)} C_p^{(p)}},$$

where Re is the Reynolds number, Pr is the Prandtl number and $\mathrm{Br}^{(i)}$ is the Brinkman number for the phase i. The Reynolds number measures the ratio of inertial force to viscous force, the Prandtl number is a material property dimensionless number, and the Brinkman number gives the ratio of the dissipation to the heat conduction. Π and λ are associated with the irreversible and reversible rates of internal energy change to the heat conduction, respectively. Λ measures the ratio of the energy required to maintain sharp gradients in the hydrogel to the magnitude of free energy per volume. T^* and P^* are associated with the temperature and pressure changes to the overall temperature and pressure, respectively, and C^* gives the ratio of the liquid density and heat capacity to the corresponding ones in the polymer phase.

Introducing these scales and non-dimensional numbers into the governing equations and constitutive relations and dropping the tildes yields

$$\frac{\partial \phi}{\partial t} + \nabla \cdot (\phi \mathbf{v}) = -\nabla \cdot [\phi(1-\phi)(\mathbf{v}^{(p)} - \mathbf{v}^{(f)})] \tag{5.95}$$

$$\nabla \cdot \mathbf{v} = 0 \tag{5.96}$$

$$\left\{\frac{\rho_0^{(p)} U^2}{l}\right\} \phi \frac{D^{(p)}\mathbf{v}^{(p)}}{Dt} = \left\{\frac{[\sigma_{\text{eff}}^{(p)}]}{l}\right\} \nabla \cdot \sigma_{\text{eff}}^{(p)} - \left\{\frac{\Delta p}{l}\right\} \phi \nabla p - \{[\zeta]U\}\zeta(\mathbf{v}^{(p)} - \mathbf{v}^{(f)}) \tag{5.97}$$

$$\left\{\frac{\rho_0^{(f)} U^2}{l}\right\} (1-\phi) \frac{D^{(f)}\mathbf{v}^{(f)}}{Dt} = -\left\{\frac{\Delta p}{l}\right\} (1-\phi)\nabla p + \left\{\frac{[\mu^{(f)}]U}{l^2}\right\} \nabla \cdot (2\mu^{(f)}\mathrm{E}^{(f)}) + \{[\zeta]U\}\zeta(\mathbf{v}^{(p)} - \mathbf{v}^{(f)}) \tag{5.98}$$

$$\begin{aligned} &\{\mathrm{Re}\,\mathrm{Pr}\}\left[\phi \frac{D^{(p)}T}{Dt} + C^*(1-\phi)\frac{D^{(f)}T}{Dt}\right] = \{1\}\nabla \cdot (k\nabla T) \\ &-\{\lambda\}(P^*p+1)[\phi\nabla \cdot \mathbf{v}^{(p)} + (1-\phi)\nabla \cdot \mathbf{v}^{(f)}] \\ &\quad + \{\mathrm{Br}^{(p)}\}\sigma^{(p)}{:}\nabla \mathbf{v}^{(p)} + \{\mathrm{Br}^{(f)}\}2\mathrm{E}^{(f)}{:}\nabla \mathbf{v}^{(f)} + \{\Pi\}(\mathbf{v}^{(f)} - \mathbf{v}^{(p)}) \cdot \pi \end{aligned} \tag{5.99}$$

The dimensional quantities and non-dimensional numbers are given for convenience in brackets {}, which will be useful for determination of the remaining unknown scales, namely U, $[p^{(\mathrm{osm})}]$, Δp and $[t_{\mathrm{heat}}]$.

5.6.2.2 Time Scales

A temperature-sensitive neutral hydrogel, which deforms due to an environmental temperature perturbation, experiences simultaneous heat and mass transfer, whence it is necessary to consider two time scales. The time scale for the deformation of the hydrogel has been found as

$$[t_{\mathrm{def}}] \sim \frac{l}{U}.$$

The time scale for the heat transfer can be estimated by returning to the dimensional energy equation and definitions of the internal energies and enthalpies, Eqs. (5.46), (5.53), (5.54), (5.67) and (5.68) and relevant scales, from which it can be inferred that the transient terms have to balance either with the convective or with the conductive energy transfer, i.e.

$$\rho_0^{(p)} C_p^{(p)} \frac{\Delta T}{[t_{\mathrm{heat}}]} \sim \max\left(\frac{[k]\Delta T}{l^2}, \rho_0^{(p)} C_p^{(p)} \frac{U\Delta T}{l}\right)$$

providing that the remaining terms are all much smaller than 1 (<< 1). It will be shown later that this condition is indeed satisfied. Furthermore, $[t_{\mathrm{heat}}]$ cannot exceed $[t_{\mathrm{def}}]$ as long as the convective energy transfer originates only from the deformation of the hydrogel itself, whence two cases can be identified for the relative magnitudes of the time scales:

$$[t_{\mathrm{heat}}] \sim \min\left(\frac{\rho_0^{(p)} C_p^{(p)} l^2}{[k]}, [t_{\mathrm{def}}]\right).$$

When the conductive heat transfer is faster than the deformation, i.e. $[t_{\mathrm{heat}}]<[t_{\mathrm{def}}]$, it follows that $[t_{\mathrm{heat}}] \sim \rho_0^{(p)} C_p^{(p)} l^2/[k]$. Therefore, thermal equilibrium is reached faster than swelling equilibrium, which is in line with experimental findings for PNIPA by Gehrke (1993), who studied heat and mass transfer experimentally in a PNIPA gel with a thermocouple. For $[t_{\mathrm{heat}}]\sim[t_{\mathrm{def}}]$, heat transfer and deformation will be of the same order, which occurs for hydrogels with a sufficiently fast deformation response rate.

5.6.2.3 Momentum Balances

Before proceeding with the identification of the remaining scales, U, $[p^{(\mathrm{osm})}]$ and Δp, it is instructive to recall that the deformation of hydrogels is considered to be induced by a change in temperature, which manifests itself as a driving force in the Cauchy stress tensor and osmotic pressure. It is expected that the low permeability of the hydrogels, $\iota_1 \sim 10^{-16} - 10^{-19}\mathrm{m}^2$ (see Table 5.2), will lead to $U \ll 1\,\mathrm{ms}^{-1}$. Now, close to the swelling equilibrium of the hydrogel, the stress is expected further to balance with the osmotic pressure, $O(p^{(\mathrm{osm})}) \sim O(\sigma_{\mathrm{eff}}^{(p)})$, whence $[p^{(\mathrm{osm})}] \sim [\sigma_{\mathrm{eff}}^{(p)}] = k_B T_\infty J_\infty^{-1}/(V_m N_x)$. As such, it is noted that the deformations originating from an applied pressure are not considered due to an external flow or externally applied loading. This in turn implies that $\Delta p \sim [p^{(\mathrm{osm})}]$. In light of this, the relevant scale for the velocity can be inferred from Eq. (5.97), by balancing the velocity term with either one of the first two terms on the RHS. Choosing the pressure term results in

$$[\zeta]U \sim \frac{[p^{(\mathrm{osm})}]}{l}$$

whence (*N.B.* $[\zeta] = [\mu^{(f)}]/[\kappa]$)

$$U = \frac{[p^{(\mathrm{osm})}][\kappa]}{l[\mu^{(f)}]} = \frac{k_B T_\infty [\kappa] J_\infty^{-1}}{[\mu^{(f)}] l N_x V_m} << 1\,\mathrm{m/s}^1$$

for the parameters considered (see Tables 5.1, 5.2 and 5.3). Had we instead chosen to scale the velocity with the inertial term on the LHS, we would have obtained $U = [\zeta] l/\rho_0^{(p)} >> 1\mathrm{m/s}^1$; a velocity, which is physically unreasonable given the low permeability of the hydrogels.

Table 5.1 Physical parameters

Physical parameters	
k_B	1.38054×10^{-23} J/K
V_m	3.3×10^{-28} m^3
R	8.314 J/mol/K
Physical parameters (liquid phase)	
c_1	-1.05×10^{-5} W/m/K^3
c_2	7.98×10^{-3} W/m/K^2
c_3	-8.38×10^{-1} W/m/K
$\rho^{(f)}$	10^3 kg/m^3
$C_p^{(f)}$	4.2×10^3 J/kg/K (Bird et al., 2002)
Physical parameters (polymer phase)	
$K^{(p)}$	2 W/m/K (estimated from Eq. (5.80) and Prokop et al., 2003)
$C_p^{(p)}$	2×10^4 J/kg/K (estimated from Eq. (5.81), Prokop et al. (2003) and Sanchez et al., 2004)
$\rho^{(p)}$	1.1×10^3 kg/m^3 (assumed for NC000, NC030, G8) 1.2×10^3 kg/m^3 (AS15, AS20 from Bae et al. (1989))

Table 5.2 Mixture properties

Physical parameters (mixture)	AS15	AS20	NC000	NC030	G8
$\Delta\Gamma(\times10^{-23}$ J K$^{-1})$	−3.50	−4.40	−3.50	−3.50	−3.71
$\Delta\Theta(\times10^{-21}$ J$)$	−8.11	−10.5	−8.67	−8.67	−9.20
$N_x(\times10^2)$	1.09	0.945	80.7	216	621
$\chi_2(\times10^{-1})$	0.889	1	3	3	0
$\phi_{ref}(\times10^{-2})$	99	99	6	1.73	0.454
$\iota_1(\times10^{-19}$ m$^2)$	1.2	1.5	40	800	1700
$\phi_0(\times10^{-1})$	3.4 (27°C) 4.2 (37°C)	4.5 (27°C) 5.4 (37°C)	0.25 (20°C) 4.7 (52°C)	0.15 (20°C) 4.6 (52°C)	0.078 (20°C) 2.6 (48°C)
$l_0(\times10^{-3}$ m$)$	0.97 (27°C) 0.90 (37°C)	0.89 (27°C) 0.84 (37°C)	3.9 (20°C) 1.5 (52°C)	3.8 (20°C) 1.2 (52°C)	3.2 (20°C) 0.98 (48°C)

Table 5.3 Geometry and operating conditions

Geometry and operating conditions	
P_∞	101.325×10^3 N/m^2
Case (i)	$R_0=l_0$ (see Table 5.2)
Case (ii)	$L=1\times10^{-3}$ m, H=1×10^{-3} m
Case (iii)	$L=1\times10^{-3}$ m, H=2×10^{-3} m, W=2×L $T_0=30$°C, $T_1=20$°C, $T_2=40$°C

Inserting the stress and pressure scales into Eq. (5.97), and noting that $\rho_p^0 U/([\zeta]l) << 1$, yield the reduced form

$$\zeta(\mathbf{v}^{(p)} - \mathbf{v}^{(f)}) = \nabla \cdot \sigma_{eff}^{(p)} - \phi \nabla p \tag{5.100}$$

The present discussion has so far been limited to the momentum equation for the polymer phase, Eq. (5.97). As seen later, however, the scales U and Δp are indeed consistent with the fluid-phase momentum equation, Eq. (5.98). Substituting U and Δp into Eq. (5.98) and dividing by $\{[\zeta]U\}$ render the inertia $(\rho_f^0 U/([\zeta]l) << 1)$ and viscous term $([\mu^{(f)}]/([\zeta]l^2) << 1)$ negligible, compared with the pressure gradient and friction, such that one can write Eq. (5.98) as

$$\zeta(\mathbf{v}^{(p)} - \mathbf{v}^{(f)}) = (1 - \phi)\nabla p \tag{5.101}$$

which upon introduction of the mixture velocity, reduces to

$$\nabla p = \frac{\mu^{(f)}}{\kappa}(\mathbf{v}^{(p)} - \mathbf{v}) \tag{5.102}$$

It is noted that, with the present definition of the friction factor, Eq. (5.77), Eq. (5.102) simplifies to Darcy's law for a stationary polymer phase $(\mathbf{v}^{(p)}=\mathbf{0}, \mathbf{v}=(1-\phi)\mathbf{v}^{(f)})$, since the mixture velocity $\mathbf{v}$ then corresponds to the fluid superficial velocity, as required by Darcy's law (Whitaker, 1999). By combining Eq. (5.100) with (5.101), Eq. (5.100) can be expressed solely in terms of the gradient of the polymer stress tensor, i.e.

$$\mathbf{v}^{(p)} - \mathbf{v} = \frac{\kappa}{\mu^{(f)}} \nabla \cdot \sigma_{\text{eff}}^{(p)} \tag{5.103}$$

Finally, by adding Eqs. (5.102) and (5.103), the mixture stress for the hydrogel can be written to behave as

$$\nabla \cdot \sigma = 0 \tag{5.104}$$

5.6.2.4 Parameters and Non-dimensional Numbers

The underlying physical, geometrical and operating conditions for the studied hydrogels can be found in Tables 5.1, 5.2 and 5.3. Typical scales for the hydrogels considered here include $l \sim 10^{-3}$ m, $\Delta T \approx 10°\text{C} \sim 50°\text{C}$, $[p^{(\text{osm})}] \sim [\sigma_{\text{eff}}^{(p)}] \approx 10^4 - 10^5/\text{Nm}^{-2}$, $[\mu^{(f)}] \approx 10^{-3}\text{kg/m/s}$, $U \approx 10^{-6} - 10^{-8}\text{m/s}$. Thence for the non-dimensional numbers, one arrives at

$$\begin{aligned} &\text{Re} \le 10^{-3}, \text{Pr} \sim 1, \text{B}r^{(f)} \le 10^{-18}, \text{B}r^{(p)} \le 10^{-7}, \Pi \le 10^{-6} \\ &\lambda \le 10^{-6}, \Lambda \sim 10^{-1}, T^* \sim 10^{-1}, P^* \le 1, C^* \sim 1 \end{aligned}$$

Furthermore, the interaction parameter

$$\chi = -\frac{\Delta\Gamma}{k_B} + \frac{\Delta\Theta}{k_B T_\infty (T^* T + 1)} + \chi_2 \phi \sim 1,$$

as assumed from the onset of the analysis. The non-dimensional osmotic pressure now reads

$$p^{(\mathrm{osm})} = -N_x J_\infty (T^* T + 1) \left[(\phi + \chi\phi^2 + \ln(1-\phi)) + \Lambda \left(\phi \nabla^2 \phi - \frac{1}{2} (\nabla\phi)^2 \right) \right] \tag{5.105}$$

In the following subsections, it will be assumed that Λ<<1, since no experimental data is available for the energy required to maintain sharp gradients.

5.6.2.5 Mass Balance

The dimensionless conservation of polymer mass now reads

$$\frac{\partial\phi}{\partial t} + \nabla \cdot (\phi \mathbf{v}) = -\nabla \cdot [\phi(1-\phi)(\mathbf{v}^{(p)} - \mathbf{v}^{(f)})] \tag{5.106}$$

and the total mixture mass balance is given by

$$\nabla \cdot \mathbf{v} = 0 \tag{5.107}$$

5.6.2.6 Energy Balance

The dissipative terms are negligible in the fluid and polymer phases, since $\mathrm{Br}^{(f,p)}$, Π<<1, as well as the reversible energy transfer, given by λ<<1. In other words, any heat dissipation stemming from the deformation or the friction between the penetrating fluid and the polymer is small compared with the overall heat transfer. The three primary mechanisms for heat transfer in the hydrogel are the convection in the fluid, the deformation of the polymer and the conduction, the ratio of which is captured by RePr:

- RePr<<1, for which the energy balance reduces to steady-state heat conduction

$$\nabla \cdot (k \nabla T) = 0 \tag{5.108}$$

- RePr∼1, for which it is necessary to consider both transient heat conduction and convection

$$\phi \frac{D^{(p)} T}{Dt} + C^* (1-\phi) \frac{D^{(f)} T}{Dt} = \frac{1}{\mathrm{Re}\,\mathrm{Pr}} \nabla \cdot (k \nabla T) \tag{5.109}$$

- RePr>>1, which would only require to solve for the heat convection in the bulk and conduction in boundary layers. This is unlikely, as it would require very fast-response hydrogels or an external velocity field, giving rise to a high convective flow inside the hydrogel, i.e. $U >> [k]/(\rho_0^{(p)} C_p^{(p)} l)$.

5.6.3 Model Formulation in Lagrangian Frame and Boundary and Initial Conditions

The governing equations have so far been formulated in an Eulerian frame. While this is a valid approach, it requires a moving mesh for the deformation even though external fields are not considered. A more suitable option is to recast the model in Lagrangian coordinates, namely the coordinates which deform with the hydrogel itself.

The continuity equation for the polymer phase in the spatial description, Eq. (5.106), reduces to

$$J = \det \mathbf{F} \tag{5.110}$$

in the corresponding material frame. We proceed by introducing the Piola identity $\nabla_X \cdot (J\mathbf{F}^{-T}) = 0$ and the transformation $\nabla_X \cdot \mathbf{P} = J\nabla \cdot \sigma$ (Holzapfel, 2000), for the right-hand side of the momentum equation, Eq. (5.104), which gives

$$\nabla_X \cdot \mathbf{P} = 0 \tag{5.111}$$

where the first Piola–Kirchhoff stress tensor of the mixture $\mathbf{P} = -Jp\mathbf{F}^{-T} + \mathbf{P}_{\text{eff}}^{(p)}$, and ∇_X denotes the gradient in the Lagrangian configuration. The second momentum equation and the conservation of the total mass, Eqs. (5.102) and (5.107), can be written in the form as

$$\mathbf{V} - \mathbf{V}^{(p)} = -\frac{\kappa}{\mu^{(f)}} J\mathbf{C}^{-1} \nabla_X p \tag{5.112}$$

$$\nabla_X \cdot \mathbf{V} = 0 \tag{5.113}$$

where $\mathbf{C}^{-1}$=$\mathbf{F}^{-1}$ $\mathbf{F}^{-T}$ is the inverse of the right Cauchy–Green tensor, $\mathbf{V}$=$J\mathbf{F}^{-1}$ $\mathbf{v}$ and $\mathbf{V}^{(p)}$=$J\mathbf{F}^{-1}$ $\mathbf{v}^{(p)}$ are the pull-backs of the mixture and polymer phase velocities, respectively. For the transformation, it is also used that the Eulerian gradient and the Lagrangian counterpart are related via the deformation gradient tensor, i.e. $\nabla p = \mathbf{F}^{-T} \nabla_X p$. The deformation of the hydrogel with time is given by $D\mathbf{U}^{(p)}/Dt = J^{-1}\mathbf{F}\mathbf{V}^{(p)} = \mathbf{v}^{(p)}$.

It is more convenient for the spherical deformation to express the relative velocity in Eq. (5.112), as a function of the stress tensor, as seen in Eq. (5.103), instead of the pressure gradient, i.e.

$$\mathbf{V} - \mathbf{V}^{(p)} = \frac{\kappa}{\mu^{(f)}} \mathbf{F}^{-1} \nabla_X \cdot \mathbf{P}_{\text{eff}}^{(p)} \tag{5.114}$$

Finally, the heat transfer in the hydrogel in the material configuration is given by

$$J[C^*(1-\phi)+\phi]\frac{DT}{Dt} = -\frac{1}{\mathrm{Re}\,\mathrm{Pr}}\nabla_X\cdot\mathbf{Q} + C^*\nabla_X T\cdot(\mathbf{V}^{(p)}-\mathbf{V}) \tag{5.115}$$

where $\mathbf{Q} = -kJ\mathbf{C}^{-1}\nabla_X T$ is the Piola–Kirchhoff conductive heat flux. The second term on the RHS originates from rewriting the material time derivative with respect to the fluid motion in terms of polymer motion.

The non-dimensional boundary and initial conditions in a Lagrangian frame for the three cases (i)–(iii) are given accordingly:

(i) *Uniform spherical deformation*:

Symmetry. For (R=0, t>0):

$$\frac{\partial T}{\partial R} = 0,\ U_R = V_R = 0 \tag{5.116}$$

Free surface. For (R=1, t>0):

$$P_{rR} = 0,\ p^f = 0,\ T = 0 \tag{5.117}$$

Initial conditions. For ($0\le R\le 1$, $t=0$):

$$p = 0,\ U_R = 0,\ T = 1 \tag{5.118}$$

(ii) *Non-uniform deformation*:

Symmetry (I):

$$\mathbf{U}\cdot\mathbf{N} = \mathbf{V}\cdot\mathbf{N} = \nabla_X T\cdot\mathbf{N} = (\mathbf{N}\cdot\nabla_X)(\mathbf{T}\cdot\mathbf{U}) = 0 \tag{5.119}$$

Free surface (II):

$$\mathbf{P}\cdot\mathbf{N} = \mathbf{0},\ p^f = 0,\ T = 0 \tag{5.120}$$

Initial conditions. For ($0\le X\le 1$, $0\le Y\le 1$, t=0):

$$p = 0,\ \mathbf{U} = \mathbf{0},\ T = 1 \tag{5.121}$$

In the Lagrangian frame, **N** and **T** are the outward unit vectors normal and tangential to the surface, respectively.

(iii) *Deformation in a temperature gradient*:

Symmetry (I):

$$\mathbf{U}\cdot\mathbf{N} = \mathbf{V}\cdot\mathbf{N} = (\mathbf{N}\cdot\nabla_X)(\mathbf{T}\cdot\mathbf{U}) = 0 \tag{5.122}$$

Free surface (II):

$$\mathbf{P} \cdot \mathbf{N} = \mathbf{0}, \ p\prime = 0 \tag{5.123}$$

Point constraint. For (X=0,Y=0):

$$\mathbf{U} = \mathbf{0} \tag{5.124}$$

Initial conditions. For ($-0.5 \leq X \leq 0.5$, $0 \leq Y \leq 2$, t=0):

$$p = 0, \ \mathbf{U} = \mathbf{0} \tag{5.125}$$

5.6.4 Numerical Implementation

The final model, given by Eqs. (5.111), (5.112) or (5.114) and (5.115), together with appropriate boundary and initial conditions as well as constitutive relations for the cases (i–iii), constitutes a highly coupled nonlinear differential algebraic system for the dependent variables $\mathbf{U}, \mathbf{V}, p$ and T. The commercial finite element solver, Femlab 3.1 (see Femlab for details), is employed to implement the derived model due to its versatility in handling general coupled nonlinear partial differential equations. Each case study is resolved as follows:

(i) *Transient uniform response to a temperature change*. The radial direction as shown in Fig. 5.23 is resolved with ~1000 elements to ensure mesh-independent solutions. The computations are carried out in a 2 GHz PC with 1 GB RAM, required around 1 min. The governing equations in the radial direction are summarized for a uniform spherical deformation as follows.
The uniform spherical deformation of hydrogels is the most extensive model development and study, as it allows a reduction in dimensionality and mathematical complexity. It only needs to resolve the radial direction because of the isotropic deformation, as shown in Fig. 5.23. Further, the radial component of the mixture velocity V_R=0, as Eq. (5.113), which in the radial direction of spherical coordinates becomes $\partial V_R/\partial R$=0, has to satisfy the boundary condition, Eq. (5.116). Hence, one can determine the fluid pressure a posteriori. We are left with solving the radial deformation U_R and temperature T, for which the governing equations, Eqs. (5.114) and (5.115) are

$$\frac{DU_R^{(p)}}{Dt} = \frac{\kappa}{\mu^{(f)} J}\left(\frac{1}{R^2}\frac{\partial}{\partial R}[R^2 P_{rR}^{(p)}] - \frac{P_{\theta\Theta}^{(p)} + P_{\varphi\Phi}^{(p)}}{R}\right) \tag{5.126}$$

$$[C^*(1-\phi)+\phi]\frac{DT}{Dt} = -\frac{1}{\mathrm{Re}\,\mathrm{Pr}\, JR^2}\frac{\partial}{\partial R}[R^2 Q_R] + C^*\frac{DU_R^{(p)}}{Dt}F_{rR}^{-1}\frac{\partial T}{\partial R} \tag{5.127}$$

$$Q_R = -kJC_{rR}^{-1}\frac{\partial T}{\partial R} \tag{5.128}$$

For slow-response hydrogels, i.e. RePr<<1, Eqs. (5.127), (5.128), (5.116) and (5.117) reduce to

$$\frac{\partial}{\partial R}[R^2 Q_R] = 0 \tag{5.129}$$

with the trivial solution T=0 for the whole sphere. In other words, as already stated previously, the heat conduction is much faster than the swelling, such that thermal equilibrium can be assumed for the whole swelling of the sphere. For equilibrium conditions, namely the steady-state deformation at a given temperature, Eq. (5.126) can be integrated once and together with the boundary conditions, Eqs. (5.116) and (5.117) for the mixture stress, one arrives at

$$P_{rR}^{(p)} = JF_{rR}^{-1} p^{(osm)} \tag{5.130}$$

or in an Eulerian frame

$$\sigma_{rR}^{(p)} = p^{(osm)} \tag{5.131}$$

The steady-state deformation is determined in Eq. (5.64), by studying the deformation gradient tensor in the radial direction for a uniform deformation. Now, the evolution from the reference to the initial state is assumed to be a uniform deformation with the initial equilibrium state. A similar expression can thus be formulated as $R(R_{\text{ref}}) = J_0^{1/3} R_{\text{ref}}$, which is substituted into Eq. (5.64) yields $r(R) = J^{1/3}R$, whence $\sigma_{rR}^{(p)} = (J_{\text{ref}}^{-1/3} - 0.5J_{\text{ref}}^{-1})$ and the radial displacement $U_R = r - R = (J^{1/3} - 1)R$. Substituting this expression for the effective stress tensor and the osmotic pressure, Eq. (5.105), into Eq. (5.131), yields a transcendental function for the polymer volume fraction at a given equilibrium temperature

$$T = \frac{1}{\Delta T}\left(-\frac{\Delta\Theta\phi^2}{k_B(\sigma_{rr}^{(p)}/N_x + \phi + \chi_2\phi^3 + \log(1-\phi)) - \Delta\Gamma\phi^2} - T_\infty\right) \tag{5.132}$$

At equilibrium, it is determined that T=0, such that the equilibrium temperature in dimensional form is

$$T_\infty = -\frac{\Delta\Theta\phi^2}{k_B(\tilde{\sigma}_{rr}^{(p)}/N_x + \phi + \chi_2\phi^3 + \log(1-\phi)) - \Delta\Gamma\phi^2} \tag{5.133}$$

The equilibrium polymer volume fraction can be found by solving this transcendental function at any given temperature, which in turn allows for the determination of the steady-state radius and deformation of the spherical hydrogel.

(ii) *Transient non-uniform response to a temperature change*. The geometry as shown in Fig. 5.24 is resolved with ~6200 elements, amounting to $\sim 4.1 \times 10^4$ degrees of freedom, to ensure mesh-independent solutions. The computations are carried out in a 2 GHz PC with 1 GB RAM, required around 70 min.

(iii) *Transient deformation in a temperature gradient*. The geometry as shown in Fig. 5.25 is resolved with ~5400 elements, amounting to $\sim 2.5 \times 10^4$ degrees of freedom, to ensure mesh-independent solutions. The computations are carried out in a 2 GHz PC with 1 GB RAM required around 10 min.

For the parameter adaptation of the free energy and stress of the hydrogels to experimental data, the transcendental steady-state uniform solution, Eq. (5.133), is solved via a nonlinear least-squares method in Matlab 6 (see Matlab for details).

5.6.5 Simulations and Discussions

The analysis presented above shows that the deformation and heat transfer can be characterized by several dimensionless parameters, the most important of which are Re and Pr. Scales for the underlying physics have also been secured, e.g. for the stress, osmotic pressure and velocities. In addition, the time scales have been identified for the deformation and heat transfer.

In the following subsections, several parameters pertaining to the free energy and stress tensor are determined from equilibrium swelling curves, after which the permeabilities are found from the swelling kinetics. We then proceed with several transient deformation studies for the three aforementioned cases, namely (i) a uniform and (ii) a non-uniform response to a step change in temperature as well as (iii) the deformation in a temperature gradient. While discussing these, it is instructive to recall the scales and dimensionless numbers, and in particular see how well they predict the behaviour of the hydrogels. We will revert to the Eulerian frame and dimensional quantities, except for the coordinates, for which we retain the dimensionless counterparts in order to better compare with the derived scales.

5.6.5.1 Temperature Dependence of Equilibrium Degree of Swelling

The equilibrium swelling curve for the hydrogel in a given solvent at various temperatures is one of the fundamental criteria for characterization and evaluation of the smart hydrogel. The simulations will reveal the degree of swelling and whether the hydrogel at hand exhibits an upper critical solution temperature (UCST) or a lower critical solution temperature (LCST). The former refers to a hydrogel that shrinks at low temperatures and swells with increasing temperature. The latter, on the other hand, exhibits the opposite behavior, where it swells at low temperatures and shrinks at higher. These changes are mainly brought about by competing polymer–solvent interactions, such as hydrogen bonding and hydrophobic interactions, which are accounted for by χ presented by Eq. (5.72), and the degree of crosslinking, N_x, which is incorporated into the stress tensor, Eq. (5.60). The

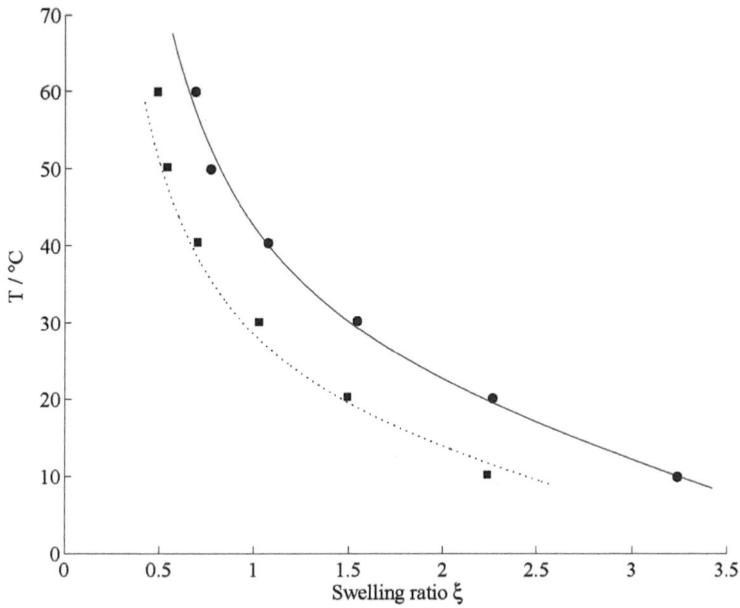

Fig. 5.27 Temperature dependence of equilibrium degree of swelling of crosslinked poly(APy-*co*-St) in water. The experimentally measured values (Bae et al., 1989) are () 20 mol% St (AS20); () 15 mole St (AS15). The model predictions are (···) AS20; (—) AS15

experimentally measured and numerically predicted swelling behaviours of the two crosslinked poly(Apy-*co*-St), AS15 and AS20, in water are shown in Fig. 5.27. The curves exhibit a moderate swelling from around 0.5 (AS20) and 0.75 (AS15) at 60°C to 2.3 and 3.3 at 10°C. The swelling equilibrium for the conventional and macroporous poly(*N*-isopropylacrylamide) gels, NC000, G8 and NC030, on the other hand, exhibit a larger degree of swelling and a clear LCST behavior, which can be inferred from Fig. 5.28. At 50°C, the three hydrogels are in shrunken state, with a swelling ratio of 4. Around 32°C, all three undergo a sudden change in volume, with NC000, NC030 and G8 reaching a swelling ratio of around 35, 60 and 80 at 15°C, respectively. These measured swelling curves are used to estimate the initial polymer volume fractions, ϕ_0, and diameters, l_0, as well as for parameter adaptation of Eq. (5.133), to determine N_x for the stress tensor, $\nabla\Theta$, $\Delta\Gamma$, χ_2 and ϕ_{ref} for the interaction parameter, χ, of the free energy, the values of which can be found in Table 5.2. The parameter adapted enthalpy, $\Delta\Theta$, and entropy, $\Delta\Gamma$, are within the range of the experimentally estimated counterparts for a PNIPA gel by Hirotsu (1987). Overall, the model for steady-state uniform swelling agrees well with the experimentally measured values, especially so for the slow-response AS15, AS20 and NC000. The model overestimates the LCST for NC000, NC030 and G8 by several degrees Celsius, whence it is not able to fully capture the temperature for the sudden change in volume. It is noted that the parameters for the stress tensor and interaction parameter have to be determined for each new hydrogel considered, since they are functions of the degree of crosslinking, type of monomers, porosity, etc.

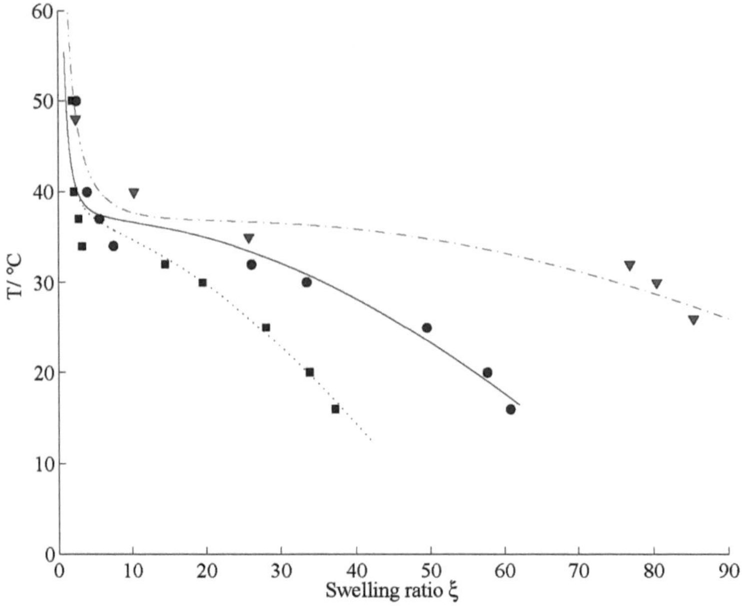

Fig. 5.28 Temperature dependence of equilibrium degree of swelling of the normal and macroporous poly(*N*-isopropylacrylamide) gels in water. The experimentally measured values are () NC000 (Cheng et al., 2003), () NC030 (Cheng et al., 2003) and () G8 (Zhang and Zhuo, 2000a, b). The model predictions are (···) NC000, (—) NC0030 and (—) G8

5.6.5.2 Permeability for Swelling and Deswelling Kinetics

The swelling and deswelling kinetics are mainly determined by the permeability of the hydrogel. A low permeability is likely to result in a slow deformation response, as the penetrating solvent experiences a high resistance inside the hydrogel, whereas a high permeability allows for easier solvent penetration. Based on the measured swelling and deswelling kinetics for AS15 and AS20, as shown in Fig. 5.29, and the deswelling response for NC000 and NC030 and G8, as depicted in Fig. 5.30, the permeabilities for these hydrogels could be estimated for uniform swelling (Case i). Clearly, the G8 hydrogel is the fastest, losing more than 80% of the absorbed water in 1 min, followed in descending order by NC030, NC000, AS20 and AS15. NC030 requires around 2 min, i.e. twice as long as the G8, to reach water retention of around 20%. The kinetic response to the change in temperature is captured by the permeabilities, which are found to be 1.7×10^{-16}, 8×10^{-17}, 4×10^{-18}, 1.5×10^{-19} and 1.2×10^{-19} m^2 for G8, NC030, NC000, AS20 and AS15, respectively. These values are of similar order of magnitude as experimental findings by Tokita and Tanaka (1991), who measured permeabilities in the order of $10^{-17}-10^{-18}$ m^2 for a poly(acrylamide) gel.

Returning to Fig. 5.29, it is found that the time is overestimated for swelling and deswelling of both AS15 and AS20. The polymer volume fraction dependence of

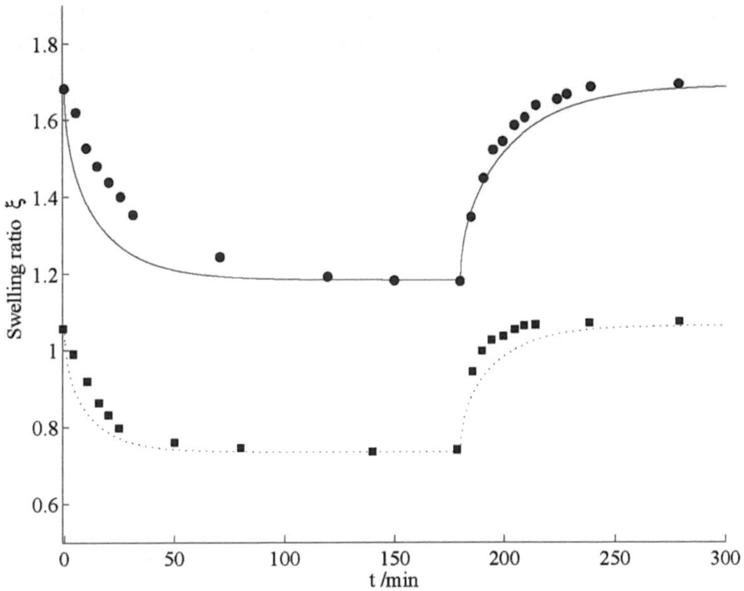

Fig. 5.29 Swelling and deswelling curves of crosslinked poly(APy-*co*-St) between 27 and 37°C. The experimentally measured values (Bae et al., 1989) are () 20 mol% St (AS20) and () 15 mol% St (AS15). The model predictions are (···) AS20 and (—) AS15

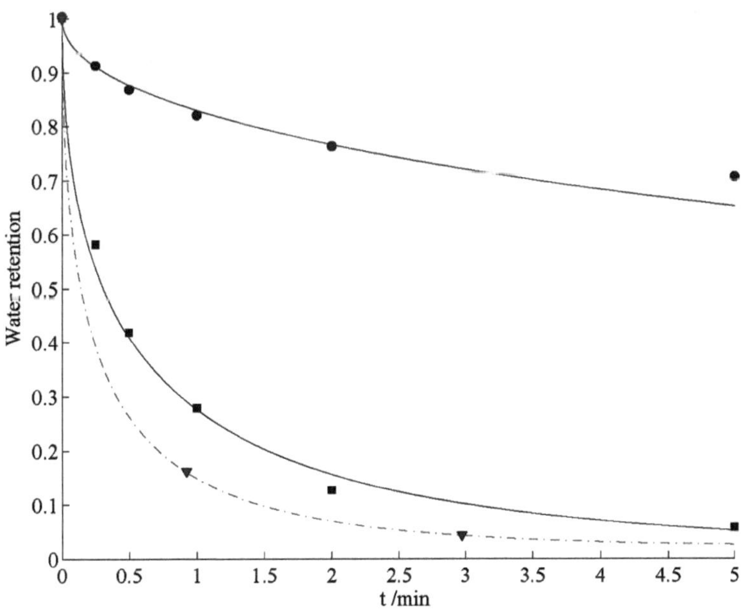

Fig. 5.30 Deswelling curves for the water retention of the normal and macroporous poly(*N*-isopropylacrylamide) gels in water. The experimentally measured values are () G8 (Zhang and Zhuo, 2000a, b), () NC030 (Cheng et al., 2003) and () NC000 (Cheng et al., 2003). The model predictions are (-··) G8, (—) NC030 and (-) NC000

the permeability, taken to be $\sim\phi^{-3/2}$, is thus not able to fully predict the kinetics. If better predictions of the kinetics of the crosslinked poly(Apy-*co*-St), AS15 and AS20, are required, we have to determine the dependency of the kinetics on the polymer volume fraction, similar to what has been done for the PNIPA gel (Tokita and Tanaka, 1991; Grattoni et al., 2001). For the macroporous gels, NC030 and G8, the assumption of constant permeability due to the high porosity seems to be valid as it is able to accurately solve the water retention during deswelling. The non-modified predictions of PNIPA gel, NC000, however, initially follow the measurements, but start to deviate from measurements at around 5 min, as seen in Fig. 5.30. The answer to the deviation can be found in the permeability: at t=0 min, the temperature is 20°C, whence NC000 is in a swollen state with a high porosity. For this state, the assumption of constant permeability is valid. As the hydrogel starts to shrink, the porosity decreases and the permeability becomes increasingly more dependent on the local polymer volume fraction.

5.6.5.3 Uniform Deformation Behaviour

As mentioned before, isothermal uniform swelling due to step changes in temperature has been studied extensively for the hydrogels, mostly for the PNIPA or derivatives thereof. In an effort to extend previous work, the present studies are conducted not only on the evolution of slow-response hydrogels, represented by AS15, but also on the macroporous fast-response hydrogel NC030 in tandem with the heat transfer. While doing so, one finds it convenient to return to the scales to aid in the interpretation. It begins with the local behaviour of the deformation in terms of the polymer volume fraction, as depicted in Fig. 5.31. Here, several features are apparent; foremost is that the deswelling behaviour of the fast-response NC030, as seen Fig. 5.31d, occurs as a sharp front in time, whereas for AS15, as seen Fig. 5.31b, the shrinking is more akin to a diffusion process, where the whole interior of the hydrogel sees a change of polymer volume fraction with time. The shrinking occurs several times faster for NC030 compared with AS15. In fact, the deformation is so rapid that there is not enough time for the solvent to penetrate deeply into the hydrogel, as it is the case for AS15. The penetrating front for NC030 widens with time, but it is only in the final stages of the shrinking, at around t>400 s, that the front has widened significantly and reached the centre of the hydrogel. While the deswelling of the fast-response NC030 is very rapid, only requiring around 5–10 min, the swelling is much slower, requiring up to 500–600 min for attaining equilibrium. Recalling the deformation time scales $[t_{\mathrm{def}}]\sim 10$ min and $[t_{\mathrm{def}}]\sim 500$ min for the deswelling and swelling of NC030, respectively, it is found that they are of the same order of magnitude as the computed deformation time. May it explain that deswelling is significantly faster than the swelling for NC030? The answer is in the affirmative: returning to the scale analysis, the deformation time scale and velocity scale in particular. It is observed that the deformation time can be expressed as $[t_{\mathrm{def}}] \sim l/U = [\mu^{(f)}]l^2/([p^{(\mathrm{osm})}][\kappa])$, where the osmotic pressure appears in the denominator. For the deswelling from 20 to 52°C, $[p^{(\mathrm{osm})}]\sim 10^4$ N/m^2 at the new equilibrium temperature 52°C, in contrast to $[p^{(\mathrm{osm})}]\sim 500$ N/m^2 at 20°C for the

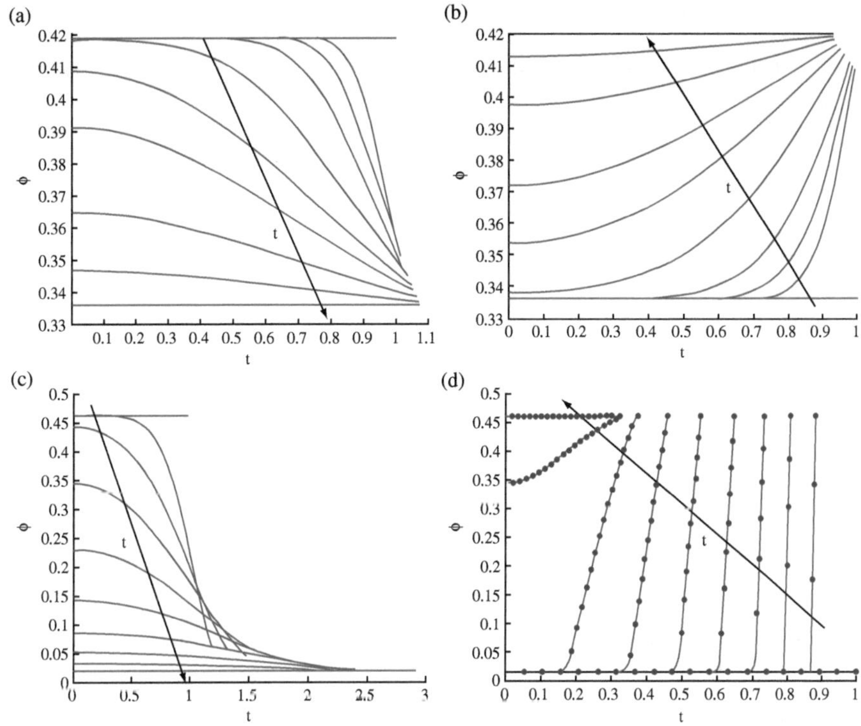

Fig. 5.31 The local polymer volume fractions along the radial direction (Eulerian frame): (**a**) swelling of AS15 from 37 to 27°C and (**b**) deswelling of AS15 from 27 to 37°C at 0, 1, 2, 3, 9, 17, 25, 42, 67, 170 min; (**c**) swelling of NC030 from 52 to 20°C at 0, 1, 3, 6, 13, 30, 67, 130, 333, 667 min; and (**d**) deswelling of NC030 from 20 to 52°C (—) and isothermally () at 52°C for 0, 5, 20, 50, 70, 100, 200, 400, 700, 1200 s

swelling from 52 to 20°C. The driving force for the former is almost two orders of magnitude higher than the latter, which explains the fast deswelling compared with the swelling. However, the deswelling of the macroporous NC030 does not necessarily correspond to a fast response, as it is inferred from Fig. 5.32, where the deswelling from 20°C for various ΔT is depicted. It is so only if the hydrogel is heated to a temperature above its LCST, i.e. in this case for $\Delta T \geq 35$°C, that the deswelling rate amounts to minutes instead of hours or more. This behaviour originates from the sudden jump in polymer volume fraction when the hydrogel passes from a swollen to a shrunken state, i.e. $[\sigma_{\text{eff}}^{(p)}] \sim [p^{(\text{osm})}] \propto J_{\infty}^{-1} = (\phi_{\infty}/\phi_{\text{ref}})$. In the swollen state, the polymer volume fraction, $\phi_{\infty} \sim 10^{-2}$, is much smaller than in the shrunken state with $\phi_{\infty} \sim 0.5$. As long as the deswelling occurs below the LCST, $J_{\infty}^{-1} \sim 1$ for swelling (*N.B.* $\phi_{\text{ref}} \sim 10^{-2}$ for NC030), as opposed to shrinking from below the LCST to above, which manifests itself in $J_{\infty}^{-1} \sim 100$. This behaviour is in line with experimental findings by Kim et al. (2003a, b), who investigated the fast deswelling response of semi-interpenetrated thermo-sensitive

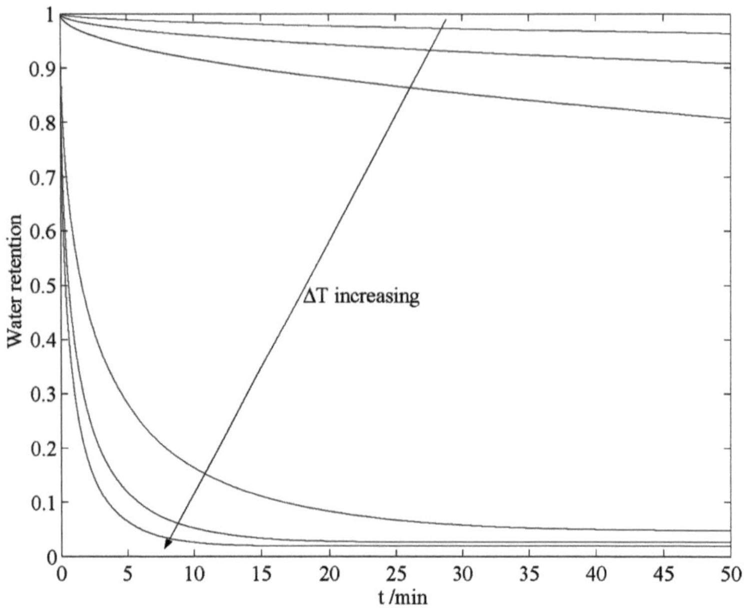

Fig. 5.32 Deswelling curves for the water retention from T_0=20°C of NC030 in water for $\Delta T = T_\infty - T_0$ =5, 10, 15, 20, 25, 30°C

poly(*N*-isopropylacrylamide)/poly(ethylene oxide) hydrogels. Furthermore, the earlier work by Barrière and Leibler (2002) shows that the surface of a hydrogel does not immediately reach its equilibrium during swelling towards a shrunken state. This delay is clearly visible in Fig. 5.31a–c for both AS15 and NC030, except for the very fast shrinking of NC030 as shown in Fig. 5.31d.

Turning the attention towards the heat transfer for AS15, as illustrated in Fig. 5.33a and b, it is known that the heat transfer occurs very quickly, only taking ~4 s to attain the new equilibrium temperature. For AS15, the heat transfer is so rapid that the hydrogel has no time to deform significantly. In other words, for the slow-response hydrogel, AS15, the heat transfer only sees the initial shape of the hydrogel. This is in line with the predicted heat transfer scale $[t_{heat}] \sim 10$ s and $RePr \sim 10^{-4}$, such that it only needs to consider the steady-state version of the heat transfer, Eq. (5.129), when solving for the deformation. For the case of a uniform step change in temperature, the steady-state solution is simply the new equilibrium temperature. A closer look at the temperature profiles in Fig. 5.32a and b reveals that the heat transfer is somewhat faster for swelling than deswelling, which may be attributed to the initial larger size of the hydrogel at the lower temperature 27°C. For the fast-response hydrogel, NC030, the heat transfer requires ~10 (swelling) and ~40 s (deswelling), as it can be inferred from Fig. 5.33c and d; the longer time for the latter stems from the initially larger radius in the swollen state at 20°C. During the swelling, heat transfer is clearly much faster than the deformation. For the deswelling from 20 to 52°C, however, the reduction in radius is apparent in the

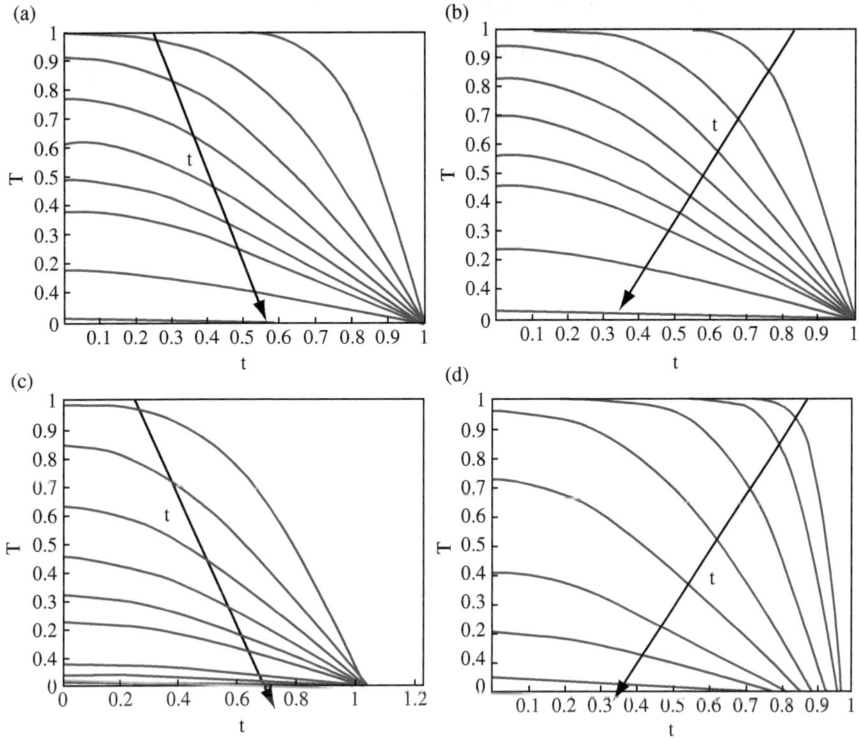

Fig. 5.33 The local temperature along the radial direction (Eulerian frame): (**a**) swelling of AS15 from 37 to 27°C and (**b**) deswelling of AS15 from 20 to 37°C at 0, 0.1, 0.3, 0.5, 0.7, 0.9, 1.1, 1.3, 1.9, 3.9 s; (**c**) swelling of NC030 from 52 to 20°C at 0, 1, 2, 2.5, 3, 3.5, 5, 6.5, 8 s; and (**d**) deswelling of NC030 from 20 to 52°C at 0, 0.5, 1, 2, 5, 9, 15, 20, 30, 40 s

temperature profiles. At the new thermal equilibrium, the radius drops by 30%. This seems to suggest that the heat transfer might have an impact on the overall deformation of the hydrogel. The fact that RePr$\sim 10^{-2}$ indicates otherwise, implying that the heat transfer equation should at leading order reduce to a steady-state conduction problem, with the new temperature as trivial solution. The question that arises is therefore whether it needs to consider transient heat transfer if the present main concern is the deformation of the hydrogel, or whether we can prescribe the new temperature from the onset. The answer can be found in the circles in Fig. 5.31d, which corresponds to an isothermal deformation based on the new equilibrium temperature. It is shown that the isothermal deformation follows its thermal counterpart closely, implying that it is unnecessary to consider transient heat transfer for NC030. An additional explanation for the agreement could be related to the moving front during deswelling: this front, which controls the deformation, is always close to or at the equilibrium temperature, as it is evident from Figs. 5.31d and 5.33d. The temperature in the core of the hydrogel can thus be expected to be less important for the deswelling for this case.

5.6.5.4 Non-uniform Deformation Behaviour

The uniform responses of both slow and fast-response hydrogels have thus far been discussed in terms of deformation and heat transfer. It is followed by studying the non-uniform deformation for the two hydrogels, AS15 and NC030, when subjected to a step change in temperature. As mentioned above in the subsections for the model formulation, the motion of the hydrogels is constrained in the normal direction, whence it is only required to resolve the deformations in the streamwise and spanwise directions, i.e. *x*- and *y*-directions, respectively.

Achilleos et al. (2000a, b, 2001) demonstrated that the polyelectrolyte slabs deform faster at the corners than at the sides for change in salt concentration. The question here is whether the temperature-sensitive hydrogels display the same deformation behaviour for change in temperature. For this purpose, the evolution of AS15 and NC030 is studied for a step change in temperature from 37 to 27°C (swelling) and from 27 to 37°C (shrinking), as depicted in Fig. 5.34. For both AS15 and NC030, most of the swelling/deswelling occur in the first 100 min after the change in temperature. Deswelling is marginally faster than the swelling for the slow-response hydrogel AS15, whereas shrinking is around four to five times faster than swelling for the fast-response hydrogel NC030, which requires around 100 min to shrink from 27°C to the new equilibrium state at 37°C, as compared with around 400–500 min for the swelling from 37 to 27°C. For both the cases RePr<<1 and

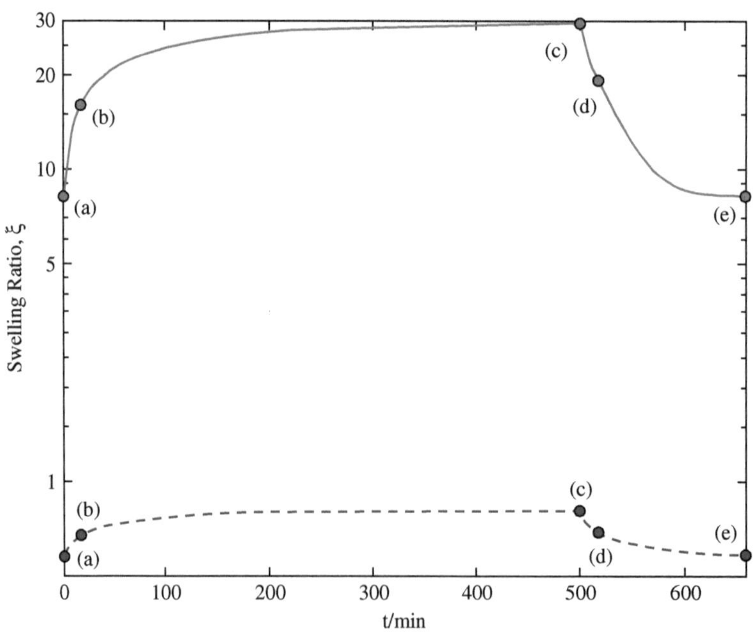

Fig. 5.34 Swelling and deswelling curves between 27 and 37°C: (—) NC030 and (—) AS15. The specific times are (**a**) 0 min, (**b**) 17 min, (**c**) 500 min, (**d**) 517 min and (**e**) 660 min

$[t_{heat}]{\sim}10$ s, whence the deformations take place at the steady-state temperature. The initial equilibrium shape at 37°C of AS15 and NC030, denoted (a) in Fig. 5.34, is illustrated in Figs. 5.35a and 5.36a, respectively. The polymer volume fraction of AS15 is $4.2{\times}10^{-1}$ and $9.2{\times}10^{-2}$ for NC030. A decrease in the temperature to 27°C results in both the hydrogels swelling towards the new equilibrium state, as shown in Figs. 5.35b and 5.36b. Here, several features are apparent. Foremost is the influx of solvent at the sides of the free surfaces: close to the symmetry lines, the flow field is more or less unidirectional towards the core of the hydrogel, but as it approaches the evolving corners, the penetrating solvent can be seen to flow not only into the hydrogel interior but also towards the corner. Clearly, the corners drag solvent with them due to the friction between the solvent and polymer phases, as

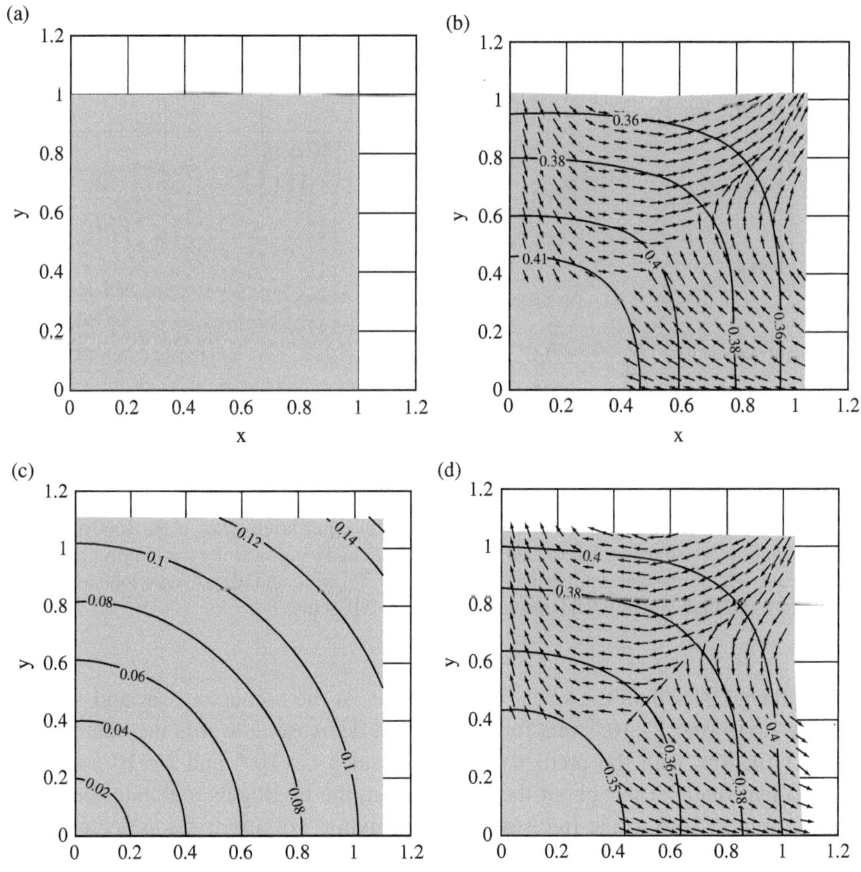

Fig. 5.35 Deformation behaviour of AS15: (**a**) initial equilibrium state at T_0 (t=0 min); (**b**) polymer volume fraction distribution and fluid velocity field during swelling (t=17 min); (**c**) deformation field, $|\mathbf{U}|$, for the new equilibrium state at T_∞ (t=500 min); and (**d**) polymer volume fraction distribution and fluid velocity field during shrinking (t=517 min)

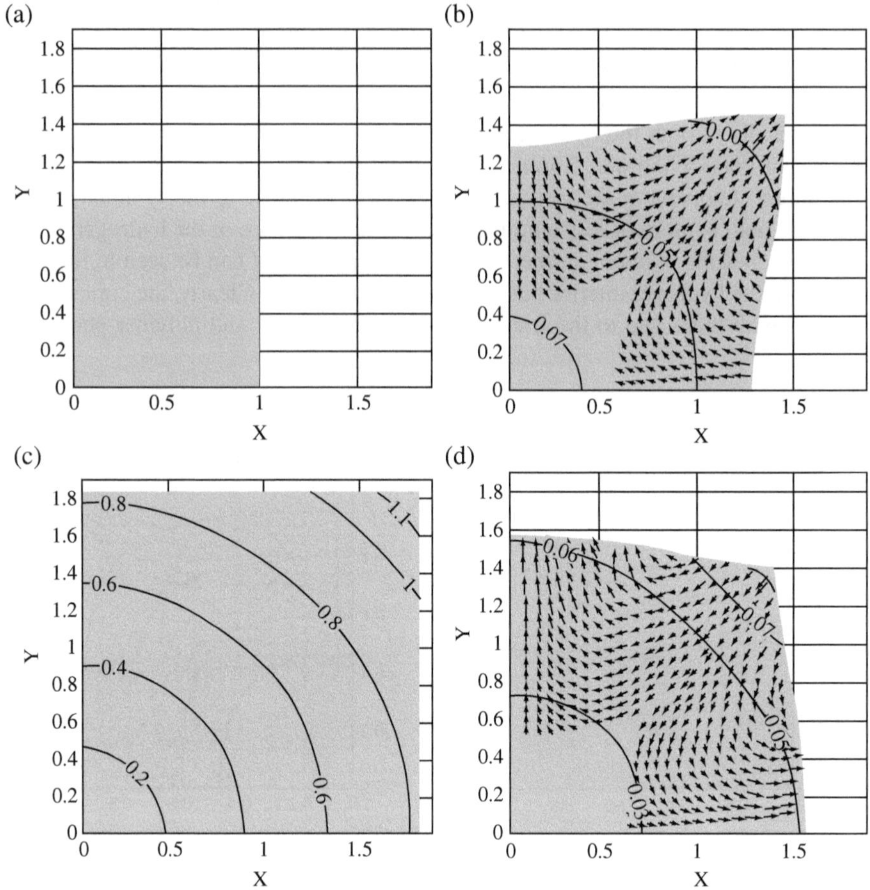

Fig. 5.36 Deformation behaviour of NC030: (**a**) initial equilibrium state at T_0 (t=0 min); (**b**) polymer volume fraction distribution and fluid velocity field during swelling (t=17 min); (**c**) deformation field, |**U**|, for the new equilibrium state at T_∞ (t=500 min); and (**d**) polymer volume fraction distribution and fluid velocity field during shrinking (t=517 min)

they deform faster than the sides. The velocities at the corners are around 4×10^{-8} m/s for AS15 and 2×10^{-7} m/s for NC030 and decreases towards the centre core, which is in line with the predicted velocity scales $U\sim10^{-8}$ and $U\sim10^{-7}$ m/s for the two hydrogels. Throughout the deformation, the hydrogels maintain the shape of the corner. Close to the free surface, the polymer volume fraction decreases as compared with the interior due to the entering solvent. As the hydrogels swell, the solvent gradually penetrates deeper into the core until an equilibrium state with a uniform polymer volume fraction is attained. The smooth rounded contour lines in the interior reminisce of the diffusion behaviour observed for the uniform swelling. The low volume fractions around 3×10^{-2} to 7×10^{-2} for NC030 indicate the macroporous nature of the hydrogel, in contrast to the denser AS15, with polymer volume fractions of around 4×10^{-1}. The higher stiffness of AS15 due to a lower N_x in the stress tensor, as seen in Eq. (5.59), restricts the deformation of the corner, and the whole hydrogel more or less keeps its initial shape as it evolves towards the new equilibrium condition, in contrast to NC030, with an N_x value ~100 times as high as AS15, for which the corner deforms significantly faster and seems to drive the swelling of the hydrogel. The new equilibrium shape, essentially a larger version of the initial state with the same form, is depicted in Figs. 5.35c and 5.36c, with a volume increase of ~10% for AS15 and ~200% for NC030. The polymer volume fractions drop to 3.4×10^{-1} and 2.8×10^{-2} for AS15 and NC030, respectively. The larger swelling ratio of the macroporous NC030 as compared with AS15 is evident in the deformation field, $|\mathbf{U}|$, which ranges from 0 to ~0.14 for AS15 and from 0 to ~1.1 for NC030. The subsequent shrinking of the hydrogels is illustrated in Figs. 5.35d and 5.36d as the temperature increases to 37°C again. Akin to the swelling, the shrinking is faster at the corners, although only marginally so for the stiffer AS15. The flow field is now reversed, with water being pushed inwards at the corner and out the sides, which is reflected in a higher polymer volume fraction close to the free surface than in the core. The flow here is also more unidirectional close to the symmetry line y=0. After the water is squeezed out during the shrinking, the hydrogels return to the initial shape, as shown in point (e) in Fig. 5.34, thus amounting to a fully reversible process.

It is well known that the macroporous hydrogels suffer from low elastic strength, while a fast-response hydrogel is desirable in many applications, due to the large pores in the network (Kim et al., 2004a, b). This mechanical weakness is reflected in the high N_x values, as presented in Table 5.1, for the macroporous hydrogels as compared with the low N_x of AS15 and AS20. During the swelling, the Cauchy stress component, σ_{xx} (N.B. σ_{yy} is symmetric to σ_{xx} in this case), at t=17 min as shown in Fig. 5.37a (point (b) in Fig. 5.34), varies between 5.8×10^4 and 7.6×10^4

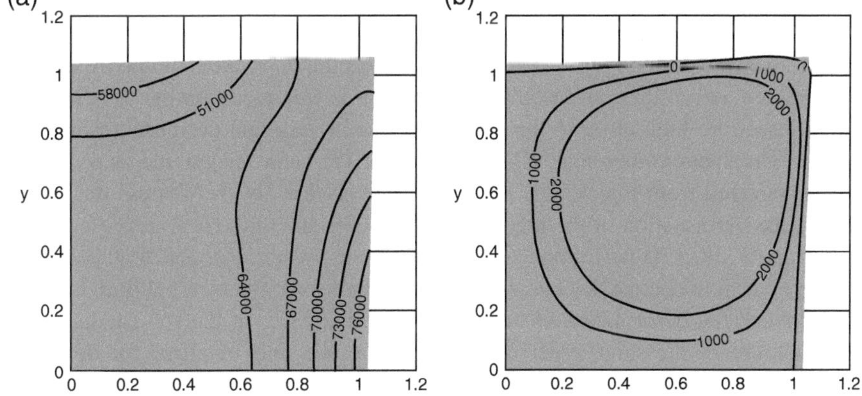

Fig. 5.37 Dimensional Cauchy stress distribution in the polymer phase for AS15 during swelling at t=17 min: (**a**) $\sigma_{xx}^{(p)}$ and (**b**) $\sigma_{xy}^{(p)}$

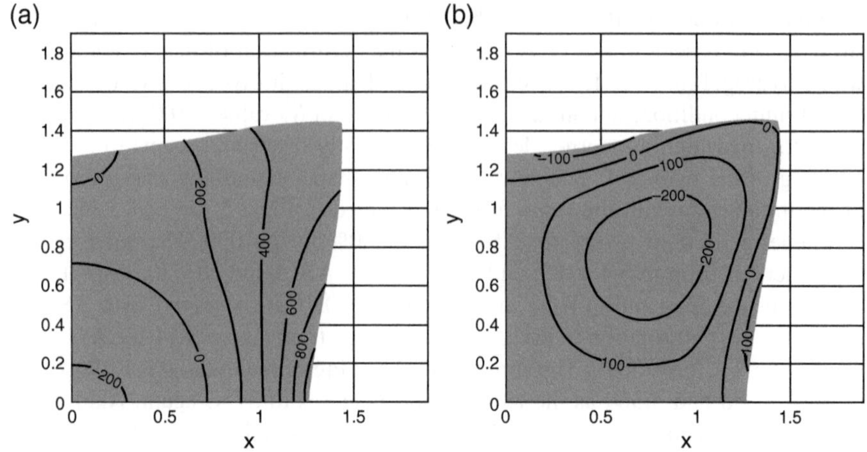

Fig. 5.38 Dimensional Cauchy stress distribution in the polymer phase for NC030 during swelling at t=17 min: (**a**) $\sigma_{xx}^{(p)}$ and (**b**) $\sigma_{xy}^{(p)}$

N/m^2. Recalling the stress scale $[\sigma_{\text{eff}}^{(p)}] \sim 10^5$N/m^2, it is found that it is of the same order of magnitude as the computed stress. The highest stress is found close to the symmetry line at y=0, which explains the slower movement of the hydrogel adjacent to the symmetry line as opposed to the corner, which experiences less stress and hence can expand faster. The movement of the corner is observed to induce a shear stress in the core of the hydrogel, as the hydrogel swells, with a σ_{xy} component (*N.B.* $\sigma_{yx}= -\sigma_{xy}$ for this case) of almost two orders of magnitude lower than σ_{xx}, as depicted in Fig. 5.37b. As mentioned in the scale analysis, the stress scale is derived mainly for the extensional stresses and therefore fails to predict the shear stress. For NC030, as seen in Fig. 5.38a, the stress components σ_{xx} is significantly lower, varying between -2×10^2 and 9×10^2 N/m^2, in agreement with the predicted stress scale $[\sigma_{\text{eff}}^{(p)}] \sim 10^3$N/m^2. Close to the core, the hydrogel is slightly compressed and further out towards the corner, it is extended. Similar to AS15, the stress levels are high close to the symmetry line at y=0 and decrease towards the corner. The shear component, σ_{xy}, is almost of the same magnitude as σ_{xx}, as it can be inferred from Fig. 5.38b. The comparatively low levels of stress that arises during the deformation of the macroporous NC030 due to the low stiffness can be expected to affect its performance in environments, where external flow is important, e.g. in flow control devices. Under such conditions, the external flow is likely to influence the deformation as soon as the fluid stresses at the solvent–hydrogel interface are of the same order of magnitude as the ones required for deformation. Clearly, the low elastic strength is not desirable as the hydrogel might be too weak to sustain its function when subject to external stress. On the other hand, AS15 is mechanically stronger but is limited by the slow-response and low swelling ratios.

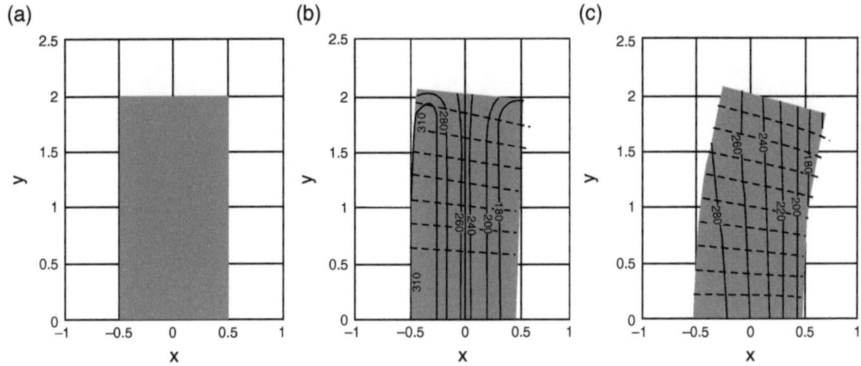

Fig. 5.39 Deformation and osmotic pressure in a temperature gradient for NC030: (**a**) initial equilibrium state at T_0 (t=0 min), (**b**) at t=0.3 min and (**c**) at t=17 min

5.6.5.5 Deformation Behaviour in a Temperature Gradient

In this subsection, the deformation of a temperature-sensitive hydrogel is addressed when immersed in a solvent subject to a temperature gradient. Previous studies (Li et al., 2004a, b; Wallmersperger et al., 2004; Zhou et al., 2002; Snita et al., 2001) demonstrated that polyelectrolyte hydrogels exhibit a bending motion when exposed to an electric field. This bending motion is able to generate a force, which can be harnessed in, e.g. artificial muscles or actuators. The origin of the bending motion is the difference in the osmotic pressure that the hydrogel experiences in the electric field. Now, considering the osmotic pressure as a function of temperature, the temperature-sensitive hydrogels are expected to behave analogously, i.e. to align itself with the gradient and deform to a curved equilibrium state. This is indeed what happens, as it can be inferred from Fig. 5.39a–c, where the NC030 bends under the influence of the temperature gradient. The hydrogel at the initial state is in equilibrium at T_0=30°C. The initial temperature is chosen such that the temperature remains at 30°C at the Origo for t>0, which allows for focusing on the bending. At t>0, a linear temperature gradient, with T=20°C and 40°C at the left and right walls (see Fig. 5.25), is imposed and the hydrogel responds with a bending motion. After around half a minute, as seen in Fig. 5.39b, the left side of the hydrogel, where the temperature is lower (≈27°C), sees an influx of solvent and undergoes swelling. The right side, in contrast, sees a higher temperature than in the initial state (≈32.5°C), whence it squeezes out solvent as it shrinks. At this time, there is already a difference of around 150 N/m^2 in osmotic pressure, $\Delta p^{(\mathrm{osm})}$, between the left and the right, which gives rise to the bending motion. The solvent velocity in the upper part of the hydrogel is $O(10^{-6}$ m/s) and decreases towards the symmetry line. This flow originates from the friction coupled with the deformed polymer phase, since the solvent flow induced by the osmotic pressure difference can only give rise to a solvent flow of $O(\kappa \Delta p^{(\mathrm{osm})}/([\mu^{(f)}]l)) \sim 10^{-8}$m/s as it can be inferred from Eq. (5.102), which is significantly lower than the observed velocity. After around

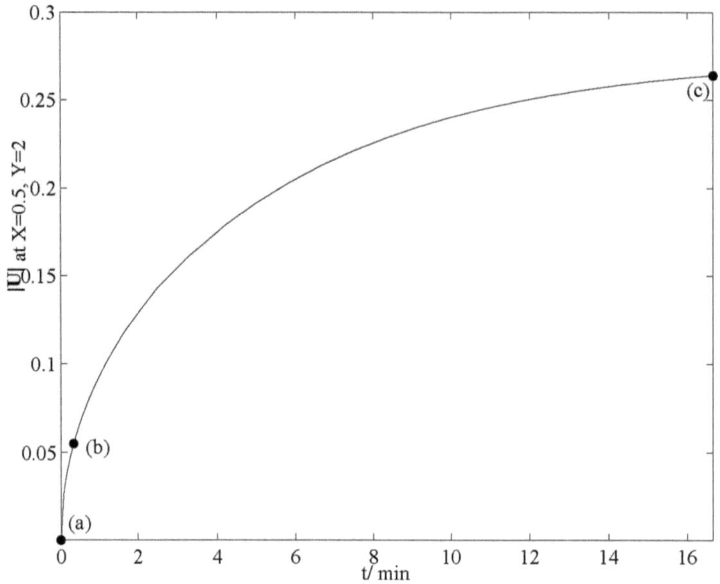

Fig. 5.40 The absolute value of the deformation, |**U**|, at the upper right corner (X=0.5, Y=2) during the bending motion for NC030: (**a**) t=0 min, (**b**) t=0.3 min, and (**c**) t=17 min

17 min, the point (c) in Fig. 5.40 implies that the deformation of the right top corner (X=0.5, Y=2) has more or less reached its final state, such that the hydrogel has attained its new equilibrium state. This final shape is shown in Fig. 5.39c, where the elastic force is balanced by the osmotic pressure gradient due to the temperature difference of around 5°C between the cooler left and warmer right sides of the hydrogel. Even though the deformation comes to an end, the solvent observed still flows from the left to the right inside the hydrogel at speed of $O(10^{-8}$ m/s), which is in line with the above predicted solvent flow stemming from the osmotic pressure difference. Similar to the non-uniform response to a step change in temperature, outlined above, $[t_{heat}]\sim 10$ s, which is sufficiently much faster than the deformation, justifying the assumption of a steady-state temperature gradient.

In summary, it is suggested that the temperature-sensitive hydrogels can undergo bending motions when subject to a temperature gradient, similar to the behaviour of polyelectrolyte hydrogels in an electric field. An increase or decrease in the temperature gradient can be expected to give rise to a larger or smaller deflection, respectively, such that the degree of bending can be controlled.

5.7 Remarks

In this chapter, the two models for simulation of the temperature-sensitive hydrogels are presented: one for steady-state simulation and the other for transient simulation.

For the steady-state simulation of swelling equilibrium of ionized thermo-sensitive hydrogels, the multiphysics model, called the multi-effect-coupling thermal-stimulus (MECtherm) model, is developed to predict the swelling equilibrium with the volume phase transition for ionized thermo-sensitive hydrogels. It is composed of steady-state nonlinear coupled Poisson–Nernst–Planck partial differential equations as well as a transcendental equation governing the swelling equilibrium state. For solution of the MECtherm model, a hierarchical Newton iteration strategy is implemented in the computational flowchart. The meshless Hermite-cloud method (Li et al., 2003) is employed for numerical simulation of the responsive hydrogels subject to changes in environmental temperature. The simulation results are in good agreement with experimental data for the relations between the temperature and volume swelling ratio. In parameter studies, the influences of various material properties and environmental conditions, such as the initial fixed charge density, crosslink density, initial volume fraction and electrolyte solution concentration, on the responsive characteristics of the thermal-sensitive hydrogels are investigated in details. The mobile ion concentrations, fixed charge density as well as the electric potential distributed in both the interior hydrogel and the exterior bathing solution are also simulated and discussed. It is concluded that the degree of swelling can be improved by increasing the initial fixed charge density, or by decreasing the electrolyte solution concentration. The present studies provide a working platform for designers of smart hydrogel-based BioMEMS devices to enhance the performance of the responsive hydrogels in BioMEMS as critically active sensing/actuating components.

For the transient simulation of swelling kinetics of slow- and fast-response temperature-sensitive neutral hydrogels, the generic biphasic model is developed for mass, momentum, heat and species transfer. The governing equations, constitutive relations and boundary conditions are derived in an Eulerian frame, followed by a non-dimensionalization and reduction via a scale analysis. The reduced model is then reformulated and solved numerically in a Lagrangian frame for a uniformly spherically deformed hydrogel and a constrained hydrogel slab. The interaction parameter, crosslinking parameter and permeabilities are parameter-adapted and validated with experimental equilibrium and transient swelling curves. The scale analysis has provided the proper scales for the mechanisms of transport and deformation, and also indicated that inertia, energy interaction between the polymer and solvent and heat dissipation are negligible for the conditions considered here. Three cases, namely (i) a uniform and (ii) a non-uniform responses to a step change in temperature and (iii) a deformation in a temperature gradient, are investigated. The first yields the information on the deformation and heat transfer during swelling and shrinking. Due to $RePr \ll 1$ for both the slow- and fast-response hydrogels, the heat transfer can be reduced to the steady-state solution, which is simply the new equilibrium temperature. For the deswelling of NC030, the solvent is found to penetrate the hydrogel as a sharp front, which would gradually increase in thickness until reaching the core of the hydrogel. For AS15 and the swelling of NC030, the deformation is more akin to a diffusion process, where the whole interior of the hydrogel experiences the solvent penetration. The second gives an insight into the non-uniform response of a hydrogel slab: the deformation is faster at the corner than the sides, due to the higher stress close to the symmetry lines. For NC030, the evolution of the corner is readily apparent, seeming to drive the deformation of the hydrogel, whereas the corner is only marginally faster for the stiffer AS15, such that AS15 more or less retains its square shape throughout. The third demonstrates the ability of the macroporous temperature-sensitive hydrogel NC030 to align itself with the temperature gradient and attain a curved equilibrium state. During the deformation as well as at equilibrium, the solvent flows from the cooler left to the warmer right sides due to differences in the osmotic pressure. In tandem with a discussion of the cases (i) and (ii), the scales predicted give good agreement with the computed quantities. In summary, the derived model provides an insight into the kinetics of temperature-sensitive neutral hydrogels and the associated phenomena, e.g. solvent flow field, stress level and heat flux. Once the necessary parameters pertaining to the free energy, stress and permeability are obtained, the model can be applied to arbitrary neutral hydrogel, allowing for extensive parameter studies as well as providing answers to questions related to, for example, whether the hydrogel can withstand the stress levels when subjected to temperature gradient. If an external flow is present, the boundary conditions of the hydrogel can easily be modified to account for the stress induced at the interface between the hydrogel and the fluid flow, whence the response of the hydrogel can be studied in flow control applications. Proposed future work may seek to incorporate additional environmental stimuli into the framework of the present model, such as pH, electric field and biochemically active molecules, with a view to provide a comprehensive model that is able to account for multiple stimuli. Furthermore, the one-way coupling assumed here for the case (iii) is only valid if the thermal properties of the hydrogel are similar to that of the surrounding solvent, e.g. for the swollen state of the macroporous NC030. For denser hydrogels, the difference in thermal properties can be expected to affect the surrounding temperature field, which calls for two-way coupling. Such a coupling can be accomplished by invoking a moving mesh for the external field, e.g. via an Arbitrary Lagrangian Eulerian (ALE) formulation, which can be coupled to the model derived here.

References

E.C. Achilleos, K.N. Christodoulou, I.G. Kevrekidis. (2001). A transport model for swelling of polyelectrolyte gels in simple and complex geometries. *Computational and Theoretical Polymer Science*, 11, 63–80.

E.C. Achilleos, R.K. Prud'homme, K.N. Christodoulou, K.R. Gee, I.G. Kevrekidis. (2000a). Dynamic deformation visualization in swelling of polymer gels. *Chemical Engineering Science*, 55, 3335–3340.

E.C. Achilleos, R.K. Prud'homme, I.G. Kevrekidis, K.N. Christodoulou, K.R. Gee. (2000b). Quantifying deformation in gel swelling experiments and simulations. *AIChE Journal*, 46, 2128–2139.

M. Andersson, A. Axelsson, G. Zacchi. (1998). Swelling kinetics of poly (*N*-isopropylacrylamide) gel. *Journal of Controlled Release*, 50, 273–281.

T. Araki, H. Tanaka. (2001). Three-dimensional numerical simulations of viscoelastic phase separation: Morphological characteristics. *Macromolecules*, 34, 1953–1963.

G.A. Athesian, N.O. Chahine, I.M. Basalo, C.T. Hung. (2004). The correspondence between equilibrium biphasic and triphasic material properties in mixture models of articular cartilage. *Journal of Biomechanics*, 37, 391–400.

R.J. Atkin, R.E. Craine. (1976). Continuum theories of mixtures: Basic theory and historical development. *The Quarterly Journal of Mechanics and Applied Mathematics*, 29, 209–244.

Y.H. Bae, T. Okano, S.W. Kim. (1989). Insulin permeation through thermo-sensitive hydrogels. *Journal of Controlled Release*, 9, 271–279.

Y.C. Bae, J.J. Shim, D.S. Soane, J.M. Prausniz. (1993). Representation of vapor–liquid and liquid–liquid equilibria for binary systems containing polymers: Applicability of an extended Flory-Huggins equation. *Journal Applied Polymer Science*, 47, 1193–1206.

J.P. Baker, L.H. Hong, H.W. Blanch, J.M. Prausnitz. (1994). Effect of initial total monomer concentration on the swelling behavior of cationic acrylamide-based hydrogels. *Macromolecules*, 27, 1446–1454.

B. Barriere, L. Leibler. (2003). Kinetics of solvent absorption and permeation through a highly swellable elastomeric network. *Journal of Polymer Science Part B: Polymer Physics*, 41, 166–182.

R.B. Bird, W.E. Stewart, E.N. Lightfoot. (2002). *Transport Phenomena*, 2nd edn. New York: John Wiley and Sons.

E. Birgersson, H. Li, S.N. Wu. (2008). Transient analysis of temperature-sensitive neutral hydrogels. *Journal of the Mechanics and Physics of Solids*, 56, 444–466.

T.M. Birshtein, V.A. Pryamitsyn. (1991). Coil-globule type transitions in polymers. 2. Theory of coil-globule transition in linear macromolecules. *Macromolecules*, 24, 1554–1560.

M.A.T. Bisschops, K. Ch, A.M. Luyben, L.A.M. van der Wielen. (1998). Generalized Maxwell-Stefan approach for swelling kinetics of Dextran gels. *Industrial & Engineering Chemistry Research*, 37, 3312–3322.

R.M. Bowen. (1980). Incompressible porous media models by use of the theory of mixtures. *International Journal of Engineering Science* 18, 1129–1148.

G. Camera-Roda, G.C. Sarti. (1990). Mass transport with relaxation in polymers. *AIChE Journal*, 36, 851–860.

J. Chen, H. Park, K. Park. (1999). Synthesis of superporous hydrogels: Hydrogels with fast swelling and superabsorbent properties. *Journal of Biomedical Materials Research*, 44, 53–62.

S.X. Cheng, J.T. Zhang, R.X. Zhuo. (2003). Macroporous poly(*N*-isopropylacrylamide) hydrogels with fast response rates and improved protein release properties. *Journal of Biomedical Materials Research*, 67, 96–103.

J.M. Chern, W.F. Lee, Y. Hsieh. (2004). Preparation and swelling characterization of poly(*N*-isopropylacrylamide)-based porous hydrogels. *Journal of Applied Polymer Science*, 92, 3651–3658.

D.T. Chung, C.O. Horgan, R. Abeyaratne. (1986). The finite deformation of internally pressurized hollow cylinders and spheres for a class of compressible elastic materials. *International Journal of Solids and Structures*, 22, 1557–1570.

I. Colombo, M. Grassi, M. Fermeglia, R. Lapasin, S. Pricl. (1996). Modelling phase transitions and sorption desorption kinetics in thermo-sensitive gels for controlled drug delivery systems. *Fluid Phase Equilibria*, 116, 148–161.

M. Doi. (1990). *Effects of Viscoelasticity on Polymer Diffusion, Springer Proceedings in Physics*, Vol. 52, Berlin/Heidelberg: Springer-Verlag.

J. Dolbow, E. Fried, H. Ji. (2004). Chemically induced swelling of hydrogels. *Journal of the Mechanics and Physics of Solids*, 52, 51–84.

J. Dolbow, E. Fried, H. Ji. (2005). A numerical strategy for investigating the kinetic response of stimulus-responsive hydrogels. *Computer Methods in Applied Mechanics and Engineering*, 194, 4447–4480.

D.T. Eddington, D.J. Beebe. (2004). Flow control with hydrogels. *Advanced Drug Delivery Reviews*, 56, 199–210.

B. Erman, P.J. Flory. (1986). Critical phenomena and transitions in swollen polymer networks and in linear macromolecules. *Macromolecules*, 19, 2342–2353.
P.J. Flory. (1953). Principles of Polymer Chemistry, Ithaca, New York: Cornell University Press.
I.Y. Galaev, B. Mattiasson. (1999). 'Smart' polymers and what they could do in biotechnology and medicine. *Trends in Biotechnology*, 17, 335–340.
D. Gawin, C.E. Majorana, B.A. Schrefler. (1999). Numerical analysis of hygro-thermal behaviour and damage of concrete at high temperature. *Mechanics of Cohesive-frictional Materials*, 4, 37–74.
S.H. Gehrke. (1993). Synthesis, equilibrium swelling, kinetics, permeability and applications of environmentally responsive gels. *Advances in Polymer Science*, 110, 81–144.
Y.K. Godovsky. (1992). *Thermophysical Properties of Polymers*, New York: Springer-Verlag.
M. Grassi, I. Colombo, R. Lapasin. (2000). Drug release from an ensemble of swellable crosslinked polymer particles. *Journal of Controlled Release*, 68, 97–113.
C.A. Grattoni, H.H. Al-Sharji, C. Yang, A.H. Muggeridge, R.W. Zimmerman. (2001). Rheology and permeability of cross-linked polyacrylamide gel. *Journal of Colloid and Interface Science*, 240, 601–607.
T. Hino, J.M. Prausnitz. (1998). Molecular thermodynamics for volume-change transitions in temperature-sensitive polymer gels. *Polymer*, 39, 3279–3283.
S. Hirotsu. (1987). Phase transition of a polymer gel in pure and mixed solvent media. *Journal of the Physical Society of Japan*, 56, 233–242.
S. Hirotsu. (1991). Softening of bulk modulus and negative Poisson's ratio near the volume phase transition of polymer gels. *Journal of Chemical Physics*, 94, 3949–3957.
S. Hirotsu. (1993). Coexistence of phases and the nature of first-order phase transition in poly-*N*-isopropylacrylamide gels. *Advances in Polymer Science*, 110, 1–26.
S. Hirotsu, Y. Hirokawa, T. Tanaka. (1987). Volume-phase transitions of ionized *N*-isopropylacrylamide gels. *Journal of Chemical Physics*, 87, 1392–1395.
A. S. Hoffman. (2002). Hydrogels for biomedical applications. *Advanced Drug Delivery Reviews*, 43, 3–12.
M.H. Holmes. (1986). Finite deformation of soft tissue: Analysis of a mixture model in uni-axial compression. *ASME Journal of Biomechanical Engineering*, 108, 372–381.
M.H. Holmes, W.M. Lai, V.C. Mow. (1985). Singular perturbation analysis of the nonlinear, flow-dependent compressive stress relaxation behavior of articular cartilage. *ASME Journal of Biomechanical Engineering*, 107, 206–218.
M.H. Holmes, V.C. Mow. (1990). The nonlinear characteristics of soft gels and hydrated connective tissues in ultrafiltration. *Journal of Biomechanics*, 23, 1145–1156.
G.H. Holzapfel. (2000). *Nonlinear Solid Mechanics, A Continuum Approach for Engineering*, Chichester, UK: John Wiley & Sons.
J.M. Huyghe, J.D. Janssen. (1999). Thermo-chemo-electro-mechanical formulation of saturated charged porous solids. *Transport in Porous Media*, 34, 129–141.
H. Ji, H. Mourad, E. Fried, J. Dolbow. (2006). Kinetics of thermally induced swelling of hydrogels. *International Journal of Solids and Structures*, 43, 1878–1907.
E. Kati. (1997). Volume-phase transition of *N*-isopropylacrylamide gels induced by hydrostatic pressure. *Journal of Chemical Physics*, 106, 3792–3797.
S.J. Kim, S.J. Park, S.I. Kim. (2003a). Synthesis and characteristics of interpenetrating polymer network hydrogels composed of poly(vinyl alcohol) and poly(*N*-isopropylacrylamide). *Reactive and Functional Polymers*, 55, 61–67.
S.J. Kim, G.Y. Yoon, Y.M. Lee, S.I. Kim. (2003b). Electrical sensitive behavior of poly(vinyl alcohol)/poly(diallyldimethylammonium chloride) IPN hydrogel. *Sensors and Actuators B: Chemical*, 88, 286–291.
S.J. Kim, C.K. Lee, S.I. Kim. (2004a). Electrical/pH responsive properties of poly(2-acrylamido-2-methylpropane sulfonic acid)/hyaluronic acid hdyrogels. *Journal of Applied Polymer Science*, 92, 1731–1736.

S.J. Kim, S.G. Yoon, S.M. Lee, S.H. Lee, S.I. Kim. (2004b). Electrical sensitivity behavior of a hydrogel composed of polymethacrylic acid/poly(Vinyl alcohol). *Journal of Applied Polymer Science*, 91, 3613–3617.

A.K. Lele, M.V. Badiger, M.M. Hirve, R.A. Mashelkar. (1995). Thermodynamics of hydrogen-bonded polymer gel-solvent systems. *Chemical Engineering Science*, 50, 3535–3542.

A.K. Lele, I. Devotta, R.A. Mashelkar. (1997). Predictions of thermoreversible volume phase transitions in copolymer gels by lattice-fluid-hydrogen-bond theory. *Journal of Chemical Physics*, 106, 4768–4772.

H. Li, J. Chen, K.Y. Lam. (2004a). Multiphysical modelling and meshless simulation of electric-sensitive hydrogels. *Journal of Polymer Science Part B: Polymer Physics*, 42, 1514–1531.

H. Li, T.Y. Ng, J.Q. Cheng, K.Y. Lam. (2003). Hermite-cloud: A novel true meshless method. *Computational Mechanics*, 33, 30–41.

Y. Li, T. Tanaka. (1990). Kinetics of swelling and shrinking of gels. *Journal of Chemical Physics*, 92, 1365–1371.

Y. Li, T. Tanaka. (1992). Phase transitions of gels. *Annual Review of Material Science*, 22, 243–276.

H. Li, Z. Yuan, K.Y. Lam, H.P. Lee, J. Chen, J. Hanes, J. Fu. (2004b). Model development and numerical simulation of electric-stimulus-responsive hydrogels subject to an externally applied electric field. *Biosensors and Bioelectronics*, 19, 1097–1107.

R.M. Luo, H. Li, K.Y. Lam. (2008). Modeling and analysis of pH-electric-stimuliresponsive hydrogels. *Journal of Biomaterials Science – Polymer Edition*, 19, 1597–1610.

S.R. Lustig, J.M. Caruthers, N.A. Peppas. (1992). Continuum thermodynamics and transport theory for polymer-fluid mixtures. *Chemical Engineering Science*, 47, 3037–3057.

L. Mao, Y. Hu, Y. Piao, X. Chen, W. Xian, D. Piao. (2005). Structure and character of artificial muscle model constructed from fibrous hydrogel. *Current Applied Physics*, 5, 426–428.

G. Maurer, J.M. Prausnitz. (1996). Thermodynamics of phase equilibrium for systems containing gels. *Fluid Phase Equilibria*, 115, 113–133.

R. Moerkerke, R. Koningsveld, H. Berghmans, K. Dusek, K. Solc. (1995). Phase transitions in swollen networks. *Macromolecules*, 28, 1103–1107.

V.C. Mow, S.C. Kuei, W.M. Lai, C.G. Armstrong. (1980). Biphasic creep and stress relaxation of articular cartilage in compression: Theory and experiments. *Journal of Biomechanical Engineering*, 102, 73–84.

P.A. Netti, F. Travascio. (2003). Coupled macromolecular transport and gel mechanics: Poroviscoelastic approach. *AIChE Journal*, 49, 1580–1596.

D.A. Nield, A. Bejan. (1998). Convection in Porous Media, 2nd ed. New York: Springer Verlag.

K. Nishizawa, T. Shirose, O. Itoh. (1981). *Disposable Diaper*, United States Patent 4269188.

R.W. Ogden. (1997). *Non-linear Elastic Deformations*, New York: Dover publications.

K.S. Oh, Y.C. Bae. (1998). Swelling behavior of submicron gel particles. *Journal of Applied Polymer Science*, 69, 109–114.

T. Okajima, I. Harada, K. Nishio, S. Hirotsu. (2002). Kinetics of volume phase transition in poly(*N*-isopropylacrylamide) gels. *Journal of Chemical Physics*, 116, 9068–9077.

E.D. Oliveira, A.F.S. Silva, R.F.S. Freitas. (2004). Contributions to the thermodynamics of polymer hydrogel systems. *Polymer*, 45, 1287–1293.

A. Onuki. (1989). Theory of pattern formation in gels: Surface folding in highly compressible elastic bodies. *Physical Review A*, 39, 5932–5948.

Y. Osada, J.P. Gong. (1993). Stimuli-responsive polymer gels and their application to chemomechanical systems. *Progress in Polymer Science*, 18, 187–226.

K. Otake, H. Inomata, M. Konno, S. Saito. (1989). A new model for the thermally induced volume phase transition of gels. *Journal of Chemical Physics*, 91, 1345–1350.

G.V. Pamuk. (2004). Controlling water dynamics in scots pine (pinus sylvestris L.) seeds before and during seedling emergence. Ph.D Thesis, Umeå University, Sweden.

O. Pohl. (2004). Now, diaper technology takes on a desert, *The New York Times, Science Desk*, July 20.
A.F. Prokop, S. Vaezy, M.L. Noble, P.J. Kaczkowski, R.W. Martin, L.A. Crum. (2003). Polyacrylamide gel as an acoustic coupling medium for focused ultrasound therapy. *Ultrasound in Medicine & Biology*, 20, 1351–1358.
Y. Qiu, K.N. Park. (2001). Environment-sensitive hydrogels for drug delivery. *Advanced Drug Delivery Reviews*, 53, 321–339.
I. Roy, N. Gupta. (2003). Smart polymeric materials: Emerging biochemical applications. *Chemistry and Biology*, 10, 1161–1171.
E. Samson, J. Marchand, J.L. Robert, J.P. Bournazel. (1999). Modelling ion diffusion mechanisms in porous media. *International Journal for Numerical Methods in Engineering*, 46, 2043–2060.
M.S. Sanchez, M.M. Prada, J.L.G. Ribelles. (2004). Thermal transitions in PHEA hydrogels by thermomechanical analysis. A comparison with DSC data. *European Polymer Journal*, 40, 329–334.
K. Sekimoto. (1991). Thermodynamics and hydrodynamics of chemical gels. *Journal de Physique II (France)*, 1, 19–36.
K. Sekimoto, M. Doi. (1991). Dynamics of interface of gels undergoing volume phase transition. *Journal de Physique II (France)*, 1, 1053–1066.
M. Shibayama, T. Tanaka. (1993). Volume Phase Transition and Related Phenomena of Polymer Gels. In: *Responsive Gels: Volume Transitions I, Advances in Polymer Science* Vol. 109, K. Dusek (Ed.) Berlin: Springer-Verlag, pp. 1–62.
H. Shirota, N. Endo, K. Horie. (1998). Volume phase transition of polymer gel in water and heavy water. *Chemical Physics*, 238, 487–494.
J. Singh, M.E. Weber. (1996). Kinetics of one-dimensional gel swelling and collapse for large volume change. *Chemical Engineering Science*, 51, 4499–4508.
D. Snita, M. Paces, J. Lindner, J. Kosek, M. Marek. (2001). Nonlinear behaviour of simple ionic systems in hydrogel in an electric field. *Faraday Discuss*, 120, 53–66.
T. Tanaka, D.J. Fillmore. (1979). Kinetics of swelling of gels. *Journal of Chemical Physics*, 70, 51214–1218.
T. Tanaka, D. Fillmore, S.T. Sun, I. Nishio, G. Swislow, A. Shah. (1980). Phase transition in ionic gels. *Physical Review Letters*, 45, 1636–1639.
T. Tanaka, L.O. Hocker, G.B. Benedek. (1973). Spectrum of light scattered from a viscoelastic gel. *Journal of Chemical Physics*, 59, 5151–5159.
R. Taylor, R. Krishna. (1993). *Multicomponent Mass Transfer*, John Wiley & Sons.
M. Tokita, T. Tanaka. (1991). Friction coefficient of polymer networks of gels. *Journal of Chemical Physics*, 95, 4613–4619.
T. Wallmersperger, B. Kroplin, R.W. Gulch. (2004). Coupled chemo-electro-mechanical formulation for ionic polymer gels – numerical and experimental investigations. *Mechanics of Materials*, 36, 411–420.
C. Wang, 1Y. Li, Z. Hu. (1997). Swelling kinetics of polymer gels. *Macromolecules*, 30, 4727–4732.
J.A. Wesselingh, A.M. Bollen. (1997). Multicomponent diffusivities from the free volume theory. *IChemE Journal Transactions*, 75A, 590–602.
J.A. Wesselingh, R. Krishna. (2000). *Mass Transfer in Multicomponent Mixtures*, Delft: Delft University Press.
S. Whitaker. (1999). *The Method of Volume Averaging*, Dordrecht: Kluwer.
B.A. Wolf. (1984). Thermodynamic theory of flowing polymer solutions and its application to phase separation. *Macromolecules*, 17, 615–618.
C.W. Wolgemuth, A. Mogilner, G. Oster. (2004). The hydration dynamics of polyelectrolyte gels with applications to cell motility and drug delivery. *European Biophysics Journal*, 33, 146–158.
S. Wu, H. Li, J.P. Chen, K.Y. Lam. (2004). Model investigation of hydrogel volume transition. *Macromolecular Theory and Simulations*, 13, 13–29.

X.Z. Zhang, R.X. Zhuo. (2000a). Preparation of fast responsive, thermally sensitive poly(*N*-isopropylacrylamide) gel. *European Polymer Journal*, 36, 2301–2303.

X.Z. Zhang, R.X. Zhuo. (2000b). Novel synthesis of temperature-sensitive poly(*N*-isopropylacrylamide) hydrogel with fast deswelling rate. *European Polymer Journal*, 36, 643–645.

X. Zhou, Y.C. Hon, S. Sun, A.F.T. Mak. (2002). Numerical simulation of the steady-state deformation of a smart hydrogel under an external electric field. *Smart Materials and Structures*, 11, 459–467.

M. Zrínyi, A. Szilágyi, G. Filipcsei, J. Fehér, J. Szalma, G. Móczár. (2001). Smart gel-glass based on the responsive properties of polymer gels. *Polymers for Advanced Technologies*, 12, 501–505.

Chapter 6
Novel Models for Smart Hydrogel Responsive to Other Stimuli: Glucose Concentration and Ionic Strength

6.1 Introduction

This chapter introduces the author's latest research work, which covers the modelling of the glucose-sensitive hydrogel and the ionic strength-sensitive hydrogel, respectively.

First, a multiphysics model, termed the multi-effect-coupling glucose-stimulus (MECglu) model, is developed for simulation of the response characteristics of soft smart hydrogel to change in environmental glucose concentration. The model considers the effect of the glucose oxidation reaction catalysed by enzymes, where the enzymes include the glucose oxidase and the catalase. It is composed of the Nernst–Planck diffusion–reaction equations for mobile species concentrations in the solvent, the Poisson equation for electric potential and nonlinear mechanical equation for large deformation of the hydrogel that arises due to the conversion of chemical energy to mechanical energy. Based on the theory of chemo-electro-mechanical coupled fields, the formulation of the fixed charge groups bound onto the crosslinked polymeric network chains is associated with the change of the ambient solution pH. The MECglu model is examined by comparison between the steady-state simulation and experimental equilibrium swelling curves published in open literature, and good agreement is achieved. The model can be employed for parameter study to ascertain the impact of various solvent and material parameters on the responsive swelling behaviour of the hydrogel. One key parameter here is the glucose concentration, which varies within the range of practical physiological glucose concentrations from 0 to 16.5 mM (300 mg/ml), in order to support the design and optimization of the insulin delivery system that is based on the glucose-sensitive hydrogels with the immobilized glucose oxidase and catalase. The model is able to predict the distributive profiles of reacting and diffusive species concentrations, the electric potential, and the displacement as well as the swelling ratio of the glucose-sensitive hydrogels when various hydrogel material properties and environmental solution conditions are considered.

Second, a chemo-electro-mechanical model, called the multi-effect-coupling ionic-strength-stimulus (MECis) model, is presented for simulation of the deformable ionic strength-sensitive hydrogels responding to the change in the ionic strength of surrounding solution. As well known, the ionic strength of a solution is

H. Li, *Smart Hydrogel Modelling*, DOI 10.1007/978-3-642-02368-2_6,

also one of important environmental stimuli to smart hydrogels, like solution pH, externally applied electric voltage or surrounding temperature. In other words, the ionic strength-sensitive hydrogel is one of important smart biomaterials, like the pH-, electrical- or thermal-sensitive hydrogels. However, extensive literature search reveals that most of studies were experimental based, and few of them were involved in theoretical model development. Furthermore, so far no study has been found for transient modelling of responsive characteristics of the ionic strength-sensitive hydrogels, although some of them considered other stimuli, such as pH, but they took the ionic strength of the surrounding solution as an environmental condition only. Therefore, it is necessarily required to provide an efficient platform for theoretically deeper understanding of responsive behaviour of the ionic strength-sensitive hydrogels, as the smart hydrogels possess the unique properties such as controllable swelling/shrinking, sorption capacities, mechanical properties, permeability and surface properties, and they have increasingly attracted attention for applications in BioMEMS, drug delivery, artificial organ and so on. Therefore, the present objective is to develop the multiphysics MECis model for simulation of the mechanism of volume transition of the smart hydrogel in response to the change in the ionic strength of bathing solution. The MECis model is composed of coupled nonlinear partial differential equations. One of them is Poisson–Nernst–Planck system, which incorporates the effect of the ionic strength through the chemical activity coefficient and describes the ionic fluxes due to diffusion and electrophoresis as well as convection in both the hydrogel and bathing solution, coupled with electrical potential distributed in whole domain. The fixed charge density is characterized by Langmuir isotherm theory with consideration of the effect of the ionic strength. The other is the mechanical finite deformation governing equation to predict how the hydrogel deforms induced by osmotic pressure and fixed charge repulsion. It is demonstrated that the MECis model is capable of efficiently simulating the distributive variations of diffusive ionic species, electrical potential and mechanical deformation for the ionic strength-sensitive hydrogels immersed in a solution with change in ionic strength, as the swelling/deswelling behaviour of the hydrogel is considerably dependent on environment conditions as well as the physical and chemical material properties of the hydrogel.

6.2 Multi-Effect-Coupling Glucose-Stimulus (MECglu) Model for Glucose-Sensitive Hydrogel

Diabetes is one of the most serious health concerns in the twenty-first century, and its worldwide prevalence is predicted to double to 366 millions by 2030 from 171 millions in 2000 (WHO, 2008). For insulin-dependent diabetes, the administration of insulin is one of major treatments to control glycaemia. In order to significantly improve the lives of diabetic patients, the systems for administration of insulin should be continuous, non-invasive or minimally invasive closed-loop systems, and they should attempt to deliver insulin in direct response to blood glucose level, and to achieve the feedback-controlled release of insulin (Traitel et al., 2000; Kim and Park, 2001; Guiseppi-Elie et al., 2005). Development of self-regulated insulin delivery systems is one of the most challenging targets in area of controlled drug delivery. Fortunately, the glucose-sensitive hydrogels, some of which incorporate the immobilized glucose oxidase (GOx) and catalase enzymes into the pH-sensitive hydrogels, are elegant and highly promising biomaterials for development of smart insulin delivery systems (Qiu and Park, 2001; Abdekhodaie and Wu, 2005; Ulijn et al., 2007).

So far various self-regulated insulin delivery systems based on the pH-sensitive hydrogels have been developed, to which the glucose oxidase sensitive to the glucose concentration is loaded for regulation of insulin release (Ishihara and Matsui, 1986; Albin et al., 1987; Parker et al., 1999; Cao et al., 2001; Baldi et al., 2003). The glucose-oxidase-loaded pH-sensitive hydrogel system is a kind of pH-responsive polymeric polybasic hydrogels, in which the glucose oxidase and the catalase are immobilized. If the system is immersed in a glucose buffer solution, the glucose in the solution diffuses into the hydrogel, and then the glucose oxidase catalyses the conversion of the glucose to the gluconic acid, thereby lowering the pH level within the hydrogel. The pH drop results in the change in the osmotic pressure because of the difference of concentrations between the hydrogel and surrounding solution, causing the deswelling of the hydrogel. Typical examples of suitable polymers commonly include hydroxyethyl methacrylate (HEMA) and *N,N*-dimethylaminoethyl methacrylate (DMA), poly(HEMA-*co*-DMAEMA) (Ishihara and Matsui, 1986; Albin et al., 1987; Traitel et al., 2000, 2003; Brahim et al., 2002), poly[(diethylaminoethyl methacrylate)-hydroxyethyl methacrylate-*graft*-(ethylene glycol)] poly(DEAEM-HEMA-*g*-EG) (Podual and Peppas, 2005), poly[(diethylaminoethyl methacrylate)-*graft*-(ethylene glycol)] poly(DEAEM-*g*-EG) (Podual and Peppas, 2005), poly(methacrylic acid-*g*-ethylene glycol) poly(MAAc-*g*-EG) (Parker et al., 1999; Cao et al., 2001), *N*-isopropylacrylamide, methacrylic acid, ethyleneglycol dimethacrylate (NIPA-MAA-EGDMA) (Zhang and Wu, 2002; Misra and Siegel, 2002; Dhanarajan and Siegel, 2005), *N,N*-dimethylacrylamide (DMAAm) (Kang and Bae, 2003), p(MPBA-*co*-AAm) (Siegel et al., 2004.) and NIPA-MAA (Suzuki and Kumagai, 2003).

Numerous experiments on the glucose-sensitive hydrogels have thus far been carried out. However, only a few theoretical studies have been made for modelling and simulation. For example, Parker and Schwartz (1987) developed a ping-pong kinetics model for analysis of the overall reaction rate of the immobilized glucose oxidase. Albin et al. (1987) simulated a cationic glucose-sensitive membrane, based on the assumption that the diffusivity of each species within the membrane is equal to that in the bulk solution and independent of the swelling or pH gradient of the membrane. By the work of Albin et al. (1987), Klumb et al. (1992) discussed various design configurations to overcome the oxygen limitation in the insulin delivery system. Gough et al. (1985, 1988) carried out steady-state and transient simulations of a cylindrical glucose sensor. However, all the models mentioned above are unable to predict the mechanical deformation of the hydrogel, where the diffusion and reaction of mobile species are simulated only with the effect of the enzymes. By formulating the mesh size of the polymeric network to characterize the volume swelling ratio, Abdekhodaie and Wu (2005) proposed a theoretical model for the swelling-dependent diffusivity of the species and the oxygen limitation for a cationic glucose-sensitive membrane, in which the diffusion of non-electrolyte species is considered only, and did not account for electrolyte species. It is thus difficult for this model to provide the distribution of electric potential. In brief, all the models discussed above fail to take account of two important effects, namely the electric potential and the ionic strength of electrolyte species in surrounding solution, and their impact on the response deformation of the glucose-sensitive hydrogels.

6.2.1 Development of the MECglu Model

Due to the shortcomings of the above models, a multiphysics model is presented here for simulation of the glucose stimulus-responsive hydrogel, which is entitled the multi-effect-coupling glucose-stimulus (MECglu) model. It is based on the theory of chemo-electro-mechanical coupled fields and considers the effect of the glucose oxidation reaction, which is catalysed by the two enzymes, the glucose oxidase and catalase. The MECglu model comprises the Nernst–Planck equations for the mobile species concentrations, the Poisson equation for the electric potential and a nonlinear mechanical governing equation for the finite deformation of the hydrogel, which captures the conversion of chemical energy to mechanical energy. The formulation of the fixed charge groups bound onto the crosslinked polymeric network chains is associated with the change of the ambient solution pH. The model consists of a system of coupled nonlinear partial differential governing equations, and it can be solved numerically with the meshless Hermite-cloud method (Li et al., 2003) for simulation of the equilibrium and kinetics deformation of smart hydrogel responding to an environmental glucose stimulus, when immersed in a glucose buffer solution. For validation of the MECglu model, a comparison for the one-dimensional equilibrium swelling of a glucose-sensitive hydrogel is carried out between steady-state computational simulations and published experimental curve data, in which good agreements are obtained. The model is applicable for parameter study on the responsive equilibrium and kinetics swelling of the hydrogel within the practical physiological glucose concentration ranging from 0 to 16.5 mM (300 mg/ml), which allows for design and optimization of the insulin delivery system based on glucose-sensitive hydrogels with the immobilized glucose oxidase and catalase. The (MECglu) model can be employed for the distributive variations of reacting and diffusive species concentrations, the electric potential, and the displacement as well as the swelling/deswelling ratio of the glucose stimulus-responsive hydrogels with various environmental conditions and material properties, such as the concentrations of glucose, oxygen, enzyme and hydrogen ion H^+ in surrounding solution, the ionic strength and pH of bathing solution, the initially fixed charge density, initial geometrical size and Young's modulus of the hydrogels.

6.2.1.1 Mechanism and Assumptions

According to the thermodynamics theory, a hydrogel swells due to the diffusive small molecules mixing with the crosslinked polymeric network, where the configuration entropy of mixture increases, while the configuration entropy of the network decreases (Birgersson et al., 2008). If the changes of the two entropies balance each other, the system of the hydrogel and solvent will reach equilibrium state (Hong et al., 2008). Based on the response mechanism of the hydrogels to change in the glucose concentration, the glucose-sensitive hydrogels can generally be classified into three categories, namely the lectin-loaded glucose-sensitive hydrogels, the acid moieties and the glucose-oxidase-loaded pH-sensitive hydrogels. For the glucose-oxidase-loaded pH-sensitive hydrogels, the mechanism of equilibrium swelling may be described in five steps: (1) the diffusion of glucose molecule into the hydrogel; (2) the enzyme reaction occurring within the hydrogel and converting the glucose into the gluconic acid; (3) the change in ionization of the fixed charged groups; (4) the generation of osmotic pressure due to the difference of the concentrations between the interior hydrogel and exterior solution and (5) the swelling of the hydrogel, as illustrated in Fig.6.1.

It is well known that the glucose oxidase (GOx) can convert the glucose into the lactone species (Glucono–δ–lactone) and the hydrogen peroxide (H_2O_2) (Whitaker, 1994). The intermediate lactone species produced undergoes a spontaneous hydrolysis to gluconic acid, i.e.,

$$\text{Glucose} + O_2 \xrightarrow{\text{GOx}} \text{Glucono} - \delta - \text{lactone} + H_2O_2 \tag{6.1}$$

$$\text{Glucono} - \delta - \text{lactone} + H_2O \rightarrow \text{Gluconic Acid} \tag{6.2}$$

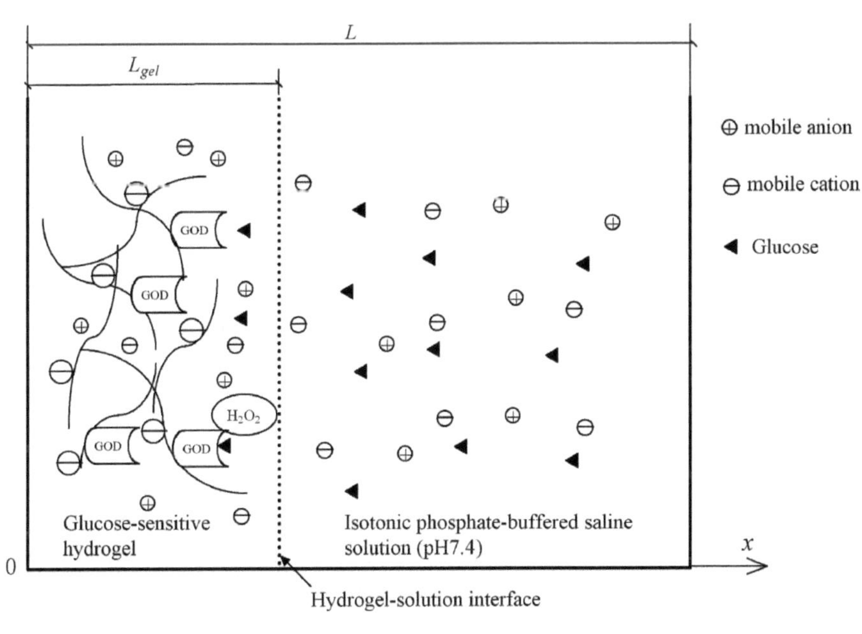

Fig. 6.1 Schematic of ion transport and reaction in the glucose-sensitive hydrogel system

The kinetics of the glucose oxidase can generally be characterized by the Michaelis–Menten kinetics, which is often approximated by the ping-pong kinetics (Parker and Schwartz, 1987). In the present MECglu model, the glucose oxidase is assumed to follow the ping-pong kinetics, namely moving from a fully oxidized state to a fully reduced form and then back to the oxidized state in a catalytic cycle. By combination of conventional and rapid reaction techniques, the mechanism of the overall reaction of the glucose oxidase can be characterized as follows (Parker and Schwartz, 1987; Kurnik et al., 1998):

$$E_0 + G \underset{k_{-1}}{\overset{k_{+1}}{\rightleftarrows}} E_B \xrightarrow{k_{+2}} E_R + \text{D - gluconic acid} \tag{6.3}$$

$$E_R + O_2 \xrightarrow{k_{+3}} E_0 + H_2O_2 \tag{6.4}$$

where E_0 denotes the oxidized form of the glucose oxidase, E_R the reduced form of the glucose oxidase, G the glucose. E_B indicates the enzyme substrate complex and E_B=E_0–G. k_{-1}, k_{+1}, k_{+2} and k_{+3} represent the constants of reaction rate of the corresponding enzyme reactions.

Usually the produced hydrogen peroxide H_2O_2 inhibits the glucose oxidase reaction (Traitel et al., 2000). In order to reduce the inhibition due to the hydrogen peroxide, the catalase is incorporated with the glucose oxidase to catalyse the dissociation of the hydrogen peroxide H_2O_2, in which the oxygen limitation is improved by adding the oxygen of 0.5 mol to the reaction catalysed by the glucose oxidase (Tse and Gough, 1987)

$$H_2O_2 \xrightarrow{\text{Catalase}} H_2O + \frac{1}{2}O_2 \tag{6.5}$$

in which it is assumed that the catalase incorporated in the hydrogel is sufficient enough to reduce all the hydrogen peroxide instantaneously.

The overall stoichiometry of the glucose oxidation is thus written as

$$\text{Glucose} + \frac{1}{2}O_2 \xrightarrow{\text{GOx/Catalase}} \text{Gluconate} + H^+ \tag{6.6}$$

The MECglu model incorporates the effects of the diffusion of mobile species and the reaction of the glucose, oxygen and gluconic acid within the glucose-oxidase-loaded pH-sensitive hydrogel. It is known that the glucose is converted to the gluconic acid after the glucose and oxygen diffuse into the hydrogel from the bathing solution. The enzyme reaction results in a drop in pH within the hydrogel, and thereby triggers the deswelling of the hydrogel. According to the above mechanism, the MECglu model makes the assumptions as follows:

(1) The hydrogel is always maintained in isothermal condition. The rates of the enzyme reactions, such as k_{-1}, k_{+1}, k_{+2} and k_{+3}, are constant in the enzyme reactions (6.3) and (6.4).

(2) The enzymes include the glucose oxidase and the catalase, and they are immobilized and distributed uniformly within the hydrogel. The activities of the enzymes are constant.

(3) No inhibition of the substrate or product of the chemical reaction occurs. The glucose oxidase follows the ping-pong kinetics, namely moving from a fully oxidized state to a fully reduced form and then back to the oxidized state in a catalytic cycle. There is sufficient catalase contained in the hydrogel so that the catalase can always reduce the hydrogen peroxide H_2O_2 to O_2 and H_2O immediately, where the hydrogen peroxide H_2O_2 is a byproduct in the glucose oxidase reaction. It is not desired that there is not sufficient catalase in the hydrogel, since the produced hydrogen peroxide then attacks the glucose oxidase, deactivates the enzyme and finally inhibits the glucose oxidase reaction (Traitel et al., 2000).

(4) The diffusivity of each mobile species or reactant inside the hydrogel, including the glucose and oxygen, is equal to its counterpart in the surrounding aqueous solution, and independent of the gel deformation, due to the macroporous nature of the hydrogel. The diffusivities are independent of the degree of swelling/deswelling and the local pH gradient. The diffusivity of the gluconic acid is also equal to that of glucose.

(5) All acid/base reactions are in local equilibrium, since they are much faster than the reaction catalysed by the glucose oxidase.

6.2.1.2 Formulation in Deformed Configuration

In order to develop the MECglu model, we should consider at least six diffusive species in the model system, namely the glucose (glu), oxygen (ox), gluconic acid (ga), hydrogen ion H^+, cation and anion species. For each diffusive species k (k= glu, ox, ga, H^+, cation and anion), the law of mass conservation is employed for characterization of the mass transport between the hydrogel and the surrounding solution, i.e.,

$$\frac{\partial}{\partial t}\left(\phi^{(w)} c_k\right) + \nabla \cdot \mathbf{n}_k = \nu_k r \; (k = \text{glu,ox,...,}N) \tag{6.7}$$

where $\phi^{(w)}$ is the volume fraction of the solvent within the hydrogel, c_k, $\mathbf{n}_k$ and ν_k indicate the concentration (mM), molar flux (mM/s) and stoichiometric coefficient in the chemical reaction for the kth diffusive species (k=glu,ox,. . .,N). N is the number of total diffusive species and r is the rate of chemical reaction representing a source term. For non-ideal ionic solution, the steady-state Nernst–Planck equations can be obtained by simplifying Eq. (6.7) for characterization of the total flux of the species k through the hydrogel membrane (Plawsky, 2001)

$$\nabla \cdot \mathbf{n}_k = \nu_k r \; (k = \text{glu,ox,...,}N) \tag{6.8}$$

$$\mathbf{n}_k = \mathbf{J}_{k(D)} + \mathbf{J}_{k(E)} + c_k \mathbf{v}^{(w)} = -\mathbf{D}_k \left[\nabla c_k + \frac{z_k F}{RT} c_k \nabla \psi + c_k \nabla(\ln \gamma_k)\right] + c_k \mathbf{v}^{(w)} \tag{6.9}$$

where $\mathbf{D}_k$, z_k and γ_κ denote the diffusivity tensor (m^2/s), valence number and chemical activity coefficient for the kth diffusive species. $\mathbf{v}^{(w)}$ is the velocity of the solvent flow, ψ is the electrostatic potential. F, R and T are the Faraday's constant (9.6487×10^4C/mol), universal gas constant (8.314 J/mol $\cdot$ K) and absolute temperature (K), respectively. On the right-hand side of Eq. (6.9), $\mathbf{J}_{k(D)}$ represents the mass flux due to the molecular diffusion, $\mathbf{J}_{k(E)}$ is the mass flux of charged species responding to the electric potential gradient and $c_k\mathbf{v}^{(w)}$ is the convection flux of the solvent due to osmosis and electro-osmosis. When the hydrogel is placed in an unstirred solution and the pores of the hydrogel are sufficiently small, the ion transport due to convection can be neglected reasonably (Nikonenko, et al., 2003).

For implementation of Eq. (6.7), the rate r of chemical reaction is required to characterize the enzyme reaction of the glucose. According to the mechanism of the ping-pong kinetics, the rate of the enzyme reaction is written as follows: (Kurnik et al., 1998)

$$r = \frac{V_{\max} c_{\text{glu}} c_{\text{ox}}}{c_{\text{ox}}(c_{\text{glu}} + K_{\text{glu}}) + K_{\text{ox}} c_{\text{glu}}} \tag{6.10}$$

where c_{glu} and c_{ox} are the concentrations of the glucose and oxygen, respectively. $V_{\max}$ is the maximum reaction velocity of the enzyme glucose oxidase and $V_{\max} = k_{+2}[E_0]$. K_{glu} and K_{ox} are the Michaelis constants for the glucose and oxygen at infinite concentrations of the other substrate and $K_{\text{glu}} = (k_{-1} + k_{+1})/k_{+1}$ and $K_{\text{ox}} = k_{+2}/k_{+3}$, respectively.

It is well known that the oxygen concentration in the body fluid always remains saturated, namely c_{ox}=0.274 mM (Abdekhodaie and Wu, 2005), and the practical physiological glucose concentration ranges from 0 to 16.5 mM (300 mg/ml). Compared with the respective Michaelis–Menten constants K_{ox}=6992 mM, $c_{\text{ox}} << K_{\text{ox}}$. If this oxygen limitation is assumed, Eq. (6.10) can be simplified as

$$r = V_{\max}\left[\frac{c_{\text{ox}}}{K_{\text{ox}}} - \left(\frac{c_{\text{ox}}}{K_{\text{ox}}}\right)^2\right] \tag{6.11}$$

In order to include the effect of the electric potential coupled with the diffusive ionic concentrations, the MECglu model incorporates the following Poisson equation defined in the whole computational domain covering both the hydrogel and surrounding medium (Luo et al., 2007a, b):

$$\nabla^2\psi = -\frac{F}{\varepsilon\varepsilon_0}\left(\sum_k z_k c_k + z_f c_f\right)(k = \text{glu,ox,...,}N) \tag{6.12}$$

where ϵ is the relative dielectric constant of surrounding medium and ϵ_0 the vacuum permittivity or dielectric constant (8.85418×10^{-12}C^2/N $\cdot$ m^2). z_{glu}=z_{ox}=0, z_f and c_f are the valence number and density of the charge groups fixed onto the polymeric hydrogel network chains, and one can have (Luo et al., 2007)

$$c_f = \frac{c_{f0}^s \cdot K \cdot \phi_0^s}{(K + c_{\text{H}})\sqrt{1 + 2F_1(\mathbf{E}) + 4F_2(\mathbf{E}) + 8F_3(\mathbf{E})}} \tag{6.13}$$

where c_{f0}^s is the concentration of the fixed charge groups at dry state, K is the dissociation constant of the fixed charge groups, ϕ_0^s is the volume fraction of the polymeric network matrix in the initial configuration and c_{H} is the concentration of hydrogen ion H^+ within the hydrogel. $F_1(\mathbf{E})$ $F_2(\mathbf{E})$ and $F_3(\mathbf{E})$ are the first, second and third invariants of Green strain tensor $\mathbf{E}$ of the apparent polymeric solid phase, respectively.

When the system considered is in equilibrium state, the swelling stress is balanced by the elastic retractive force exerted by the polymeric chain network of the hydrogel in order to maintain the current hydration state. Then the equilibrium of linear momentum leads to the mechanical governing equation as follows:

$$\nabla \cdot \sigma + \rho \mathbf{b} - \rho \dot{\mathbf{V}} = \mathbf{f} \tag{6.14}$$

where $\boldsymbol{\sigma}$ is the Cauchy stress tensor of the hydrogel, $\mathbf{u}$ is the displacement vector, ρ is the density of hydrogel membrane, $\mathbf{b}$ is body force, $\dot{\mathbf{V}}$ is the acceleration and $\mathbf{f}$ is the external loading. If the steady-state simulation is carried out ($\rho\dot{\mathbf{V}}$=0), neither the external force ($\mathbf{f}$=0) nor the body force ($\mathbf{b}$=0) is considered.

6.2.1.3 Formulation in Reference Configuration

So far the governing equations of the MECglu model have been formulated in the deformed configuration, where a moving mesh is required due to the deformation of the hydrogel. Therefore it is necessary to recast the model in the undeformed configuration. As such, the Nernst–Planck and Poisson equations are rewritten as follows:

$$\nabla_X \cdot \mathbf{N}_k = J v_k r \; (k = \text{glu,ox,...,}N) \tag{6.15}$$

$$\mathbf{N}_k = -J D_k \mathbf{C}^{-1} \left[\nabla_X c_k + \frac{z_k F}{RT} c_k \nabla_X \psi \right] \tag{6.16}$$

$$r = V_{\max} \left[\frac{c_{\mathrm{ox}}}{K_{\mathrm{ox}}} - \left(\frac{c_{\mathrm{ox}}}{K_{\mathrm{ox}}} \right)^2 \right] \tag{6.17}$$

$$\nabla_X \cdot \left(J \mathbf{C}^{-1} \nabla_X \psi \right) = -\frac{FJ}{\varepsilon \varepsilon_0} \left(\sum_k z_k c_k + z_f c_f \right) \tag{6.18}$$

$$c_f = \frac{c_{f0}^s \cdot K \cdot \phi_0^s}{(K + c_H)\sqrt{1 + 2F_1(\mathbf{E}) + 4F_2(\mathbf{E}) + 8F_3(\mathbf{E})}} \tag{6.19}$$

where $\mathbf{N}_k$ is the Piola–Kirchhoff molar flux, J=det($\mathbf{F}$), $\mathbf{F}$ is the deformation gradient tensor and C^{-1} is the inverse of the right Cauchy–Green tensor and $\mathbf{C}^{-1}=\mathbf{F}^{-1}\mathbf{F}^{-\mathrm{T}}$.

For incorporation of the effect of mechanical equilibrium on swelling/deswelling of the hydrogel into the MECglu model, the mechanical governing equations for large deformation based on a total Lagrangian description are given as

$$\nabla_X \cdot \mathbf{P} = 0 \tag{6.20}$$

$$\mathbf{u} = \mathbf{u}* \quad \text{in} \quad \Gamma_{u*} \tag{6.21}$$

$$\mathbf{P} \cdot \mathbf{N} = \mathbf{s}* \quad \text{in} \quad \Gamma_{s*} \tag{6.22}$$

where **P** is the first Piola–Kirchhoff stress tensor that is a kind of expatriate, living partially in the deformed configuration **x** and partially in the reference configuration **X**. **u** is the displacement vector from the initial configuration **X** to the deformed configuration **x** so that **x**=**X**+**u**. **u**$*$ is the specified displacement vector on boundary portion Γ_{u*}, **s**$*$ is the surface traction vector on boundary Γ_{s*}, **N** is the unit outward normal vector.

For a polymeric porous mixture, the first Piola–Kirchhoff stress tensor **P** can be written as (Li et al., 2007)

$$\mathbf{P} = -J^{-1}\mathbf{F}p_{\text{osmotic}}(X)\mathbf{I} + \mathbf{S}\mathbf{F}^{\text{T}} \tag{6.23}$$

where **I** is identity tensor, **S**=**C**:**E** and **C** is the material moduli tensor. p_{osmotic} is the osmotic pressure and computed by

$$p_{\text{osmotic}} = RT\sum_k (c_k - c_k^*) \tag{6.24}$$

where c_k^* is the boundary value of the ionic concentration and c_k is the diffusive ionic concentration within the whole domain for the *k*th ionic species.

In this section, one-dimensional steady-state simulations are carried out with the MECglu model for analysis of the equilibrium response of a cylindrical glucose-sensitive hydrogel immersed in the glucose buffer solution. The deformation of hydrogel is confined in axial direction, and thus it swells in the radial direction only. Due to the constraints, the equilibrium swelling of the cylindrical hydrogel may thus be modelled approximately as a one-dimensional uniaxial problem along the radius direction of the hydrogel. Furthermore, it is assumed that the surrounding bath is well mixed, such that the concentrations of the surrounding solution are constant. As a result, the present one-dimensional axisymmetrical computational domain consists of the hydrogel radius representing the interior cylindrical hydrogel and the glucose buffer solution along the radius direction representing the surrounding glucose buffer medium. The computational domain of Eqs. (6.15) with (6.16) to (6.17) and (6.18) with (6.19) is defined as covering the whole domain, including both the hydrogel and surrounding solution, for the diffusive species concentrations and electric potential. However, the computational domain of Eqs. (6.20), (6.21) and (6.22) for the hydrogel displacement is defined as covering the hydrogel region only.

For the axisymmetrical problem, the Neumann type of the electrochemical boundary conditions is imposed at the centre of the cylindrical hydrogel to ensure the continuity inside the hydrogel

$$\frac{\partial c_k}{\partial X} = 0, \frac{\partial \psi}{\partial X} = 0, (k = \text{glu}, \text{ox}, ..., N) \quad \text{at} \quad X = 0 \tag{6.25}$$

and Dirichlet boundary conditions are imposed at the solution edges

$$c_{\text{ox}} = c^*_{\text{ox}}, c_{\text{glu}} = c^*_{\text{glu}}, c_{\text{ga}} = 0, c_k = c^*_k \psi = 0 \quad \text{at} \quad X = L \tag{6.26}$$

At the interface between the hydrogel and the solution, one can have

$$(\lambda + 2\mu)\left[\frac{du}{dX} + \frac{1}{2}\left(\frac{du}{dX}\right)^2\right] = p_{\text{osmotic}}, X = L_{\text{gel}} \tag{6.27}$$

In addition, a point constraint is required at the centre of the hydrogel to prevent the hydrogel from undergoing rigid body motion

$$u = 0 \quad \text{at} \quad X = 0 \tag{6.28}$$

Formulation of the MECglu model has thus far been completed, and it is composed of the coupled nonlinear partial differential equations, namely the Nernst–Planck equation (6.15) with the rate of the enzyme reaction of the glucose (6.17), the Poisson equation (6.18) with the density of the fixed charge groups (6.19) and the mechanical equations (6.20), (6.21) and (6.22). In numerical implementation of the MECglu model, there are several mathematical challenges, such as the multi-energy domains associated with the coupled nonlinear partial differential equations, the computational domain remeshing due to moving interfaces and the localized high gradient over the hydrogel–solution interfaces. A strong-form meshless technique termed the Hermite-cloud method (Li et al., 2003), which combines with an iterative computing technique for the nonlinear partial differential equations, is employed for numerical solution of the MECglu model for the equilibrium response of the glucose-sensitive hydrogels. Following the computational flowchart shown in Fig.6.2, the distribution of the fixed charge concentration is computed first by Eq. (6.19) according to the given boundary conditions and input parameters. By the computed c_f, the Nernst–Planck equations (6.15) coupled with the Poisson equation (6.18) are solved numerically with a Newton iterative technique for distribution of the converged mobile species c_k and electric potential ψ. The converged concentration c_k is then substituted into the mechanical equilibrium governing equation (6.20) for the corresponding displacement u of the hydrogel, representing the swelling/deswelling of the glucose-sensitive hydrogel. Because of the deformation u, the fixed charge density c_f is redistributed within the hydrogel, and a new iteration with the remeshing domains is required. Such a computational loop is carried out until all the independent variables including c_k, ψ, c_f and u converge.

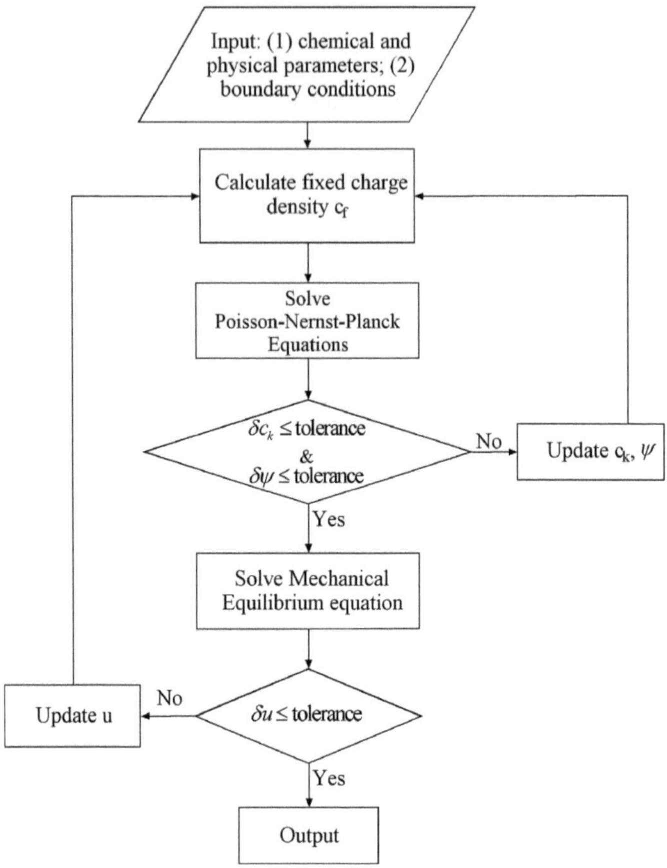

Fig. 6.2 Computational flowchart of the MECglu model

6.2.2 Model Validation with Experiment

For simulation of the equilibrium response of the glucose-sensitive hydrogels, the MECglu model is validated here by comparison between the simulation results and experimental curve data published (Kang and Bae, 2003), as shown in Figs. 6.3, 6.4 and 6.5, in which the experiment was conducted by Kang and Bae (2003) for a typical glucose-sensitive hydrogel, the poly(*N*,*N*-dimethylacrylamide) gel (PDMAAm) covalently immobilized the glucose oxidase (GOx) and the catalase, which was immersed in an isotonic phosphate buffered saline (PBS) solution with pH 7.4. As well known, the pH-sensitive hydrogel can exhibit a volume transition in a narrow range of pH, which is associated with the pK_a of the pendant ionizable groups attached onto the crosslinked polymeric network chains (Peppas et al., 2000). Similarly, the PDMAAm hydrogel contains the sulfonic acid $-SO_2N-H^-$ attached onto the polymeric network chains so that it can also demonstrate the volume

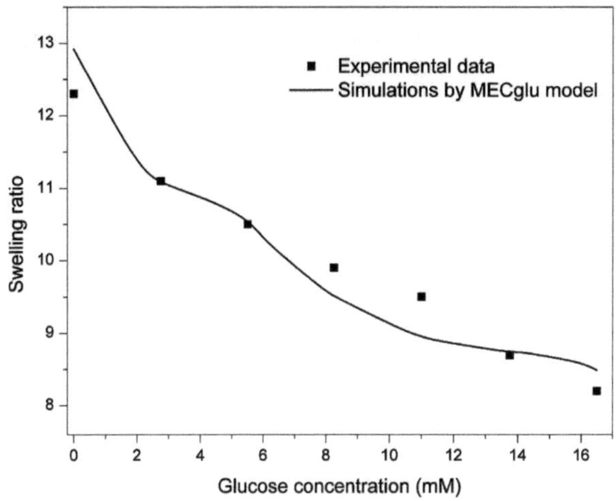

Fig. 6.3 Comparison of the glucose-dependent swelling of the glucose-sensitive hydrogel between the MECglu model predictions and experimental data (Kang and Bae, 2003) in the glucose buffer solution at 37°C

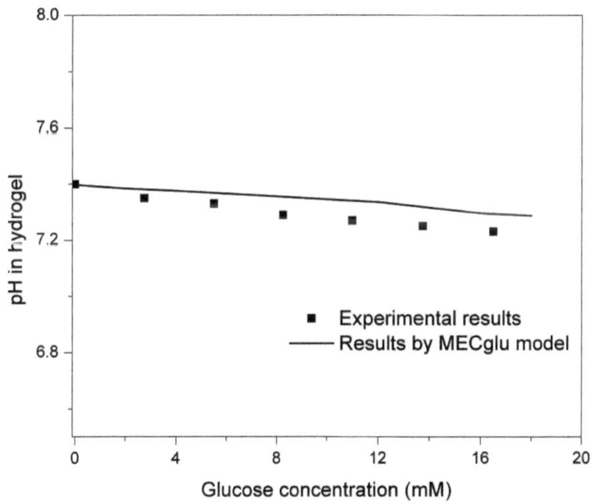

Fig. 6.4 Comparison of the average pH, ⟨pH⟩, within the hydrogel strip between the MECglu model predictions and experimental data (Kang and Bae, 2003) in the glucose buffer solution at 37°C

transition in the dynamics and equilibrium swelling in response to the change in environmental pH. The ionization of the PDMAAm hydrogel occurs when the environmental pH is higher than the pK_a of the ionizable sulfonic group (Kang and Bae, 2002). By design of appropriate pK_a, the PDMAAm hydrogel can be designed to control the transition of solubility and swelling at a desired pH level of about 7.4.

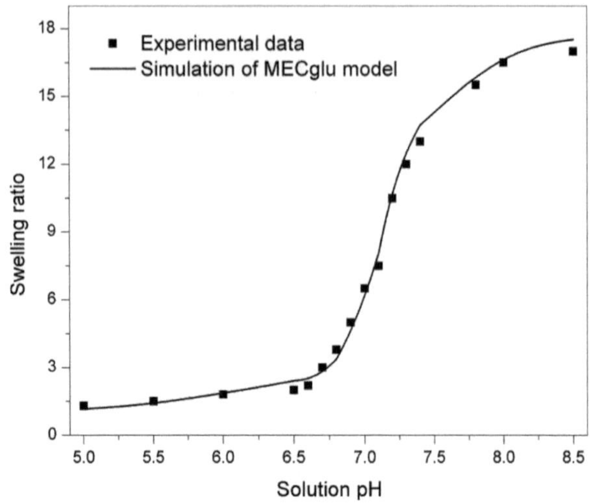

Fig. 6.5 Comparison of the swelling ratio of the hydrogel between the MECglu model predictions and experimental data (Kang and Bae, 2003) in the pH buffer solution at 37°C

The glucose oxidase immobilized within the hydrogel catalyses the glucose diffusive from the surrounding solution and converts it to the gluconic acid. The hydrogen ion species produced from the enzyme reaction thereby lower the pH level within the hydrogel and alter the ionization of the sulfonic acid $-SO_2N-H^-$ of PDMAAm and the diffusion of other ionic species. This will develop a concentration difference over the interface between the interior hydrogel and the exterior surrounding solution, which generates an osmotic pressure over the interface. This in turn results in the change of swelling deformation of the hydrogel.

Table 6.1 summarizes the input parameters required by the MECglu model for computational simulation. As mentioned early, the oxygen concentration in the body fluid always remains saturated, namely c_{ox}=0.274 mM, since the oxygen pressure in the capillary blood is higher than that in the tissue (Guyyon, 1991). In addition, the experiments reveal that Young's modulus of the hydrogel varies with the surrounding pH value (Kidoaki et al., 2001; Johnson et al., 2002; Isayava et al., 2002). In the present simulation with the MECglu model, the variation of Young's modulus as function of solution pH is characterized by a hyperbolic tangent curve, which consists of three segments: the constant Young's modulus of 1.45 MPa if pH<6.5 and 0.64 MPa if pH>7.5, and the third segment formulated by a hyperbolic function defined in the range of pH 6.5–7.5 (Kidoaki et al., 2001; Sudipto et al., 2002).

The comparison for the glucose-dependent swelling of the hydrogel is demonstrated in Fig. 6.3 between the simulation results and the experimental counterpart (Kang and Bae, 2003). It is observed that the swelling ratio of the hydrogel decreases with increment of the glucose concentration from 0 to 16.5 mM, which is associated with the drop of pH within the hydrogel and involves the conversion of the glucose to the gluconic acid by the enzyme glucose oxidase. Meanwhile, the catalase reaction can prevent the accumulation of peroxide and partially regenerate the oxygen. The

Table 6.1 Input data required by the MECglu model

Faraday's constant	$F = 9.645 \times 10^4$C/mol
Universal gas constant	$R = 8.314$ J/mol·K
Absolute temperature	T=310 K
Vacuum permittivity or dielectric constant (Li et al., 2005)	$\varepsilon_0 = 8.854 \times 10^{-12}$C^2/N·m^2
Relative dielectric constant of surrounding medium (Li et al., 2005)	ϵ=80
Michaelis–Menten constants (Abdekhodaie and Wu, 2005)	$K_{\text{ox}} = 6.992 \times 10^3$ mM; $K_g = 0.6178$ mM
Maximum reaction velocity	$V_{\max}(\text{mol/s} \cdot \text{cm}^3) = 860(\text{s}^{-1}) \times c_{\text{enz}}(\text{mol/cm}^3)$
Equilibrium dissociation constant (Kang and Bae, 2002)	$K = 10^{-6.17}$mM
Young's modulus (Kidoaki et al., 2001; Sudipto et al., 2002)	E=0.6∼1.5 MPa
Initial water volume fraction (Li et al., 2004a, b)	$\phi_0^w = 0.8$
Initially fixed charge density and valence number of the fixed charge groups	$c_f^0 = 10$ mM; $z_f = -1$
Boundary conditions	$c^* = c_{\text{Na}^+} = c_{\text{Cl}^-} = 138.0$ mM; $V_e = 0$V
Geometry of hydrogel strip	$L = 4000\ \mu$m; $L_{\text{gel}}^0 = 600\ \mu$m
Diffusion coefficients for dilute aqueous solution (25°C)	
Hydrogen ion (Plawsky, 2001)	$D_H = 9.31 \times 10^{-9}$ m^2/s
Gluconic acid and glucose (Albin et al., 1987; Klumb and Horbett, 1992)	$D_a = D_g = 6.75 \times 10^{-10}$ m^2/s
Oxygen (Albin et al., 1987; Klumb and Horbett, 1992)	$D_{\text{ox}} = 2.29 \times 10^{-9}$ m^2/s

important product from the reaction is the hydrogen ion H^+, which lowers the local pH level, makes a change in the osmotic pressure and then the hydrogel shrinks. It is also shown that the theoretically predicted swelling deformation coincides quantitatively and qualitatively with the experimental results, where the swelling ratio of the hydrogel decreases almost linearly from 12 to 8 at the same rate for both the simulation and experimental curves.

If an average pH is defined as

$$\langle \text{pH} \rangle = (\text{V}_{\text{gel}})^{-1} \int_{\text{V}_{\text{gel}}} \text{pH}dV,$$

Fig. 6.4 illustrates that the reduction of the numerically simulated ⟨pH⟩ within the hydrogel from 7.4 to 7.28 is slightly larger than the experimental ⟨pH⟩ that varies from 7.4 to 7.24, where the relative discrepancy is less than 0.5%. Therefore, it is examined that the MECglu model can provide a robust simulation, not only to support the experimental phenomena and the trends of a glucose-sensitive hydrogel system but also to predict the variation of average pH, ⟨pH⟩, achieved in the glucose-sensitive hydrogel.

Figure 6.5 is plotted for comparison of the pH-dependent swelling of the SDM-PDMAAm hydrogel, where a gradual increase of swelling ratio is found as pH increases. It is also observed that the simulation results marked by solid line agree qualitatively and quantitatively with the published experimental data marked by squares (Kang and Bae, 2003). The ionization of fixed charge groups is altered by changing the environmental pH. The figure also clearly shows the volume transition in the range of pH from 6.5 to 8.0, in which the swelling ratio of the hydrogel increases with pH level. Since the polymeric complexes are broken and the fixed charge groups of the PDMAAm hydrogel are ionized, the osmotic pressure increases and the hydrogel swells further. If the pH is lower than 6.5 around the pK_a of the sulfonic acid $-SO_2N - H^-$, the volume change of the hydrogel is small, that is, the hydrogel polymeric networks collapse and the swelling ratio is small. A likely reason is that the interaction between the ionized and un-ionized fixed charge groups at the solid state prevents the solubilization of crosslinked polymeric network until the ionized groups become dominant (Kang and Bae, 2001). It is also known from the figure that the simulation results are slightly larger than the experimental data in the range of pH higher than 7.0, which may be attributable to the immobilized glucose oxidase and catalase. The incorporation of enzymes may make the hydrogel thicker and stronger (Dhanarajan and Siegel, 2005).

Based on the above comparisons and discussions, it is concluded that the simulations by the MECglu model agree well with experimental findings. Therefore, the MECglu model can provide the information to predict the swelling/deswelling characteristics of a glucose stimulus-responsive hydrogel, and to support the design and optimization of the glucose-sensitive hydrogel-based bioengineering devices, such as the self-regulated insulin delivery systems.

6.3 Multi-Effect-Coupling Ionic-Strength-Stimulus (MECis) Model for Ionic Strength-Sensitive Hydrogel

Ionic strength of a solution is generally defined as

$$I = \frac{1}{2}\sum_{k=1}^{N} c_k z_k^2,$$

where c_k denotes the ionic concentration (mM) and z_k the valence/charge number for the kth ionic compound species and N is the number of total ionic compound species in that solution. As one of the important characteristics of such an electrolyte solution with dissolved ionic species, the ionic strength is a measure of the concentrations of ionic species in which electrolyte solution containing multiple ionic compounds, which dissociate into ions when dissolved in water.

The ionic strength-sensitive hydrogels are a class of environmental stimuli-responsive hydrogels, like pH-, electrical- or thermal-sensitive hydrogels. They possess the swelling or shrinking behaviour in response to the variation of ionic strength of surrounding solution. Literature search shows that numerous experimental studies were carried out to synthesize the ionic strength-sensitive hydrogels for investigation of various characteristics of the hydrogels. For instance, early studies include the experiment conducted by Ohmine and Tanaka (1982) who observed the ionized acrylamide gels undergoing a discrete phase transition of the equilibrium volume subject to change in the ionic strength of a solution, and explained the experimental phenomenon by Flory–Huggins model and Donnan theory. Other early studies include the experiment conducted by Liu et al. (1995) who studied the swelling characteristics of sulfonate gels placed in water and various aqueous solutions with different ionic strengths. Jeon et al. (1998) studied the effect of the ionic strength with different initial monomer concentrations on the swelling behaviour of polyelectrolyte gels. Dhara et al. (1999) synthesized the interpenetrating polymeric network mixture for the effect of component contents on the swelling or deswelling of the hydrogels, which demonstrated the deformation dependence on the pH and ionic strength of swelling medium. Later studies include the experiment conducted by Zhao and Moore (2001) who developed a technique for preparation of the pH and ionic strength-sensitive hydrogels with capability of fast response time. Belma and Banu (2005) prepared a series of acrylamide and *N*-vinylimidazole, named poly(AAm-*co*-NVI), for the effect of the ionic strength on the hydrogel swelling. Caykara and Dogmus (2005) studied the swelling and shrinking behaviour of poly(acrylamide-*co*-itaconic acid) hydrogels in water and aqueous NaCl solutions and drew the conclusion that the equilibrium swelling ratio of the hydrogels increases drastically with the ionic strength of bath solution. Caykara and Ayçiçek (2005) synthesized the ionic poly[(*N*,*N*-diethylaminoethylmethacrylate)-*co*-(*N*-vinyl-2-pyrrolidone)] hydrogels and examined the responsive characteristics to external stimuli including the ionic strength. Recent studies include the experiment conducted by Abd El-Mohdy (2007) who prepared a novel highly swelling hydrogel by grafting crosslinked polyacrylamide (PAM) chains onto carboxymethylcellulose (CMC) and showed the ionic strength-dependent swelling behaviour of the hydrogel. Lin et al. (2007) explored the effect of the ionic strength on the tensile properties of the hydrogel. Liu et al. (2007) investigated the responsive swelling of acrylamide/maleic acid copolymer sensitive to the pH and ionic strength. In brief, all the experimental studies found demonstrate the significant influence of the ionic strength of environmental solution on the swelling/shrinking behaviour of the hydrogels.

In terms of theoretical modelling, probably Flory (1953) was the first who introduced the contribution of the ionic strength of surrounding solution into the swelling of hydrogels. However, Flory's theory could not handle the situation well when the ionic strength is treated as an environmental stimulus to the hydrogel. Literature search reveals that few studies focus on the model development for the ionic strength-sensitive hydrogels. Most of them used the existing models for explanation of experiments concerning the influence of the ionic strength on the deformation characteristics of the hydrogels. For example, Hooper et al. (1990) used Donnan theory and a model concerning the swelling equilibrium of neutral and ionized polyacrylamide gels in water or aqueous salt solution for analysis of the effects of the ionic strength and fixed charge density on swelling equilibrium, which was based on the extension of framework they proposed earlier (Prange et al., 1989). Other early studies include a simple Flory model with Donnan theory employed by Baker et al. (1994, 1995), where the polyampholyte hydrogel with negative fixed charge was synthesized and measured for analysis of swelling equilibrium of the smart hydrogel responding to the changes in the solution ionic strength and fixed charge density (Baker et al., 1992). English et al. (1996) qualitatively compared the experimental results with theoretical predictions by extending the Flory–Huggins theory with a quasi-lattice screened Coulombic term for analysis of swelling ratio of the ionized hydrogel responsive to the ionic strength. Okay et al. (1998) utilized the Flory–Rehner theory for computation of swelling ratio of the anionic hydrogel in salt solution with the ionic strength ranging from 10^{-5} to 1 M, and compared with the experimental results, where the theoretical simulation captured the experimental swelling in some limited cases of the fixed charge densities. Later studies include the model presented by Caykara et al. (2000), which was based on the work by Brannon-Peppas and Peppas (1991), to study the equilibrium swelling of the ionic strength-sensitive hydrogel prepared by containing triprotic acid moieties. Peppas et al. (2000) theoretically expressed the swelling ratio associated with the ionic strength for the anionic and cationic hydrogels. Recent studies include the model presented by Caykara et al. (2003), which was based on the Flory–Huggins thermodynamics theory, the Donnan theory and the James-Guth phantom network theory (1943), for investigation of the volume swelling ratio of the ionic strength-sensitive hydrogels that they synthesized. In summary, the ionic strength of environmental solution was considered as boundary conditions in most of the theoretical studies mentioned above, and not incorporated into the governing equations of the models. These models were either oversimplified so that they were applicable in limited cases only or too complicated to use.

6.3.1 Development of the MECis Model

In this section, a multiphysics model is developed, and called the multi-effect-coupling ionic-strength-stimulus (MECis) model, for simulation of the swelling or shrinking characteristics of the ionic strength-sensitive hydrogels. The MECis model is composed of coupled nonlinear partial differential equations. One of them is the Nernst–Planck flux system, characterizing the mechanism of the ionic species transport due to diffusion and electrophoresis as well as convection in the hydrogel and surrounding solution, in which the effect of the ionic strength is incorporated through the chemical activity coefficient. Considering the influence of the electrical potential on the ionic flux, the Poisson equation is coupled with Nernst–Planck equations to form the Poisson–Nernst–Planck (PNP) system. The fixed charge density is characterized by Langmuir isotherm theory with consideration of the effect of the ionic strength. The other is the mechanical finite deformation governing equation simulating the deformed hydrogel induced by the osmotic pressure and fixed charge repulsion. In this section, after several assumptions the MECis model is formulated first by the Eulerian description, and it then is transformed to the Lagrangian coordinates for more convenient computation of large deformation.

It is noted that the MECis model has significant advantage over the previous models discussed above. For example, the effect of the ionic strength of bathing solution is associated with the fixed charge density by the Langmuir adsorption isotherm theory, and also incorporated into the Poisson–Nernst–Planck (PNP) governing equations. The fixed charge density plays an important role in the swelling ratio due to the osmotic pressure and the repulsive force between fixed charges, which is also included in the MECis model. The MECis model is a good tool for simulation of responsive characteristics of the ionic strength-sensitive hydrogels, with consideration of large deformation formulated in Lagrangian coordinates.

6.3.1.1 Assumptions

Development of the present MECis model is based on the assumptions as follows:

- The system consisting of the hydrogel and surrounding solution is maintained in isothermal condition
- Material properties of the hydrogel are isotropic and homogeneous
- Diffusivity of each species is constant everywhere and independent of the swelling or concentration gradient of the hydrogel
- All the binding reactions are in local equilibrium and the chemical reaction in the solution is neglected

6.3.1.2 Formulation with Eulerian Description

When the ionized hydrogel with fixed charged groups is immersed into a solution, the concentration gradient, electrical potential, chemical potential or pressure develop, which cause the flux of ionic species that is governed by several basic laws.

The diffusive flux of ionic species in ideal mixture can be characterized by Fick's law (Fick, 1855). In the system other than ideal solution or mixture, however, the driving source for diffusion of each species is the gradient of chemical potential of the corresponding species. Thus the Fick's law is used as

$$j_{k(D)} = -\frac{D_k c_k}{RT} \nabla \mu_k \tag{6.29}$$

where $\mathbf{j}_{k(D)}$(mol/m^2·s), $\mathbf{D}_k$(m^2/s) and c_k(mol/m^3) denote the molar flux, the diffusivity tensor and the concentration of the kth ionic species, respectively. R(J/mol·K) and T(K) are the universal gas constant and absolute temperature, respectively. The negative sign in front of the right-hand side of Eq. (6.29) is due to the ionic flow down the concentration gradient.

Usually the chemical potential μ_k(J/mol) of the species k in liquid mixture is expressed in terms of the activity a_k, and given by

$$\mu_k = \mu_k^0 + RT \ln a_k \tag{6.30}$$

where μ_k^0 is the chemical potential of the kth ion species in a reference configuration, for example, in a pure liquid of the kth ion species. The activity of the kth species a_k is given by the concentration c_k and chemical activity coefficient γ_κ of the species k as

$$a_k = c_k \gamma_k \tag{6.31}$$

The diffusive flux of the species k due to chemical potential is thus written as

$$\begin{aligned} j_{k(D)} &= -\frac{D_k c_k}{RT} \nabla \mu_k = -D_k c_k (\nabla \ln a_k) = -D_k c_k (\nabla \ln c_k + \nabla \ln \gamma_k) \\ &= -D_k c_k \left(\frac{1}{c_k} \nabla c_k + \nabla \ln \gamma_k \right) = -D_k (\nabla c_k + c_k \nabla \ln \gamma_k) \end{aligned} \tag{6.32}$$

In addition, the electrostatic force exists with or without the externally applied electric potential, which moves the ionic species for the migration flux due to the gradient of electrical potential. Thus the flux due to the electrophoresis is given as

$$j_{k(E)} = -z_k \mu_k c_k \nabla \psi \tag{6.33}$$

where $\mathbf{j}_{k(E)}$(mol/m^2·s), z_k and $\boldsymbol{\mu}_k$(m^2/V·s) represent the migration flux due to the gradient of electrical potential, the valence/charge number and the mobility tensor of the kth ionic species, respectively. ψ(V) is the electrical potential.

The diffusion and drift experience the same frictional resistance in solution. Therefore, based on the Nernst–Einstein equation, the relation between mobility $\boldsymbol{\mu}_k$ and the diffusion coefficient tensor $\mathbf{D}_k$can be written as

$$\mu_k = \frac{D_k F}{RT} \tag{6.34}$$

where F(C/mol) is the Faraday constant. Substituting Eq. (6.34) into Eq. (6.33), the migration flux due to the electrophoresis is rewritten as

$$j_{k(E)} = -\frac{D_k F}{RT} z_k c_k \nabla \psi \tag{6.35}$$

Apart from that, if the effect of convection of solvent flow on the flux of the species k is considered, one can have

$$j_{k(C)} = c_k v_c \tag{6.36}$$

where $\mathbf{j}_{k(C)}$(mol/m^2·s) is the flux due to the solvent convection and $\mathbf{v}_c$(m/s) is the velocity of solvent flow.

Therefore, the total flux of the kth ionic species in the hydrogel and solution consists of the diffusion/migration flux and convection flux. The former is the flux due to chemical diffusion and electrophoresis, and the latter results from the convection of fluid or bulk flow. The total flux can thus be described as

$$\begin{aligned} j_k &= \mathbf{j}_{k(d)} + \mathbf{j}_{k(c)} = \left(j_{k(D)} + j_{k(E)}\right) + j_{k(C)} \\ &= -D_k\left(\nabla c_k + c_k \nabla \ln \gamma_k\right) - \frac{D_k F}{RT} z_k c_k \nabla \psi + c_k v_c \\ &= -D_k\left(\nabla c_k + c_k \nabla \ln \gamma_k + \frac{F}{RT} z_k c_k \nabla \psi\right) + c_k v_c \end{aligned} \tag{6.37}$$

where $\mathbf{j}_{k(d)} = \mathbf{j}_{k(D)} + \mathbf{j}_{k(E)}$ and $\mathbf{j}_{k(c)}$ are the diffusive and convective fluxes, respectively. In formulation of Eq. (6.37), the flux driven by pressure is generally neglected, which is included only when high acceleration such as in centrifuges exists in the system. The flux resulting from thermal diffusion is also excluded for small temperature gradient since it is negligibly small, relative to other fluxes.

Consider a control volume $V(t)$ with a closed region. The surface of the control volume is denoted by $S(t)$. Based on the law of mass conservation, the material derivative of concentration with respect to time t for the ionic species k in the control volume $V(t)$ is given by

$$\frac{D}{Dt}\int_{V(t)} c_k dV = -\int_{S(t)} j_{k(d)} \cdot n dS + \int_{V(t)} g_{k(V)} dV \tag{6.38}$$

where $\mathbf{n}$ is the outward unit vector normal to the surface and thus there is a negative sign in the first term of right-hand side. $g_{k(V)}$ is the generation/consumption rate of the kth ionic species due to reaction and is expressed by $g_{k(V)} = v_k r$, in which v_k is stoichiometric coefficient and r is reaction ratio.

Equation (6.38) characterizes the accumulation of the kth species in the control volume $V(t)$, which is the sum of the diffusion across the control surface and the generation or consumption in the control volume. Since the fluid within the hydrogel or external solution move due to the convection, the moving velocity $\mathbf{v}_c$ of the control volume should be considered. By the Reynolds transport theorem, the material derivative of the integral is written by (Belytschko et al., 2001)

$$\begin{aligned} \frac{D}{Dt}\int_{V(t)} c_k dV &= \int_{V(t)} \left(\frac{Dc_k}{Dt} + c_k \nabla \cdot \mathbf{v}_c\right) dV = \int_{V(t)} \left(\frac{\partial c_k}{\partial t} + \nabla \cdot (c_k \mathbf{v}_c)\right) dV \\ &= \int_{V(t)} \frac{\partial c_k}{\partial t} dV + \int_{S(t)} c_k v_c \cdot n dS \end{aligned} \tag{6.39}$$

Substituting Eq. (6.39) into Eq. (6.38) yields the conservation equation as

$$\int_{V(t)} \left(\frac{Dc_k}{Dt} + c_k \nabla \cdot \mathbf{v}_c\right) dV = -\int_{S(t)} j_{k(d)} \cdot n dS + \int_{V(t)} v_k r dV \tag{6.40}$$

The diffusion across the control surface could be expressed by the divergence theorem, i.e., the sum of all sources and sinks gives the flow out of the control volume,

$$\int_{S(t)} j_{k(d)} \cdot n dS = \int_{V(t)} \nabla \cdot j_{k(d)} dV \tag{6.41}$$

By Eqs. (6.40) and (6.41) one has

$$\int_{V(t)} \left(\frac{Dc_k}{Dt} + c_k \nabla \cdot \mathbf{v}_c + \nabla \cdot j_{k(d)} - \nu_k r \right) dV = 0 \tag{6.42}$$

The control volume is chosen arbitrarily so that the integrand could be vanished. Equation (6.42) is thus simplified to

$$\frac{Dc_k}{Dt} + c_k \nabla \cdot \mathbf{v}_c + \nabla \cdot j_{k(d)} - \nu_k r = 0 \tag{6.43}$$

Substituting Eq. (6.37) into Eq. (6.43) leads to

$$\frac{Dc_k}{Dt} + c_k \nabla \cdot \mathbf{v}_c = \nabla \cdot \left[D_k \left(\nabla c_k + c_k \nabla \ln \gamma_k + \frac{F}{RT} z_k c_k \nabla \psi \right) \right] + \nu_k r \tag{6.44}$$

where the term $\ln\gamma_\kappa$, associated with the chemical activity coefficient γ_κ for the ionic species k, is characterized by

$$\ln \gamma_k = \begin{cases} -\ln 10 A z_k^2 \sqrt{I} & I \in [0, 0.02] \\ -\dfrac{\ln 10 A z_k^2 \sqrt{I}}{1 + a_k B \sqrt{I}} & I \in [0.02, 0.1] \\ -\ln 10 \left(\frac{A z_k^2 \sqrt{I}}{1 + a_k B \sqrt{I}} - CI \right) & I > 0.1 \end{cases} \tag{6.45}$$

which results from Sinko (2006)

$$\log \gamma_k = \begin{cases} -A z_k^2 \sqrt{I} & I \in [0, 0.02] \\ -\dfrac{A z_k^2 \sqrt{I}}{1 + a_k B \sqrt{I}} & I \in [0.02, 0.1] \\ -\left(\dfrac{A z_k^2 \sqrt{I}}{1 + a_k B \sqrt{I}} - CI \right) & I > 0.1 \end{cases} \tag{6.46}$$

where A is a factor depending on the temperature T and dielectric constant ϵ of medium, $A = 0.5F^2q(2\varepsilon RT)^{-3/2}$ (Samson and Marchand, 1999). The factor B is a constant associated with the nature of solvent and temperature, $B = (2F^2/(\varepsilon RT))^{1/2}$ (Samson and Marchand, 1999). z_k is the valence number of the ion species k. a_k is called the mean effective ionic diameter or the ion size parameter, which is the mean distance of approach of the ions. *CI* is an empirical term (Sinko, 2006).

Equation (6.44) is the well-known Nernst–Planck molar flux equation, which characterizes the mass transport of mobile species in the system of the hydrogel immersed in a solution and demonstrates the variation of concentration of the ionic species k due to the fluxes of chemical potential, electrophoresis, convection and the chemical reaction.

In the present analysis, the effect of spatial charge should be included and it could be described in several possible ways, for example, the constant field where the electric potential varies linearly across the system, the constant current where the electroneutrality is assumed and Poisson equation where the electrical potential is governed by a partial differential equation. It is ascertained that the electroneutrality condition and the constant field solution are in fact the limited cases of the Poisson equation (MacGillivray, 1968; MacGillivray and Hare, 1969). The electroneutrality condition is satisfied only if the concentrations are high, while the constant field solution is applicable specifically for low concentrations. Therefore, the Poisson equation is incorporated in the MECis model to predict the spatial charge inside and outside of the hydrogel. Formulation of the Poisson equation starts from Gauss's law for magnetism as

$$\nabla \cdot B = 0 \tag{6.47}$$

where **B** denotes magnetic field. By defining the magnetic potential as **A** (V·s/m), and using Helmholtz's theorem, the magnetic field **B** can be written as

$$B = \nabla \times A \tag{6.48}$$

Here Faraday's law gives

$$\nabla \times E = -\frac{\partial B}{\partial t} \tag{6.49}$$

where **E**(N/C) denotes electric field and t(s) is time. Substituting Eq. (6.48) into Eq. (6.49) gives

$$\nabla \times \left(E + \frac{\partial A}{\partial t}\right) = 0 \tag{6.50}$$

in which the vector field, $\mathbf{E}+\partial\mathbf{A}/\partial t$, is curl free, and thus it can be written in terms of electric potential ψ

$$E + \frac{\partial A}{\partial t} = -\nabla\psi \tag{6.51}$$

By assumption of the absence of changing the magnetic field, which is usually reasonable for the system of the hydrogel immersed in solution, Eq. (6.51) yields

$$E = -\nabla\psi \tag{6.52}$$

Gauss's law for electricity in a differential control volume yields

$$\nabla \cdot D = \rho \tag{6.53}$$

where $\mathbf{D}$(C/m^2) is the electric displacement field and ρ(C/m^3) is the charge density in the control volume. By the assumption that the hydrogel and solution are isotropic and homogeneous, the displacement field is given by

$$D = \varepsilon E \tag{6.54}$$

where ϵ(C^2/N/m^2) is the permittivity of the medium. Substituting Eq. (6.54) into Eq. (6.53) gives

$$\nabla \cdot E = \frac{\rho}{\varepsilon} \tag{6.55}$$

By Eqs. (6.52) and (6.55), one has

$$\nabla^2 \psi = -\frac{\rho}{\varepsilon} \tag{6.56}$$

Equation (6.56) is the well-known Poisson equation. In order to solve the Poisson equation for the electric potential, the distribution of the charge density is required. If the charge density is zero, Eq. (6.56) reduces to a Laplace equation. If the charge density follows a Boltzmann distribution, Eq. (6.56) becomes Poisson–Boltzmann equation.

The permittivity of the medium ϵ describes the interaction between the electric field and dielectric medium. It is characterized by the ability of a material to polarize in response to the electric field. Usually it is defined as the vacuum permittivity ϵ_0 multiplying with a relative permittivity ϵ_r that is determined by material or medium considered

$$\varepsilon = \varepsilon_r \varepsilon_0 \tag{6.57}$$

where ϵ_r represents the relative permittivity of the material or medium considered and ϵ_0 the vacuum permittivity.

The total charge density includes the fixed charge density ρ_f within the hydrogel and the mobile charge density ρ_m in both the hydrogel and solution medium

$$\rho_f = N_A q z_f c_f = F z_f c_f \tag{6.58}$$

$$\rho_m = N_A q \sum_k z_k c_k = F \sum_k z_k c_k \tag{6.59}$$

where N_A(mol^{-1}) is Avogadro number, q(C) is elementary charge, z_f denotes the valence number of the fixed charge. The total charge density is thus given by

$$\rho = \rho_f + \rho_m = F\left(\sum_k z_k c_k + z_f c_f\right) \tag{6.60}$$

Substituting Eqs. (6.57) and (6.60) into Eq. (6.56) leads to

$$\nabla^2\psi = -\frac{F}{\varepsilon_r\varepsilon_0}\left(\sum_k z_k c_k + z_f c_f\right) \tag{6.61}$$

The Poisson equation approximates the electric potential ψ as a function of the ion–ion and ion–fixed charge interactions. It is examined that the electroneutrality condition and constant field are in fact the special cases of the Poisson equation (MacGillivray, 1968; MacGillivray and Hare, 1969).

On the other hand, the fixed charges play an important role in swelling/deswelling of the charged hydrogel, since large volume transition is attributed to the charged groups bound onto the polymer network chains, which is strongly dependent on the dissociation constant or binding constant. The charged groups of the hydrogel adsorb the mobile ions diffusing from the solution to reduce the fixed charge density and raise the difference of ionic concentrations between the internal hydrogel and the external solution, which leads to the increase of the osmotic pressure. The simplest model for the adsorption under equilibrium conditions was developed by Langmuir using the kinetic theory, in which a monolayer surface is considered with a specific number of binding sites, and each site can adsorb one molecule. In the Langmuir monolayer adsorption theory, it is assumed that binding at a site has no influence on the neighbouring sites. The adsorption process based on Langmuir monolayer theory can be described by

$$\mathrm{MS} \underset{K_a}{\overset{K_d}{\rightleftarrows}} \mathrm{M} + \mathrm{S} \tag{6.62}$$

where MS denotes the occupied fixed charge site in the hydrogel. M is the mobile ion bound onto the polymeric chains in the interstitial solution, S is the unoccupied fixed charge site. K_d(mol/m^3) is the dissociation constant and K_a(m^3/mol) is the association constant. The dissociation constant is defined as

$$K_d = \frac{[\mathrm{M}][\mathrm{S}]}{[\mathrm{MS}]} \tag{6.63}$$

where [M] is the concentration of mobile ions bound onto the polymeric chains in the interstitial solution. [S] and [MS] are the densities of unbinding fixed charge and binding fixed charge, respectively. They are written as

$$[\mathrm{M}] = c_b;\ \ [\mathrm{S}] = c_f;\ \ [\mathrm{MS}] = c_f^0 - c_f \tag{6.64}$$

where c_b is the concentration of the mobile ions which are bound to the polymeric network chains, c_f denotes the concentration of fixed charge after binding and c_f^0 is the initial concentration of fixed charge before binding. Substituting them into Eq. (6.63) gives

$$K_d = \frac{c_b c_f}{c_f^0 - c_f} \Rightarrow c_f = \frac{K_d c_f^0}{K_d + c_b} \tag{6.65}$$

The total concentration of the fixed charge groups within the hydrogel at the relaxed state, $c_{f,s}^0$, is defined as

$$c_{f,s}^0 = \frac{n_f}{V^s} \tag{6.66}$$

where n_f(mol) is the mole of the fixed charge within the hydrogel and V^s(m^3) is the volume of the hydrogel at dry state. Following similar definition, the concentration of the fixed charge at initial state is defined as

$$c_f^0 = \frac{n_f}{V} \tag{6.67}$$

where V is the volume of the hydrogel at initial state before binding and

$$V = V^s + V^w \tag{6.68}$$

Based on Eqs. (6.66), (6.67) and (6.68), the initial concentration of the fixed charge before binding is written as

$$c_f^0 = \frac{n_f}{V} = \frac{n_f}{V^s + V^w} = \frac{n_f}{V^s} \frac{1}{1 + V^w/V^s} = \frac{c_{f,s}^0}{1 + H} \tag{6.69}$$

where H is called the local hydration of the hydrogel, and defined as

$$H = \frac{V^w}{V^s} \tag{6.70}$$

The dissociation constant of the fixed charge groups is associated with the ionic strength of interstitial solution by (Sinko, 2006)

$$pK_d = pK_{d0} - \frac{A(2n-1)\sqrt{I}}{1+\sqrt{I}} \tag{6.71}$$

where pK_{d0} denotes the intrinsic dissociation constant in the form of minus logarithm with respect to proton and pK_d is the apparent dissociation constant. n is the absolute charge value of ions (e.g. equal to 1). By removing the logarithm, Eq. (6.71) can be converted to

$$K_d = K_{d0} \exp\left[A \ln(10)(2n-1)\sqrt{I}/(1+\sqrt{I})\right] \tag{6.72}$$

As a result, the concentration of the fixed charge within the hydrogel at current state is finally derived as

$$c_f = \frac{c_{f,s}^0}{1+H} \frac{K_{d0} \exp\left[A \ln(10)(2n-1)\sqrt{I}/(1+\sqrt{I})\right]}{K_{d0} \exp\left[A \ln(10)(2n-1)\sqrt{I}/(1+\sqrt{I})\right] + c_b} \tag{6.73}$$

For characterization of the deformation of the ionic strength-sensitive hydrogel, the mechanical equilibrium governing equation is required, in which the swelling force is balanced by the elastic retractive force exerted by the crosslinked polymeric network solid matrix of the hydrogel in order to maintain the current hydration state. The balance of linear momentum leads to the equation of motion as

$$\frac{D}{Dt}\int_{V(t)} \rho \mathbf{v} dV = \int_{S(t)} \mathbf{t} dS + \int_{V(t)} \rho \mathbf{b} dV \tag{6.74}$$

where ρ is the density of the hydrogel, $\mathbf{v}$ is the velocity field, $\mathbf{b}$ is the body force vector, $\mathbf{t}$ is the traction field on the surface of the hydrogel, which could be represented by the Cauchy stress $\boldsymbol{\sigma}$

$$\mathbf{t} = \sigma \cdot \mathbf{n} \tag{6.75}$$

where $\mathbf{n}$ is the outward unit normal vector on the surface of the hydrogel. Using divergence theorem, based on the law of the mass conservation, and considering the arbitrariness of the control volume, Eq. (6.74) becomes

$$\rho \frac{D\mathbf{v}}{Dt} = \nabla \cdot \sigma + \rho \mathbf{b} \tag{6.76}$$

In the present analysis of swelling behaviour of the hydrogel immersed in a solution, the body force $\mathbf{b}$ and velocity field $\mathbf{v}$ are generally neglected, and it is assumed that the environmental load is imposed slowly so that the inertial effect is also negligible. As such, the motion equation is simplified to

$$\nabla \cdot \sigma = 0 \tag{6.77}$$

which means the constant stress throughout the hydrogel.

Here, the Cauchy stress of the hydrogel $\boldsymbol{\sigma}$ is decomposed into two components

$$\sigma = \sigma^r + \sigma^a \tag{6.78}$$

where $\boldsymbol{\sigma}^r$ is a reactive stress determined by the fluid constraint and $\boldsymbol{\sigma}^a$ is an active stress determined by the constitutive equation. For the multiphase hydrogel mixture, the reactive stress is formulated by (Lai et al., 1991)

$$\sigma^r = -(p + T_c)\mathbf{I} \tag{6.79}$$

where p is the pressure, T_c is the chemical expansion stress which may be neglected in present analysis and $\mathbf{I}$ is unit tensor. The active stress could be written as

$$\sigma^a = \lambda(trE)\mathbf{I} + 2\mu\mathbf{E} \tag{6.80}$$

where $\mathbf{E}$ is strain tensor and $tr\mathbf{E}$ is the trace of the strain tensor. λ and μ are Lamè coefficients of the polymeric network solid matrix, and they are expressed by Young's modulus E_y and Poisson's ratio v in the form of

$$\lambda = \frac{\nu E_Y}{(1+\nu)(1-2\nu)} \tag{6.81}$$

$$\mu = \frac{E_Y}{2(1+\nu)} \tag{6.82}$$

The constitutive equation (6.80) is associated with the elasticity of the hydrogel polymeric network and the osmotic pressure due to the concentration difference between the interior hydrogel and exterior solution. In addition, there is the atomic-level force called the repulsive force that also contributes to the stress field of the hydrogel. The repulsive force is attributed to the interaction between the fixed charge repulsions within the hydrogel. Considering the driving force due to the repulsive force between the fixed charges, the pressure is composed of two components, the osmotic pressure and the repulsion between the fixed charges, namely

$$p = p_{\text{osmotic}} + p_{\text{repulsion}} \tag{6.83}$$

and thus the constitutive equation (6.80) is extended to

$$\sigma = -(p_{\text{osmotic}} + p_{\text{repulsion}})\mathbf{I} + \lambda(trE)\mathbf{I} + 2\mu\mathbf{E} \tag{6.84}$$

where $p_{\text{repulsion}}$ represents the stress due to the charge repulsion and p_{osmotic} the osmotic pressure. On the basis of the assumption of the mean field theory, the mechanical equation can be rewritten as

$$\nabla \cdot \left[-(p_{\text{osmotic}} + p_{\text{repulsion}})\mathbf{I} + \lambda(trE)\mathbf{I} + 2\mu\mathbf{E}\right] = 0 \tag{6.85}$$

It is well known that the osmotic pressure is a hydrostatic pressure resulting from the difference in ionic concentrations between the hydrogel and the bathing solution, and it is calculated by

$$p_{\text{osmotic}} = RT\sum_k (c_k - c_{k0}) \tag{6.86}$$

where c_{k0} and c_k are the concentrations of the kth ion species in buffer solution and within the hydrogel, respectively.

In order to characterize the nominal repulsive stress, the mesh size is required as a characteristic length between the fixed charges, representing the average distance between the consecutive crosslinks and serving as an indicator of the screening effect of the network on solute diffusion (Peppas et al., 1985; Canal and Peppas, 1989). By the Debye–Huckel model and considering the Debye shielding effect, the potential energy of all charges within the hydrogel in the reference configuration is given as (Sinko, 2006)

$$\varphi_0 = \frac{N_0 q^2}{4\pi\varepsilon_0 r_0} e^{-r_0/\kappa_0} \tag{6.87}$$

where ϕ_0 is the potential energy, N_0 is the number of charges and r_0 is the mesh size in initial configuration. K_0 is called Debye length and defined as

$$\kappa_0 = \sqrt{\frac{\varepsilon_0 \varepsilon_r k_B T}{2 N_A q^2 I_h}} \tag{6.88}$$

where k_B is Boltzmann constant and I_h is the ionic strength within the hydrogel. After swelling or shrinking, the mesh size becomes r from r_0, and the number of fixed charge becomes N from N_0. Variation of potential energy from initial state to current state is thus given as

$$\Delta\varphi = \frac{q^2}{4\pi\varepsilon_0}\left[\frac{N_0}{r_0}\exp\left(-\frac{r_0}{\kappa_0}\right) - \frac{N}{r}\exp\left(-\frac{r}{\kappa}\right)\right] \tag{6.89}$$

where the mesh size after swelling/shrinking of the hydrogel can be calculated by

$$r = Q^{1/3} r_0 \tag{6.90}$$

where Q is the swelling ratio of the hydrogel and defined as

$$Q = \frac{V}{V_0} \tag{6.91}$$

The numbers of charges, N_0 in the initial state and N in the current state, could be calculated by

$$N_0 = c_f^0 V_0 N_A; \; N = c_f V N_A \tag{6.92}$$

Therefore, the nominal repulsive stress is defined as

$$\begin{aligned} p_{\text{repulsion}} &= \frac{\Delta\varphi}{V} = \frac{1}{V}\frac{q^2}{4\pi\varepsilon_0}\left[\frac{N_0}{r_0}\exp\left(-\frac{r_0}{\kappa_0}\right) - \frac{N}{r}\exp\left(-\frac{r}{\kappa}\right)\right] \\ &= \frac{Fq}{4\pi\varepsilon_0}\left[\frac{c_f^0}{Q r_0} e^{-r_0/\kappa_0} - \frac{c_f}{r} e^{-r/\kappa}\right] \end{aligned} \tag{6.93}$$

6.3.1.3 Formulation with Lagrangian Description

The governing equations of the MECis model developed in the above section are formulated in the deformed configuration state. However, it is applicable only for small deformation of the hydrogel, and implementation of numerically computational simulation requires the movement of the mesh to capture deformation of the hydrogel. In general, the hydrogels are able to absorb the solvent from the bath solution from 10 to 20% up to thousands of times compared with the dry weight

(Hoffman, 2002). This makes the hydrogel deform largely. Therefore, it is necessary to formulate the MECis model based on the nonlinear deformation theory.

For description of the largely deformed hydrogel in the reference or undeformed configuration, the Lagrangian coordinates, which are also called the material coordinates, are required and denoted here by **X**. On the other hand, the Eulerian coordinates, which are also called the spatial coordinate, are employed to describe the deformed configuration and denoted here by **x**, where **x**=**X** at t=0. The displacement **u** of any point marked at the hydrogel is thus defined as the difference between the current and reference positions of that point, i.e.

$$\mathbf{u} = \mathbf{x} - \mathbf{X} \tag{6.94}$$

As usual, the deformation gradient tensor is defined as

$$F = \frac{\partial \mathbf{x}}{\partial \mathbf{X}} = \nabla_X \mathbf{x} = \mathbf{I} + \nabla \mathbf{u} \tag{6.95}$$

and the Jacobian between the current and reference configurations is given as

$$J = \left| \frac{\partial x_i}{\partial X_I} \right| = \det \mathbf{F} \tag{6.96}$$

where x_i and X_I are the components of the position vectors **x** and **X**, respectively. det**F** is the determinant of the deformation gradient tensor **F**. The Green strain tensor is defined as

$$\mathbf{E} = \frac{1}{2}(\mathbf{F}\mathbf{F}^T - \mathbf{I}) \tag{6.97}$$

The del operators can be transformed from the reference configuration to the deformed configuration or vice versa by

$$\nabla_x = \mathbf{F}^{-\mathrm{T}} \cdot \nabla_X \tag{6.98}$$

$$\nabla_X = \mathbf{F}^{\mathrm{T}} \cdot \nabla_x \tag{6.99}$$

where $\mathbf{F}^{-\mathrm{T}}$ is the transformation for the inverse of the deformation gradient tensor.

For the mass transport of mobile ion species in the deformed configuration, Eq. (6.37) is rewritten below for the diffusive flux of the ionic species k,

$$j_{k(d)} = -D_k \left(\nabla_x c_k + c_k \nabla_x \ln \gamma_k + \frac{F}{RT} z_k c_k \nabla_x \psi \right) \tag{6.100}$$

By transforming Eq. (6.100) into the reference configuration via Eq. (6.98), the gradients of ionic species concentration, electrical potential and chemical activity coefficient are given as

$$\nabla_x c_k(\mathbf{x},t) = \mathbf{F}^{-\mathrm{T}} \nabla_X c_k(\mathrm{X},t) \tag{6.101}$$

$$\nabla_x \psi(\mathbf{x},t) = \mathbf{F}^{-\mathrm{T}} \nabla_X \psi(\mathbf{X},t) \tag{6.102}$$

$$\nabla_x \ln \gamma_k(\mathbf{x},t) = \mathbf{F}^{-\mathrm{T}} \nabla_X \ln \gamma_k(\mathbf{X},t) \tag{6.103}$$

For convenience, the Piola–Kirchhoff species diffusion is defined as

$$\mathbf{J}_{k(d)} = J\mathbf{F}^{-1}\mathbf{j}_{k(d)} \tag{6.104}$$

Substituting Eqs. (6.100), (6.101), (6.102) and (6.103) into Eq. (6.104), the Piola–Kirchhoff species diffusion is given by

$$\begin{aligned} \mathbf{J}_{k(d)} &= J\mathbf{F}^{-1}\left[-D_k\mathbf{F}^{-\mathrm{T}}\left(\nabla_X c_k + c_k\nabla_X \ln \gamma_k + \frac{F}{RT} z_k c_k \nabla_X \psi\right)\right] \\ &= -J\mathbf{F}^{-1}D_k\mathbf{F}^{-\mathrm{T}}\left(\nabla_X c_k + c_k\nabla_X \ln \gamma_k + \frac{F}{RT} z_k c_k \nabla_X \psi\right) \end{aligned} \tag{6.105}$$

In terms of the Nernst–Planck molar flux equation, the Piola–Kirchhoff species flux (6.105) should be used for formulation of the Nernst–Planck equation in the reference configuration. Considering the control surface and volume, S_0 and V_0, in the reference configuration, the mass conservation in the reference configuration can be obtained by transforming Eq. (6.38) to

$$\frac{D}{Dt}\int\limits_{V_0} Jc_k dV = -\int\limits_{S_0} JF^{-1} j_{k(d)} \cdot \mathbf{N} dS + \int\limits_{V_0} Jg_{k(V)} dV \tag{6.106}$$

where $\mathbf{N}$ is the outward unit vector of the control surface S_0 and the transformation uses Nanson's law. Based on the definition of the Piola–Kirchhoff species diffusion, the mass conservation in the Lagrangian coordinates is written as

$$\frac{D}{Dt}\int\limits_{V_0} Jc_k dV = -\int\limits_{S_0} J_{k(d)} \cdot N dS + \int\limits_{V_0} Jg_{k(V)} dV = -\int\limits_{V_0} J\nabla_X \cdot J_{k(d)} dV + \int\limits_{V_0} Jg_{k(V)} dV \tag{6.107}$$

where the material derivative with respect to time in the Lagrangian coordinates is derived as (Belytschko et al., 2001)

$$\frac{D}{Dt}\int\limits_{V_0} Jc_k dV = \int\limits_{V_0}\left(\frac{\partial c_k}{\partial t}J + Jc_k\nabla_x \cdot \mathbf{v}_c\right)dV = \int\limits_{V_0}\left(\frac{\partial c_k}{\partial t}J + c_k\nabla_X \cdot \left(J\mathbf{F}^{-1}\mathbf{v}_c\right)\right)dV \tag{6.108}$$

in which the following divergence relation between the Eulerian and Lagrangian descriptions is used:

$$\nabla_X \cdot \left(J\mathbf{F}^{-1}\mathbf{v}\right) = J\nabla_x \cdot \mathbf{v} \tag{6.109}$$

By Eqs. (6.105), (6.106), (6.107) and (6.108), the Nernst–Planck equation in the Lagrangian coordinates is given as

$$\begin{aligned}&\frac{\partial c_k}{\partial t} + c_k J^{-1}\nabla_X \cdot \left(J\mathbf{F}^{-1}\mathbf{v}_c\right)\\&= J^{-1}\nabla_X \cdot \left[J\mathbf{F}^{-1}D_k\mathbf{F}^{-\mathrm{T}}\left(\nabla_X c_k + c_k\nabla_X \ln\gamma_k + \frac{F}{RT}z_k c_k \nabla_X \psi\right)\right] + \nu_k r\end{aligned} \tag{6.110}$$

In the present MECis model, the diffusivity of substance remains constant everywhere in each direction, such that the diffusivity tensor $\mathbf{D}_k$ can be reduced to a scalar D_k, and then the term $\mathbf{F}^{-1}$ $\mathbf{F}^{-\mathrm{T}}$ becomes the inverse of Cauchy–Green stress tensor $\mathbf{C}^{-1}$. Therefore, the Nernst–Planck equation can be expressed as

$$\begin{aligned}&\frac{\partial c_k}{\partial t} + c_k J^{-1}\nabla_X \cdot \left(J\mathbf{F}^{-1}\mathbf{v}_c\right)\\&= J^{-1}\nabla_X \cdot \left[JD_k\mathbf{C}^{-1}\left(\nabla_X c_k + c_k\nabla_X \ln\gamma_k + \frac{F}{RT}z_k c_k \nabla_X \psi\right)\right] + \nu_k r\end{aligned} \tag{6.111}$$

In terms of the spatial charge, the Poisson equation is also transformed into the reference configuration. Considering Eqs. (6.102) and (6.109), we have

$$\begin{aligned}\nabla_x \cdot (\nabla_x\psi) &= J^{-1}\nabla_X \cdot \left(J\mathbf{F}^{-1}\nabla_x\psi\right) = J^{-1}\nabla_X \cdot \left(J\mathbf{F}^{-1}\mathbf{F}^{-\mathrm{T}}\nabla_X\psi\right)\\&= J^{-1}\nabla_X \cdot \left(J\mathbf{C}^{-1}\nabla_X\psi\right)\end{aligned} \tag{6.112}$$

Substituting Eqs. (6.112) into (6.61) yields the Poisson equation in the reference configuration as follows:

$$J^{-1}\nabla_X \cdot \left(J\mathbf{C}^{-1}\nabla_X\psi\right) = -\frac{F}{\varepsilon_r\varepsilon_0}\left(\sum_k z_k c_k + z_f c_f\right) \tag{6.113}$$

In addition, the fixed charge density in the large deformation of the hydrogel is derived, which is based on the derivation in the above section. Since the hydrogel is composed of three phases, the polymeric network matrix solid phase, interstitial water phase and ionic species phase, the volume fractions of the three phases are defined as

$$\phi^s = \frac{V^s}{V};\ \phi^w = \frac{V^w}{V};\ \phi^i = \frac{V^i}{V} \tag{6.114}$$

Relatively compared with the volume fractions of the polymeric matrix and interstitial water, the volume fraction of ionic species is very small and it is neglected here, thus $\phi^s + \phi^w \approx 1$. Therefore, the volume fractions of the polymeric matrix solid phase and interstitial water phase can be written as

$$\phi^s = \frac{V^s}{V^w + V^s} = \frac{1}{1+H} \tag{6.115}$$

$$\phi^w = \frac{V^w}{V^s + V^w} = \frac{H}{1+H} \tag{6.116}$$

As such

$$\phi^w = 1 - \phi^s = 1 - \frac{V^s}{V} = 1 - \frac{V^s}{V_0}\frac{V_0}{V} = 1 - \phi_0^s \cdot J^{-1} \tag{6.117}$$

where V_0 is the volume of the hydrogel mixture in the reference configuration and ϕ_0^s is the volume fraction of the polymeric matrix solid phase in the reference configuration. $J=dV/dV_0$ is the volume ratio of the apparent polymeric solid phase, and it can be calculated with the three invariants of Green strain tensor $\mathbf{E}$ as

$$J = \det \mathbf{F} = \sqrt{1 + 2F_1(\mathbf{E}) + 4F_2(\mathbf{E}) + 8F_3(\mathbf{E})} \tag{6.118}$$

where $F_1(\mathbf{E})$, $F_2(\mathbf{E})$ and $F_3(\mathbf{E})$ are the first, second and third invariants of Green strain tensor. With Eqs. (6.116) and (6.117), the hydration is written as

$$H = \frac{J - \phi_0^s}{\phi_0^s} \tag{6.119}$$

As a result, the concentration of fixed charge within the hydrogel at current state with large deformation is obtained as

$$\begin{aligned} c_f &= \frac{c_{f,s}^0}{(1+H)}\frac{K_d}{(K_d + c_b)} \\ &= \frac{c_{f,s}^0 \phi_0^s K_{d0} \exp\left[A \ln(10)(2n-1)\sqrt{I}/(1+\sqrt{I})\right]}{\left(K_{d0} \exp\left[A \ln(10)(2n-1)\sqrt{I}/(1+\sqrt{I})\right] + c_b\right)\sqrt{1+2F_1(\mathbf{E})+4F_2(\mathbf{E})+8F_3(\mathbf{E})}} \end{aligned} \tag{6.120}$$

In terms of the mechanical deformation of the hydrogel, the balance of linear momentum in the Lagrangian description can be derived by converting the mechanical equation in the Eulerian form

$$\frac{D}{Dt}\int_{V_0} \rho_0 \mathbf{v} dV = \int_{S_0} J\sigma \mathbf{F}^{-T}\mathbf{N} dS + \int_{V_0} \rho_0 \mathbf{b} dV \tag{6.121}$$

where J is the determinant of the deformation gradient tensor and $\mathbf{N}$ is the unit outward normal vector in the reference configuration of the hydrogel. ρ_0 is the density of the hydrogel at initial state, and by the law of mass conservation in the Lagrangian description $\rho_0 = J\rho$. V_0 and S_0 are the control volume and surface in the reference configuration, respectively. The first Piola–Kirchhoff stress tensor is defined as

$$\mathbf{P} = J\mathbf{F}^{-1} \cdot \sigma \tag{6.122}$$

Therefore, Eq. (6.121) yields

$$\rho_0 \frac{\partial \mathbf{v}}{\partial t} = \nabla_X \cdot \mathbf{P} + \rho_0 \mathbf{b} \tag{6.123}$$

Similarly, the first Piola–Kirchhoff stress is also decomposed into two components

$$\mathbf{P} = \mathbf{P}^r + \mathbf{P}^a \tag{6.124}$$

where $\mathbf{P}^r$ and $\mathbf{P}^a$ are the reactive and active stresses, respectively. The active stress $\mathbf{P}^a$ is associated with the second Piola–Kirchhoff stress tensor $\mathbf{S}$,

$$\mathbf{S} = \mathbf{F}^{-1}\mathbf{P}^a \tag{6.125}$$

The second Piola–Kirchhoff stress $\mathbf{S}$ is determined by the constitutive equation of the hydrogel

$$\mathbf{S} = \mathbf{DE} \tag{6.126}$$

where $\mathbf{D}$ is the material moduli tensor. For the porous hydrogel mixture with the osmotic pressure and repulsive stress, the first Piola–Kirchhoff stress is given as

$$\mathbf{P} = \mathbf{P}^r + \mathbf{P}^a = -J\mathbf{F}^{-1}\left(p_{\text{osmotic}} + p_{\text{repulsion}}\right)\mathbf{I} + \mathbf{SF}^{\mathrm{T}} \tag{6.127}$$

Based on the same assumptions made in the Eulerian descriptions, which exclude the effects of the external, body and inertial forces, the mechanical governing equation is obtained as

$$\nabla_X \cdot \left[\mathbf{SF}^{\mathrm{T}} - J\mathbf{F}^{-1}\left(p_{\text{osmotic}} + p_{\text{repulsion}}\right)\mathbf{I}\right] = 0 \tag{6.128}$$

6.4 Remarks

So far the two models have been developed in this chapter, which are the multi-effect-coupling glucose-stimulus (MECglu) model for the glucose-sensitive hydrogel and the multi-effect-coupling ionic-strength-stimulus (MECis) model for the ionic strength-sensitive hydrogel.

The MECglu model as a multiphysics simulation framework is applicable for simulation of the equilibrium and kinetics performance of the glucose-sensitive hydrogel according to its sensitivity to the glucose concentration in microenvironmental solution. Effect of the glucose oxidation reaction catalysed by the enzymes is incorporated in the model, where the enzymes include the glucose oxidase and catalase that are immobilized within the hydrogel. The MECglu model also incorporates the effect of chemo-electro-mechanical coupled multi-energy domains and the conversion of the chemical energy into mechanical energy. The model consists of the Nernst–Planck equation for mobile species concentrations, the Poisson equation for electric potential and a nonlinear mechanical governing equation for the finite deformation of the hydrogel when subjected to the environmental glucose stimulus. Density of the fixed charge groups bound onto the crosslinked polymeric network chains is formulated and associated with the change in ambient solution pH. The MECglu model is examined by numerical comparison of the equilibrium swelling degree of the glucose-sensitive hydrogel between the model predictions and experimental data, with the meshless Hermite-cloud method for one-dimensional steady-state simulations. The model can be employed for design and optimization of an insulin delivery system that is based on a glucose-sensitive hydrogel, for example, for prediction of the swelling/deswelling of the smart hydrogel responding to the practical physiological glucose concentration ranging from 0 to 16.5 mM (300 mg/ml). It can also provide the helpful equilibrium and kinetics information, for instance, the distributions of reacting and diffusive species concentrations, the distributive electric potential, as well as the displacement or swelling ratio of the glucose-sensitive hydrogels responding to the stimulus of the solution glucose when immersed in a glucose buffer solution.

The MECis model developed here is applicable for simulation of the equilibrium and kinetics characteristics of the ionic strength-sensitive hydrogel, when placed in a solution with the change in the ionic strength. It is composed of the Nernst–Planck flux equation based on the law of mass conservation for describing the flux of mobile ionic species, Poisson equation predicting the spatial charge distribution and mechanical motion equation characterizing the deformation distribution of the hydrogel. The effect of the ionic strength of surrounding solution is incorporated not only into the Nernst–Planck governing equation but also into the constitutive relation of the fixed charge density. The MECis model is presented in the Eulerian coordinates first, and then converted into the Lagrangian coordinates to capture the large deformation of the ionic strength-sensitive hydrogel. The model considers the effects of chemical, electrical and mechanical multi-energy coupled domains on the responsive behaviour due to the complex mechanism of swelling or shrinking of the charged hydrogel subject to the stimuli of the environmental ionic strength

References

M.J. Abdekhodaie, X.Y. Wu. (2005). Modelling of a cationic glucose-sensitive membrane with consideration of oxygen limitation. *Journal of Membrane Science*, 254, 119–127.

H.L. Abd El-Mohdy. (2007). Water sorption behavior of CMC/PAM hydrogels prepared by γ-irradiation and release of potassium nitrate as agrochemical. *Reactive and Functional Polymers*, 67, 1094–1102.

G. Albin, T.A. Horbett, S.R. Miller, N.L. Ricker. (1987). Theoretical and experimental studies of glucose sensitive membranes. *Journal of Controlled Release*, 6, 267–291.

J.P. Baker, H.W. Blanch, J.M. Prausnitz. (1995). Swelling properties of acrylamide-based ampholytic hydrogels: Comparison of experiment with theory. *Polymer*, 36, 1061–1069.

J.P. Baker, L.H. Hong, H.W. Blanch, J.M. Prausnitz. (1994). Effect of initial total monomer concentration on the swelling behavior of cationic acrylamide-based hydrogels. *Macromolecules*, 27, 1446–1454.

J.P. Baker, D.R. Stephens, H.W. Blanch, J.M. Prausnitz. (1992). Swelling equilibria for acrylamide-based polyampholyte hydrogels. *Macromolecules*, 25, 1955–1958.

A. Baldi, Y. Gu, P. Loftness, R.A. Siegel. (2003). A hydrogel-actuated environmentally-sensitive microvalve for active flow control. IEEE/ASME Journal of Microelectromechanical Systems, 12, 613–621.

I.S.I.K. Belma, D. Banu. (2005). Swelling behavior of poly (acrylamide-*co*-*N*-vinylimidazole) hydrogels under different environment conditions. *Journal of Applied Polymer Science*, 96, 1783–1788.

T. Belytschko, W.K. Liu, B. Moran. (2001). *Nonlinear Finite Elements for Continua and Structures*, New York: John Wiley and Sons.

E. Birgersson, Hua Li, S.N. Wu. (2008). Transient analysis of temperature-sensitive neutral hydrogels. *Journal of the Mechanics and Physics of Solids*, 56(2), 444–466.

S. Brahim, D. Narinesingh, A. Guiseppi-Elie. (2002). Bio-smart hydrogels: Co-joined molecular recognition and signal transduction in biosensor fabrication and drug delivery. *Biosensors and Bioelectronics*, 17, 973–981.

L. Brannon-Peppas, N.L. Peppas, (1991). Equilibrium swelling behavior of pH-sensitive hydrogels. *Chemical Engineering Science*, 46, 715–722.

T. Canal, N.A. Peppas. (1989). Correlation between mesh size and equilibrium degree of swelling of polymeric networks. *Journal of Biomedical Materials Research*, 23, 1183–1193.

X. Cao, S. Lai, L.J. Lee. (2001). Design of a self-regulated drug delivery device. *Biomedical Microdevices*, 3, 109–118.

T. Caykara, I. Aycicek. (2005). External stimuli-responsive characteristics of ionic poly[(*N*,*N*-diethylaminoethyl methacrylate)-*co*-(*N*-vinyl-2-pyrrolidone)] hydrogels. *Macromolecular Materials Engineering*, 290, 468–474.

T. Caykara, U. Bozkaya, O. Kantoglu. (2003). Network structure and swelling behavior of poly(acrylamide/crotonic acid) hydrogels in aqueous salt solution. *Journal of Polymer Science Part B: Polymer Physics*, 41, 1656–1664.

T. Caykara, M. Dogmus. (2005). Swelling-shrinking behavior of poly(acrylamide-*co*-itaconic acid) hydrogels in water and aqueous NaCl solutions. *Journal of Macromolecular Science, Part A*, 42, 105–111.

T. Caykara, C. Ozyurek, O. Kantoglu, O. Guven. (2000). Equilibrium swelling behavior of pH- and temperature-sensitive poly(*N*-vinyl 2-pyrrolidone-g-citric acid) polyelectrolyte hydrogels. *Journal of Polymer Science Part B: Polymer Physics*, 38, 2063–2071.

A.P. Dhanarajan, R.A. Siegel. (2005). Time-dependent permeabilities of hydrophobic, pH-sensitive hydrogels exposed to pH gradients. *Macromolecular Symposia*, 227, 105–114.

D. Dhara, C.K. Nisha, P.R. Chatterji. (1999). Super absorbent hydrogels: Interpenetrating networks of poly (acrylamide-*co*-acrylic.acid) and poly (vinyl alcohol): Swelling behavior and structural parameters. *Journal of Macromolecular Science: Pure and Applied Chemistry*, A36, 197–210.

A.E. English, S. Mafe, J.A. Manzanares, X. Yu, A.Y. Grosberg, T. Tanaka. (1996). Equilibrium swelling properties of polyampholytic hydrogels. *Journal of Chemistry and Physics*, 104, 8713–8720.

A. Fick. (1855). On liquid diffusion. *Philosophical Magazine*, 10, 31–39.

P.J. Flory. (1953). *Principles of Polymer Chemistry*, Ithaca, New York: Cornell University Press.

D.A. Gough, J.Y. Lusisano, P.H.S. Tse. (1985). Two dimensional enzyme electrode sensor for glucose. *Analytical Chemistry*, 57, 2351–2357.

A. Guiseppi-Elie, S. Brahim, G. Slaughter, K.R. Ward. (2005). Design of a subcutaneous implantable biochip for monitoring of glucose and lactate. *IEEE Sensors Journal*, 5, 345–355.

A.C. Guyyon. (1991). *Textbook of Medical Physiology*, 8th edn. Philadelphia: W.B. Saunders Company, pp. 433–443.
A. S. Hoffman. (2002). Hydrogels for biomedical applications. *Advanced Drug Delivery Reviews*, 43, 3–12.
W. Hong, X.H. Zhao, J.X. Zhou, Z. Suo. (2008). A theory of coupled diffusion and large deformation in polymeric gels. *Journal of the Mechanics and Physics of Solids*, 56, 1779–1793.
H.H. Hooper, J.P. Baker, H.W. Blanch, J.M. Prausnitz. (1990). Swelling equilibria for positively ionized polyacrylamide hydrogels. *Macromolecules*, 23, 1096–1104.
I.S. Isayava, S.A. Yankovshi, J.P. Kennedy. (2002). Novel amphiphilic membranes of poly(*N,N*-dimethylacrylamide) crosslinked with octa-methacrylate-telechelic polyisobutylene stars. *Polymer Bulletin*, 48, 475–482.
K. Ishihara, K. Matsui. (1986). Glucose-responsive insulin release from polymer capsule. *Journal of Polymer Science: Polymer Letters Edition*, 24, 413–417.
C.H. Jeon, E.E. Makhaeva, A.R. Khokhlov. (1998). Swelling behavior of polyelectrolyte gels in the presence of salts. *Macromolecular Chemistry and Physics*, 199, 2665–2670.
B.D. Johnson, D.J. Niedermaier, W.C. Crone, J. Moorthy, D.J. Beebe. (2002). Mechanical properties of a pH sensitive hydrogel, *Proceedings of the 2002 Annual Conference of Society for Experimental Mechanics,*Milwaukee, Wisconsin.
S.I. Kang, Y.H. Bae. (2001). pH-induced volume-phase transition of hydrogels containing sulfonamide side group by reversible crystal formation. *Macromolecules*, 34, 8173–8178.
S.I. Kang, Y.H. Bae. (2002). pH-induced solubility transition of sulfonamide-based polymers. *Journal of Controlled Release*, 80, 145–155.
S.I. Kang, Y.H. Bae. (2003). A sulfonamide based glucose-responsive hydrogel with covalently immobilized glucose oxidase and catalase. *Journal of Controlled Release*, 86, 115–121.
S. Kidoaki, Y. Nakayama, T. Matsuda. (2001). Measurement of the interaction forces between proteins and iniferter-based graft-polymerized surfaces with an atomic force microscope in aqueous media. *Langmuir*, 17, 1080–1087.
J.J. Kim, K. Park. (2001). Modulated insulin delivery from glucose-sensitive hydrogel dosage forms. *Journal of Controlled Release*, 77, 39–47.
L.A. Klumb, T.A. Horbett. (1992). Design of insulin delivery device based on glucose-sensitive membrane. *Journal of Controlled Release*, 18, 59–80.
R.T. Kurnik, B. Berner, J. Tamada, R.O. Potts. (1998). Design and simulation of a reverse lontophoretic glucose monitoring device. *Journal of electrochemistry Society*, 145, 4119–4125.
W.M. Lai, J.S. Hou, V.C. Mow. (1991). A triphasic theory for the swelling and deformation behaviors of articular cartilage. *ASME Journal of Biomechanical Engineering*, 113, 245–258.
H. Li, J. Chen, K.Y. Lam. (2004). Multiphysical modelling and meshless simulation of electric-sensitive hydrogels. *Journal of Polymer Science Part B: Polymer Physics*, 42, 1514–1531.
H. Li, R.M. Luo, K.Y. Lam. (2007). Modelling and simulation of deformation of hydrogels responding to electric stimulus. *Journal of Biomechanics*, 40, 1091–1098.
H. Li, T.Y. Ng, J.Q. Cheng, K.Y. Lam. (2003). Hermite-cloud: A novel true meshless method. *Computational Mechanics*, 33, 30–41.
H. Li, Z. Yuan, K.Y. Lam, H.P. Lee, J. Chen, J. Hanes, J. Fu. (2004). Model development and numerical simulation of electric-stimulus-responsive hydrogels subject to an externally applied electric field. *Biosensors and Bioelectronics*, 19, 1097–1107.
Z. Lin, W. Wu, J. Wang, X. Jin. (2007). Studies on swelling behaviors, mechanical properties, network parameters and thermodynamic interaction of water sorption of 2-hydroxyethyl methacrylate/novolac epoxy vinyl ester resin copolymeric hydrogels. *Reactive and Functional Polymers*, 67, 789–797.
H. Liu, M. Zhen, R. Wu. (2007). Ionic-strength- and pH-responsive poly[acrylamide-*co*-(maleic acid)] hydrogel nanofibers. *Macromolecular Chemistry and Physics*, 208, 874–880.
R.M. Luo, Hua Li, K.Y. Lam. (2007a). Modelling and simulation of chemo-electro-mechanical behavior of pH-electric-sensitive hydrogel. *Analytical and Bioanalytical Chemistry*, 389, 863–873.

R.M. Luo, Hua Li, K.Y. Lam. (2007b). Coupled chemo-electro-mechanical simulation for smart hydrogels that are responsive to an external electric field. *Smart Materials and Structures*, 16(4), 1185–1191.

J.Y. Lusisano, D.A. Gough. (1988). Transient response of the two dimensional glucose sensor. *Analytical Chemistry*, 60, 1272–1281.

A.D. MacGillivray. (1968). Nernst–Planck equation and the electroneutrality and Donnan equilibrium assumptions. *Journal of Chemical Physics*, 48, 2903–2907.

A.D. MacGillivray, D. Hare. (1969). Applicability of goldman's constant field assumption to biological systems. *Journal of Theoretical Biology*, 25, 113–126.

G.P. Misra, R.A. Siegel. (2002). New mode of drug delivery: Long term autonomous rhythmic hormone release across a hydrogel membrane. *Journal of Controlled Release*, 81, 1–6.

V. Nikonenko, K. Lebedev, J.A. Manzanares, G. Pourcelly. (2003). Modelling the transport of carbonic acid anions through anion-exchange membranes. *Electrochimica Acta*, 48, 3639–3650.

I. Ohmine, T. Tanaka. (1982). Salt effects on the phase transition of ionic gels. *Journal of Chemistry and Physics*, 77, 5725–5729.

O. Okay, S.B. Sariisik, S.D. Zor. (1998). Swelling behavior of anionic acrylamide-based hydrogels in aqueous salt solutions: Comparison of experiment with theory. *Journal of Applied Polymer Science*, 70, 567–575.

R.S. Parker, F.J. Doyle III, N.A. Peppas. (1999). A model-based algorithm for blood glucose control in type I diabetic patients. *IEEE Transactions on Biomedical Engineering*, 46, 148–157.

J.W. Parker, C.S. Schwartz. (1987). Modelling the kinetics of immobilized glucose oxidase. *Biotechnology and Bioengineering*, 30, 724–735.

N.A. Peppas, P. Bures, W. Leobandung, H. Ichikawa. (2000). Hydrogels in pharmaceutical formulations. *European Journal of Pharmaceutics and Biopharmaceutics*, 50, 27–46.

J.L. Plawsky. (2001). *Transport Phenomena Fundamentals*, New York: Marcel Dekker Inc.

K. Podual, N.A. Peppas. (2005). Relaxational behavior and swelling-pH master curves of poly[(diethylaminoethyl methacrylate)-graft-(ethylene glycol)] hydrogels. *Polymer International*, 54, 581–593.

M.M. Prange, H.H. Hooper, J.M. Prausnitz. (1989). Thermodynamics of aqueous systems containing hydrophilic polymers or gels. *AIChE Journal*, 35, 803–813.

Y. Qiu, K.N. Park. (2001). Environment-sensitive hydrogels for drug delivery. *Advanced Drug Delivery Reviews*, 53, 321–339.

E. Samson, J. Marchand. (1999). Numerical solution of the extended Nernst–Planck model. *Journal of Colloid and Interface Science*, 215, 1–8.

R.A. Siegel, Y.D. Gu, A. Baldi, B. Ziaie. (2004). Novel swelling/shrinking behaviors of glucose-binding hydrogels and their potential use in a microfluidic insulin delivery system. *Macromolecular Symposia*, 207, 249–256.

P.J. Sinko. (2006). *Martin's Physical Pharmacy and Pharmaceutical Sciences*, Pennsylvania: Lippincott Williams & Wilkins.

K.D. Sudipto, N.R. Aluru, B. Johnson, W.C. Crone, D.J. Beebe, J. Moore. (2002). Equilibrium swelling and kinetics of pH-responsive hydrogels: Models, experiments, and simulations. *Journal of Microelectromechanical Systems*, 11, 544–555.

H. Suzuki, A. Kumagai. (2003). A disposable biosensor employing a glucose-sensitive biochemomechanical gel. *Biosensor and Bioelectronics*, 18, 1289–1297.

T. Traitel, Y. Cohen, J. Kost. (2000). Characterization of glucose-sensitive insulin release systems in simulated in vivo conditions. *Biomaterials*, 21, 1679–1687.

T. Traitel, J. Kost, S.A. Lapidot. (2003). Modelling ionic hydrogels swelling: Characterization of the Non-steady state. *Biotechnology and Bioengineering*, 84, 20–28.

P.H.S. Tse, D.A. Gough. (1987). Time-dependent inactivation of immobilized glucose oxidase and catalase. *Biotechnology and Bioengineering*, 29, 705–713.

R.V. Ulijn, N. Bibi, V. Jayawarna, P.D. Thornton, S.J. Rodd, R.J. Mart, A.M. Smith, J.E. Gough. (2007). Bioresponsive hydrogels. *Materials Today*, 10, 40–48.

J.R. Whitaker. (1994). Principle of Enzymology for the Food Science, 2nd ed. New York: Marcel Dekker Inc.

S. Whitaker. (1999). *The Method of Volume Averaging*, Dordrecht: Kluwer.
K. Zhang, X.Y. Wu. (2002). Modulated insulin permeation across a glucose sensitive polymeric composite membrane. *Journal of Controlled Release*, 80, 169–181.
B. Zhao, J.S. Moore. (2001). Fast pH- and ionic strength-responsive hydrogels in microchannels. *Langmuir*, 17, 4758–4763.

Chapter 7
Simulation of Controlled Drug Release from Non-Swellable Micro-Hydrogel Particle

7.1 Introduction

In this chapter, a transient model is presented for simulation of drug delivery from non-swellable micro-hydrogel particle, where the nifedipine controlled release from the spherical chitosan microgel is investigated numerically. The model developed mathematically takes into account both the drug dissolution and drug diffusion through the continuous matrix of spherical microgel. The influences of several important microgel and drug parameters on drug release are evaluated, which include the microsphere mean radius, equivalent drug saturation concentration, drug dissolution rate and drug diffusion coefficient.

7.2 Formulation of Model

Nifedipine is a poorly water-soluble drug with solubility less than 10 mg/l (Liu et al., 2000). As one of well-known calcium channel blockers, nifedipine is most commonly used for treatment of hypertension, a chronic disease that influences 10–20% of the global population and induces cardiovascular complication (Hombreiro et al., 2003). However, many serious adverse effects due to the nifedipine released immediately are revealed, such as the hypotension, myocardial ischaemia or infarction, ventricular fibrillation and cerebral ischaemia (Mansoor and von Hagel Keefer, 2002). Given the seriousness of the reported adverse events and the lack of any clinical documentation attesting to a benefit, Food and Drug Administration (FDA) of the USA concluded that the use of nifedipine released immediately for hypertensive emergencies is neither safe nor effective, and thus it should not be used (Grossman et al., 1996).

So far microsphere system for controlled drug release has increasingly attracted research attention. Controlled nifedipine release was investigated experimentally by various polymer-based microgels, such as the spherical chitosan microgel, and Eudragit microcapsule and poly(DL-lactide-*co*-glycolide acid) microsphere. However, few theoretical efforts are made to develop mathematical model for numerical simulation of the nifedipine release process due to complexity. The

H. Li, *Smart Hydrogel Modelling*, DOI 10.1007/978-3-642-02368-2_7,

objective of this chapter is thus to present a model with consideration of the mechanisms of drug dissolution and diffusion for numerical investigation of the nifedipine release from the chitosan microgel.

In general, the initial drug loading concentration C_0 in spherical microgel is larger than the drug saturation concentration C_s. This may be achieved either by preparation of a solution and total evaporation of the solvent or by partial evaporation or phase inversion (Harland et al., 1988). When the polymeric microgel is placed into a well-stirred release medium, four steps of the mass transfer take place consequently as follows (Hombreiro et al., 2003): (1) drug dissolution within the microgel; (2) drug diffusion within the matrix of microgel; (3) drug diffusion through the unstirred liquid boundary layer on the surface of the microgel and (4) drug diffusion and convection within the release medium. Since the convective transport within the environmental medium is usually very fast when compared with the diffusive mass, the effect of convective transport can be neglected when the overall rate of drug release from the polymeric microgel is considered. Therefore, it may reasonably be assumed that the drug dissolution and diffusion through the continuous matrix of the spherical microgel control the drug release in a well-stirred release medium.

Kinetics of drug release from the microgel with radius R may be characterized by the partial differential governing equation as (Harland et al., 1988)

$$\frac{\partial C(r,t)}{\partial t} = D\left(\frac{\partial^2 C(r,t)}{\partial r^2} + \frac{2}{r}\frac{\partial C(r,t)}{\partial r}\right) + k(\varepsilon C_s - C(r,t)) \tag{7.1}$$

and the following boundary and initial conditions for simulation of the drug release process in a well-stirred release medium

$$\begin{aligned} &\frac{\partial C(r,t)}{\partial r} = 0 \quad \text{at } r = 0 \ \ t > 0 \\ &C(r,t) = 0 \quad \text{at } r = R \ \ t > 0 \\ &C(r,t) = \varepsilon C_s \quad \text{at } 0 < r < R \ \ t = 0 \end{aligned} \tag{7.2}$$

where $C(r,t)$(g/cm^3) is the drug concentration at the radial position r within the microgel system at the release time t, D(cm^2/s) is the drug diffusion coefficient, k (s^{-1}) is the first-order drug dissolution rate, and ε is a parameter for the polymeric network meshes of microgel and it is related directly to the crosslinking density of the polymeric microsphere. If C_s (g/cm^3) is defined as the drug saturation concentration in the system, εC_s(g/cm^3) is referred to as the equivalent drug saturation concentration in microgel with a network mesh parameter ε.

The first term on the right-hand side of Eq. (7.1) is well known as Fick's second law of diffusion for a spherical system (Crank, 1975), which describes the drug diffusion release process in the microgel due to the continuous drug dissolution. The second term on the right-hand side of Eq. (7.1) corresponds to the potential rate-limiting drug dissolution process (Harland et al., 1988). It is observed that

Eq. (7.1) is reduced to the classic Fick's diffusion equation, if the initial drug loading concentration C_0 is smaller than the drug saturation concentration C_s. Although the drug diffusion coefficient D in the polymeric microgel may be solvent concentration dependent, usually one reasonably assumes an approximate constant D for simplicity.

It is also assumed that the drug is distributed uniformly throughout the microgel with the equivalent drug saturation concentration εC_s at initial state. Under perfect sink conditions, the release medium is considered to be well stirred, such that the drug concentration outside of microgel is further assumed to be constant and equal to zero.

Defining dimensionless parameters, $\xi = r/R$ as dimensionless radius, $\tau = Dt/R^2$ as dimensionless Fourier time, $\beta = kR^2/D$ as dimensionless dissolution/diffusion number and $\overline{C}(\xi,\tau) = 1 - C(r,t)/\varepsilon C_s$ as dimensionless concentration, which indicates the non-dimensional drug concentration additionally required to reach saturation dissolution, the partial differential governing equation (7.1) and boundary and initial conditions (7.2) can be rewritten in the dimensionless forms as follows:

$$\frac{\partial \overline{C}(\xi,\tau)}{\partial \tau} = \frac{\partial^2 \overline{C}(\xi,\tau)}{\partial \xi^2} + \frac{2}{\xi}\frac{\partial \overline{C}(\xi,\tau)}{\partial \xi} - \beta \overline{C}(\xi,\tau) \tag{7.3}$$

$$\begin{aligned} \frac{\partial \overline{C}(\xi,\tau)}{\partial \xi} &= 0 \qquad \text{at } \xi = 0 \;\; \tau > 0 \\ \overline{C}(\xi,\tau) &= 1 \qquad \text{at } \xi = 1 \;\; \tau > 0 \end{aligned} \tag{7.4}$$

$$\overline{C}(\xi,\tau) = 0 \;\; \text{at } 0 < \xi < 1 \;\; \tau = 0 \tag{7.5}$$

Solving the above governing equation together with boundary and initial conditions, $\overline{C}(\xi,\tau)$ is simulated and then the drug concentration $C(r, t)$ is found. According to Fick's first law (Robert, 1996), the flux $J = J(r, t)$, the rate of drug transfer per unit area of section, is considered as

$$J(r,t) = -D\frac{\partial C(r,t)}{\partial r} \tag{7.6}$$

The rate of drug release from the microgel is thus formulated as (Robert, 1996)

$$\frac{\partial M_t}{\partial t} = AJ(r,t)\,|_{r=R} \tag{7.7}$$

where A is the area of microgel with radius R, M_t denotes the amount of drug released after time t and it can be calculated by simply integrating Eq. (7.7)

$$\begin{aligned} M_t &= \int_0^t AJ(r,\bar{t})|_{r=R} d\bar{t} = 4\pi R^2 \int_0^t \left(-D\frac{\partial C(r,\bar{t})}{\partial r}|_{r=R}\right) d\bar{t} \\ &= -4\pi R^2 D \int_0^t \frac{\partial C(r,\bar{t})}{\partial r}\,|_{r=R}\, d\bar{t} \end{aligned} \tag{7.8}$$

7.3 Numerical Implementation

For simulation of the kinetics of drug release, the non-dimensional transient partial differential governing equation (7.3) is discretized first in spatial domain by Hermite-cloud method (Li et al., 2003), and then discretized in time domain by linear interpolation technique, namely for

$$\frac{\partial \overline{C}(\xi,\tau)}{\partial \tau} = \frac{\partial^2 \overline{C}(\xi,\tau)}{\partial \xi^2} + \frac{2}{\xi}\frac{\partial \overline{C}(\xi,\tau)}{\partial \xi} - \beta \overline{C}(\xi,\tau) \tag{7.9}$$

by Hermite-cloud method (Li et al., 2003) for spatial discretization, one has

$$\overline{C}(\xi,\tau) = \sum_{n=1}^{N_T} N_n(\xi)\overline{C}_n(\tau) + \sum_{m=1}^{N_S}\left(\xi - \sum_{n=1}^{N_T} N_n(\xi)\xi_n\right) M_m(\xi)\overline{C}_{\xi m}(\tau) \tag{7.10}$$

$$\overline{C}_\xi(\xi,\tau) = \frac{\partial \overline{C}(\xi,\tau)}{\partial \xi} = \sum_{m=1}^{N_S} M_m(\xi)\overline{C}_{\xi m}(\tau) \tag{7.11}$$

$$\frac{\partial^2 \overline{C}(\xi,\tau)}{\partial \xi^2} = \sum_{n=1}^{N_T} N_{n,\xi\xi}(\xi)\overline{C}_n(\tau) \tag{7.12}$$

Substituting Eqs. (7.10), (7.11) and (7.12) into Eq. (7.9), the governing equation for drug release is discretized at ξ_i in spatial domain as

$$\begin{aligned}\frac{\partial \overline{C}(\xi_i,\tau)}{\partial \tau} &= \sum_{n=1}^{N_T} \left(N_{n,\xi\xi}(\xi_i) - \beta N_n(\xi_i)\right)\overline{C}_n(\tau) \\ &+ \left(\frac{2}{\xi_i} - \beta\left(\xi_i - \sum_{n=1}^{N_T} N_n(\xi_i)\xi_n\right)\right)\sum_{m=1}^{N_S} M_m(\xi_i)\overline{C}_{\xi m}(\tau)\end{aligned} \tag{7.13}$$

By the linear interpolation technique, the weighted average of the time derivative $\partial \overline{C}/\partial \tau$ is approximated at two consecutive time steps as follows (Reddy, 1993):

$$(1-\lambda)\frac{\partial \overline{C}(\xi_i,\tau)}{\partial \tau} + \lambda\frac{\partial \overline{C}(\xi_i,\tau+\Delta\tau)}{\partial \tau} = \frac{\overline{C}(\xi_i,\tau+\Delta\tau) - \overline{C}(\xi_i,\tau)}{\Delta\tau} \tag{7.14}$$

where λ is a weighted coefficient ($0 \le \lambda \le 1$).

Substituting Eqs. (7.10) and (7.13) into Eq. (7.14), and considering the auxiliary condition (7.11), the governing equation discretized in both spatial and time domains for drug release with time iteration is finally derived, and reduced to a set of discrete algebraic equations in the following matrix form:

$$\begin{bmatrix} [G_{ij}^{11}]_{N_T\times N_T} & [G_{ij}^{12}]_{N_T\times N_S} \\ [G_{ij}^{21}]_{N_S\times N_T} & [G_{ij}^{22}]_{N_S\times N_S} \end{bmatrix} \begin{Bmatrix} \{\overline{C}_i(\tau+\Delta\tau)\}_{N_T\times 1} \\ \{\overline{C}_{\xi i}(\tau+\Delta\tau)\}_{N_S\times 1} \end{Bmatrix} = \begin{bmatrix} [G_{ij}^{*11}]_{N_T\times N_T} & [G_{ij}^{*12}]_{N_T\times N_S} \\ [G_{ij}^{*21}]_{N_S\times N_T} & [G_{ij}^{*22}]_{N_S\times N_S} \end{bmatrix} \begin{Bmatrix} \{\overline{C}_i(\tau)\}_{N_T\times 1} \\ \{\overline{C}_{\xi i}(\tau)\}_{N_S\times 1} \end{Bmatrix} \tag{7.15}$$

where

$$[G_{ij}^{11}] = [(1+\lambda\beta\Delta\tau)N_j(\xi_i) - \lambda\Delta\tau N_{j,\xi\xi}(\xi_i)]$$

$$[G_{ij}^{12}] = \left[-\frac{2\lambda\Delta\tau}{\xi_i} + (1+\lambda\beta\Delta\tau)\left(\xi_i - \sum_{n=1}^{N_T} N_n(\xi_i)\xi_n\right)\right] M_j(\xi_i)$$

$$[G_{ij}^{21}] = [G_{ij}^{*21}] = [N_{j,\xi}(\xi_i)]$$

$$[G_{ij}^{22}] = [G_{ij}^{*22}] = \left[\left(-\sum_{n=1}^{N_T} N_{n,\xi}(\xi_i)\xi_n\right) M_j(\xi_i)\right]$$

$$[G_{ij}^{*11}] = [(1-\lambda)\Delta\tau N_{j,\xi\xi}(\xi_i) + (1-\beta(1-\lambda)\Delta\tau)N_j(\xi_i)]$$

$$[G_{ij}^{*12}] = \left[\left(\frac{2(1-\lambda)\Delta\tau}{\xi_i} + (1-\beta(1-\lambda)\Delta\tau)\left(\xi_i - \sum_{n=1}^{N_S} N_n(\xi_i)\xi_n\right)\right) M_j(\xi_i)\right] \tag{7.16}$$

7.4 Comparison with Experiment

The experimentally measured data of the nifedipine release for the spherical nifedipine-loaded chitosan microgel exposed to phosphate buffer (pH 7.4), achieved by Filipovic et al. (1996) through the chitosan microgel preparation and characterization and nifedipine release determination, are simulated numerically by the present model. The series of B samples (B1–B5) are taken and the corresponding microgel radii R are listed in Table 7.1 (Filipovic et al., 1996). The totally loaded

Table 7.1 Experimental and identified parameters of nifedipine microgels

	Experimental data (Filipovic et al., 1996)		Identified parameters		
Sample	R ($\times 10^{-4}$ cm)	M_∞ ($\times 10^{-13}$ g)	D ($\times 10^{-11}$ cm^2/s)	k ($\times 10^{-7}$ s^{-1})	εC_s ($\times 10^{-6}$ g/cm^3)
B1	12.10	0.20	0.40	7.0	1.225
B2	13.90	0.24	0.40	7.0	1.225
B3	13.05	0.20	0.35	7.0	1.033
B4	12.20	0.16	0.30	7.0	0.823
B5	14.50	0.32	0.40	7.0	1.225

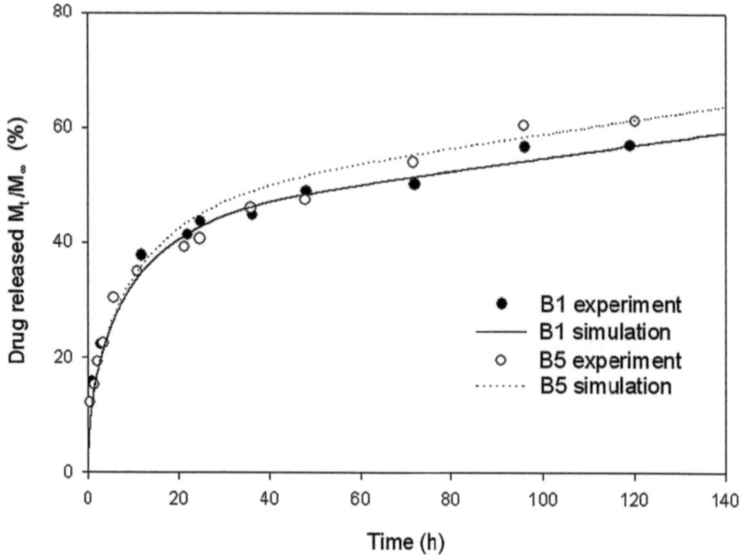

Fig. 7.1 Rate of nifedipine release from chitosan microgels with different radii R

drug mass M_∞ listed in Table 7.1 is calculated by the mass of drug-loaded microgel m(g), total drug content d(%), the mean radius of dry microgel R(cm) and the volume of the dissolution medium V(cm^3), which are extracted from the experimental data. By assuming that the drug is released and dissolved in the dissolution medium,

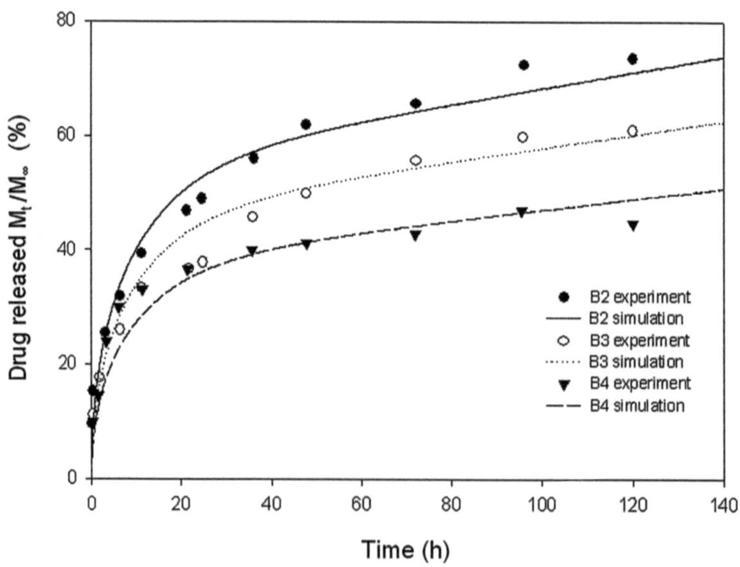

Fig. 7.2 Rate of nifedipine release from chitosan microgels with different network mesh parameter ε

$M_\infty = (4\pi R^3/3) \times (md/V)$. It is noted that such an assumption will bring about unpredictable error in M_∞. However, since only the ratio M_t/M_∞ is concerned here, the inaccuracy would not influence the prediction. The comparisons of the model with experimental measurement of nifedipine release published in open literature are shown in Figs. 7.1 and 7.2, and discussed in Sect. 7.5.1.

7.5 Parameter Studies by Transient Simulation

For analysis of the influences of several important microgel and drug parameters on the kinetics of drug release, including the microsphere radius R, equivalent saturation concentration εC_s, drug dissolution rate k and drug diffusion coefficient D, the present model is employed for simulation of kinetics of the drug release. When the sensitivity study on one of the parameters is carried out, other parameters remained the same. Parameter studies are conducted in Sects. 7.5.2, 7.5.3, 7.5.4 and 7.5.5 with support of Figs. 7.3, 7.4, 7.5 and 7.6, where the experimental data of drug release from B5 sample are taken as a comparative reference:. Before that, as mentioned before, the model is validated first by comparison with the published experimental data of nifedipine release in Sect. 7.5.1.

7.5.1 *Identification of Physical Parameters*

Figure 7.1 illustrates the kinetics of in vitro drug release from the microgels with different microsphere radii and initially loaded drug amounts. As time increases,

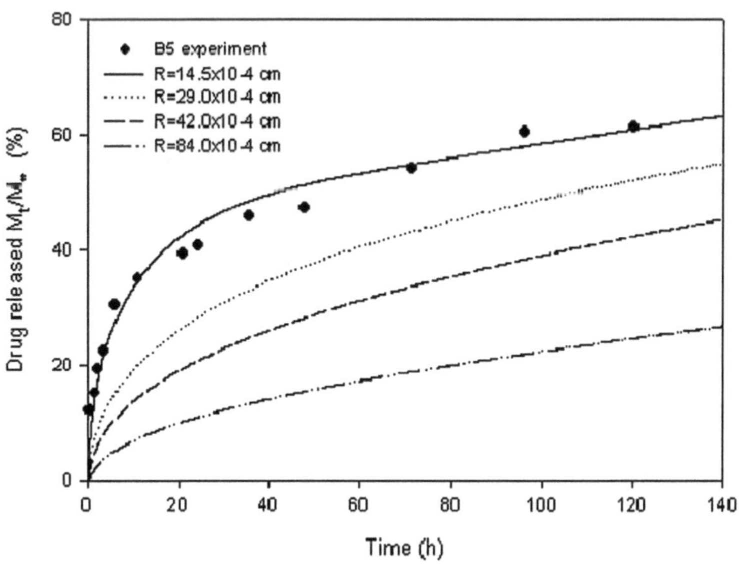

Fig. 7.3 Effect of the microsphere radius R on the rate of nifedipine release from chitosan microgels when $D = 0.40 \times 10^{-11}(\text{cm}^2/\text{s})$, $k = 7.0 \times 10^{-7}(\text{s}^{-1})$, $\varepsilon C_s = 1.225 \times 10^{-6}(\text{g/cm}^3)$

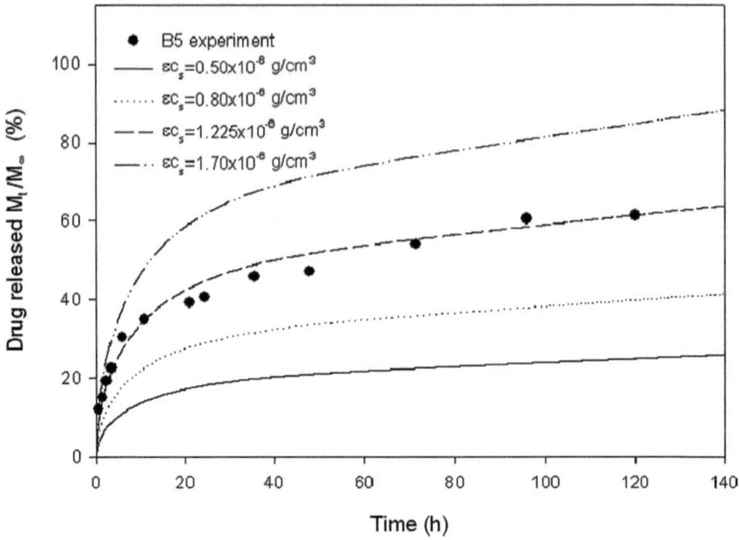

Fig. 7.4 Effect of the equivalent drug saturation concentration εC_s on the rate of nifedipine release from chitosan microgels when $D = 0.40 \times 10^{-11}(\text{cm}^2/\text{s})$, $R = 14.5 \times 10^{-4}(\text{cm})$, $k = 7.0 \times 10^{-7}(\text{s}^{-1})$

the drug release amount increases rapidly at early state, followed by a gradual drug release. B5 has larger microsphere radius R and higher total loaded drug amount M_∞ (see Table 7.1). It is found that nifedipine release from B5 is faster than that from B1. Good agreement is achieved between numerically fitted results with the

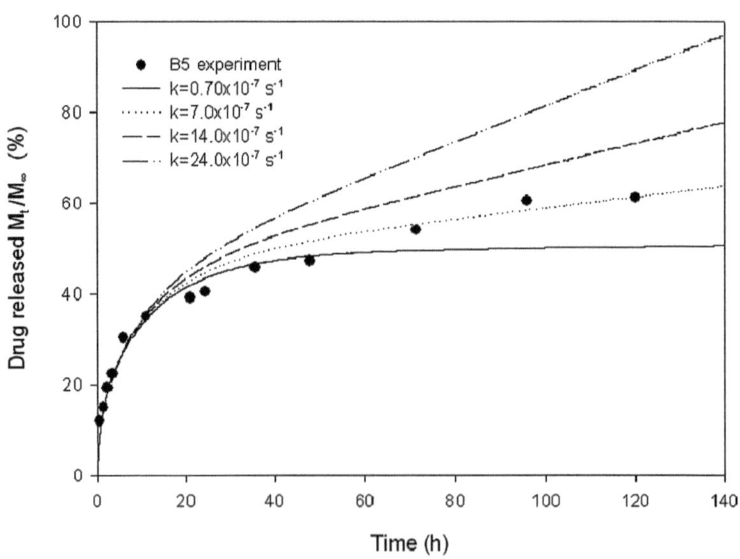

Fig. 7.5 Effect of the drug dissolution rate k on the rate of nifedipine release from chitosan microgels for $R = 14.5 \times 10^{-4}(\text{cm})$, $D = 0.40 \times 10^{-11}(\text{cm}^2/\text{s})$, $\varepsilon C_s = 1.225 \times 10^{-6}(\text{g/cm}^3)$

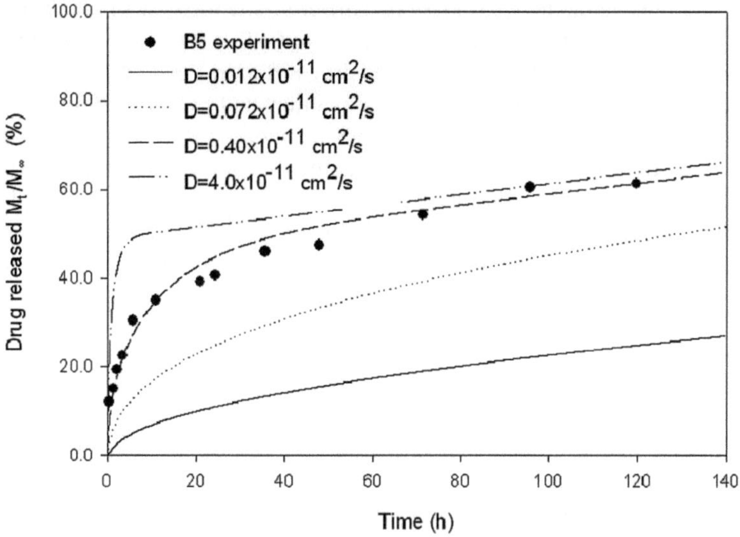

Fig. 7.6 Effect of the drug diffusion coefficient D on the rate of nifedipine release from chitosan microgels when $R = 14.5 \times 10^{-4}$(cm), $k = 7.0 \times 10^{-7}(\mathrm{s}^{-1})$, $\varepsilon C_s = 1.225 \times 10^{-6}(\mathrm{g/cm}^3)$

present model and experimental data published for both B1 and B5. It is seen that the model successfully captures the effect of microsphere radius R. The values of the equivalent drug saturation concentration εC_s, drug dissolution rate k and diffusion coefficient D are identified by best fitting the computed results to the experimental data. The identified εC_s, k and D are summarized in Table 7.1. The D value found is smaller than the reported value of nifedipine in crosslinked hydrogel of polyacrylamide-grafted guar gum (Soppimath et al., 2001). Generally the D value for various drugs in polymeric hydrogels ranges from 10^{-6} to 10^{-9} (cm^2/s). Several effects may contribute to the extremely low D value of nifedipine in the studied microgels. First, the solubility of nifedipine in the release medium is very low, about 11 (μg/ml), which implies relatively large partition coefficient of nifedipine between the polymeric hydrogel and the release medium. Second, the microgel that is loaded with a high content of drug tends to absorb less water than those containing a lower content of drug so that the diffusion process is retarded. Last, it is reported that the increase in drug content will increase the crystallinity of the drug, and thus slow down the release of such a crystalline drug (Soppimath et al., 2000).

Figure 7.2 demonstrates the kinetics of nifedipine release from chitosan microgels with different network mesh parameters ε. These microgels are formed with the same nifedipine amount but different glutaraldehyde reaction time. With increasing glutaraldehyde reaction time, the crosslinking degree of the microgels increases, which results in a decrease in the network mesh parameter ε of the microgels. This further causes the decrease in the equivalent saturation concentration εC_s. The fitting represents the experimental results well. The corresponding εC_s and D are identified by best fitting the simulation results to the experimental data, and they are tabulated in Table 7.1. The diffusion coefficient D studied is dependent on the crosslinking density, and thus on the network mesh parameter ε. The increase of εC_s from 0.823 to 1.225 enlarges the D value from 0.30×10^{-11} to $0.40 \times 10^{-11}(\mathrm{cm}^2/\mathrm{s})$. Increasing the network mesh parameter ε enlarges the drug diffusion coefficient D, which is in consistence with the experimental findings by Pillay and Fassihi (1999).

In brief, Figs. 7.1 and 7.2 validate the present model with capability of predicting well the nifedipine release from chitosan microgels with different microsphere conditions. It can capture the characteristics of the various important physical parameters affecting the kinetics of nifedipine release. Therefore, it is concluded that this model provides a suitable simulation platform to provide deeper insight into the effects of the microgel and drug parameters on drug release from the microgels.

7.5.2 Influence of Mean Radius of Micro-Hydrogel Particle

Figure 7.3 shows the influence of microsphere mean radius R on the controlled drug release, where the drug diffusion coefficient $D = 0.40 \times 10^{-11}(\mathrm{cm}^2/\mathrm{s})$, drug dissolution rate $k = 7.0 \times 10^{-7}(\mathrm{s}^{-1})$ and drug equivalent saturation concentration $\varepsilon C_s = 1.225 \times 10^{-6}(\mathrm{g}/\mathrm{cm}^3)$. The microsphere radius R in the figure ranges from 9.5×10^{-4} to 17.0×10^{-4} (cm). With increasing microsphere radius R, the overall drug release rate becomes slower. It is noted that a slight change in the microsphere radius R results in remarkable alteration of nifedipine release rate. A smaller microgel has larger specific surface area of contact with the release medium and facilitates the drug diffusion through the continuous matrix of microgel into the release medium if compared with larger microgel. Decreasing the microsphere radius$R$ increases both the initial fast release rate and the following gradual release rate. This implicates that the variation of microsphere radius$R$ influences both the drug dissolution and diffusion processes.

7.5.3 Influence of Equivalent Drug Saturation Concentration

Influence of the equivalent drug saturation concentration εC_s on drug release kinetics is shown in Fig. 7.4, where the equivalent drug saturation concentration εC_s ranges from 0.5×10^{-6} to 1.7×10^{-6} (g/cm^3). The drug release remarkably increases with the porosity. An increase of the network mesh parameter ε of microgels, i.e. a decrease of the crosslinking density of microgels, enlarges the equivalent drug saturation concentration εC_s and the drug diffusion coefficient D simultaneously, which results in an increase in both the drug dissolution and diffusion rates. These led to the increase of nifedipine release rate synergetically.

7.5.4 Influence of the First-Order Drug Dissolution Rate

Figure 7.5 illustrates the influence of the drug dissolution rate k on drug release, where the dissolution rate k ranges from 0.70×10^{-7} to $24.0 \times 10^{-7}(\mathrm{s}^{-1})$. It is manifest that alteration of the drug dissolution rate k has insignificant effect on initial drug release rate. However, after a period of release time, the drug release rate increases with increasing dissolution rate k. This indicates that different mechanisms control different stages of drug release. At the initial stage, diffusion through continuous matrix of microgel predominantly affects the drug release. After a period of drug release, the drug dissolution becomes significant on the drug release rate.

7.5.5 Influence of Drug Diffusion Coefficient

Figure 7.6 is plotted for discussion of the influence of the drug diffusion coefficient D on the drug release, where$D$ ranges from 0.012×10^{-11} to $4.0\times10^{-11}(\mathrm{cm}^2/\mathrm{s})$. It is observed that the diffusion process controls significantly the early drug release. As the diffusion coefficient D increases, the initial drug release rate increases distinctly, and the drug release amount becomes a linear function of time at relatively short time. However, after a certain period of release time, the drug release reaches a constant level. This indicates that the drug release with lower diffusion coefficient D is characterized by the diffusion mechanism, whereas the diffusion process cannot fully control the drug release with higher diffusion coefficient D.

7.6 Remarks

The present model has been examined well by comparison with the experimental nifedipine release from the spherical chitosan microgels, which is characterized by the mechanisms of both the drug dissolution and diffusion through the continuous matrices of microgels. It provides a computer-based platform for better understanding of the underlying mechanisms of drug release via microsphere hydrogel system and represents a highly efficient tool for analysis of the influences of important parameters of the drug and the microgels, such as the diffusion coefficient, the micro-spherical radius and the network mesh parameter. Consequently, it can be applied for design and optimization of the microhydrogel-based controlled drug release system.

References

J. Crank. (1975). *The Mathematics of Diffusion*, 2nd edn. Oxford: Clarendon Press.

J. Filipovic-Grcic, M. Becirevic-Lacan, N. Skalko, I. Jalsenjak. (1996). Chitosan microspheres of nifedipine and nifedipine-cyclodextrin inclusion complexes. *International Journal of Pharmaceutics*, 135, 183–190.

E. Grossman, F.H. Messerli, T. Grodzicki, P. Kowey. (1996). Should a moratorium be placed on sublingual nifedipine capsules given for hypertensive emergencies and pseudoemergencies. *Journal of the American Medical Association*, 276, 1328–1331.

R.S. Harland, C. Dubernet, J.P. Benoit, N.A. Peppas. (1988). A model of dissolution-controlled, diffusional drug release from non-swellable polymeric microspheres. *Journal of Controlled Release*, 7, 207–215.

M. Hombreiro-Perez, J. Siepmann, C. Zinutti, A. Lamprecht, N. Ubrich, M. Hoffman, R. Bodmeier, P. Maincent. (2003). Non-degradable microparticles containing a hydrophilic and/or a lipophilic drug: Preparation, characterization and drug release modelling. *Journal of Controlled Release*, 88, 413–428

L.X. Liu, J. Ku, G. Khang, B. Lee, J.M. Rhee, H.B. Lee. (2000). Nifedipine controlled delivery by sandwiched osmotic tablet system. *Journal of Controlled Release*, 68, 145–156.

H. Li, T.Y. Ng, J.Q. Cheng, K.Y. Lam. (2003). Hermite-cloud: A novel true meshless method. *Computational Mechanics*, 33, 30–41.

A.F. Mansoor, L.A. von Hagel Keefer. (2002). The dangers of immediate-release nifedipine for hypertensive crises. *Pharmacy and Therapeutics*, 27, 362–365.

V. Pillay, R. Fassihi. (1999). A new method for dissolution studies of lipid-filled capsules employing nifedipine as a model drug, *Pharmaceutical Research*, 16, 333–338.

J.N. Reddy. (1993). An Introduction to the Finite Element Method, 2nd ed. New York: McGraw-Hill.

A.A. Robert. (1996). Physical Chemistry, 2nd ed. New York: Wiley.

K.S. Soppimath, R.A. Kulkarni, T.M. Sminsbhavi. (2000). Controlled release of antihypertensive drug from the interpenetrating network poly(vinyl alcohol)-guar gum hydrogel microspheres. *Journal of Biomaterials Science – Polymer Edition*, 11, 27–44.

K.S. Soppimath, A.R. Kulkarni, T.M. Aminabhavi. (2001). Chemically modified polyacrylamide-g-guar gum-based crosslinked anionic microgels as pH-sensitive drug delivery systems: Preparation and characterization. *Journal of Controlled Release*, 75, 331–345.

References

M.E. Byrne, K. Park, N.A. Peppas. (2002). Molecular imprinting within hydrogels. Advanced Drug Delivery Reviews, 34, 149–161.

K. Dusek. (1993a). Advances in Polymer Science, Vol. 109, Berlin/Heidelberg: Springer-Verlag.

K. Dusek. (1993b). Advances in Polymer Science, Vol. 110, Berlin/Heidelberg: Springer-Verlag.

M.E. Freeman, M.J. Furey, B.J. Love, J.M. Hampton. (2000). Friction wear and lubrication of hydrogels as synthetic articular cartilage. Wear, 241, 129–135.

P. Gupta, K. Vermani, S. Garg. (2002). Hydrogels: From controlled release to pH-responsive drug delivery. Drug Discovery Today, 7, 569–579.

R. Haigh, S. Rimmer, N.J. Fullwood. (2000). Synthesis and properties of amphiphilic networks 1: The effect of hydration and polymer composition on the adhesion of immunoglobulin

I.S. Isayeva, S.A. Yankovski, J.P. Kennedy. (2002). Novel amphiphilic membranes of poly(N,N-dimethyl acrylamide) crosslinked with octa-methacrylate-telechelic polyisobutylene stars. Polymer Bulletin, 48, 475–482.

B. Jeong, A. Gutowska. (2002). Lessons from nature: Stimuli-responsive polymers and their biomedical applications. Trends in Biotechnology, 20, 305–311.

R.H. Liu, Q. Yu, D.J. Beebe. (2002). Fabrication and characterization of hydrogel-based microvalves. Journal of Microelectromechanical Systems, 11, 45–53.

F.L. Mi, C.Y. Kuan, S.S. Shyu, S.T. Lee, S.F. Chang. (2000). The study of gelation kinetics and chain-relaxation properties of glutaraldehyde-cross-linked chitosan gel and their effects on microspheres preparation and drug release. Carbohydrate Polymers, 41, 389–396.

M. Micic, Y. Zheng, V. Moy, X.H. Zhang, M. Andreopoulos, R.M. Leblanc. (2002). Comparative studies of surface topography and mechanical properties of a new photo-switchable PEG-based hydrogel. Colloids and Surfaces B: Biointerfaces, 27, 147–158.

T. Miyata, T. Uragami, K. Nakamae. (2002). Biomolecule-sensitive hydrogels. Advanced Drug Delivery Reviews, 54, 79–98.

C. Ozyurek, T. Caykara, O. Kantoglu, O. Guven. (2000). Characterization of network structure poly(N-vinyl 2-pyrrolidone/acrylic acid) polyelectrolyte hydrogels by swelling measurements. Journal of Polymer Science Part B: Polymer Physics, 38, 3309–3317.

N.A. Peppas, E.W. Merrill. (1976). Poly(vinyl alcohol) hydrogels: Reinforcement of radiation-crosslinked network by crystallization. Journal of Polymer Science Part A: Polymer Chemistry, 14, 441–457.

K.M. Raju, M.P. Raju, Y.M. Mohan. (2002). Synthesis and water absorbency of crosslinked superabsorbent polymers. Journal of Applied Polymer Science, 85, 1795–1801.

M. Sen, O. Güven. (2000). Prediction of the swelling behavior of amphiphilic hydrogels and the determination of average molecular weight between cross-links. Computational and Theoretical Polymer Science, 11, 475–482.

K.S. Soppimath, T.M. Aminabhavi, A.M. Dave, S.G. Kumbar, W.E. Rudzinski. (2002). Stimulus-responsive "smart" hydrogels as novel drug delivery systems. Drug Development and Industrial Pharmacy, 28, 957–974.

H. Li, *Smart Hydrogel Modelling*, DOI 10.1007/978-3-642-02368-2_BM2,

J.A. Wesselinght, A.M. Bollen. (1997). Multicomponent diffusivities from the free volume theory. IChemE Journal Transactions, 75A, 590–602.

M.J. Whitcombe, E.N. Vulfson. (2001). Imprinted polymers. Advanced Materials, 13, 467–478.

M. Wu, J. Chen, Q. Qian, N. Bao. (2002). Thermo-responsive interpenetrating polymer networks composed of AAc/AAm/NMA by radiation grafting. Journal of Radioanalytical and Nuclear Chemistry, 252, 531–535.

J. O'M. Bockris, B.E. Conway, E. Yeager (Eds.) (1983). Comprehensive Treatise of Electrochemistry, Vol. 6, Electrodics: Transport, New York: Plenum Press.

L. Zhang, W.R. Seitz. (2002). A pH sensor based on force generated by pH-dependent polymer swelling. Analytical and Bioanalytical Chemistry, 373, 555–559.

S. Sun, A.F.T. Mak. (2001). The dynamical response of a hydrogel fiber to electrochemical stimulation. *Journal of Polymer Science Part B: Polymer Physics*, 39, 236–246.

Acknowledgements

The author would like to sincerely acknowledge the use of figures and tables reproduced from the following sources.

E. Birgersson, H. Li, S.N. Wu. (2008). Transient analysis of temperature-sensitive neutral hydrogels. *Journal of the Mechanics and Physics of Solids*, 56(2), 444–466.

H. Li. (2009). Kinetics of smart hydrogels responding to electric field: A transient deformation analysis. *International Journal of Solids and Structures*, 46(6) 1326–1333.

H. Li, J. Chen, K.Y. Lam. (2004). Multiphysical modelling and meshless simulation of electric-sensitive hydrogels. *Journal of Polymer Science – Part B: Polymer Physics*, 42(8), 1514–1531.

H. Li, J. Chen, K.Y. Lam. (2006). A transient simulation to predict the kinetic behavior of hydrogels responsive to electric stimulus. *Biomacromolecules*, 7(6), 1951–1959.

H. Li, J. Chen, K.Y. Lam. (2007). Transient simulation of kinetics of electric-sensitive hydrogels. *Biosensors and Bioelectronics*, 22(8), 1633–1641.

H. Li, R.M. Luo, E. Birgersson, K.Y. Lam. (2007). Modelling of multiphase smart hydrogels responding to pH and electric voltage coupled stimuli. *Journal of Applied Physics*, 101(11), 114905-(1∼7).

H. Li, R.M. Luo, K.Y. Lam. (2007). Modelling and simulation of deformation of hydrogels responding to electric stimulus. *Journal of Biomechanics*, 40(5), 1091–1098.

H. Li, R.M. Luo, K.Y. Lam. (2007). Modelling of ionic transport in electric-stimulus-responsive hydrogels. *Journal of Membrane Science*, 289(1–2), 284–296.

H. Li, R.M. Luo, K.Y. Lam. (2009). Multiphysics modelling of electrochemomechanically smart microgels responsive to coupled pH/electric stimuli. *Macromolecular Bioscience*, 9(3), 287–297.

H. Li, R.M. Luo, E. Birgersson, K.Y. Lam. (2009). A chemo-electro-mechanical model for simulation of responsive deformation of glucose-sensitive hydrogels with the effect of enzyme catalysis. *Journal of the Mechanics and Physics of Solids*, 57(2), 369–382.

H. Li, T.Y. Ng, Y.K. Yew, K.Y. Lam. (2005). Modelling and simulation of the swelling behavior of pH-stimulus-responsive hydrogels. *Biomacromolecules*, 6(1), 109–120.

H. Li, T.Y. Ng, Y.K. Yew, K.Y. Lam. (2007). Meshless modelling of pH-sensitive hydrogels subjected to coupled pH and electric field stimuli: Young modulus effects and case studies. *Macromolecular Chemistry and Physics*, 208(10), 1137–1146.

H. Li, X.G. Wang, Z.J. Wang, K.Y. Lam. (2005). Multiphysics modelling of volume phase transition of ionic hydrogels responsive to thermal stimulus. *Macromolecular Bioscience*, 5(9), 904–914.

H. Li, Z.J. Wang, X.G. Wang, K.Y. Lam. (2005). Simulation of the influences of bathing solution and crosslink density on the swelling equilibrium of ionic thermo-sensitive hydrogels. *Biophysical Chemistry*, 118(2–3), 57–68.

H. Li, X.G. Wang, G.P Yan, K.Y. Lam, S.X. Cheng, T. Zou, R.X. Zhuo. (2005). A novel multiphysic model for simulation of swelling equilibrium of ionized thermal-stimulus responsive hydrogels. *Chemical Physics*, 309(2–3), 201–208.

H. Li, G.P. Yan, S.N. Wu, Z.J. Wang, K.Y. Lam. (2004). Numerical simulation of controlled nifedipine release from chitosan microgels. *Journal of Applied Polymer Science*, 93(4), 1928–1937.

H. Li, Y.K. Yew. (2009). Simulation of soft smart hydrogels responsive to pH stimulus: Ionic strength effect and case studies. *Materials Science and Engineering C*, 29(7), 2261–2269.

H. Li, Y.K. Yew, T.Y. Ng. (2008). Computational analysis of the influence of initial fixed charge density on pH-sensitive hydrogels. *Modelling and Simulation in Materials Science and Engineering*, 16(8), 085004-(1∼20).

H. Li, Y.K. Yew, T.Y. Ng. (2009). Modelling for analysis of the effect of Young's modulus on soft active hydrogels subject to pH stimulus. *Smart Materials and Structures*, 18(4), 045010-(1∼10).

H. Li, Y.K. Yew, T.Y. Ng, K.Y. Lam. (2005). Meshless steady-state analysis of chemo-electro-mechanical coupling behavior of pH-sensitive hydrogel in buffered solution. *Journal of Electroanalytical Chemistry*, 580(1), 161-172.

H. Li, Z. Yuan, K.Y. Lam, H.P. Lee, J. Chen, J. Hanes, J. Fu. (2004). Model development and numerical simulation of electric-stimulus-responsive hydrogels subject to an externally applied electric field. *Biosensors and Bioelectronics*, 19(9), 1097–1107.

R.M. Luo, Hua Li, E. Birgersson, K.Y. Lam. (2008). Modelling of electric-stimulus-responsive hydrogels immersed in different bathing solutions. *Journal of Biomedical Materials Research Part A*, 85A(1), 248–257.

R.M. Luo, Hua Li, K.Y. Lam. (2007). Modelling and simulation of chemo-electro-mechanical behavior of pH-electric-sensitive hydrogel. *Analytical and Bioanalytical Chemistry*, 389(3), 863–873.

R.M. Luo, Hua Li, K.Y. Lam. (2007). Coupled chemo-electro-mechanical simulation for smart hydrogels that are responsive to an external electric field. *Smart Materials and Structures*, 16(4), 1185–1191.

R.M. Luo, Hua Li, K.Y. Lam. (2009). Modelling the effect of environmental solution pH on the mechanical characteristics of glucose-sensitive hydrogels. *Biomaterials*, 30(4), 690–700.

T.Y. Ng, Hua Li, Y.K. Yew, K.Y. Lam. (2007). Effects of initial-fixed charge density on pH-sensitive hydrogels subjected to coupled pH and electric field stimuli: A meshless analysis. *ASME Journal of Biomechanical Engineering*, 129(2), 148–155.

Y.K. Yew, T.Y. Ng, Hua Li, K.Y. Lam. (2007). Analysis of pH and electrically controlled swelling of hydrogel-based micro-sensors/actuators. *Biomedical Microdevices*, 9(4), 487–499.

Index

MIX
Papier aus verantwortungsvollen Quellen
Paper from responsible sources
FSC® C105338

If you have any concerns about our products,
you can contact us on
ProductSafety@springernature.com

In case Publisher is established outside the EU,
the EU authorized representative is:
Springer Nature Customer Service Center GmbH
Europaplatz 3, 69115 Heidelberg, Germany

Printed by Libri Plureos GmbH
in Hamburg, Germany